# Studies in Logic

Volume 11

# Foundations of the Formal Sciences V

## Infinite Games

Volume 1
Proof Theoretical Coherence
Kosta Dosen and Zoran Petric

Volume 2
Model Based Reasoning in Science and Engineering
Lorenzo Magnani, editor

Volume 3
Foundations of the Formal Sciences IV: The History of the Concept of the Formal Sciences
Benedikt Löwe, Volker Peckhaus and Thoralf Räsch, editors

Volume 4
Algebra, Logic, Set Theory. Festschrift für Ulrich Felgner zum 65. Geburtstag
Benedikt Löwe, editor

Volume 5
Incompleteness in the Land of Sets
Melvin Fitting

Volume 6
How to Sell a Contradiction: The Logic and Metaphysics of Inconsistency
Francesco Berto

Volume 7
Fallacies — Selected Papers 1972-1982
John Woods and Douglas Walton, with a Foreword by Dale Jacquette

Volume 8
A New Approach to Quantum Logic
Kurt Engesser, Dov M. Gabbay and Daniel Lehmann

Volume 9
Handbook of Paraconsistency
Jean-Yves Béziau, Walter Carnielli and Dov Gabbay, editors

Volume 10
Automated Reasoning in Higher-Order Logic. Set Comprehension and Extensionality in Church's Type Theory
Chad E. Brown

Volume 11
Foundations of the Formal Sciences V: Infinite Games
Stefan Bold, Benedikt Löwe, Thoralf Räsch and Johan van Benthem, editors

Studies in Logic Series Editor
Dov Gabbay
dov.gabbay@kcl.ac.uk

# Foundations of the Formal Sciences V

## Infinite Games

edited by

Stefan Bold
Benedikt Löwe
Thoralf Räsch
Johan van Benthem

ISBN 978-1-904987-75-8
College Publications
Scientific Director: Dov Gabbay
Managing Director: Jane Spurr
Department of Computer Science
King's College London
Strand, London WC2R 2LS, UK

Original cover design by Richard Fraser
Cover produced by orchid creative www.orchidcreative.co.uk

---

# CONTENTS

# Preface

Infinity can feature in games in various forms: we can play games of infinite length, with infinitely many players, or allow for infinitely many moves or strategies. Games of infinite length have been thoroughly investigated by mathematicians and have played a central rôle in mathematical logic. However, their applications go far beyond mathematics: they feature prominently in theoretical computer science, philosophical *Gedankenexperiments*, as limit cases in economical applications, and in many other fields. The conference "*Foundations of the Formal Sciences V: Infinite Games*" (FotFS V) focused on games of infinite length, but was open to include other notions of infinity in games. It brought together researchers from the various areas that employ infinitary game techniques to develop cross-cultural bridges. This volume contains selected and fully refereed papers presented at the conference.

FotFS V was held from November 26th to 29th, 2004 at the Rheinische Friedrich-Wilhelms-Universität Bonn, as the fifth conference in the series *Foundations of the Formal Sciences* after FotFS I in Berlin (May 1999), FotFS II in Bonn (November 2000), FotFS III in Vienna (September 2001), and FotFS IV in Bonn (February 2003).[1] The conference series aims at bringing together researchers from various areas who are interested in the foundations of formal reasoning. The interdisciplinary character of the conference series makes it necessary for authors to explain their material to non-specialists and get them interested in the research problems of other areas. This emphasis on catalyzing joint research between representatives of various communities is inherited by the proceedings volumes of the series. The proceedings volumes of the first four conferences appeared in the journal *Synthese* (Volume 133, Number 1/2), the book series *Trends in Logic* (Volumes 17 and 23) and *Studies in Logic* (Volume 3), respectively. After FotFS V, we held another successful meeting in the series, FotFS VI, on the topic of "Reasoning about Probabilities and Probabilistic Reasoning" in Amsterdam (May 2007). The next conference, FotFS VII is planned for October 2008 in Brussels.

**The conference.**

The conference FotFS V was held at the *Mathematisches Institut* at the *Rheinische Friedrich-Wilhelms-Universität Bonn*, organized in collaboration with the *Institute for Logic, Language and Computation* (ILLC) at the

[1]We would like to point out a rather embarrassing misprint in the proceedings volume of FotFS IV: in the preface, we claim that the conference took place in February 2004.

*Universiteit van Amsterdam.* This collaboration was also represented by the group of organizers (Stefan Bold, Boudewijn de Bruin, Peter Koepke, Benedikt Löwe, Thoralf Räsch, Johan van Benthem) which consisted of representatives of both institutes. The main financial sponsor of the conference was the *Deutsche Forschungsgemeinschaft* (4851/197/04). During the preparatory phase of the conference, we had the unreserved support of our host institute for which we would like to extend our gratitude. For the conference itself, we had a reliable staff of helpers consisting of Irina Arndt, Dorothée Baoues, Patrick Braselmann, Alexander Gilbers, Philipp Hieronymi, Bernhard Irrgang, Tikitu de Jager, Jorrit Kirsten, Michael Klein, Alexander Rothkegel, Karen Seidel, Jip Veldman, Thomas Waltke, and Florian Werne. The following is a list of all presentations scheduled for our conference:

Samson Abramsky, Oxford UK: *Game Semantics and Infinite Games*

Alessandro Andretta, Torino: *Wadge-determinacy and the semi-linear ordering principle*

Dietmar Berwanger, Aachen: *Rationality and regularity in multi-player games*

Julian Bradfield, Edinburgh UK: *IF Logic and Fixpoint Logic*

Matthias Blonski, Frankfurt: *Games in Aggregated Form*

Adam Brandenburger, New York NY: *Admissibility in Games*

Jérémie Cabessa, Lausanne: *An infinite game on omega-semigroups*

Francien Dechesne, Tilburg / Eindhoven: *Game theoretical semantics and infinity*

Samson de Jager, Amsterdam: *Solving asymmetric combinatorial games*

Natasha Dobrinen, State College PA: *Infinite Games on Boolean Algebras*

Jacques Duparc, Lausanne: *Wadge games and sets of reals recognized by simple machines: An $\omega$-power of a finite context-free language which is Borel above* $\boldsymbol{\Delta}^0_\omega$

Johannes Emrich, Erlangen: *A dialogical decision procedure for predicate logic*

Angel Garrido, Madrid: *Matrix Theory and Artificial Intelligence*

Lorenz Halbeisen, Belfast UK: *Infinite Games in Forcing Constructions*

Christoph Heinatsch, Münster: *The determinacy strength of* $\boldsymbol{\Pi}^1_2$*-comprehension*

Ian Hodkinson, London UK: *Games in algebraic logic: axiomatisations and beyond*

Josef Hofbauer, London UK: *Evolutionary dynamics for infinite games*

Kevin Kelly, Pittsburgh PA: *Ockham's Razor as the Most Efficient Strategy in the Infinite Truth-finding Game of Science*

Gleb A. Koshevoy, Moscow: *Cooperative games with infinite number of players, Projective systems, and Cores*

Kenneth A. Presting, Chapel Hill NC: *The Valuation of Choices in The Logic of Decision*

Sandra Quickert, Edinburgh: *Games and Model Checking*

Hamid Sabourian, Cambridge UK: *Stability and equilibrium selection in evolutionary game theory*

Victor Selivanov, Novosibirsk: *Memoir on Wadge degrees of regular star-free $\omega$-languages*

Marion Scheepers, Boise ID: *Infinite Games in Topology*

Robert Simon, London UK: *Games of incomplete information, ergodic theory, and the measurability of equilibria*

Brian Skyrms, Irvine CA: *Replicator Dynamics and the Unreasonable Effectiveness of Cheap Talk in Evolutionary Games*

Fernando Tohmé, Bahía Blanca: *A New Set-Theoretic Foundation, Based on Determinacy, for Economic Theory*

Boaz Tsaban, Jerusalem: *Game theory of generalized selection hypotheses*

Jouko Väänänen, Helsinki: *A new game in infinitary logic*

Stefano Vannucci, Siena: *On concept lattices of coalitional game forms*

Yde Venema, Amsterdam: *Automata, logic and games: a coalgebraic perspective*

Philip Welch, Bristol UK: *A game theoretic approach to quasi-inductive definitions*

Enrico Zoli, Firenze: *The role of Wolfgang Schmidt's $(\alpha, \beta)$-games in Descriptive Set Theory*

**This volume.**

The spirit of the conference that brought together researchers from various fields interested in notions of infinity in games is well reflected in this proceedings volume. A large number of anonymous referees helped to ensure the scientific quality of this volume while keeping in mind its intended wide audience and watching the quality of exposition in the submitted papers. Some of the referee reports were exceptionally detailed and came with many helpful suggestions to improve the presentation. We would like to gratefully acknowledge the support of Jane Spurr of College Publications. The final typesetting meeting in Amsterdam was supported by grant DN 62-630 of the *Nederlandse Organisatie voor Wetenschappelijk Onderzoek.*

Köln/Amsterdam/Bonn
November 2007 S. B. B. L. Th. R. J.F.A.K. v. B.

# Schedule

<table>
<tr><th>Time</th><th>Friday,<br>November 26</th><th>Saturday,<br>November 27</th><th>Sunday,<br>November 28</th><th>Monday,<br>November 29</th></tr>
<tr><td>9:30–10:50</td><td rowspan="9"></td><td>Kelly</td><td>Andretta</td><td>Scheepers</td></tr>
<tr><td>10:30–10:50</td><td>Break</td><td>Break</td><td>Break</td></tr>
<tr><td>10:50–11:10</td><td>Presting</td><td rowspan="2">Duparc</td><td rowspan="2">Tsaban</td></tr>
<tr><td>11:10–11:30</td><td>Tohmé</td></tr>
<tr><td>11:30–11:50</td><td>Dechesne</td><td rowspan="3">Abramsky</td><td rowspan="2">Halbeisen</td></tr>
<tr><td>11:50–12:10</td><td>Garrido</td></tr>
<tr><td>12:10–12:30</td><td>Conference Photo</td><td>Zoli</td></tr>
<tr><td>12:30–12:50</td><td rowspan="4">Lunch Break</td><td rowspan="4">Lunch Break</td><td>Break</td></tr>
<tr><td>12:50–13:30</td><td>Dobrinen</td></tr>
<tr><td>13:50–14:10</td><td>Opening</td><td>Closing</td></tr>
<tr><td>14:10–14:30</td><td rowspan="2">Skyrms</td><td rowspan="16"></td></tr>
<tr><td>14:30–15:10</td><td rowspan="2">Sabourian</td><td>Welch</td></tr>
<tr><td>15:10–15:30</td><td>Break</td><td>Heinatsch</td></tr>
<tr><td>15:30–15:50</td><td rowspan="2">Riedel</td><td>Break</td><td>Break</td></tr>
<tr><td>15:50–16:10</td><td rowspan="2">Blonski</td><td>de Jager</td></tr>
<tr><td>16:10–16:30</td><td rowspan="2">Venema</td><td>Väänänen</td></tr>
<tr><td>16:30–16:50</td><td rowspan="2">Brandenburger</td><td rowspan="2">Selivanov</td></tr>
<tr><td>16:50–17:10</td><td>Break</td></tr>
<tr><td>17:10–17:30</td><td rowspan="3">Hodkinson</td><td>Vannucci</td><td>Break</td></tr>
<tr><td>17:30–17:50</td><td>Break</td><td>Quickert</td></tr>
<tr><td>17:50–18:10</td><td rowspan="3">Simon</td><td rowspan="2">Bradfield</td></tr>
<tr><td>18:10–18:30</td><td rowspan="4"></td></tr>
<tr><td>18:30–18:50</td><td>Break</td></tr>
<tr><td>18:50–19:10</td><td rowspan="2"></td><td>Cabessa</td></tr>
<tr><td>19:10–19:30</td><td>Berwanger</td></tr>
</table>

# Conference Photo

Stefan **Bold**, Benedikt **Löwe**,
Thoralf **Räsch**, Johan **van Benthem** (*eds.*)
**Foundations of the Formal Sciences V**
Infinite Games

# The SLO Principle and the Wadge Hierarchy

ALESSANDRO ANDRETTA*

Dipartimento di Matematica
Università di Torino
Via Carlo Alberto 10
10123 Torino, Italy
alessandro.andretta@unito.it

ABSTRACT. The Semi-Linear Ordering principle for continuous functions ($\mathsf{SLO}^{\mathrm{W}}$) is one of the most useful tools for studying the Wadge hierarchy, and says that $A$ is continuously reducible to $B$ or the complement of $B$ is continuously reducible to $A$, for any two sets $A, B \subseteq \mathbb{R}$. We survey some old and new results on $\mathsf{SLO}^{\mathrm{W}}$ and Wadge degrees.

## 1 Introduction

Game theoretic methods figure prominently in set theory and other parts of mathematical logic. Most of the games studied by set theorists are of the following form: there are two players (usually denoted by **I** and **II**) which take turns in choosing elements from a non-empty set $X$, with **I** playing first:

| | | | | | |
|---|---|---|---|---|---|
| **I** | $a_0$ | | $a_1$ | | $\cdots$ |
| **II** | | $b_0$ | | $b_1$ | $\cdots$ |

The game lasts $\omega$ many rounds with **I** playing first in each round, and at any finite stage of the game either player knows exactly the opponent's previous moves. (In this paper $\omega$ is the first transfinite ordinal, identified with $\{0, 1, 2, \dots\}$, the set of all natural numbers.) If the sequence constructed

*The author would like to thank the organizers of the fifth FotFS meeting for the invitation to talk on the Wadge hierarchy, and Alberto Marcone for suggestions that greatly simplify the exposition in Section 5.

by the two players by the end of the game

$$\langle a_0, b_0, a_1, b_1, \ldots \rangle \tag{1}$$

belongs to a set $C \subseteq {}^{\omega}X$ specified at the outset, then **I** wins; otherwise **II** wins. (The set ${}^{\omega}X$ is the collection of all $f : \omega \to X$.) The set $C$ is called the **pay-off set**, and the above game is denoted by $\mathcal{G}(C)$. Games of this form are usually called **infinite length, perfect-information, zero-sum, two players games on** $X$. A **strategy for I** in $\mathcal{G}(C)$ is a recipe[1] (*i.e.*, a function on sequences) that tells **I** how to answer to **II**'s moves; it is **winning** if the sequence (1) obtained following such strategy is in $C$, no matter what **II** plays. (The definition of (winning) strategy for **II** is similar.) A game is **determined** if one of the two players has a winning strategy—clearly the two players cannot have both winning strategies since otherwise, pitting them against each other, a sequence which is both inside and outside the pay-off set would result.

Mycielski and Steinhaus in [MycSte62] started a systematic investigation of the so-called **determinacy hypotheses**, *i.e.*, statements of the form

$$\forall C \subseteq {}^{\omega}X \text{ the game } \mathcal{G}(C) \text{ is determined.} \tag{$\mathsf{AD}_X$}$$

If $X$ is a singleton, then $\mathsf{AD}_X$ is trivially true, and if $X \subseteq Y$ or, more generally, if there is an injection from $X$ into $Y$, then $\mathsf{AD}_Y \implies \mathsf{AD}_X$. In [MycSte62] it is shown that $\mathsf{AD}_n \iff \mathsf{AD}_\omega$ for every $2 \leq n < \omega$, and the statement $\mathsf{AD}_\omega$ is called the **Axiom of Determinacy**, and it is denoted by AD. The word *axiom* is probably a misnomer, since AD is not meant to be an intuitively true fact about sets. In fact it contradicts a widely accepted mathematical postulate—the **Axiom of Choice**, AC—as it implies that all sets are Lebesgue measurable [MycSwi64]. In other words, AD is false in ZFC, the Zermelo-Fraenkel set theory ZF augmented with AC.

On the other hand, if the set $C$ is not too complex, then the game $\mathcal{G}(C)$ can be shown to be determined in ZFC: Gale and Stewart showed in [GalSte53] that for any set $X$, if $C$ is a closed subset of ${}^{\omega}X$, then $\mathcal{G}(C)$ is determined, and this was later extended by Martin [Mar75] to all Borel subsets of ${}^{\omega}X$. The topology on ${}^{\omega}X$ is given by the complete metric

$$d(f,g) = \begin{cases} 2^{-n} & \text{if } f \restriction n = g \restriction n \text{ and } f(n) \neq g(n), \\ 0 & \text{if } f = g. \end{cases} \tag{2}$$

The Axiom of Choice is essential for these results, since the Gale-Stewart theorem (the determinacy of all games with pay-off closed in ${}^{\omega}X$, for any

[1] The formal definition of strategy for will be given in Section 2.2.

$X$) implies AC. Therefore, although the statement that *all* games on $\omega$ are determined contradicts AC, the statement that all *closed* games on *any* $X$ is equivalent to AC over ZF. On the other hand, by a deep result of Woodin [Woo88] building on previous work of Martin and Steel [MarSte89], assuming ZFC and the existence of large enough cardinals, it can be shown that AD holds in $L(\mathbb{R})$, the smallest transitive model of ZF containing all reals and all ordinals. In particular, AD is consistent with ZF, modulo large cardinals. Woodin has shown that a similar result holds for $\mathsf{AD}_{\mathbb{R}}$, the **Axiom of Real Determinacy**, which is much stronger than AD (see [Woo99]). These results cannot be extended to $\mathsf{AD}_X$ for other uncountable $X$, since,*e.g.*, $\mathsf{AD}_{\omega_1}$ and $\mathsf{AD}_{\mathscr{P}(\mathbb{R})}$ are inconsistent with ZF. It is worth mentioning that the collection of all determined sets depends on the amount of strong axioms of set theory that one is willing to accept: if we restrict ourselves to ZFC alone, then we should diet on Borel sets only, but if we are willing to assume the existence of large cardinals, we might consider projective sets, or sets in $L(\mathbb{R})$, or even sets in larger inner models for determinacy [Woo99].

The main reason for studying AD and $\mathsf{AD}_{\mathbb{R}}$ is that they have interesting consequences in descriptive set theory. This area of mathematics, lying at the crossroads between analysis, topology and set theory, arose from the labors of the analysts of the turn of the 20th century (Baire, Borel, Lebesgue, Lusin, Suslin, to name a few) to put integration theory on firm grounds. After World War II the subject was rejuvenated by the introduction of methods from mathematical logic, and it is nowadays one of the most active areas of set theory. Modern descriptive set theory can be roughly defined as the study of definable subsets of Polish spaces, *i.e.*, separable completely metrizable spaces. Since any two uncountable Polish spaces are Borel-isomorphic, and since many set-theoretic questions are invariant under Borel isomorphism, it is customary in set theory to dub any uncountable Polish space $\mathbb{R}$. In set theory it is often desirable to deal with a version of $\mathbb{R}$ which is transparent from a combinatorial point of view, so we will work with the **Baire space** ${}^{\omega}\omega$ and with the **Cantor space** ${}^{\omega}2$: as they are zero-dimensional, they are homeomorphic to their $n$-th power ($1 \leq n \leq \omega$) so that pairs, tuples, or $\omega$-sequences can be continuously coded into a single element—something that cannot be done with the real line $(-\infty, +\infty)$ studied in Calculus.[2] In this paper, $\mathbb{R}$ will always stand for the Baire space,

$$\mathbb{R} \stackrel{\text{def}}{=} {}^{\omega}\omega \,,$$

but on a couple of occasions the Cantor space ${}^{\omega}2$ will also be considered.

[2]For a more compelling reason why the real line is not suited for our purposes, see Example 3.

The concept of "definable set" used in our characterization of modern descriptive set theory is a bit vague, and it is taken to mean: set in a reasonable family of sets, like Borel, projective, *etc.* In any case, such family should be immune from all pathologies (sets which are non-Lebesgue measurable, or do not have the property of Baire, *etc.*) spawned by the Axiom of Choice. A standard requirement is that such family should be closed under continuous pre-images, which brings us to the subject matter of the present paper, the **Wadge hierarchy**.

**Definition 1.** For $A$, $B$ subsets of $\mathbb{R}$, say that $A$ is **reducible** to $B$ if $A$ is the pre-image of $B$, *i.e.*,

$$A = f^{-1}(B)$$

for some $f : \mathbb{R} \to \mathbb{R}$, which is called a **reduction** of $A$ to $B$. If the reduction $f$ is continuous, then $A$ is said to be **Wadge reducible** to $B$, in symbols

$$A \leq_{\mathrm{W}} B\,. \tag{3}$$

The intuition behind this definition is that

$$\text{the set } A \text{ is at most as complex as the set } B. \tag{4}$$

Complexity here refers to some intuitive pre-ordering of topological complexity: we require that no set is strictly simpler than its complement. Thus if (3) holds, any upper bound for the complexity of $B$ yields an upper bound for the complexity of $A$ and—conversely—a lower bound for the complexity of $A$ yields a lower bound for the complexity of $B$. For example, if $B$ is Borel (or $\mathbf{F}_\sigma$, closed, *etc.*), then also $A$ must be Borel ($\mathbf{F}_\sigma$, closed, *etc.*), and if $A$ is *not* Borel ($\mathbf{F}_\sigma$, closed, *etc.*) then same can be said of $B$.

Although the notion of continuous pre-image harks back to the dawn of general topology, and had been used in a variety of situations, it was William W. Wadge, a PhD student of John Addison in Berkeley in the late-60s–early-70s, that first studied the structural properties of the relation $\leq_{\mathrm{W}}$ *per se.* Wadge's breakthrough relied on a reformulation of the reducibility relation (3) in terms of games (see Section 2.3 below), and it was during that time that infinite games and AD and other determinacy hypotheses were becoming a standard tool in descriptive set theory. In Wadge's own words:

> Yet nowhere (to our knowledge) is the relation $A = f^{-1}(B)$ for some continuous $f$ ever explicitly defined and studied as a partial order, not even in exhaustive work such as Kuratowski (1958) or Sierpiński (1952). In the latter, Sierpiński discusses preimage in general, continuous image and homeomorphic image, but not (explicitly) continuous preimage, which is perhaps the most natural. One possible explanation is that the investigation of $\leq$

> naturally involves infinite games, and it is only recently that game methods have been fully understood and appreciated. [Wad83, p. 3][3]

Wadge proved a simple, but fundamental result, now known as Wadge's Lemma, which has ushered a slew of new results in descriptive set theory.

**Lemma 2 (Wadge's Lemma).** Assume AD. Then for all $A, B \subseteq \mathbb{R}$

$$A \leq_{\mathrm{W}} B \quad \vee \quad \mathbb{R} \setminus B \leq_{\mathrm{W}} A \,. \tag{5}$$

This implies the converse of the principle (4): if $A$ is simpler than (or as complex as) $B$ from the topological perspective, and if $A \leq_{\mathrm{W}} B$ does not hold, then $\mathbb{R} \setminus B \leq_{\mathrm{W}} A$ and hence $\mathbb{R} \setminus B$ is simpler than (or as complex as) $A$, from which we could conclude that $B$ is strictly simpler than its complement, contradicting our basic assumption on the topological complexity of sets. Note that Wadge's Lemma fails if the ambient space $\mathbb{R}$ is taken to be real line $(-\infty, +\infty)$ or any other connected Polish space, as the following example (taken from [Woo99, Remark 9.26, p. 624]) shows:

**Example 3.** Let $A = \mathbb{Q}$, let $C \subset [0, 1]$ be the usual 1/3-Cantor set and let $B \subseteq C$. We claim that there is no continuous reduction of $A$ to $B$: since $B$ can be taken to be not in $\mathbf{G}_\delta$, and since $A \in \mathbf{F}_\sigma$, then $\mathbb{R} \setminus B \not\leq_{\mathrm{W}} A$, and this will witness the failure of Wadge's Lemma for $A$ and $B$. Towards a contradiction, suppose $A = f^{-1}(B)$ for some continuous $f$. Then

$$\mathbb{R} = \mathrm{Cl}(\mathbb{Q}) = \mathrm{Cl}\big(f^{-1}(B)\big) \subseteq f^{-1}\big(\mathrm{Cl}(B)\big) \subseteq f^{-1}(C)\,,$$

where Cl denotes the closure. Thus $\mathrm{ran}(f) \subseteq C$. But $\mathrm{ran}(f)$ is connected, and since the connected components of $C$ are its points, it follows that $f$ is constant, and hence it cannot be a reduction of $A$ to $B$.

Wadge's Lemma says that under AD the pre-order $\leq_{\mathrm{W}}$ is *semi*-linear, in the following sense: if $A$ is not Wadge reducible to $B$ then

$$\neg B \stackrel{\text{def}}{=} \mathbb{R} \setminus B$$

is reducible to $A$. For this reason (5) is called the Semi-Linear Ordering principle for continuous reductions, and it is denoted by $\mathsf{SLO}^{\mathrm{W}}$. The equivalence classes of the induced equivalence relation

$$A \equiv_{\mathrm{W}} B \quad \Longleftrightarrow \quad A \leq_{\mathrm{W}} B \wedge B \leq_{\mathrm{W}} A$$

are called **Wadge degrees**. The pre-order $\leq_{\mathrm{W}}$ on $\mathscr{P}(\mathbb{R})$ induces a partial order on the Wadge degrees which is denoted by $\leq$, and Wadge's Lemma

[3] Kuratowski (1958) and Sierpiński (1952) are, respectively, the monographs [Kur58] and [Sie52], and $\leq$ is our $\leq_{\mathrm{W}}$.

amounts to say that every anti-chain in $\leq$ is of length 2. By 1972 Wadge had carried out a thorough analysis of the structure of Wadge degrees of Borel sets (these result appeared much later in his PhD dissertation [Wad83]), and in the subsequent years many descriptive set theorists (Martin, Steel, Kechris, Louveau, to name a few) further developed the theory of the Wadge degrees. Some of these results are reported in papers in the Cabal volumes [KecMos78, KecMarMos81, KecMarMos83, KecMarSte88], some other are buried in PhD theses [Wad83, Van77, Ste77], but—to our knowledge—there is no single place that gives an overview of the results in this area of set theory. The ambition of the present paper is to fill this gap in the literature by surveying some classical and some recent results—other papers that provide some general information on the Wadge hierarchy are [Van78b, And03, And06]. Since all of the results presented here have appeared somewhere else (albeit in form of PhD theses), most of them are stated without proof, but with a reference to the sources. The only exceptions are the results in Section 5, which are new and presented in detail. In Section 2 some standard facts and terminology in descriptive set theory are recalled—for all undefined concepts, the reader is referred to the standard monographs [Kec95, Mos80]. Still this is *not* to meant to be a scholarly account of the theory of Wadge degrees, and many important modern developments are completely ignored: for example no attempt was made to cover the work by Louveau and Saint-Raymond on the strength of Wadge Determinacy of Borel sets [LouSai88, Lou83] or recent applications of Wadge reducibility to automata theory and languages [Dup03, DupFinRes01, Sel03].

## 2 Basic facts about Wadge degrees

Firstly, let us briefly recall the definitions and notations that will be used throughout this paper.

### 2.1 Notation

The set ${}^{Y}X$ is the collection of all functions from $Y$ to $X$. We endow ${}^{\omega}X$ with the product topology, where $X$ is taken to be discrete. The resulting space is completely metrizable (via the complete metric given in (2)) and it is separable if and only if $X$ is countable. The set ${}^{<\omega}X = \bigcup_n {}^{n}X$ is the collection of all finite sequences from $X$. If $s, t \in {}^{<\omega}X$ then $s^\frown t \in {}^{<\omega}X$ is the finite sequence obtained by concatenating $s$ with $t$; when dealing with sequences of length 1, say $\langle x \rangle$, we shall write $x^\frown t$ and $s^\frown x$ rather than $\langle x \rangle^\frown t$ or $s^\frown \langle x \rangle$. The notion of concatenation can be defined also when $s \in {}^{<\omega}X$ and $f \in {}^{\omega}X$: in this case $s^\frown f \in {}^{\omega}X$.

A **tree on** $X$ is a set $T \subseteq {}^{<\omega}X$ which closed under initial segments, *i.e.*, $t \in T \wedge s \subseteq t \implies s \in T$; the elements of $T$ are called **nodes**. In particular

${}^{<\omega}X$ is a tree on $X$. A node $t \in T$ is **terminal** if and only if there is no $s \in T$ such that $t \subset s$; $\mathbf{tn}(T)$ is the set of all terminal nodes. A tree without terminal nodes is said to be **pruned**. A **branch** through a tree $T$ on $X$ is an $f \in {}^{\omega}X$ such that $f \restriction n \in T$ for all $n \in \omega$. The **body** of a tree $[T]$ is the set of all branches through $T$. If $T$ is a pruned non-empty tree on $X$, then $[T] \neq \varnothing$ by DC—see below for the definition of DC. For $n \in \omega$,

$$x^{(n)} = \underbrace{\langle x, x, \ldots \rangle}_{n}$$

is the sequence of $n$-many consecutive $x$'s where $n \leq \omega$, with the understanding that $x^{(0)} = \varnothing$. If $\varphi : {}^{<\omega}\omega \to {}^{<\omega}\omega$ is such that

(i) $s \subseteq t \implies \varphi(s) \subseteq \varphi(t)$, and
(ii) $\lim_{n \to \infty} \mathrm{lh}(\varphi(x \restriction n)) = \infty$, for all $x \in \mathbb{R}$,

then a continuous function $f_\varphi : \mathbb{R} \to \mathbb{R}$, $f_\varphi(x) = \bigcup_n \varphi(x \restriction n)$ is defined, and, conversely, any continuous function from $\mathbb{R}$ to $\mathbb{R}$ is of the form $f_\varphi$. Since $\varphi$ can be coded by a real, we fix from now on an $\mathbb{R}$-parametrization of the set of all continuous functions,

$$\{\boldsymbol{f}_x \mid x \in \mathbb{R}\}. \tag{6}$$

If (ii) is strengthened to $\mathrm{lh}(\varphi(s)) = \mathrm{lh}(s)$, then the resulting function $f_\varphi$ is said to be **Lipschitz**. Thus $f$ is Lipschitz if and only if

$$\forall n \in \omega \, (x \restriction n = y \restriction n \implies f(x) \restriction n = f(y) \restriction n),$$

or, equivalently,

$$d(f(x), f(y)) \leq d(x, y).$$

Since Lipschitz functions are continuous, in analogy with (6) it is possible to find an $\mathbb{R}$-parametrization

$$\{\boldsymbol{\ell}_x \mid x \in \mathbb{R}\} \tag{7}$$

of the set of all Lipschitz functions. Moreover it is not hard to arrange this so that the map

$$\mathbb{R} \times \mathbb{R} \to \mathbb{R}, \qquad (x, y) \mapsto \boldsymbol{\ell}_x(y) \tag{8}$$

is continuous—see [AndHjoNee07]. Finally, a Lipschitz function $f : \mathbb{R} \to \mathbb{R}$ such that

$$x \restriction n = y \restriction n \implies f(x) \restriction n + 1 = f(y) \restriction n + 1,$$

is said to be a **contraction**. Let $A \subseteq \mathbb{R}$. The **complement** of $A$ is $\neg A = \mathbb{R} \setminus A$. If $s \in {}^{<\omega}\omega$, then

$$s^\frown A = \{s^\frown x \mid x \in A\}.$$

The set

$$A_{\lfloor s \rfloor} = \{x \mid s^\frown x \in A\} \tag{9}$$

is called the **localization** of $A$ at $s$, and clearly $(s^\frown A)_{\lfloor s \rfloor} = A$. If $A_n \subseteq \mathbb{R}$, let

$$\bigoplus\nolimits_n A_n = \bigcup\nolimits_n n^\frown A_n$$

and let $A \oplus B = \bigoplus_n A_n$ where $A_{2n} = A$ and $A_{2n+1} = B$.

### 2.2 Determinacy and Choice

Given a non-empty set $X$ and a $C \subseteq {}^{\omega}X$, we define a **game** $\mathcal{G}(C)$ in which two players **I** and **II** alternately pick elements from $X$, with **I** playing first. Let $a_0, a_1, \ldots$ and $b_0, b_1, \ldots$ be the sequences of **moves** by **I** and **II**, respectively. A play of $\mathcal{G}(C)$ lasts $\omega$-many rounds so that at the end of the two players have cooperatively constructed a sequence $\langle a_0, b_0, a_1, b_1, \ldots \rangle$ as in (1). Then **I** wins iff such sequence in in $C$; otherwise **II** wins. A **strategy for I** is a function

$$\sigma : {}^{<\omega}X \to X$$

and it is said to be winning if for any sequence $\langle b_0, b_1, \ldots \rangle$ played by **II**, letting $a_0 = \sigma(\varnothing)$, $a_1 = \sigma(b_0)$, $a_2 = \sigma(b_0, b_1)$, $\ldots$, the resulting (1) sequence is in $C$. A strategy for **II** is a function

$$\tau : {}^{<\omega}X \setminus \{\varnothing\} \to X,$$

and it is winning if for any sequence $\langle a_0, a_1, \ldots \rangle$ played by **I**, letting $b_0 = \tau(a_0)$, $b_1 = \tau(a_0, a_1)$, $\ldots$, the sequence (1) is not in $C$.

A game $\mathcal{G}(C)$ is **determined** iff one (and only one) of the two players has a winning strategy, and $\mathsf{AD}_X$ is the statement that all games on $X$ are determined.

The Axiom of Determinacy $\mathsf{AD}$ (that is $\mathsf{AD}_\omega$) contradicts the Axiom of Choice, $\mathsf{AC}$, since it implies that:

(LM) every set of reals is Lebesgue measurable,

(BP) every set of reals has the property of Baire,

(PSP) every set of reals is countable, or else it contains a perfect subset,[4]

and it is well-known that any one of these three properties contradicts $\mathsf{AC}$—see [Kan03, p. 373–377]. There are weak forms of $\mathsf{AC}$, though, that are consistent with $\mathsf{AD}$.

[4] This is known as the perfect set property.

**Definition 4.** The **Axiom of Countable Choice over** $X$, $\mathsf{AC}_\omega(X)$, is the statement: for any family $\{A_n \mid n \in \omega\}$ of non-empty subsets of $X$, there is a choice function $n \mapsto a_n \in A_n$.

The **Axiom of Dependent Choice over** $X$, $\mathsf{DC}(X)$, is the statement: for any relation $R \subseteq X \times X$ such that $\forall x \in X\, \exists y \in X\, (x\, R\, y)$, and for any $x_0 \in X$, there is a sequence $x_1, x_2, \ldots$ such that $x_n\, R\, x_{n+1}$, for all $n \in \omega$. The axioms $\mathsf{AC}_\omega$ and $\mathsf{DC}$ are the global versions of $\mathsf{AC}_\omega(X)$ and $\mathsf{DC}(X)$, namely:

$$\forall X\, \mathsf{AC}_\omega(X) \quad \text{and} \quad \forall X\, \mathsf{DC}(X)\,.$$

The axiom $\mathsf{AC}_\omega(\mathbb{R})$ implies that the countable union of countable subsets of $\mathbb{R}$ is countable and that $\omega_1$ is regular. It is well-known that $\mathsf{AD} \implies \mathsf{AC}_\omega(\mathbb{R})$ [Kan03, Proposition 27.10], while it is open wether the same is true for $\mathsf{DC}(\mathbb{R})$:

**Problem 5.** Does $\mathsf{AD}$ imply $\mathsf{DC}(\mathbb{R})$?

By a theorem of Kechris' [Kec84], this is true if $\mathrm{V} = \mathrm{L}(\mathbb{R})$ is assumed, and since $\mathsf{DC}(\mathbb{R}) \implies \mathsf{DC}$ holds in $\mathrm{L}(\mathbb{R})$, it follows that $\mathsf{AD}$ is consistent with $\mathsf{DC}$. By unpublished work of Woodin, the consistency strength of the statement $\mathsf{AD} + \neg\mathsf{DC}(\mathbb{R})$ is higher than that of $\mathsf{AD}$, if consistent at all.

### 2.3 Lipschitz and Wadge games

Wadge introduced two important games on $\omega$. The **Lipschitz game for** $A, B \subseteq \mathbb{R}$

$$G_{\mathrm{L}}(A, B)$$

is the game in which players **I** and **II** play natural numbers, $a_i$ and $b_i$, respectively,

| **I** | $a_0$ | | $a_1$ | | $\cdots$ |
|---|---|---|---|---|---|
| **II** | | $b_0$ | | $b_1$ | $\cdots$ |

and player **II** wins just in case

$$a = \langle a_i \mid i \in \omega \rangle \in A \iff b = \langle b_i \mid i \in \omega \rangle \in B\,.$$

It is not hard to see that $G_{\mathrm{L}}(A, B)$ is a game on $\omega$ according to our definition, *i.e.*, it is of the form $\mathcal{G}(C)$ for an appropriate $C \subseteq \mathbb{R}$. A strategy for **II** is—essentially—a Lipschitz function, while a strategy for **I** is a contraction. Hence

$$\begin{aligned} \textbf{I} \text{ wins } G_{\mathrm{L}}(A, B) &\iff \neg B \text{ is reducible to } A \text{ via a contraction, and} \\ \textbf{II} \text{ wins } G_{\mathrm{L}}(A, B) &\iff A \text{ is reducible to } B \text{ via a Lipschitz map.} \end{aligned}$$

When $A$ is reducible to $B$ via a Lipschitz function we say that $A$ is **Lipschitz reducible** to $B$ and write

$$A \leq_{\mathrm{L}} B\,.$$

In order to illustrate the power of game theoretic techniques we pause for a simple example.

**Example 6.** Suppose $A$ is a closed set and $B$ is not open, and let us show that $A \leq_{\mathrm{L}} B$. If $A = \varnothing$, then this must hold, since $B$ cannot be equal to $\mathbb{R}$, so we may assume that $A = [T]$ with $T$ a non-empty pruned tree on $\omega$. By case assumption there is an $x \in B \setminus \mathrm{Int}(B)$. Then **II** wins $G_{\mathrm{L}}(A, B)$ as follows:

> As long as **I** plays inside $T$, then **II** enumerates $x$. If at some round $n$ **I** leaves $T$, then **II** picks a real $x_n \supset x{\restriction}n$ such that $x_n \notin B$, and starts following $x_n$ rather than $x$. (Such $x_n$ exists as $x$ is not in the interior of $B$.)

Thus if **I**'s play is in $[T]$, then **II**'s play will be $x \in B$, and **I**'s play is not in $[T]$, then **II**'s play will be $x_n \notin B$, where $n$ is the round when **I** left the tree $T$.

The **Wadge game** is similar to the Lipschitz game, but **II** has the further option of *passing* (*i.e.*, not playing) at any round, with the proviso that he must play infinitely often. Formally, **II** can play at any round a non-integer move $\mathsf{p}$, which means *pass*: if the sequence of his moves is definitely equal to $\mathsf{p}$, then he loses; otherwise consider the sequence of his moves once the $\mathsf{p}$ are dropped: it is infinite and of the form $b = \langle b_n \mid n \in \omega \rangle$, where the $b_n$'s are the actual moves on **II**, and **II** wins iff

$$a \in A \quad\Longleftrightarrow\quad b \in B\,,$$

where $a$ is the sequence of moves of **I**. Thus the Wadge game can be thought as the Lipschitz game with player **II** taking short naps. Defined this way, $G_{\mathrm{W}}$ is not a game on $\omega$ because of the $\mathsf{p}$ move: one possible way of construing it as a honest game on $\omega$ is to have **II** play 0 to mean $\mathsf{p}$, and $n+1$ to mean $n$, but we will never think of $G_{\mathrm{W}}$ this way. Notice that

$$\begin{aligned} \textbf{I} \text{ wins } G_{\mathrm{W}}(A,B) &\quad\Longrightarrow\quad \neg B \text{ is reducible to } A \text{ via a contraction,} \\ \textbf{II} \text{ wins } G_{\mathrm{W}}(A,B) &\quad\Longleftrightarrow\quad A \leq_{\mathrm{W}} B\,, \end{aligned}$$

and therefore **I**'s burden is much heavier in the Wadge game than in the Lipschitz game.

We can now easily prove Lemma 2, Wadge's Lemma: by determinacy, one of the two players has a winning strategy in the game $G_{\mathrm{L}}(A, B)$. If **II** has a winning strategy, then $A \leq_{\mathrm{L}} B$, and if **I** has a winning strategy, then $\neg B \leq_{\mathrm{L}} A$. Therefore AD implies that

$$A \leq_{\mathrm{W}} B \quad \vee \quad \neg B \leq_{\mathrm{W}} A\,. \tag{10}$$

and as Lipschitz functions are continuous, this can be weakened to

$$A \leq_{\mathrm{L}} B \quad \vee \quad \neg B \leq_{\mathrm{L}} A\,. \tag{11}$$

A set which is reducible to its complement is said to be **Wadge self-dual**; otherwise it is said to be **Wadge non-self-dual**. It is easy to check that any clopen set other than $\mathbb{R}$ or $\varnothing$ is self-dual. For an example of non-self-dual set take $\mathbb{R}$ or $\varnothing$ or, less trivially, an open or closed set which is not clopen. Note that Wadge's Lemma implies that a self-dual set is comparable with any other set.

## 2.4 Lipschitz versus Wadge degrees

Two sets are **Wadge equivalent**, in symbols

$$A \equiv_{\mathrm{W}} B\,,$$

iff $A \leq_{\mathrm{W}} B$ and $B \leq_{\mathrm{W}} A$. Note that $A \leq_{\mathrm{W}} B$ if and only if $\neg A \leq_{\mathrm{W}} \neg B$ and hence

$$A \leq_{\mathrm{W}} \neg A \iff A \equiv_{\mathrm{W}} \neg A\,.$$

A $\equiv_{\mathrm{W}}$-class is called a **Wadge degree** and

$$[A]_{\mathrm{W}} = \{B \mid B \equiv_{\mathrm{W}} A\}$$

is the Wadge degree of $A$. Since the notion of self-duality is invariant under $\equiv_{\mathrm{W}}$ we can speak of (non-)self-dual Wadge degrees. All these notions can be recast for Lipschitz reductions: in this case we shall speak of Lipschitz degrees, and since Lipschitz functions are continuous, every Wadge degree is union of Lipschitz degrees. It is easy to check that

$$[\mathbb{R}]_{\mathrm{W}} = [\mathbb{R}]_{\mathrm{L}} = \{\mathbb{R}\} \quad \text{and} \quad [\varnothing]_{\mathrm{W}} = [\varnothing]_{\mathrm{L}} = \{\varnothing\}$$

are minimal among Wadge degrees, and among Lipschitz degrees. The strict part of the pre-orders $\leq_{\mathrm{L}}$ and $\leq_{\mathrm{W}}$ are defined by

$$\begin{aligned} A <_{\mathrm{L}} B &\iff A \leq_{\mathrm{L}} B \wedge B \not\leq_{\mathrm{L}} A, \\ A <_{\mathrm{W}} B &\iff A \leq_{\mathrm{W}} B \wedge B \not\leq_{\mathrm{L}} A\,. \end{aligned}$$

**Lemma 7.** For any $A, B \subseteq \mathbb{R}$,

$$\begin{aligned} A <_{\mathrm{L}} B &\implies \neg A <_{\mathrm{L}} B\,, \\ A <_{\mathrm{W}} B &\implies \neg A <_{\mathrm{W}} B\,. \end{aligned} \tag{12}$$

*Proof.* Since $B \not\leq_{\mathrm{L}} A$ then $\neg A \leq_{\mathrm{L}} B$ by (10), so it is enough to check that $B \equiv_{\mathrm{L}} \neg A$ does not hold: otherwise $A \leq_{\mathrm{L}} B$ and transitivity imply that $A \leq_{\mathrm{L}} \neg A$ and therefore $A \equiv_{\mathrm{L}} \neg A \equiv_{\mathrm{L}} B$, contradicting $A <_{\mathrm{L}} B$. The case for $<_{\mathrm{W}}$ is identical. q.e.d.

Wadge's Lemma says—essentially—two things:

(i) the antichains in the Wadge degrees have length at most 2 and are of the form $\{[A]_{\mathrm{W}}, [\neg A]_{\mathrm{W}}\}$ with $A$ non-self-dual, and
(ii) if we identify every non-self-dual degree with its dual (*i.e.*, if we collapse each anti-chain to a single object), then the resulting order is linear.

And similarly for the Lipschitz degrees. Donald A. Martin, building on previous work of Leonard Monk, showed in 1972 that—under the identification as in (ii) above—a well-order is obtained both in the case of Wadge and Lipschitz degrees.

**Theorem 8 (Martin–Monk).** Assume AD. Then there is no sequence $\langle A_n \mid n < \omega \rangle$ of sets of reals such that either $\forall n \in \omega\,(A_{n+1} <_{\mathrm{L}} A_n)$ or $\forall n \in \omega\,(A_{n+1} <_{\mathrm{W}} A_n)$.

**Corollary 9.** Assume AD and DC($\mathbb{R}$). Then the ordering on the Wadge and Lipschitz degrees is well-founded.

*Proof.* Since each Wadge degree is a union of Lipschitz degrees, it is enough to prove the result for Lipschitz degrees. Towards a contradiction, suppose $\mathcal{A} \neq \varnothing$ is a collection of Lipschitz degrees without a minimal element and let $A$ be a set such that $[A]_{\mathrm{L}} \in \mathcal{A}$. Consider the tree

$$T = \big\{\langle x_0, \ldots, x_n \rangle \in {}^{<\omega}\mathbb{R} \mid \forall i < n\,\big(\boldsymbol{\ell}_{x_{i+1}}(A) <_{\mathrm{L}} \boldsymbol{\ell}_{x_i}(A)\big) \wedge \forall i \leq n\,([\boldsymbol{\ell}_{x_i}(A)]_{\mathrm{L}} \in \mathcal{A})\big\}\,,$$

searching for an infinite descending chain $\langle [\boldsymbol{\ell}_{x_i}(A)]_{\mathrm{L}} \mid i \in \omega \rangle$ in $\mathcal{A}$. (The $\boldsymbol{\ell}_x$'s are defined in (7).) By case assumption on $\mathcal{A}$, $T$ is a pruned tree on $\mathbb{R}$, and hence by DC($\mathbb{R}$) it must have an infinite branch, contrarily to Theorem 8. q.e.d.

Since the Lipschitz (Wadge) degrees are essentially well-ordered, we will speak of the **Lipschitz** (**Wadge**) **hierarchy**, and $\|A\|_{\mathrm{L}}$ ($\|A\|_{\mathrm{W}}$) is the ordinal of $A$ in the well-founded pre-order $\leq_{\mathrm{L}}$ ($\leq_{\mathrm{W}}$). For technical reasons, it is convenient to start counting from 1 rather than 0; in other words

$$\|\mathbb{R}\|_{\mathrm{L}} = \|\mathbb{R}\|_{\mathrm{W}} = \|\varnothing\|_{\mathrm{L}} = \|\varnothing\|_{\mathrm{W}} = 1\,.$$

Since the Lipschitz and Wadge degrees are essentially well-ordered, we will speak of the Lipschitz or Wadge hierarchy, respectively. A **successor degree** is a degree which has an immediate predecessor, *i.e.*, its rank is a successor ordinal different from 1. A degree which is not minimal (*i.e.*, different from $\{\mathbb{R}\}$ or $\{\varnothing\}$) and not a successor is called **limit**.

Recall (Open Problem 5) that it is not known whether AD implies DC($\mathbb{R}$). Obviously an affirmative answer would simplify the hypotheses of Corollary 9, and to the best of our knowledge, the following weakening of Problem 5 seems also to be open:

**Problem 10.** Is DC($\mathbb{R}$) needed for Corollary 9? In other words: does AD imply the well-foundedness of the Wadge and Lipschitz hierarchies?

*Proof of Theorem 8.* Any $<_{\mathrm{W}}$-descending sequence is a $<_{\mathrm{L}}$-descending sequence, so it is enough to prove the Theorem for $<_{\mathrm{L}}$. Towards a contradiction suppose $A_{n+1} <_{\mathrm{L}} A_n$ for all $n$. By (12) for Lipschitz reducibility, $\neg A_{n+1} <_{\mathrm{L}} A_n$ must hold, so by Wadge's Lemma player **I** wins both $G_{\mathrm{L}}(A_n, A_{n+1})$ and $G_{\mathrm{L}}(A_n, \neg A_{n+1})$: for the sake of definiteness fix winning strategies $\sigma_n^0$ and $\sigma_n^1$ for **I** in these games. For $z \in {}^{\omega}2$ let

$$\mathcal{G}_n^z = \begin{cases} G_{\mathrm{L}}(A_n, A_{n+1}) & \text{if } z(n) = 0\,, \\ G_{\mathrm{L}}(A_n, \neg A_{n+1}) & \text{if } z(n) = 1\,. \end{cases}$$

The games $\{\mathcal{G}_n^z \mid n \in \omega\}$ will be played simultaneously with **I** using $\sigma_n^{z(n)}$ in $\mathcal{G}_n^z$ and **II** copying **I**'s moves from the next game, $\mathcal{G}_{n+1}^z$. To see how this is done consider the diagram of Figure 1, where $\xrightarrow{\sigma_n}$ is the strategy $\sigma_n^{z(n)}$ and $\Longrightarrow$ is the copying strategy. **I** starts by filling-in the first column, then **II** copies and fills-in the second column. Now **I** can use his strategies to fill-in the third column. And so on. Let $a_n^z = \langle a_n^z(i) \mid i \in \omega \rangle$ be the real played by **I** is $\mathcal{G}_n^z$. Then $a_{n+1}^z$ is also the real played by **II** in $\mathcal{G}_n^z$. So we have the following

$$\begin{aligned} z(n) = 0 &\implies \left(a_n^z \in A_n \iff a_{n+1}^z \notin A_{n+1}\right) \\ z(n) = 1 &\implies \left(a_n^z \in A_n \iff a_{n+1}^z \in A_{n+1}\right). \end{aligned} \tag{13}$$

Notice that if $z$ and $w$ are eventually equal, that is for some $m_0$

$$\forall m \geq m_0 \, (z(m) = w(m))\,,$$

then

$$\forall m \geq m_0 \, (a_m^z = a_m^w)\,.$$

**Claim 11.** Suppose $z, w \in {}^{\omega}2$ are such that $\exists! k \in \omega \, (z(k) \neq w(k))$. Then

$$a_0^z \in A_0 \iff a_0^w \notin A_0\,.$$

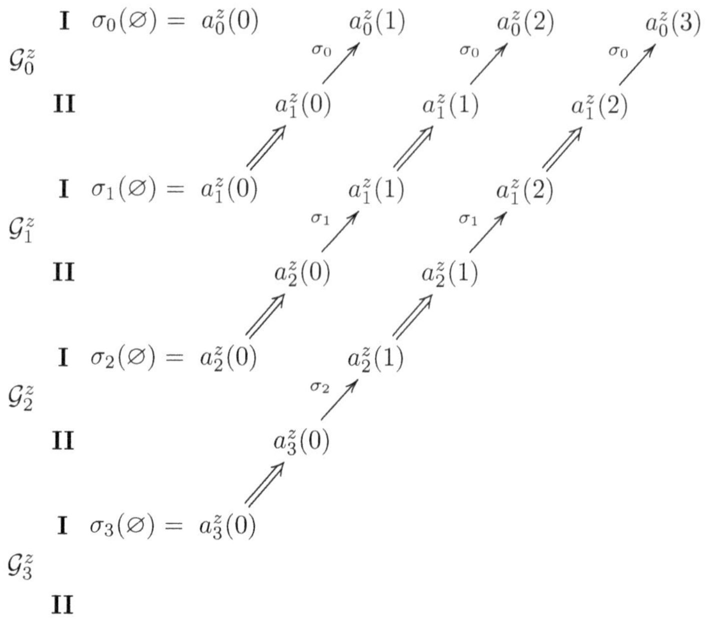


Figure 1. The Martin-Monk diagram

*Proof of Claim 11.* As $z$ and $w$ agree after $k$, then $a_{k+1}^z = a_{k+1}^w$. By (13) above, $a_k^z \in A_k \iff a_k^w \notin A_k$. As $z{\restriction}k = w{\restriction}k$ and again by repeated applications of (13), $a_0^z \in A_0 \iff a_0^w \notin A_0$. q.e.d.

Therefore the set

$$F = \{z \in {}^{\omega}2 \mid a_0^z \in A_0\}$$

has the property that

$$\forall z, w \in {}^{\omega}2\, (\exists! k\; z(k) \neq w(k) \implies (z \in F \iff w \notin F))\,. \tag{14}$$

Any subset of the Cantor set that satisfies (14) is called a **flip-set** since the truth value of "$z \in F$" flips between true and false every time a coordinate of $z$ is changed. No flip-set can have the property of Baire or can be Lebesgue measurable[5] and therefore the existence of $F$ contradicts AD.

[5] The argument in [Kec95, Theorem 8.46 and Exercise 8.50] proves it only for the Baire property, but the same proof, using the fact that if a set $X$ is Lebesgue measurable then $\mu(X \cap \boldsymbol{N}_s) = 0$ or $\mu(X \cap \boldsymbol{N}_s) = \mu(\boldsymbol{N}_s)$ for some $s$, does it for the measure case.

Having reached a contradiction, the proof of Theorem 8 is complete.

q.e.d. (Theorem 8)

### 2.5 The structure of the Wadge hierarchy

Suppose $A_0 <_{\mathrm{L}} A_1 <_{\mathrm{L}} A_2 <_{\mathrm{L}} \ldots$. Then each $A_i$ is Lipschitz reducible to $\bigoplus_n A_n$, since **II** wins $G_{\mathrm{L}}(A_i, \bigoplus_n A_n)$ by playing $i$ and then copying **I**'s moves, so $[\bigoplus_n A_n]_{\mathrm{L}}$ is an upper-bound of the $[A_n]_{\mathrm{L}}$'s. In fact it is the least upper-bound: if $B <_{\mathrm{L}} \bigoplus_n A_n$ then **I** wins $G_{\mathrm{L}}(\bigoplus_n A_n, B)$ via some strategy $\sigma$. If $i = \sigma(\varnothing)$ is **I**'s first move, then $\sigma$ induces a winning strategy for **II** in $G_{\mathrm{L}}(\neg B, A_i)$ and therefore $\neg B \leq_{\mathrm{L}} A_i$. Since $A_i <_{\mathrm{L}} \neg A_n$ for $i < n$ by an analogue of (12), we have $B \leq_{\mathrm{L}} A_n$ for all sufficiently large $n$'s. Moreover

$$\bigoplus\nolimits_n A_n \leq_{\mathrm{L}} \neg(\bigoplus\nolimits_n A_n) = \bigoplus\nolimits_n \neg A_n$$

since **II** wins $G_{\mathrm{L}}(\bigoplus_n A_n, \bigoplus_n \neg A_n)$ as follows: if **I** plays $n$, then **II** plays $n+1$ and then follows any winning strategy for the game $G_{\mathrm{L}}(A_n, \neg A_{n+1})$ (which exists by (12)).

We have stated the above results for the Lipschitz hierarchy, but the proofs go through *verbatim* for the Wadge hierarchy. To recap: any increasing $\omega$-sequence of Wadge/Lipschitz degrees has a self-dual least upper bound given by the $\bigoplus$-construction.

**Lemma 12.** Assume AD. If $[A]_{\mathrm{L}}$ is self-dual then, for any $k \in \omega$, the degree $[k^\frown A]_{\mathrm{L}}$ is self-dual and is the least Lipschitz degree above $[A]_{\mathrm{L}}$.

*Proof.* $A \leq_{\mathrm{L}} k^\frown A$ is clear. Fix $\tau$ a winning strategy for **II** in $G_{\mathrm{L}}(A, \neg A)$. Then **I** wins $G_{\mathrm{L}}(k^\frown A, A)$ by playing $k$ and then following $\tau$. Therefore **II** cannot have a winning strategy and hence $k^\frown A \not\leq_{\mathrm{L}} A$. Suppose $B <_{\mathrm{L}} k^\frown A$ and let $\sigma$ be a winning strategy for **I** in $G_{\mathrm{L}}(k^\frown A, B)$. Then **II** can use $\sigma$ to win $G_{\mathrm{L}}(\neg B, A)$, showing thus that $\neg B \leq_{\mathrm{L}} A$, hence $B \leq_{\mathrm{L}} \neg A$ and therefore $B \leq_{\mathrm{L}} A$. q.e.d.

Recall from (9) the notion of localization.

**Corollary 13.** Assume AD. If $A$ is Lipschitz self-dual, then $A_{\lfloor n \rfloor} <_{\mathrm{L}} A$ for any $n$.

*Proof.* Notice that $A_{\lfloor n \rfloor} \leq_{\mathrm{L}} n^\frown A_{\lfloor n \rfloor} \leq_{\mathrm{L}} A$: if $A_{\lfloor n \rfloor}$ is self-dual then the first inequality is strict by Lemma 12; if $A_{\lfloor n \rfloor}$ is non-self-dual and $A \leq_{\mathrm{L}} A_{\lfloor n \rfloor}$, then $A \equiv_{\mathrm{L}} A_{\lfloor n \rfloor}$ and hence $A$ would be non-self-dual, a contradiction. q.e.d.

Using Lemma 12 at successor stages and by taking least upper bounds (the $\bigoplus$-construction) at limit stages, it follows that any self-dual $[A]_{\mathrm{L}}$ is followed by $\omega_1$-many self-dual degrees. By the preceding arguments, if $\operatorname{cof}(\|A\|_{\mathrm{L}}) = \omega$, then $A$ is self-dual. Conversely, if $[A]_{\mathrm{L}}$ is a limit degree and self-dual, then by Corollary 13

$$\forall n \left(A_{\lfloor n \rfloor} <_{\mathrm{L}} A\right).$$

If $\operatorname{cof}(\|A\|_{\mathrm{L}}) > \omega$, let $B$ be a set such that $\|A_{\lfloor n \rfloor}\|_{\mathrm{L}} \leq \|B\|_{\mathrm{L}} < \|A\|_{\mathrm{L}}$, for all $n$: then $A = \bigoplus_n A_{\lfloor n \rfloor} \leq_{\mathrm{L}} B$, a contradiction. In other words, we have shown that any limit Lipschitz degree of uncountable cofinality must be non-self-dual. Therefore the Lipschitz hierarchy looks like this:

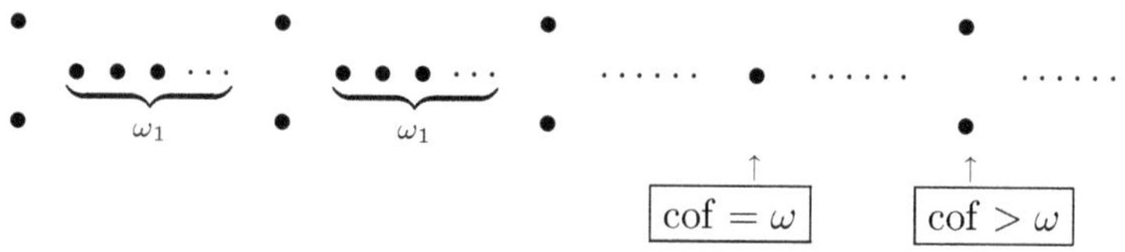

(15)

John Steel and Robert Van Wesep—at the time graduate students in Berkeley—used the technique of Theorem 8 (the so-called Martin–Monk method) to prove the following:

**Theorem 14 (Steel–Van Wesep).** Assume AD. Then

$$A \leq_{\mathrm{W}} \neg A \implies A \leq_{\mathrm{L}} \neg A.$$

For a proof see [Van78b, Theorem 3.1].

**Corollary 15.**
Assume AD and that $A \not\leq_{\mathrm{W}} \neg A$. Then $B \leq_{\mathrm{W}} A \implies B \leq_{\mathrm{L}} A$.

It is easy to check that each $\omega_1$-block of consecutive self-dual Lipschitz degrees is contained in a single self-dual Wadge degree, and by the Steel–Van Wesep Theorem a Wadge self-dual degree is made-up exactly of one $\omega_1$-block of self-dual Lipschitz degrees. Therefore in the Wadge hierarchy self-dual degrees and non-self-dual pairs alternate, with self-dual degrees at limit stages of countable cofinality, and non-self-dual pairs at levels of uncountable cofinality:

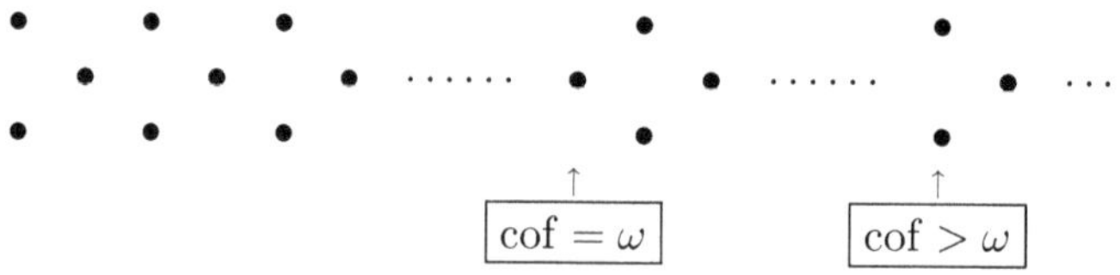

(16)

The bottom of the hierarchy is occupied by the non-self-dual pair $[\mathbb{R}]_{\mathrm{W}}$ and $[\varnothing]_{\mathrm{W}}$. Immediately above there is the self-dual degree of all clopen sets (except for $\mathbb{R}$ and $\varnothing$), and above that the non-self-dual pair of all open and all closed sets (except for the clopen sets); the $\mathbf{F}_\sigma$ and $\mathbf{G}_\delta$ sets appear at level $\omega_1$.

## 2.6 The length of the Wadge hierarchy

The ordinal

$$\Theta = \sup\{\alpha \mid \exists f\,(f : \mathbb{R} \twoheadrightarrow \alpha)\},$$

the supremum of the ordinals which are surjective images of $\mathbb{R}$, figures prominently in the more abstract parts of descriptive set theory. Under AD, it is a very large cardinal—see [Kan03, p. 396–399]. The following result from [Sol78] explains why $\Theta$ is important to us.

**Theorem 16.** The length of the Wadge hierarchy is $\Theta$.

*Proof.* We start with a few preliminaries. Let

$$\begin{aligned} &\mathrm{J} : \mathscr{P}(\mathbb{R}) \to \mathscr{P}(\mathbb{R}) \\ &\mathrm{J}(A) = \{0^\frown x \mid \boldsymbol{f}_x(0^\frown x) \notin A\} \cup \{1^\frown x \mid \boldsymbol{f}_x(1^\frown x) \in A\}, \end{aligned} \tag{17}$$

where $\{\boldsymbol{f}_x \mid x \in \mathbb{R}\}$ as in (6). If $\mathrm{J}(A) \leq_{\mathrm{W}} A$ via some $\boldsymbol{f}_{x_0}$, then $0^\frown x_0 \in \mathrm{J}(A)$ iff $\boldsymbol{f}_{x_0}(0^\frown x_0) \notin A$: a contradiction. Similarly $\mathrm{J}(A) \leq_{\mathrm{W}} \neg A$ is impossible, and therefore

$$A, \neg A <_{\mathrm{W}} \mathrm{J}(A)\,. \tag{18}$$

We are now ready to prove the theorem. Fix an $A \subseteq \mathbb{R}$. For every $\beta \leq \|A\|_{\mathrm{W}}$ there is a $B \leq_{\mathrm{W}} A$ such that $\beta = \|B\|_{\mathrm{W}}$ and hence there is an $x \in \mathbb{R}$ such that $B = \boldsymbol{f}_x^{-1}(A)$. Therefore the map

$$\mathbb{R} \twoheadrightarrow \|A\|_{\mathrm{W}} + 1, \qquad x \mapsto \|\boldsymbol{f}_x^{-1}(A)\|_{\mathrm{W}}$$

witnesses that $\|A\|_{\mathrm{W}} < \Theta$.

Conversely, let $\alpha < \Theta$. We shall construct sets of real $A_\nu$ $(\nu < \alpha)$ such that

$$\nu < \xi \implies A_\nu <_{\mathrm{W}} A_\xi\,,$$

and hence $\alpha \leq \sup\{\|A\|_{\mathrm{W}} \mid A \subseteq \mathbb{R}\}$. Being $\alpha < \Theta$ arbitrary, this will complete the proof. Fix $f : \mathbb{R} \twoheadrightarrow \alpha$ and let

$$A_\nu = \mathrm{J}(\{x \oplus y \in \mathbb{R} \mid f(x) < \nu \wedge y \in A_{f(x)}\}),$$

where

$$x \oplus y\,(m) = \begin{cases} x(n) & \text{if } m = 2n, \\ y(n) & \text{if } m = 2n+1. \end{cases}$$

If $\nu < \xi < \alpha$, then let $x_0 \in \mathbb{R}$ be such that $f(x_0) = \nu$: then $z \mapsto x_0 \oplus z$ witnesses that $A_\nu \leq_{\mathrm{W}} \{x \oplus y \in \mathbb{R} \mid f(x) < \xi \wedge y \in A_{f(x)}\}$, and therefore $A_\nu <_{\mathrm{W}} A_\xi$ by (18). q.e.d.

Since $\Theta$ is an uncountable cardinal, then $\omega_1 \cdot \Theta = \Theta$, hence the length of the Lipschitz hierarchy is also $\Theta$.

## 2.7 The Wadge hierarchy in the Cantor space

Some of the results discussed so far hold also for the Cantor space (with the same proof). For example, if $\leq_{\mathrm{W}}$ and $\leq_{\mathrm{L}}$ denote continuous and Lipschitz reducibility in ${}^{\omega}2$, and AD is assumed:

- Wadge's Lemma holds for subsets of the Cantor space,
- $\leq_{\mathrm{L}}$ and $\leq_{\mathrm{W}}$ are well-founded,
- the Steel–Van Wesep Theorem 14,
- $\{{}^{\omega}2\}$ and $\{\varnothing\}$ are the two bottom degrees both in the Wadge and in the Lipschitz hierarchies,
- in the Wadge hierarchy, the self-dual and non-self-dual pairs alternate,
- the length of the Lipschitz and Wadge hierarchies is $\Theta$.

Since the $\bigoplus_n$-construction cannot be carried out in the space ${}^{\omega}2$, many results have to be changed accordingly: for example

- each self-dual Wadge degree is the union of an $\omega$-chain of Lipschitz self-dual degrees,
- all limit levels of the Lipschitz and Wadge hierarchies consist of a non-self-dual pair.

In other words, the Lipschitz degrees of the Cantor space look like this:

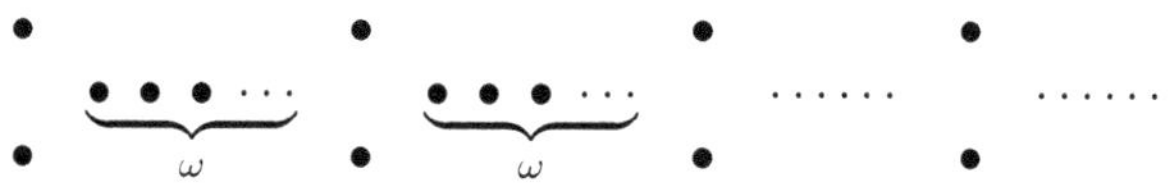


and the Wadge degrees of the Cantor space look like this:

# 3 Boldface pointclasses

A **pointclass** is a non-empty, proper subset of $\mathscr{P}(\mathbb{R})$; if moreover it is closed under continuous pre-images, then it is said to be a **boldface** pointclass. (We will only be concerned with boldface pointclasses.) Boldface pointclasses are denoted with boldface capital Greek letters like $\boldsymbol{\Gamma}$, $\boldsymbol{\Lambda}$, $\boldsymbol{\Delta}$, *etc.*, and by Wadge's Lemma they are either of the form

$$\boldsymbol{\Gamma} = \{B \subseteq \mathbb{R} \mid B \leq_{\mathrm{W}} A\} \tag{19}$$

or

$$\boldsymbol{\Gamma} = \{B \subseteq \mathbb{R} \mid B <_{\mathrm{W}} A\}\,. \tag{20}$$

The **dual** of $\boldsymbol{\Gamma}$ is the pointclass

$$\breve{\boldsymbol{\Gamma}} = \{\neg A \mid A \in \boldsymbol{\Gamma}\}\,.$$

A pointclass is self-dual if it coincides with its dual, otherwise it is non-self-dual—for example: the classical Borel pointclasses $\boldsymbol{\Sigma}^0_\alpha$ $(0 < \alpha < \omega_1)$ and the projective pointclasses $\boldsymbol{\Sigma}^1_n$ $(1 \leq n)$ are non-self-dual (and so are their duals $\boldsymbol{\Pi}^0_\alpha$ and $\boldsymbol{\Pi}^1_n$), while the collection of all clopen sets $\boldsymbol{\Delta}^0_1$ and the collection of all Borel sets $\boldsymbol{\Delta}^1_1$ are self-dual. If $\boldsymbol{\Gamma}$ is non-self-dual, then it must be as in (19) with $A$ non-self-dual, and, conversely, if $A$ is non-self-dual then $\{B \subset \mathbb{R} \mid B \leq_{\mathrm{W}} A\}$ is non-self-dual. By Wadge's Lemma, if $\boldsymbol{\Gamma}, \boldsymbol{\Lambda}$ are boldface pointclasses then either $\boldsymbol{\Gamma} \subseteq \boldsymbol{\Lambda}$ or else $\breve{\boldsymbol{\Lambda}} \subseteq \boldsymbol{\Gamma}$, and by the results in the preceding section, $\mathsf{AD} + \mathsf{DC}(\mathbb{R})$ yields a complete description of all boldface pointclasses.

## 3.1 Universal sets and the separation property

A set $U \subseteq \mathbb{R} \times \mathbb{R}$ is **universal** for $\boldsymbol{\Gamma}$ if

$$\boldsymbol{\Gamma} = \{U_{\{x\}} \mid x \in \mathbb{R}\}\,,$$

where $U_{\{x\}}$ is the $x$-section of $U$, that is

$$U_{\{x\}} = \{y \mid (x, y) \in U\}\,.$$

$\boldsymbol{\Gamma}$ is said to have a universal set if there is a $U \subseteq \mathbb{R} \times \mathbb{R}$ which is universal for $\boldsymbol{\Gamma}$ and the image of $U$ under the standard homeomorphism $\mathbb{R} \times \mathbb{R} \cong \mathbb{R}$ is in $\boldsymbol{\Gamma}$. It is not hard to see that all $\boldsymbol{\Sigma}^0_\alpha$'s and all $\boldsymbol{\Sigma}^1_n$'s (and their duals) have a universal set [Kec95, Theorems 22.3 and 37.7].

**Theorem 17.** Assume $\mathsf{AD}$ and let $\boldsymbol{\Gamma}$ be a non-self-dual pointclass. Then $\boldsymbol{\Gamma}$ has a universal set.

*Proof.* Fix $A \in \boldsymbol{\Gamma} \setminus \breve{\boldsymbol{\Gamma}}$: by Wadge's Lemma $\boldsymbol{\Gamma} = \{B \mid B \leq_{\mathrm{W}} A\}$ with $A$ non-self-dual. We claim that

$$\{\boldsymbol{\ell}_x^{-1}(A) \mid x \in \mathbb{R}\} = \boldsymbol{\Gamma}\,.$$

The inclusion from left-to-right follows from the fact that Lipschitz functions are continuous; conversely, if $B \leq_{\mathrm{W}} A$ but $B \not\leq_{\mathrm{L}} A$, then $A \leq_{\mathrm{L}} \neg B$ by (11), hence $B \equiv_{\mathrm{W}} \neg B \equiv_{\mathrm{W}} A$ would be self-dual, contrarily to our assumption. Therefore

$$U = \{(x, y) \mid \boldsymbol{\ell}_x(y) \in A\}$$

is universal for $\boldsymbol{\Gamma}$. q.e.d.

A non-self-dual boldface pointclass $\boldsymbol{\Gamma}$ has the **separation property** if given any two disjoint $A, B \in \boldsymbol{\Gamma}$ there is a $C \in \boldsymbol{\Gamma} \cap \breve{\boldsymbol{\Gamma}}$ such that $A \subseteq C$ and $C \cap B = \varnothing$. A theorem of Sierpiński [Kec95, Theorem 22.16] says that the $\boldsymbol{\Pi}^0_\alpha$'s have the separation property. Lusin's Separation Theorem [Kec95, Theorem 14.7] says that $\boldsymbol{\Sigma}^1_1$ has the separation property, and by Moschovakis' Second Periodicity Theorem—assuming all projective sets are determined—the same is true of $\boldsymbol{\Sigma}^1_{2n+1}$ and $\boldsymbol{\Pi}^1_{2n}$ [Kec95, Corollary 39.9]. Using the Martin–Monk technique, Steel [Ste81b] and Van Wesep [Van78a] proved that

**Theorem 18.** Assume $\mathsf{AD}$ and let $\boldsymbol{\Gamma}$ be a non-self-dual pointclass different from $\{\mathbb{R}\}$ and $\{\varnothing\}$. Then exactly one among $\boldsymbol{\Gamma}$ and $\breve{\boldsymbol{\Gamma}}$ has the separation property.

### 3.2 Infinitary Boolean operations

Another fact that holds true of the classical Borel and projective pointclasses is that they are obtained via some "construction procedure" from open sets. To be more specific: an **$\omega$-ary Boolean operation** is a function

$$F : {}^{\omega}\mathscr{P}(\mathbb{R}) \to \mathscr{P}(\mathbb{R})$$

such that for some fixed set $T \subseteq {}^{\omega}2$ (called the truth table of $F$) such that the truth of the statement "$x \in F(\langle A_0, A_1, \dots\rangle)$" depends on whether the sequence $\langle \chi_{A_0}(x), \chi_{A_1}(x), \dots\rangle$ is in $T$, that is:

$$x \in F(\langle A_n \mid n \in \omega\rangle) \iff \langle \chi_{A_n}(x) \mid n \in \omega\rangle \in T\,.$$

It is not hard to see that the $\boldsymbol{\Sigma}^0_\alpha$'s and the $\boldsymbol{\Sigma}^1_n$'s are of the form

$$\{F(\langle A_n \mid n \in \omega\rangle) \mid A_n \in \boldsymbol{\Sigma}^0_1\}\,, \tag{21}$$

for some carefully chosen $F$. (For the $\mathbf{\Sigma}^1_1$ use the fact that analytic sets are obtained by applying Suslin's operation $(\mathscr{A})$ to closed sets—see [Kec95, Theorem 25.7].) A pointclass that can be written as in (21) is said to be generated from open sets by an $\omega$-ary Boolean operation.

The next result generalizes the previous observation to all non-self-dual boldface pointclasses, and it is obtained by knitting together results from [Ste77, Van77, Wad83].

**Theorem 19 (Radin, Steel, Van Wesep, Wadge).** Assume $\mathsf{AD} + \mathsf{DC}(\mathbb{R})$ and let $\mathbf{\Gamma}$ be a non-self-dual boldface pointclass different from $\{\mathbb{R}\}$ and $\{\varnothing\}$. Then $\mathbf{\Gamma}$ is generated from open sets by an $\omega$-ary Boolean operation.

Since an $\omega$-ary Boolean operation $F$ is completely determined by its truth table $T = T_F$ (which is a subset of the Cantor space ${}^{\omega}2$), the Wadge degree of $F$ is defined to be the Wadge degree (in ${}^{\omega}2$) of $T_F$. All $\omega$-ary operations used to construct $\mathbf{\Sigma}^0_\alpha$ from open sets have a Borel Wadge degree (*i.e.*, their truth table is a Borel set): for example the operations of countable unions $\langle A_n \mid n < \omega \rangle \mapsto \bigcup_n A_n$ and countable intersections $\langle A_n \mid n < \omega \rangle \mapsto \bigcap_n A_n$ have truth tables $\{x \in {}^{\omega}2 \mid \exists n\, x(n) = 1\}$ and $\{x \in {}^{\omega}2 \mid \forall n\, x(n) = 1\}$ respectively, which are open and closed, respectively. On the other hand the Wadge degree of operation $(\mathscr{A})$ is the degree of complete $\mathbf{\Sigma}^1_1$ sets, and therefore it has a complexity just beyond the realm of Borel sets. John Steel has conjectured (personal communication) the following:

**Conjecture 20.** Assume $\mathsf{AD} + \mathsf{DC}(\mathbb{R})$ and suppose $\mathbf{\Gamma}$ is non-self-dual and closed under countable unions and intersections. Then either $\mathbf{\Gamma}$ or $\breve{\mathbf{\Gamma}}$ are closed under operation $(\mathscr{A})$.

For more results on general boldface pointclasses see [Kec77, Ste81a, KecSolSte81, Lou83, Bec88].

## 4 How much determinacy do you really need?

All the results in the preceding sections are stated under the "blanket assumption" that the Axiom of Determinacy (possibly augmented with $\mathsf{DC}(\mathbb{R})$) holds true. A closer inspection shows that the key determinacy hypotheses used in those proofs are that

$$\text{all Lipschitz games are determined,} \qquad (\mathsf{AD}^{\mathrm{L}})$$

or that

$$\text{all Wadge games are determined,} \qquad (\mathsf{AD}^{\mathrm{W}})$$

and in fact, in several proofs, only the statements

$$\forall A, B \subseteq \mathbb{R}\,(A \leq_{\mathrm{W}} B \vee \neg B \leq_{\mathrm{W}} A)\,, \tag{$\mathsf{SLO}^{\mathrm{W}}$}$$

and

$$\forall A, B \subseteq \mathbb{R}\,(A \leq_{\mathrm{L}} B \vee \neg B \leq_{\mathrm{L}} A)\,, \tag{$\mathsf{SLO}^{\mathrm{L}}$}$$

were used, where the acronym SLO stands for *semi-linear ordering.* It is straightforward to check that

$$\begin{array}{ccccc} & \nearrow\!\!\!\!\nearrow & \mathsf{AD}^{\mathrm{L}} & \Longrightarrow & \mathsf{SLO}^{\mathrm{L}} \\ \mathsf{AD} & & & & \Downarrow \\ & \searrow\!\!\!\!\searrow & \mathsf{AD}^{\mathrm{W}} & \Longrightarrow & \mathsf{SLO}^{\mathrm{W}} \end{array} \tag{22}$$

and it is quite natural to ask whether any of the implications above can be reversed. A conjecture, probably due to Robert Solovay, states that—under an additional assumption—all arrows above can be reversed:

**Conjecture 21.** Assume $\mathrm{V} = \mathrm{L}(\mathbb{R})$. Then $\mathsf{SLO}^{\mathrm{W}} \implies \mathsf{AD}$.

This conjecture is wide open and can be phrased as saying that for sets in $\mathscr{P}(\mathbb{R}) \cap \mathrm{L}(\mathbb{R})$ the Axiom of Determinacy and the Semi-Linear Ordering principle for continuous maps are equivalent. A plausible scenario for verifying it, is to prove the conjecture for larger and larger pointclasses contained in $\mathscr{P}(\mathbb{R}) \cap \mathrm{L}(\mathbb{R})$ using core model theory. For example, Leo Harrington (for $n = 1$, [Har78]) and Greg Hjorth (for $n = 2$, [Hjo96]) showed that

**Theorem 22.** Assume $\mathsf{SLO}^{\mathrm{W}}$ for all sets in $\mathbf{\Pi}^1_n$, where $n = 1, 2$. Then all $\mathbf{\Pi}^1_n$ games are determined.

Unfortunately, there are challenging difficulties to extend core-model techniques at higher projective levels, hence the statement of Theorem 22 is open for $n \geq 3$. On the other hand, by results of Alain Louveau and Jean Saint Raymond [LouSai88], $\mathsf{SLO}^{\mathrm{W}}$ for Borel sets is provable in second order arithmetic, while, by a celebrated result of Harvey Friedman [Fri71], the determinacy of all Borel games is not a theorem of $\mathsf{ZFC}$ −Replacement: probably, though, these results should not be construed as evidence for the falsity of the Conjecture, but rather that *any* proof that AD follows from $\mathsf{SLO}^{\mathrm{W}}$ should be indirect enough that it cannot be scaled-down to the realm of Borel sets. Another way to give some evidence that the conjecture is true, is to prove that the usual regularity properties for the reals follow already from $\mathsf{SLO}^{\mathrm{W}}$. To the best of our knowledge, only two such results are known: in [And03] it is shown that

**Theorem 23.** Assume $\mathsf{SLO}^{\mathrm{W}}$. Then $\mathsf{AC}_\omega(\mathbb{R})$, the Axiom of Countable Choices over $\mathbb{R}$, holds.

The second well-known consequence of determinacy which is known to follow from $\mathsf{SLO}^{\mathrm{W}}$ is the $\mathsf{PSP}$, the Perfect Subset Property, *i.e.*, the statement that each subset of $\mathbb{R}$ is either countable, or else it contains a perfect subset.

**Theorem 24 (Wadge).** $\mathsf{SLO}^{\mathrm{W}} \implies \mathsf{PSP}$.

*Proof.* Let $X \subseteq \mathbb{R}$ be arbitrary and let

$$G = \{x \in \mathbb{R} \mid \forall n \, \exists m > n \; x(m) = 0\}$$

be the set of all sequences containing infinitely many 0's. By $\mathsf{SLO}^{\mathrm{W}}$ either

(i) $X \leq_{\mathrm{W}} \neg G$, or else
(ii) $G \leq_{\mathrm{W}} X$.

Suppose (i) holds. Since $G$ is $\mathbf{G}_\delta$, then $X$ is $\mathbf{F}_\sigma$, *i.e.*, $X = \bigcup_n C_n$ where each $C_n \subseteq \mathbb{R}$ is closed. By the Cantor-Bendixson theorem (which is provable in $\mathsf{ZF}$) each $C_n$ is either countable, or it contains a perfect set: if some $C_n$ contains a perfect set, then so does $X$; otherwise each $C_n$ is countable, hence by Theorem 23 and the remark following it, $X$ is countable.

Suppose now (ii) holds, and let $f : \mathbb{R} \to \mathbb{R}$ be continuous and such that $f^{-1}(X) = G$. Let also $\varphi : {}^{<\omega}\omega \to {}^{<\omega}\omega$ be such that $f = f_\varphi$. We will construct a complete binary tree $T$ such that $[T] \subseteq G$ and $f \restriction [T]$ is injective, and therefore $f([T])$ will be a perfect subset of $X$. Recall that two finite sequences $u, t$ are **incompatible**, in symbols: $u \perp t$, if $u \nsubseteq t$ e $t \nsubseteq u$. We will construct a map ${}^{<\omega}2 \to {}^{<\omega}\omega$, $w \mapsto s_w$ such that

- $z \subset w \implies s_z \subset s_w$,
- $z \perp w \implies s_z \perp s_w$.

The required tree will be $T = \{t \in {}^{<\omega}\omega \mid \exists w \in {}^{<\omega}2 \, (t \subseteq s_w)\}$.

Set $s_\varnothing = \varnothing$. Since $\vec{0} \in G$ and $\vec{1} \notin G$, then $f(\vec{0}) \in X$ and $f(\vec{1}) \notin X$, hence there are $n, m$ such that $\varphi(0^{(n)}) \perp \varphi(1^{(m)})$. Let $s_0 = 0^{(n)}$ and $s_1 = 1^{(m)} {}^\frown 0$: since $\varphi$ is monotone $\varphi(s_0) \perp \varphi(s_1)$. Arguing as before, $f(s_i {}^\frown \vec{0}) \neq f(s_i {}^\frown \vec{1})$, for $i = 0, 1$, so there are $n', m'$ such that $\varphi(s_{i,0}) \perp \varphi(s_{i,0})$, where $s_{i,0} = s_i {}^\frown 0^{(n')}$ and $s_{i,0} = s_i {}^\frown 1^{(m')} {}^\frown 0$. We can now repeat the argument for $s_{0,0}, s_{0,1}, s_{1,0}, s_{1,1}$, constructing all $s_w$ for $w$ of length 3, and so on. q.e.d.

It is open wether the other regularity properties follow from the Semi-Linear Ordering principle, for example:

**Problem 25.** Show that $\mathsf{LM}$, the statement that all sets of reals are Lebesgue measurable, and $\mathsf{BP}$, the statement that all sets of reals have the property of Baire, both follow from $\mathsf{SLO}^{\mathrm{W}}$.

Even the non-existence of a flip-set (a fact implied by either $\mathsf{LM}$ or $\mathsf{BP}$—see page 14) is not known to follow from $\mathsf{SLO}^{\mathrm{W}}$:

**Problem 26.** Show that $\mathsf{SLO}^{\mathrm{W}}$ implies that there are no flip-sets.

Although we do not know how to prove determinacy from $\mathsf{SLO}^{\mathrm{W}}$, the other arrows in the diagram (22) can be reversed.

**Theorem 27.** Assume $\mathsf{DC}(\mathbb{R})$. Suppose that there are no flip-sets (a fact which follows from either $\mathsf{LM}$ or $\mathsf{BP}$). Then

$$\mathsf{AD}^{\mathrm{W}} \iff \mathsf{AD}^{\mathrm{L}} \iff \mathsf{SLO}^{\mathrm{L}} \iff \mathsf{SLO}^{\mathrm{W}} .$$

The proof is contained in the union of [And03] and [And06]: in the first paper the equivalence among the first three statements is shown, while in the second paper the result is extended to $\mathsf{SLO}^{\mathrm{W}}$ as well.

In view of this result, the expression "Wadge determinacy" denotes anyone of: $\mathsf{AD}^{\mathrm{L}}$, $\mathsf{AD}^{\mathrm{W}}$, $\mathsf{SLO}^{\mathrm{L}}$, and $\mathsf{SLO}^{\mathrm{W}}$. The import of Theorem 27 is that $\mathsf{SLO}^{\mathrm{W}}$ (together with $\mathsf{DC}(\mathbb{R})$ and, say, $\mathsf{BP}$) is enough to prove the results in Sections 2 and 3; in particular, it is enough to prove that the structure of the Wadge degrees is as in (16). Clearly, a proof of Open Problem 26 would allow us to remove one assumption in the statement of the theorem.

## 5 Other spaces

The Baire and Cantor spaces are of the form ${}^{\omega}X$, where $X$ is discrete and ${}^{\omega}X$ is given the product topology: a complete compatible metric is given by (2), and ${}^{\omega}X$ is separable if and only if $X$ is finite or countable. A further generalization is given by the spaces of the form $[T]$, where $T$ is a pruned tree on some non-empty set $X$. The relations of continuous and Lipschitz pre-image in $[T]$ are denoted by the symbols $\leq_{\mathrm{W}}^{[T]}$ and $\leq_{\mathrm{L}}^{[T]}$. The notions of Wadge and Lipschitz games can be adapted to the spaces $[T]$ by requiring that both players maintain their positions in $T$, the first player to break this rule resulting in an immediate loss: the two games are denoted by $G_{\mathrm{W}}^{T}$ and $G_{\mathrm{L}}^{T}$. The Semi-Linear Ordering principle can be stated for $[T]$, or more generally, for any topological space $\mathcal{X}$:

$$\forall A, B \subseteq \mathcal{X} \left( A \leq_{\mathrm{W}}^{\mathcal{X}} B \quad \vee \quad \mathcal{X} \setminus B \leq_{\mathrm{W}}^{\mathcal{X}} A \right), \qquad (\mathsf{SLO}(\mathcal{X}))$$

where $\leq_{\mathrm{W}}^{\mathcal{X}}$ denotes the relation of continuous reducibility in $\mathcal{X}$. In order to study $\mathsf{SLO}(\mathcal{X})$ on different $\mathcal{X}$'s, it is convenient to introduce the following terminology: the **Wadge structure**[6] of the space $\mathcal{X}$ is the model-theoretic structure

$$\langle \mathscr{P}(\mathcal{X}), \neg^{\mathcal{X}}, \leq_{\mathrm{W}}^{\mathcal{X}} \rangle$$

where $\neg^{\mathcal{X}}$ is a 1-ary operation of taking complements, $\neg^{\mathcal{X}} A = \mathcal{X} \setminus A$. Suppose $\mathcal{X}$ is a **retract** of $\mathcal{Y}$, that is, $\mathcal{X}$ is a closed subset of $\mathcal{Y}$ and there is a continuous surjection $\pi : \mathcal{Y} \twoheadrightarrow \mathcal{X}$ which is the identity on $\mathcal{X}$. (The map $\pi$ is called a **retraction** of $\mathcal{Y}$ onto $\mathcal{X}$.) We claim that the map $\Phi : \mathscr{P}(\mathcal{X}) \to \mathscr{P}(\mathcal{Y})$, $\Phi(A) = \pi^{-1}(A)$ is an embedding of the Wadge structure of $\mathcal{X}$ into the Wadge structure of $\mathcal{Y}$: it is easy to check that it is an injective map that preserves complementation, so it is enough to show that

$$A \leq_{\mathrm{W}}^{\mathcal{X}} B \iff \Phi(A) \leq_{\mathrm{W}}^{\mathcal{Y}} \Phi(B)\,.$$

If $f : \mathcal{X} \to \mathcal{X}$ is continuous and $A = f^{-1}(B)$, then $f \circ \pi : \mathcal{Y} \to \mathcal{X} \subseteq \mathcal{Y}$ witnesses $\Phi(A) \leq_{\mathrm{W}}^{\mathcal{Y}} \Phi(B)$; conversely, if $g : \mathcal{Y} \to \mathcal{Y}$ is continuous and $\Phi(A) = g^{-1}(\Phi(B))$, then $\pi \circ g{\restriction}\mathcal{X} : \mathcal{X} \to \mathcal{X}$ witnesses $A \leq_{\mathrm{W}}^{\mathcal{X}} B$. Therefore we have shown

**Proposition 28.** If $\pi : \mathcal{Y} \twoheadrightarrow \mathcal{X}$ is a retraction, then the map $A \mapsto \pi^{-1}(A)$ is an embedding of the Wadge structure of $\mathcal{X}$ into the Wadge structure of $\mathcal{Y}$. In fact

$$\mathsf{SLO}(\mathcal{Y}) \implies \mathsf{SLO}(\mathcal{X})\,.$$

(The last statement is immediate since $\mathsf{SLO}$ is a $\forall$-formula in the language for Wadge structures, and therefore downward absolute.) Since $S \subseteq T$ implies that $[S]$ is a retract of $[T]$ (see [Kec95, Proposition 2.8]), then $\mathsf{SLO}^{\mathrm{W}}$ implies $\mathsf{SLO}({}^{\omega}2)$ and, more generally, $\mathsf{SLO}([T])$ for any pruned tree $T$ on a countable set.

If $T$ is a tree on $X$, then $\mathsf{AD}_X$ —the statement that all 0-sum, perfect information, two-player games on $X$ are determined—implies $\mathsf{SLO}([T])$. If $X = \mathbb{R}$, then $\mathsf{SLO}({}^{\omega}\mathbb{R})$ follows from $\mathsf{AD}_{\mathbb{R}}$ which, by unpublished results of Woodin, is consistent modulo large cardinals. On the other hand $\mathsf{AD}_{\omega_1}$ is inconsistent with $\mathsf{ZF}$ [Kan03, Exercise 27.12], and the main result of this section is that the same is true of $\mathsf{SLO}({}^{\omega}\omega_1)$, that is to say: there are $A, B \subseteq {}^{\omega}\omega_1$ such that $A \not\leq_{\mathrm{W}}^{{}^{\omega}\omega_1} B$ and ${}^{\omega}\omega_1 \setminus B \not\leq_{\mathrm{W}}^{{}^{\omega}\omega_1} A$.

**Theorem 29.** The principle $\mathsf{SLO}({}^{\omega}\omega_1)$ is inconsistent with $\mathsf{ZF}$.

*Proof.* The proof is an elaboration of the ideas behind Theorem 23—see [And03, Theorem 3, p. 77]. Suppose $\mathsf{SLO}({}^{\omega}\omega_1)$ holds, and let $\pi : {}^{\omega}\omega_1 \twoheadrightarrow {}^{\omega}\omega$ be the retraction

$$\pi(x)(n) = \begin{cases} x(n) & \text{if } x(n) < \omega, \\ 2 & \text{otherwise.} \end{cases}$$

[6]The definition of Wadge structure and Proposition 28 are due to Alberto Marcone.

(The rationale for the 2 in the formula above is explained in (23) below.) Then, by Proposition 28, $\mathsf{SLO}^{\mathrm{W}}$ follows and hence Theorem 23 holds. We will show that there exist an $\omega_1$ sequence of reals, contradicting $\mathsf{PSP}$ and hence Theorem 24. In order to simplify the notation, let's agree that $\mathcal{G}$ is the Wadge game on the space ${}^{\omega}\omega_1$, *i.e.*, $\mathcal{G} = G^{T}_{\mathrm{W}}$ where $T = {}^{<\omega}\omega_1$. Similarly $\leq_{\mathrm{W}}^{{}^{\omega}\omega_1}$ and $<_{\mathrm{W}}^{{}^{\omega}\omega_1}$ are abbreviated with $\preceq$ and $\prec$. For $X \subseteq {}^{\omega}\omega_1$, let

$$X^{\triangledown} = \bigcup_{n<\omega} 0^{(n)}{}^\frown 1^\frown X \qquad \text{and} \qquad X^{\circ} = \{\vec{0}\} \cup X^{\triangledown}\,.$$

Notice that $X \preceq X^{\triangledown}$ and $X \preceq X^{\circ}$ via the function $x \mapsto 1^\frown x$. If $X \subseteq {}^{\omega}2$, then $X^{\triangledown}, X^{\circ} \subseteq {}^{\omega}2$ and

$$\pi^{-1}(X^{\triangledown}) = X^{\triangledown} \quad \wedge \quad \pi^{-1}(X^{\circ}) = X^{\circ}\,. \tag{23}$$

(This is the point of the specific definition of $\pi$.) For any $x \in {}^{\omega}2$ let $E_x$ be the binary relation on $\omega$ given by

$$n\, E_x\, m \Longleftrightarrow x(\langle n, m\rangle) = 1$$

where $\langle \cdot, \cdot \rangle : \omega \times \omega \to \omega$ is a standard bijection, *e.g.*, $\langle n, m\rangle = 2^n(2m+1)$. As usual, $\mathrm{WO}_\alpha$ is the set of all reals coding a well-order of type $\alpha \geq \omega$, *i.e.*,

$$\mathrm{WO}_\alpha = \{x \in {}^{\omega}2 \mid \langle \omega, E_x\rangle \cong \alpha\}\,.$$

By a theorem of Jacques Stern [Ste78]

$$W_\alpha \overset{\text{def}}{=} \mathrm{WO}_{\omega^\alpha} \in \boldsymbol{\Sigma}^0_{2\alpha+2} \setminus \boldsymbol{\Sigma}^0_{2\alpha+1}\,.$$

Since $W^{\circ}_\alpha$ is a countable union of sets in $\boldsymbol{\Sigma}^0_{2\alpha+2}$, then $\mathbb{R} \setminus W^{\circ}_\alpha \in \boldsymbol{\Pi}^0_{2\alpha+2}$, and therefore

$$\alpha < \beta \implies (W_\beta \not\leq_{\mathrm{W}} \mathbb{R} \setminus W^{\circ}_\alpha,\ \mathbb{R} \setminus W_\alpha,\ W^{\triangledown}_\alpha)\,. \tag{24}$$

Let

$$A = \bigcup_{1 \leq \alpha < \omega_1} \alpha^\frown W^{\circ}_\alpha \quad \text{and} \quad B = \bigcup_{\substack{h \in \omega \\ 1 \leq \alpha < \omega_1}} 0^{(h)}{}^\frown \alpha^\frown W^{\triangledown}_\alpha\,.$$

We claim that

$$A \not\preceq B \quad \wedge \quad B \not\preceq {}^{\omega}\omega_1 \setminus A\,,$$

and hence $\mathsf{SLO}({}^{\omega}\omega_1)$ fails.

**Lemma 30.** $B \not\preceq {}^{\omega}\omega_1 \setminus A$.

*Proof of Lemma 30.* Suppose otherwise, and let $\tau$ be a winning strategy for **II** in $\mathcal{G}(B, {}^{\omega}\omega_1 \setminus A)$. Let **I** play 0's as long as **II** passes: since $\tau$ is winning, there is a first round $h$ when **II** plays some ordinal number $\bar{\alpha}$, that is $\tau(0^{(h)}) = \bar{\alpha}$. Let **I** answer $\bar{\alpha}+1$, so that after this inning is over, the positions of the two players are

$$\begin{array}{llll} \mathbf{I} & \overbrace{0\cdots\cdots0}^{h} & & \bar{\alpha}+1 \\ \mathbf{II} & \underbrace{\mathsf{p}\cdots\mathsf{p}}_{h-1} & \bar{\alpha} & \quad i \end{array}$$

where $i \in \omega_1 \cup \{\mathsf{p}\}$ and $\mathsf{p}$ denotes "passing". Since $\tau$ is winning, **II** can use $\tau$ to win what's left of the game. In other words, recalling the notion of localization of a set at a given sequence (9):

$$W^{\nabla}_{\bar{\alpha}+1} = B_{\lfloor 0^{(h)} \frown \bar{\alpha}+1 \rfloor} \preceq ({}^{\omega}\omega_1 \setminus A)_{\lfloor \langle \bar{\alpha}, i\rangle \rfloor}\,,$$

and this last set is

$$({}^{\omega}\omega_1 \setminus A)_{\lfloor \langle \bar{\alpha}, i\rangle \rfloor} = \begin{cases} {}^{\omega}\omega_1 \setminus W^{\circ}_{\bar{\alpha}} & \text{if } i \in \{\mathsf{p}, 0\}, \\ {}^{\omega}\omega_1 \setminus W_{\bar{\alpha}} & \text{if } i = 1, \\ {}^{\omega}\omega_1 & \text{if } i > 1. \end{cases}$$

The third case is impossible, since the only $X$ such that $X \preceq {}^{\omega}\omega_1$ is ${}^{\omega}\omega_1$ itself and $W^{\nabla}_{\bar{\alpha}+1} \neq {}^{\omega}\omega_1$. Therefore

$$W^{\nabla}_{\bar{\alpha}+1} \preceq {}^{\omega}\omega_1 \setminus W^{\circ}_{\bar{\alpha}} \quad \vee \quad W^{\nabla}_{\bar{\alpha}+1} \preceq {}^{\omega}\omega_1 \setminus W_{\bar{\alpha}}\,.$$

Proposition 28 and (23) imply that

$$W^{\nabla}_{\bar{\alpha}+1} \leq_{\mathrm{W}} \mathbb{R} \setminus W^{\circ}_{\bar{\alpha}} \quad \vee \quad W^{\nabla}_{\bar{\alpha}+1} \leq_{\mathrm{W}} \mathbb{R} \setminus W_{\bar{\alpha}}\,,$$

and since $W_{\bar{\alpha}+1} \leq_{\mathrm{W}} W^{\nabla}_{\bar{\alpha}+1}$ either of these contradicts (24).

q.e.d. (Lemma 30)

Suppose now $A \preceq B$ and let $\tau$ be a winning strategy for **II** in $\mathcal{G}(A, B)$. If **I** plays $\alpha \frown \vec{0} \in A$, then **II** cannot respond with $\vec{0}$ since $\vec{0} \notin B$, so for any $\alpha \geq 1$ let $m_\alpha \in \omega$ be least such that $\tau(\alpha \frown 0^{(m_\alpha)}) = \gamma_\alpha > 0$, and let $h_\alpha$ be the number of 0's played so far; that is, after the $m_\alpha$th inning is completed, the sequences of actual moves[7] played by **I** and **II** are $\alpha \frown 0^{(m_\alpha)}$ and $0^{(h_\alpha)} \frown \gamma_\alpha$,

[7] That is, disregarding the p's.

respectively. (Notice that $h_\alpha < m_\alpha$ is possible, since $\tau$ can pass.) If $\gamma_\alpha < \alpha$ then

$$\begin{aligned} W_\alpha &\preceq W^\circ_\alpha \\ &= A_{\lfloor \alpha^\frown 0^{(m_\alpha)} \rfloor} \\ &\preceq B_{\lfloor 0^{(h_\alpha)}{}^\frown \gamma_\alpha \rfloor} \\ &= W^\nabla_{\gamma_\alpha} \end{aligned}$$

and hence, by Proposition 28, $W_\alpha \leq_{\mathrm{W}} W^\nabla_{\gamma_\alpha}$, contradicting (24).

Therefore $\alpha \leq \gamma_\alpha$, for all $1 \leq \alpha < \omega_1$. Since $\alpha^\frown \vec{0} \in A$ but $0^{(h_\alpha)}{}^\frown \gamma_\alpha^\frown \vec{0} \notin B$, player **II** cannot go on playing 0's forever, and arguing as before, the first non-zero ordinal played must be a 1. Let $g(\alpha) \in W_{\gamma_\alpha}$ be such that

$$0^{(h_\alpha)}{}^\frown \gamma_\alpha{}^\frown 0^{(i)}{}^\frown 1^\frown g(\alpha)$$

is $\tau$'s answer to **I** playing $\alpha^\frown \vec{0}$. Let $C \subseteq \omega_1$ be unbounded and such that the map $\alpha \mapsto \gamma_\alpha$ is injective. Since the $W_\alpha$'s are disjoint, then the map $C \to \mathbb{R}$, $\alpha \mapsto g(\alpha)$, yields an $\omega_1$ sequence of distinct reals, contradicting PSP.

q.e.d. (Theorem 29)

Let us make a few remarks about the proof. Recall that a set $A \subseteq {}^\omega X$ is $\mathbf{\Sigma}^1_1$ if it is the projection of a closed subset of ${}^\omega X \times \mathbb{R}$, and it is $\mathbf{\Delta}^1_1$ if it and its complement ${}^\omega X \setminus A$ are both in $\mathbf{\Sigma}^1_1$. Every Borel set of ${}^\omega X$ is $\mathbf{\Delta}^1_1$ and if $X$ is countable the converse holds by a classical theorem of Suslin [Kec95, Theorem 14.11]. For $X$ uncountable Suslin's theorem fails and $\mathbf{\Delta}^1_1$ is strictly larger than the collection of Borel sets, and an alternative definition of $\mathbf{\Delta}^1_1$ is the following: it is the smallest $\mathcal{D} \subseteq \mathscr{P}({}^\omega X)$ containing all open sets, closed under complements, countable unions, and **open-separated unions**, *i.e.*, for any family $A_i \in \mathcal{D}$ $(i \in I)$ such that there are open *disjoint* sets $U_i \supseteq A_i$, then $\bigcup_{i \in I} A_i \in \mathcal{D}$. Working in ZFC, Martin showed in [Mar90] that all $\mathbf{\Delta}^1_1$ games with pay-off in ${}^\omega X$ are determined and therefore under AC, no pair of $\mathbf{\Delta}^1_1$ sets can witness the failure of $\mathsf{SLO}({}^\omega \omega_1)$. On the other hand, the statement that

$$\forall X\, \forall C \subseteq {}^\omega X\, (C \text{ is closed} \implies \mathcal{G}(C) \text{ is determined})$$

implies AC. Therefore Martin's theorem implies AC and hence it is not provable in ZF. Notice that Theorem 29 *does not* give an explicit pair of sets $A, B \subseteq {}^\omega \omega_1$ such that

$$\mathsf{ZF} + \mathsf{AC}_\omega(\mathbb{R}) \vdash \mathcal{G}(A, B) \text{ is not determined,}$$

since $\mathcal{G}$ is a game on $\omega_1$, and by results of Harrington and Kechris [HarKec81, p. 133–134] there may be no definable instance of the failure of $\mathsf{AD}_{\omega_1}$. What Theorem 29 *does* show is that there is an explicit pair of $\mathbf{\Delta}^1_1$ sets $A$ and $B$ such that (provably in $\mathsf{ZF}+\mathsf{AC}_\omega(\mathbb{R})$) $B \not\preceq {}^\omega\omega_1 \setminus A$ and such that $A \preceq B$ implies the existence of an $\omega_1$ sequence of distinct reals: this contradicts PSP, but it is a consequence of AC, and in fact $\mathsf{ZFC} \vdash A \preceq B$. This is completely analogous to the proof that $\mathsf{AD}_{\omega_1}$ is inconsistent: consider the game in which **I** plays an ordinal $\alpha \geq \omega$ in the first round and **II** constructs a $z \in {}^\omega 2$ by playing 0's and 1's in the remaining rounds (**I**'s moves are irrelevant after the first round) with **II** winning just in case $z \in \mathrm{WO}_\alpha$. Then **I** cannot have a winning strategy and AC implies **II** has a winning strategy; but any strategy for **II** yields an $\omega_1$-sequence of distinct reals.

## 6 Other reducibilities

So far we have only considered continuous and Lipschitz reductions, but obviously one can consider the reducibility notion

$$A \leq_{\mathcal{F}} B \iff \exists f \in \mathcal{F}\left(A = f^{-1}(B)\right)$$

where $\mathcal{F}$ is a "reasonable" collection of functions from $\mathbb{R}$ to $\mathbb{R}$: for example we require that $\mathcal{F}$ contains the identity function (so that $\leq_{\mathcal{F}}$ is reflexive), that it is closed under composition (so that $\leq_{\mathcal{F}}$ is transitive), and that it contains all Lipschitz functions (so that $\leq_{\mathrm{L}}$ refines $\leq_{\mathcal{F}}$). Moreover the notions of Semi-Linear Ordering principle for functions in $\mathcal{F}$ ($\mathsf{SLO}^{\mathcal{F}}$), $\mathcal{F}$-degree $[A]_{\mathcal{F}}$, *etc.* can be defined accordingly. The exact definition of what "reasonable" means is not important here—a precise definition is given in [AndMar03, Definition 2, p. 178], where such $\mathcal{F}$ is dubbed **amenable**. The point is that $\mathcal{F}$ can be taken to be the collection of all Lipschitz functions, of all continuous functions, of all Borel function. Another interesting amenable $\mathcal{F}$ is the set of all **$\mathbf{\Delta}^0_\alpha$-functions**, where $f : \mathbb{R} \to \mathbb{R}$ is $\mathbf{\Delta}^0_\alpha$ if and only if

$$A \in \mathbf{\Delta}^0_\alpha \implies f^{-1}(A) \in \mathbf{\Delta}^0_\alpha,$$

or, equivalently, if

$$A \in \mathbf{\Sigma}^0_\alpha \implies f^{-1}(A) \in \mathbf{\Sigma}^0_\alpha.$$

Thus a function is $\mathbf{\Delta}^0_1$ if and only if it is continuous. It is not hard to check that a function is $\mathbf{\Delta}^0_2$ if and only if the pre-image of an open set is $\mathbf{\Delta}^0_2$ (see [And06]) so every $\mathbf{\Delta}^0_2$-function is Baire-class 1. The converse, however, is not true since the pre-image of a $\mathbf{\Sigma}^0_2$ via a Baire-class 1 function is, in general, a $\mathbf{\Sigma}^0_3$ set. The $\mathbf{\Delta}^0_2$-functions have been investigated, albeit with different names, in the literature: for example in [JayRog82] are called *first level Borel functions* and in [Kir92] are called *Baire one star functions.*

Notice that if $f$ is Baire-class $\alpha$ then $f$ is a $\boldsymbol{\Delta}^0_{\alpha\cdot\omega}$-function, so the hierarchy of $\boldsymbol{\Delta}^0_\alpha$-functions yields an alternative stratification of the Borel functions. When $\mathcal{F}$ is the collection of all $\boldsymbol{\Delta}^0_\alpha$-functions or of all $\boldsymbol{\Delta}^1_1$-functions (*i.e.*, Borel functions) we replace the "$\mathcal{F}$" in $\leq_{\mathcal{F}}$, $[A]_{\mathcal{F}}$, $\mathsf{SLO}^{\mathcal{F}}$, *etc.* with $\boldsymbol{\Delta}^0_\alpha$ or $\boldsymbol{\Delta}^1_1$. In [AndMar03] it is shown[8] that

**Theorem 31.** Assume $\mathsf{SLO}^{\boldsymbol{\Delta}^1_1} + \mathsf{DC}(\mathbb{R})$. Suppose also that there are no flip-sets. Then $\leq_{\boldsymbol{\Delta}^1_1}$ is well-founded and the structure of the $\boldsymbol{\Delta}^1_1$ degrees is analogous to the one of the Wadge degrees (see (16)): self-dual degrees and non-self-dual pairs of degrees alternate, with self-dual degrees at limit levels of countable cofinality, and non-self-dual pairs at limit levels of uncountable cofinality.

The length of the $\boldsymbol{\Delta}^1_1$ hierarchy is $\Theta$: to see this repeat the proof of Theorem 16 using an $\mathbb{R}$-parametrization $\{\boldsymbol{g}_x \mid x \in \mathbb{R}\}$ of all Borel functions instead of $\{\boldsymbol{f}_x \mid x \in \mathbb{R}\}$.

Every $\boldsymbol{\Delta}^1_1$ degree is the union of Wadge degrees: if $\mathsf{SLO}^{\mathrm{W}}$ is assumed, any non-self-dual $\boldsymbol{\Delta}^1_1$ degree is a (necessarily non-self-dual) Wadge degree, while any self-dual $\boldsymbol{\Delta}^1_1$ degree is the union of several consecutive Wadge degrees, for example: the least self-dual $\boldsymbol{\Delta}^1_1$ degree is the union of all Wadge degree of Borel sets other than $\varnothing$ and $\mathbb{R}$. In other words: if

$$\boldsymbol{\beta} \stackrel{\text{def}}{=} \|A\|_{\mathrm{W}}\,, \qquad A \in \boldsymbol{\Sigma}^1_1 \setminus \boldsymbol{\Delta}^1_1$$

or, equivalently, if $\boldsymbol{\beta}$ is the length of the Wadge hierarchy restricted to Borel sets, then the least self-dual $\boldsymbol{\Delta}^1_1$ degree is the union of $\boldsymbol{\beta}$ consecutive Wadge degrees. (The actual value of $\boldsymbol{\beta}$ is revealed in Theorem 32 below.) The next self-dual $\boldsymbol{\Delta}^1_1$ degree, *i.e.*, the one containing the sets of the form $A \oplus \neg A$ with $A \in \boldsymbol{\Sigma}^1_1 \setminus \boldsymbol{\Pi}^1_1$, is obtained by stacking together an even larger (*i.e.*, more than $\boldsymbol{\beta}$) block of consecutive Wadge degrees, and so on. This should be contrasted with the case of the Wadge-vs-Lipschitz hierarchies where each self-dual Wadge degree is the union of $\omega_1$ consecutive Lipschitz degrees.

The length of the Wadge hierarchy restricted to $\boldsymbol{\Delta}^0_\alpha$ or $\boldsymbol{\Delta}^1_1$ was explicitly computed in [Wad83]: Recall that $\varepsilon(\alpha)$ is the $\alpha$th fixed point of the map $\gamma \mapsto \omega^\gamma$ so that $\varepsilon : \mathrm{Ord} \to \{\gamma \mid \omega^\gamma = \gamma\}$ is the enumerating function and

$$\varepsilon(0) = \sup\{\omega, \omega^\omega, \omega^{\omega^\omega}, \dots\}\,.$$

If $F : \mathrm{Ord} \to \mathrm{Ord}$ is increasing and continuous at limits, then so is $F' : \mathrm{Ord} \to \mathrm{Ord}$, the function enumerating all fixed points of $F$. Thus we can

[8] The reader should be warned that in [AndMar03] the pre-order $\leq_{\boldsymbol{\Delta}^1_1}$ is denoted by $\leq_{\mathbf{B}}$ and the $\boldsymbol{\Delta}^1_1$ degrees are called Borel-Wadge degrees.

define

$$\begin{aligned}
\varepsilon^{(0)} &= \varepsilon\,, \\
\varepsilon^{(\alpha+1)} &= (\varepsilon^{(\alpha)})'\,, \\
\varepsilon^{(\lambda)} &= \text{the function enumerating } \left\{\gamma \mid \forall \alpha < \lambda \left(\varepsilon^{(\alpha)}(\gamma) = \gamma\right)\right\}, \\
&\quad \text{if } \lambda \text{ is limit.}
\end{aligned}$$

Note that

$$\forall \alpha < \omega_1 \left(\varepsilon^{(\alpha)}(\omega_1) = \omega_1 = \varepsilon^{(\omega_1)}(0)\right).$$

Wadge in [Wad83] proved that:

**Theorem 32.**

(a) Let $\gamma_0 = 1$ and $\gamma_{i+1} = \omega_1^{\gamma_i}$. Then the length of the Wadge hierarchy restricted to $\mathbf{\Delta}^0_{n+1}$ sets ($n \geq 1$) is $\gamma_n$.

(b) For any $\alpha \geq \omega$, consider its Cantor normal form

$$\alpha = \omega^{1+\nu_0} + \omega^{1+\nu_1} + \cdots + \omega^{1+\nu_n} + k$$

where $\nu_0 \geq \cdots \geq \nu_n$ and $k \in \omega$. Then the length of the Wadge hierarchy restricted to $\mathbf{\Delta}^0_\alpha$ sets is

$$\varepsilon^{(\nu_0)}\left(\varepsilon^{(\nu_1)}(\cdots(\varepsilon^{(\nu_n)}(\omega_1 + \omega_1 \cdot \gamma_k))\cdots)\right).$$

(c) The length of the Wadge hierarchy restricted to $\mathbf{\Delta}^1_1$, *i.e.*, the Borel sets, is

$$\beta = \varepsilon^{(\omega_1)}(1)\,.$$

Theorem 31 is similar to Theorem 27, as it asserts that a certain Semi-Linear Ordering principle implies structural results for the relevant hierarchy. But the analogy stops here. Theorem 27 is a result in the metamathematics of descriptive set theory, asserting that the structural properties of the Wadge hierarchy which hold under AD, follow from seemingly weaker principles, and that these principles are all equivalent. On the other hand the structural property of the $\mathbf{\Delta}^1_1$ hierarchy stated in Theorem 31 were not previously known under any hypothesis. Another important difference between the two results is that there is no known analogue of the game $G_{\mathrm{W}}$ for the $\mathbf{\Delta}^1_1$ hierarchy: the arguments in [AndMar03] are non-trivial generalizations of the ones in [And03]. Clearly, if $\mathcal{F}$ is an amenable collection of Borel functions, then

$$\mathsf{SLO}^{\mathrm{W}} \implies \mathsf{SLO}^{\mathcal{F}} \implies \mathsf{SLO}^{\mathbf{\Delta}^1_1},$$

and because of the similarities between Theorems 27 and 31, it is tempting to formulate the following

**Conjecture 33.** Assume $\mathsf{DC}(\mathbb{R})$ and that there are no flip-sets. Then

$$\mathsf{SLO}^{\boldsymbol{\Delta}^1_1} \Longleftrightarrow \mathsf{SLO}^{\mathrm{W}}.$$

Compounding this with Conjecture 21, we have a stronger

**Conjecture 34.** Assume $\mathrm{V} = \mathrm{L}(\mathbb{R})$. Then

$$\mathsf{SLO}^{\boldsymbol{\Delta}^1_1} \Longrightarrow \mathsf{AD}.$$

Towards proving Conjecture 33, in [And06] it is shown that

**Theorem 35.** Assume $\mathsf{DC}(\mathbb{R})$ and that there are no flip-sets. Then

$$\mathsf{SLO}^{\boldsymbol{\Delta}^0_2} \Longleftrightarrow \mathsf{SLO}^{\mathrm{W}}.$$

Therefore if any two sets $A$ and $B$ can be compared using $\boldsymbol{\Delta}^0_2$ functions, *i.e.*, $A \leq_{\boldsymbol{\Delta}^0_2} B$ or $\neg B \leq_{\boldsymbol{\Delta}^0_2} A$, then they can be compared using continuous functions. In [And06] it is also shown that $\mathsf{SLO}^{\boldsymbol{\Delta}^0_2}$ implies that the structure of the $\boldsymbol{\Delta}^0_2$ degrees behaves like the structure of Wadge degrees, that is: self-dual degrees and non-self-dual pairs of degrees alternate, with self-dual degrees at limit levels of countable cofinality, and non-self-dual pairs at limit levels of uncountable cofinality. Thus the $\boldsymbol{\Delta}^0_2$ hierarchy looks like this:

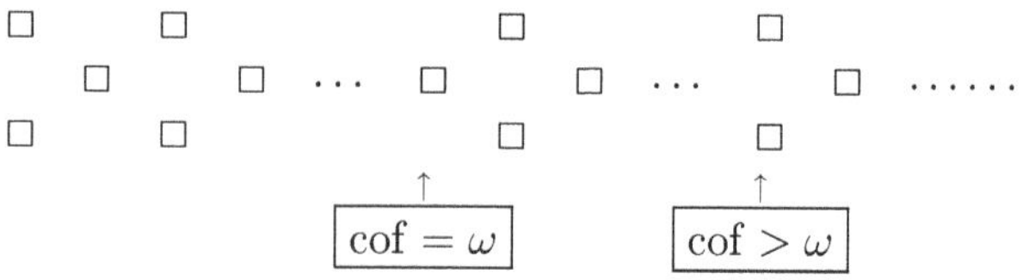


Moreover $\mathsf{SLO}^{\boldsymbol{\Delta}^0_2}$ implies that any non-self-dual $\boldsymbol{\Delta}^0_2$ degree is a Wadge degree and every self-dual $\boldsymbol{\Delta}^0_2$ is a obtained by collapsing a block of consecutive Wadge degrees: if $[A]_{\boldsymbol{\Delta}^0_2}$ is self-dual and $A$ is of minimal Wadge rank among sets in the $\boldsymbol{\Delta}^0_2$ degree, then $[A]_{\boldsymbol{\Delta}^0_2}$ is obtained by coalescing the next $\|A\|_{\mathrm{W}} \cdot \omega_1$ Wadge degrees after $[A]_{\mathrm{W}}$. The next diagram summarizes the situation: the Wadge degrees are denoted by $\bullet$ while the $\boldsymbol{\Delta}^0_2$ degrees are denoted by $\square$

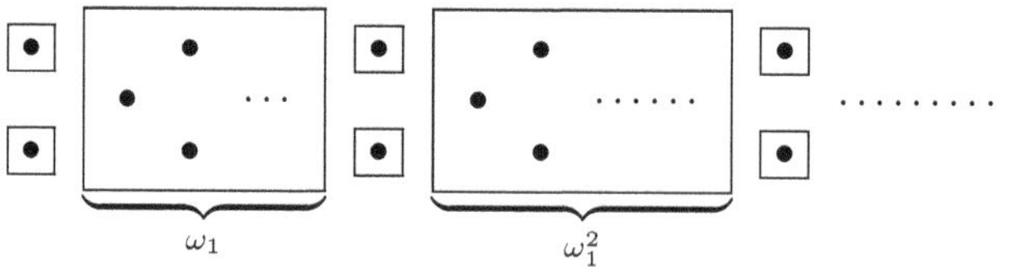


(25)

The $\boldsymbol{\Delta}^0_2$ hierarchy is intimately related to the cardinality[9] $|\boldsymbol{\Gamma}|$ of boldface pointclasses. Say that $\boldsymbol{\Gamma}$ is **cardinality pointclass** just in case $|\boldsymbol{\Gamma}'| < |\boldsymbol{\Gamma}|$, for any $\boldsymbol{\Gamma}' \subset \boldsymbol{\Gamma}$. The first few cardinality pointclasses are: $\{\varnothing\}$, $\{\mathbb{R}\}$, $\boldsymbol{\Delta}^0_1$. On the other hand $|\boldsymbol{\Sigma}^0_1| = |\mathbb{R}| = |\boldsymbol{\Delta}^0_1|$, so neither $\boldsymbol{\Sigma}^0_1$ nor $\boldsymbol{\Pi}^0_1$ is a cardinality pointclass. In fact, it can be shown that $|\boldsymbol{\Delta}^0_2| = |\mathbb{R}|$, so there is no cardinality pointclass $\boldsymbol{\Gamma}$ such that $\boldsymbol{\Delta}^0_1 \subset \boldsymbol{\Gamma} \subseteq \boldsymbol{\Delta}^0_2$.

**Lemma 36.** Assume PSP. Then $|\mathbb{R}| < |\boldsymbol{\Sigma}^0_2|$ and therefore $\boldsymbol{\Sigma}^0_2$ and $\boldsymbol{\Pi}^0_2$ are cardinality pointclasses.

*Proof.* By a theorem of Tarski's [Tar39]

$$|X| < |\mathscr{P}_{\mathrm{WO}}(X)| ,$$

where $\mathscr{P}_{\mathrm{WO}}(X) = \{Y \subseteq X \mid Y \text{ is well-orderable}\}$—for a beautiful proof of this result see [Kan97, p. 294]. By PSP

$$\mathscr{P}_{\mathrm{WO}}(\mathbb{R}) = \{Y \subseteq \mathbb{R} \mid Y \text{ is countable }\}$$

and since every countable set of reals is $\mathbf{F}_\sigma$ we are done. q.e.d.

This result implies that $|\boldsymbol{\Sigma}^0_1| < |\boldsymbol{\Sigma}^0_2|$, and Greg Hjorth showed in [Hjo98] that, assuming AD,

$$\alpha < \beta \implies |\boldsymbol{\Sigma}^0_\alpha| < |\boldsymbol{\Sigma}^0_\beta| . \tag{26}$$

In [Hjo02] he extended (26) to the projective hierarchy by showing that AD implies

$$|\boldsymbol{\Delta}^1_1| < |\boldsymbol{\Sigma}^1_1| < |\boldsymbol{\Sigma}^1_2| < \ldots \tag{27}$$

and hence $\boldsymbol{\Sigma}^1_1$ is also a cardinality pointclass. On the other hand (26) and (27) do not imply that $\boldsymbol{\Sigma}^0_\alpha$ or $\boldsymbol{\Sigma}^1_n$ are cardinality pointclasses, when $\alpha \geq 3$ and $n \geq 2$. In [AndHjoNee07] a characterization of all cardinality pointclasses is given. First a definition: a pointclass $\boldsymbol{\Gamma}$ is **$\boldsymbol{\Delta}^0_2$-closed** if and only if $A \in \boldsymbol{\Gamma} \wedge B \leq_{\boldsymbol{\Delta}^0_2} A \implies B \in \boldsymbol{\Gamma}$.

**Theorem 37.** Assume AD + DC($\mathbb{R}$) and let $\boldsymbol{\Gamma}$ be a boldface pointclass containing $\boldsymbol{\Delta}^0_2$.

(a) If $\boldsymbol{\Gamma}$ is non-self-dual, then $\boldsymbol{\Gamma}$ is a cardinality pointclass if and only if $\boldsymbol{\Gamma}$ is $\boldsymbol{\Delta}^0_2$-closed.
(b) If $\boldsymbol{\Gamma}$ is self-dual, then $\boldsymbol{\Gamma}$ is a cardinality pointclass if and only if either

[9] Since we do not assume AC, $|X| < |Y|$ means that there is an injective function from $X$ into $Y$, but no injective function from $Y$ into $X$.

- $\mathbf{\Gamma} = \bigcup_{\alpha<\lambda} \mathbf{\Gamma}_\alpha$, where $\lambda$ is limit and the $\mathbf{\Gamma}_\alpha$'s are non-self-dual, $\subset$-increasing, and $\mathbf{\Delta}^0_2$-closed; or else
- $\mathbf{\Gamma} = \{B \subseteq \mathbb{R} \mid B \leq_{\mathrm{W}} \bigoplus_n A_n\}$, where the $A_n$'s are non-self-dual, $<_{\mathrm{W}}$-increasing, and the pointclasses $\{B \subseteq \mathbb{R} \mid B \leq_{\mathrm{W}} A_n\}$ are $\mathbf{\Delta}^0_2$-closed.

In particular: $\mathbf{\Sigma}^0_\alpha$ ($\alpha \geq 2$), $\mathbf{\Delta}^0_\beta$ ($\beta \geq 3$), $\mathbf{\Sigma}^1_n$, and $\mathbf{\Delta}^1_n$ are all cardinality pointclasses.

It is worth pointing out that any pointclass $\mathbf{\Gamma}$ with a universal set (such as the $\mathbf{\Sigma}^0_\alpha$'s and $\mathbf{\Sigma}^1_n$'s) is the surjective image of $\mathbb{R}$ hence AC implies it is of cardinality $2^{\aleph_0}$.

Going back to the $\mathbf{\Delta}^0_2$-hierarchy, it is open whether the results in [And06] can be generalized to the $\mathbf{\Delta}^0_\alpha$ hierarchy, for $\alpha > 2$. Note that requiring $\forall \alpha < \omega_1\ \mathsf{SLO}^{\mathbf{\Delta}^0_\alpha}$ is weaker than requiring $\mathsf{SLO}^{\mathbf{\Delta}^1_1}$, but a proof that

$$\forall \alpha < \omega_1 (\mathsf{SLO}^{\mathbf{\Delta}^0_\alpha} \implies \mathsf{SLO}^{\mathbf{\Delta}^0_1})$$

would perhaps suggest a method for proving Conjecture 33, as $\mathsf{SLO}^{\mathbf{\Delta}^0_1}$ is just another name for $\mathsf{SLO}^{\mathrm{W}}$. The proofs in [And06] suggest that the $\mathbf{\Delta}^0_3$ hierarchy should arise from the $\mathbf{\Delta}^0_2$ hierarchy much like the $\mathbf{\Delta}^0_2$ hierarchy is arises from the Wadge hierarchy (see diagram (25)). We also believe that further generalizations to larger $\alpha$'s should ultimately provide an alternative proof of Theorem 32. To appreciate the difficulties of such generalizations we must take a closer look at the proofs in [And06]: they rely on a game-theoretic reformulation (due to Van Wesep) of $\mathbf{\Delta}^0_2$ reducibility, known as back-track reducibility. In the **back-track game** $G_{\mathrm{bt}}(A, B)$ the two players play integers, with player **II** having the privilege of passing or back-tracking (*i.e.*, erasing all his moves) at any round and the obligation of not back-tracking infinitely often and of playing infinitely often. Therefore at the end of the game the two players will have produced two reals $a$ and $b$, and **II** wins if and only if $a \in A \iff b \in B$. If **II** has a winning strategy in $G_{\mathrm{bt}}(A, B)$, then $A$ is said to be **back-track reducible** to $B$. (For more on the game $G_{\mathrm{bt}}$ the reader is referred to [And03, And06, Ste77, Van77].) Let us call any function $f : \mathbb{R} \to \mathbb{R}$ arising as a strategy for **II** in a back-track game a back-track function: by [And06] these are exactly those functions for which there exists a countable partition of $\mathbb{R}$ into closed sets $C_n$ ($n \in \omega$) such that each $f \restriction C_n$ is continuous. Since every $\mathbf{\Sigma}^0_2$ subset of the Baire space can be decomposed in a countable disjoint union of closed sets, the $C_n$'s above can be indifferently taken to be $\mathbf{\Sigma}^0_2$ or $\mathbf{\Delta}^0_2$. In [JayRog82] it is proved that

**Theorem 38 (Jayne-Rogers).** Let $f : \mathbb{R} \to \mathbb{R}$ be a function such that each $f \restriction C_n$ is continuous, where $\{C_n \mid n \in \omega\}$ is a partition of $\mathbb{R}$ into closed sets. Then $f$ is $\mathbf{\Delta}^0_2$. Hence $f$ is a back-track function if and only if $f$ is $\mathbf{\Delta}^0_2$.

Any extension of Theorem 35 and the other results in [And06] to the $\mathbf{\Delta}^0_\alpha$ hierarchy ($\alpha \geq 3$) will probably use some sort of generalization of Theorem 38 to $\mathbf{\Delta}^0_\alpha$ functions. And even if a completely different way of obtaining such an extension exists, a generalization of (or lack thereof) Theorem 38 is an interesting question *per se*. Very little is known in this direction: the only (negative) result is that the obvious generalization

> $f$ is $\mathbf{\Delta}^0_{\alpha+1}$ iff there is a partition $\{C_n \mid n \in \omega\}$ of $\mathbb{R}$ into $\mathbf{\Pi}^0_\alpha$ sets such that each $f \restriction C_n$ is continuous

is *false* when $\alpha \geq \omega$: in [Chi+91] a function $f : \mathbb{R} \to \mathbb{R}$ which is Baire class 1 (hence $\mathbf{\Delta}^0_\omega$, and *a fortiori* $\mathbf{\Delta}^0_{\omega+1}$) is constructed so that for *any* countable partition $\{C_n \mid n \in \omega\}$ there is an $n_0$ such that $f \restriction C_{n_0}$ is discontinuous.

We conclude this paper with another interesting—albeit vague—question: Is there a Wadge-style game for higher levels of reducibility, like $\mathbf{\Delta}^0_\alpha$-reducibility ($\alpha \geq 3$)? Brian Semmes has recently (spring 2005) devised a game that captures the collection of all function for which there is a partition of $\mathbb{R}$ into $\mathbf{\Delta}^0_3$ sets $\{A_n \mid n \in \omega\}$ such that $f \restriction A_n$ is $\mathbf{\Delta}^0_2$. Thus Semmes' result comes close to solve the question for $\alpha = 3$, but the question for higher $\alpha$'s is wide open.

**Note added in proof.**

Luca Motto Ros in [Mot07] has shown that the structure of the $\mathbf{\Delta}^0_\alpha$-degrees is similar to the one of the Wadge degrees, extending the results of [And06] and [AndMar03]. Furthermore his analysis can be extended to non-Borel (*e.g.*, $\mathbf{\Sigma}^1_n$) families of reductions.

## References.


[And03] Alessandro **Andretta**, Equivalence Between Wadge and Lipschitz Determinacy, **Annals of Pure and Applied Logic** 123 (2003), p. 163–192

[And06] Alessandro **Andretta**, More on Wadge Determinacy, **Annals of Pure and Applied Logic** 144 (2006), p. 2–32

[AndHjoNee07] Alessandro **Andretta**, Greg **Hjorth**, and Itay **Neeman**, Effective Cardinals of Boldface Pointclasses, **Journal of Mathematical Logic** 7 (2007), p. 35–82

[AndMar03] Alessandro **Andretta** and Donald A. **Martin**, Borel-Wadge Degrees, **Fundamenta Mathematicæ** 177 (2003), p. 175–192

[Bec88] Howard **Becker**, More Closure Properties of Pointclasses, *in:* [KecMarSte88, p. 31–36]

[Chi+91] Jacek **Chicoń**, Michał **Morayne**, Janusz **Pawlikowsky**, and Slawomir **Solecki**, Decomposing Baire Functions, **Journal of Symbolic Logic** 56 (1991), p. 1273–1283

[Dup03] Jacques **Duparc**, A Hierarchy of Deterministic Context-Free $\omega$-Languages, **Theoretical Computer Science** 290 (2003), p. 1253–1300

[DupFinRes01] Jacques **Duparc**, Olivier **Finkel**, and Jean-Pierre **Ressayre**, Computer Science and the Fine Structure of Borel Sets, **Theoretical Computer Science** 257 (2001), p. 85–105

[Fri71] Harvey **Friedman**, Higher Set Theory and Mathematical Practice, **Annals of Mathematical Logic** 2 (1971), p. 325–357

[GalSte53] David **Gale** and Frank M. **Stewart**, Infinite Games with Perfect Information, *in:* [KuhTuc53, p. 245–266]

[Har78] Leo **Harrington**, Analytic Determinacy and $0^{\#}$, **Journal of Symbolic Logic** 43 (1978), p. 685–693

[HarKec81] Leo **Harrington** and Alexander S. **Kechris**, On Determinacy of Games on Ordinals, **Annals of Mathematical Logic** 20 (1981), p. 109–154

[Hjo96] Greg **Hjorth**, $\mathbf{\Pi}^1_2$ Wadge Degrees, **Annals of Pure and Applied Logic** 77 (1996), p. 53–74

[Hjo98] Greg **Hjorth**, An Absoluteness Principle for Borel Sets, **Journal of Symbolic Logic** 63 (1998), p. 663–693

[Hjo02] Greg **Hjorth**, Cardinalities in the Projective Hierarchy, **Journal of Symbolic Logic** 67 (2002), p. 1351–1372

[JayRog82] John E. **Jayne** and C. Ambrose **Rogers**, First Level Borel Functions and Isomorphism, **Journal de Mathématiques Pures et Appliquées** 61 (1982), p. 177–205

[Kan97] Akihiro **Kanamori**, The Mathematical Import of Zermelo's Well-Ordering Theorem, **Bulletin of Symbolic Logic** 3 (1997), p. 281–311

[Kan03] Akihiro **Kanamori**, The Higher Infinite: Large Cardinals in Set Theory from Their Beginnings, 2nd edition, Springer 2003 [Springer Monographs in Mathematics]

[Kec77] Alexander S. **Kechris**, Classifying Projective-like Hierarchies, **Bulletin of the Greek Mathematical Society** 18 (1977), p. 254–275

[Kec84] Alexander S. **Kechris**, The Axiom of Determinacy Implies Dependent Choices in L($\mathbb{R}$), **Journal of Symbolic Logic** 49 (1984), p. 161–173

[Kec95] Alexander S. **Kechris**, Classical Descriptive Set Theory, Springer 1995 [Graduate Texts in Mathematics 156]

[KecMarMos81] Alexander S. **Kechris**, Donald A. **Martin**, and Yiannis N. **Moschovakis** (*eds.*), Cabal Seminar 77-79, Proceedings of the Caltech-UCLA Logic Seminar 1977-1979, Springer 1981 [Lecture Notes in Mathematics 839]

[KecMarMos83] Alexander S. **Kechris**, Donald A. **Martin**, and Yiannis N. **Moschovakis** (*eds.*), Cabal Seminar 79-81, Proceedings of the Caltech-UCLA Logic Seminar 1979-1981, Springer 1983 [Lecture Notes in Mathematics 1019]

[KecMarSte88] Alexander S. **Kechris**, Donald A. **Martin**, and John R. **Steel** (*eds.*), Cabal Seminar 81-85, Proceedings of the Caltech-UCLA Logic Seminar 1981-1985, Springer 1988 [Lecture Notes in Mathematics 1333]

[KecMos78] Alexander S. **Kechris** and Yiannis N. **Moschovakis** (*eds.*), Cabal Seminar 76-77, Proceedings of the Caltech-UCLA Logic Seminar 1976-1977, Springer 1978 [Lecture Notes in Mathematics 689]

[KecSolSte81] Alexander S. **Kechris**, Robert M. **Solovay**, and John R. **Steel**, The Axiom of Determinacy and the Prewellordering Property, *in:* [KecMarMos81, p. 101–125]

[Kir92] Bernd **Kirchheim**, Baire One Star Functions, **Real Analysis Exchange** 18 (1992), p. 385–399

[KuhTuc53] Harold W. **Kuhn** and Albert W. **Tucker** (*eds.*), Contributions to the Theory of Games, volume 2, Princeton University Press 1953 [Annals of Mathematics Studies 28]

[Kur58] Casimir **Kuratowski**, Topologie, 4th edition (1st edition 1933), Państwowe Wydawnictwo Naukowe 1958 [Monografie Matematyczne 20]

[Lou83] Alain **Louveau**, Some Results in the Wadge Hierarchy of Borel Sets, *in:* [KecMarMos83, p. 28–55]

[LouSai88] Alain **Louveau** and Jean **Saint-Raymond**, The Strength of Borel Wadge Determinacy, *in:* [KecMarSte88, p. 1-30],

[Mar75] Donald A. **Martin**, Borel Determinacy, **Annals of Mathematics** 102 (1975), p. 363–371

[Mar90] Donald A. **Martin**, An Extension of Borel Determinacy, **Annals of Pure and Applied Logic** 49 (1990), p. 279–293

[MarSte89] Donald A. **Martin** and John R. **Steel**, A Proof of Projective Determinacy, **Journal of the American Mathematical Society** 2 (1989), p. 71–125

[Mos80] Yiannis N. **Moschovakis**, Descriptive Set Theory, North Holland 1980 [Studies in Logic and the Foundations of Mathematics 100]

[Mot07] Luca **Motto Ros**, General reducibilities for sets of reals, *PhD thesis*, Politecnico di Torino, 2007

[MycSte62] Jan **Mycielski** and Hugo **Steinhaus**, A Mathematical Axiom Contradicting the Axiom of Choice, **Bulletin de l' Académie Polonaise des Sciences, Série des Sciences Mathématiques, Astronomiques et Physiques** X (1962), p. 1–3

[MycSwi64] Jan **Mycielski** and Stanisław **Swierczkowski**, On the Lebesgue Measurability and the Axiom of Determinateness, **Fundamenta Mathematicæ** 54 (1964), p. 67–71

[Sel03] Victor **Selivanov**, Wadge Degrees of $\omega$-Languages of Deterministic Turing Machines, **Theoretical Informatics and Applications** 37 (2003), p. 67–83

[Sie52] Wacław **Sierpiński**, General Topology, University of Toronto Press 1952

[Sol78] Robert M. **Solovay**, The Independence of DC from AD, *in:* [KecMos78, p. 171–184],

[Ste77] John R. **Steel**, Determinateness and Subsystems of Analysis, *PhD thesis*, University of California at Berkeley, 1977

[Ste81a] John R. **Steel**, Closure Properties of Pointclasses, *in:* [KecMarMos81, p. 147–163],

[Ste81b] John R. **Steel**, Determinateness and the Separation Property, **Journal of Symbolic Logic** 46 (1981), p. 41–44

[Ste78] Jacques **Stern**, Évaluation du rang de Borel des certains ensenbles, **Comptes Rendus Hebdomaires des Séances de l'Académie des Sciences** 286 (1978), p. 855–857

[Tar39] Alfred **Tarski**, On Well-Ordered Subsets of Any Set, **Fundamenta Mathematicæ** 32 (1939), p. 176–183

[Van77] Robert A. **Van Wesep**, Subsystems of Second-Order Arithmetic and Descriptive Set Theory Under the Axiom of Determinateness, *PhD thesis*, University of California at Berkeley, 1977

[Van78a] Robert A. **Van Wesep**, Separation Principles and the Axiom of Determinateness, **Journal of Symbolic Logic** 43 (1978), p. 77–81

[Van78b] Robert A. **Van Wesep**, Wadge Degrees and Descriptive Set Theory, *in:* [KecMos78, p. 151-170]

[Wad83] William W. **Wadge**, Reducibility and Determinateness on the Baire Space, *PhD thesis*, University of California at Berkeley, 1983

[Woo88] W. Hugh **Woodin**, Supercompact Cardinals, Sets of Reals, and Weakly Homogeneous Trees, **Proceedings of the National Academy of Sciences of the United States of America** 85 (1988), p. 6587–6591

[Woo99] W. Hugh **Woodin**, The Axiom of Determinacy, Forcing Axioms, and the Nonstationary Ideal, de Gruyter 1999 [de Gruyter Series in Logic and its Applications 1]

**Received**: April 13th, 2005;
**In revised version**: June 19th, 2005;
**Accepted by the editors**: July 15th, 2005.

FotFS V

Stefan **Bold**, Benedikt **Löwe**,
Thoralf **Räsch**, Johan **van Benthem** (*eds.*)
**Foundations of the Formal Sciences V**
Infinite Games

# The Complexity of Independence-Friendly Fixpoint Logic

JULIAN BRADFIELD AND STEPHAN KREUTZER

Laboratory for Foundations of Computer Science
University of Edinburgh
King's Buildings
Mayfield Road
Edinburgh EH9 3JZ, United Kingdom
`jcb@inf.ed.ac.uk`

Computing Laboratory
Oxford University
Wolfson Building
Parks Road
Oxford OX1 3QD, United Kingdom
`kreutzer@comlab.ox.ac.uk`

ABSTRACT. We study the complexity of model-checking for the fixpoint extension of Hintikka and Sandu's independence-friendly logic. We show that this logic captures EXPTIME; and by embedding PFP, we show that its combined complexity is EXPSPACE-hard, and moreover the logic includes second order logic (on finite structures).

## 1 Introduction

In everyday life we often have to make choices in ignorance of the choices made by others that might have affected our choice. With the popularity of the agent paradigm, there is much theoretical and practical work on logics of knowledge and belief in which such factors can be explicitly expressed in designing multi-agent systems. However, ignorance is not the only reason for making independent choices: in mathematical writing, it is not uncommon to assert the existence of a value for some parameter uniformly in some earlier mentioned parameter.

Hintikka and Sandu [HinSan96] introduced a logic, called Independence-friendly (IF) logic, in which such independent choices can be formalized by independent quantification. Some of the ideas go back some decades, for IF logic can also be viewed as an alternative account of branching quantifiers (Henkin quantifiers) in terms of games of imperfect information. Independent quantification is a subtle concept, with many pitfalls for the unwary. It is also quite powerful: it has long been known that it has existential second-order power. In previous work [BraFrö02], the first author and Fröschle applied the idea of independent quantification to modal logics, where it has natural links with the theory of true concurrency; this prompted some consideration of fixpoint versions of IF modal logics, since adding fixpoint operators is the easiest way to get a powerful temporal logic from a simple modal logic. This led the first author to an initial investigation [Bra03] of the fixpoint extension of first-order IF logic, which we call IF-LFP . It turned out that fixpoint IF logic is not trivial to define, and appears to be very expressive, with the interaction between fixpoints and independent quantification giving a dramatic increase in expressive power. In [Bra03], only some fairly simple complexity results were obtained; in this paper, we obtain much stronger results about the model-checking complexity of IF-LFP . For the data complexity, we show that not only is IF-LFP EXPTIME-complete, but it captures EXPTIME; and for the combined complexity, we obtain an EXPSPACE hardness result. This latter result is obtained by an embedding of partial fixpoint logic into IF-LFP, which shows that on finite structures IF-LFP even includes second-order logic, a much stronger result than the first author previously conjectured.

## 2 IF-FOL and IF-LFP

### 2.1 Syntax

First of all, we state one important **notational convention**: to minimize the number of parentheses, we take the scope of all quantifiers and fixpoint operators to extend as far to the right as possible.

Now we define the syntax of first-order IF logic. Here we use the version of Hodges [Hod97], and we confine the 'independence-friendly' operators to the quantifiers; in the full logic, one can also specify conjunctions and disjunctions that are independent, but these are not necessary for our purposes — their addition changes none of our results.

**Definition 1.** As for FOL, IF-FOL has proposition ($P, Q$ *etc.*), relation ($R, S$ *etc.*), function ($f, g$ *etc.*) and constant ($a, b$ *etc.*) symbols, with given arities. It also has individual *variables* $v, x$ *etc.* We write $\vec{x}, \vec{v}$ *etc.* for tuples of variables, and similarly for tuples of other objects; we use concatenation of symbols to denote concatenation of tuples with tuples or objects.

For formulae $\varphi$ and terms $t$, the (meta-level) notations $\varphi[\vec{x}]$ and $t[\vec{x}]$ mean that the free variables of $\varphi$ or $t$ are included in the variables $\vec{x}$, without repetition. The notions of 'term' and 'free variable' are as for FOL.

We assume equality $=$ is in the language, and atomic formulae are defined as usual by applying proposition or relation symbols to individual terms or tuples of terms. The free variables of the formula $R(\vec{t})$ are then those of $\vec{t}$. The compound formulae are given as follows:

- **Conjunction and disjunction.** If $\varphi[\vec{x}]$ and $\psi[\vec{y}]$ are formulae, then $(\varphi \vee \psi)[\vec{z}]$ and $(\varphi \wedge \psi)[\vec{z}]$ are also formulae, where $\vec{z}$ is the union of $\vec{x}$ and $\vec{y}$.
- **Quantifiers.** If $\varphi[\vec{y}, x]$ is a formula, $x$ a variable, and $W$ a finite set of variables, then $(\forall x/W.\, \varphi)[\vec{y}]$ and $(\exists x/W.\, \varphi)[\vec{y}]$ are formulae. If $W$ is empty, we write just $\forall x.\, \varphi$ and $\exists x.\, \varphi$.
- **Game negation.** If $\varphi[\vec{x}]$ is a formula, so is $({\sim}\varphi)[\vec{x}]$.
- **Flattening.** If $\varphi[\vec{x}]$ is a formula, so is $({\downarrow}\varphi)[\vec{x}]$.
- **(Negation.** $\neg\varphi$ is an abbreviation for ${\sim}{\downarrow}\varphi$.)

**Definition 2.** IF-FOL$^+$ is the logic in which $\sim$, $\downarrow$ and $\neg$ are applied only to atomic formulae.

### 2.2 Traditional semantics

In the independent quantifiers the intention is that $W$ is the set of independent variables, whose values the player is not allowed to know at this choice point: thus the classical Henkin quantifier $\begin{smallmatrix}\forall x\, \exists y\\ \forall u\, \exists v\end{smallmatrix}$, where $x$ and $y$ are independent of $u$ and $v$, can be written as $\forall x/\varnothing.\, \exists y/\varnothing.\, \forall u/\{x, y\}.\, \exists v/\{x, y\}$. This notion of independence is the reason for saying that IF logic is natural in mathematical English: statements such as "For every $x$, and for all $\varepsilon > 0$, there exists $\delta$, depending only on $\varepsilon$ ..." can be transparently written as $\forall x, \varepsilon > 0.\, \exists\delta/x.\, \ldots$ in IF logic.

If one then plays the Hintikka evaluation game (otherwise known as the model-checking game) with this additional condition, which can be formalized by requiring strategies to be uniform in the 'unknown' variables, one gets a game semantics of imperfect information, and defines a formula to be true if and only if Eloise has a winning strategy.

These games are not determined, so it is *not* the case that Abelard has a winning strategy iff the formula is not true. For example, $\begin{smallmatrix}\forall x\\ \exists y\end{smallmatrix}.x = y$ (or $\forall x.\, \exists y/\{x\}.\, x = y$) is untrue in any structure with more than one element, but Abelard has no winning strategy.

An alternative interpretation of the logic, dating from the early work on branching quantifiers, and one that is easier to handle mathematically in straightforward cases, is via Skolem functions with limited arguments. In

FOL, the first order sentence $\forall x. \exists y. x = y$ is converted via Skolemization to the existential second-order sentence $\exists f : \mathbb{N} \to \mathbb{N}. \forall x. x = f(x)$. In this procedure, the Skolem function always takes as arguments all the universal variables currently in scope. By allowing Skolem functions to take only some of the arguments, we get a similar translation of IF-FOL: for example, $\forall x. \exists y/\{x\}. x = y$ becomes $\exists f : 1 \to \mathbb{N}. \forall x. x = f()$. It can be shown that these two semantics are equivalent, in that an IF-FOL sentence is true in the game semantics if and only if its Skolemization is true.

It is also well known that IF-FOL$^+$ is equivalent to existential second-order logic (in the cases where this matters, 'second-order' here means function quantification rather than set quantification). This is because the Skolemization process can be inverted: given an ESO sentence, it can be turned into an IF-FOL sentence (or equivalently, a sentence with Henkin quantifiers). We shall make use of this procedure later. Details can be found in [Wal70, End70], but here let us illustrate it by a standard example that demonstrates the power of IF logic. Consider the sentence 'there is an injective endofunction that is not surjective'. This is true only in infinite domains, and therefore not first-order expressible. It can be expressed directly in ESO as

$$\exists f. (\forall x_1, x_2. f(x_1) = f(x_2) \Rightarrow x_1 = x_2) \wedge (\exists c. \forall x. f(x) \neq c)$$

which for the sake of reducing complexity below we will simplify to

$$\exists f. \exists c. \forall x_1, x_2. (f(x_1) = f(x_2) \Rightarrow x_1 = x_2) \wedge f(x_1) \neq c.$$

The basic manoeuvre for talking about functions in IF-FOL is to replace $\exists f. \forall x$ by $\forall x. \exists y$, so that $y$ plays the rôle of $f(x)$. In FOL, this works only if there is just one application of $f$; but in IF-FOL, we can do it for two (or more) applications of $f$: we write $\forall x_1. \exists y_1$, and then we write an independent $\forall x_2/\{x_1, y_1\}. \exists y_2/\{x_1, y_1\}$. Now in order to make sure that these two $(x_i, y_i)$ pairings represent the same $f$, the body of the translated formula is given a clause $(x_1 = x_2) \Rightarrow (y_1 = y_2)$. Applying this procedure to the ESO sentence above and optimizing a bit, we get

$$\forall x_1, x_2. \exists y_1/x_2. \exists y_2/x_1. \exists c/\{x_1, x_2\}. (y_1 = y_2 \Leftrightarrow x_1 = x_2) \wedge y_1 \neq c.$$

The 'game negation' $\sim$ corresponds to swapping the roles of the two players in the game. In ordinary logic, with perfect information, this corresponds exactly to classical negation; but in IF-FOL it does not. The issue of negation is still controversial, particularly for open formulae; but one approach, taken by [Hod97], is to define the 'flattening' operator $\downarrow$, which smashes its argument down to a classical formula by removing all the uniformity constraints imposed by imperfect information. Then classical negation is defined to be game negation applied to a flattened formula.

### 2.3 Trump semantics

The game semantics is how Hintikka and Sandu originally interpreted IF logic. Later on, the trump semantics of Hodges [Hod97], with variants by others, gave a Tarski-style semantics, equivalent to the original. This semantics is as follows:

**Definition 3.** Let a structure $A$ be given, with constants, propositions and relations interpreted in the usual way. A *deal* $\vec{a}$ for $\varphi[\vec{x}]$ or $\vec{t}[\vec{x}]$ is an assignment of an element of $A$ to each variable in $\vec{x}$. Given a deal $\vec{a}$ for a tuple of terms $\vec{t}[\vec{x}]$, let $\vec{t}(\vec{a})$ denote the tuple of elements obtained by evaluating the terms under the deal $\vec{a}$.

If $\varphi[\vec{x}]$ is a formula and $W$ is a subset of the variables in $\vec{x}$, two deals $\vec{a}$ and $\vec{b}$ for $\varphi$ are $\simeq_W$*-equivalent* ($\vec{a} \simeq_W \vec{b}$) if and only if they agree on the variables not in $W$. A $\simeq_W$*-set* is a non-empty set of pairwise $\simeq_W$-equivalent deals.

The denotation $[\![\varphi]\!]$ of a formula is a pair $(T, C)$ where $T$ is the set of *trumps*, and $C$ is the set of *cotrumps*. If $\varphi$ has $n$ free variables, then $T, C \subseteq \wp(\wp(A^n))$ – that is, a (co)trump is a set of deals.

- If $(R(\vec{t}))[\vec{x}]$ is atomic, then a non-empty set $D$ of deals is a trump if and only if $\vec{t}(\vec{a}) \in R$ for every $\vec{a} \in D$; $D$ is a cotrump if and only if it is non-empty and $\vec{t}(\vec{a}) \notin R$ for every $\vec{a} \in D$.
- $D$ is a trump for $(\varphi \wedge \psi)[\vec{x}]$ if and only if $D$ is a trump for $\varphi[\vec{x}]$ and $D$ is a trump for $\psi[\vec{x}]$; $D$ is a cotrump if and only if there are cotrumps $E, F$ for $\varphi, \psi$ such that every deal in $D$ is an element of either $E$ or $F$.
- $D$ is a trump for $(\varphi \vee \psi)[\vec{x}]$ if and only if it is non-empty and there are trumps $E$ of $\varphi$ and $F$ of $\psi$ such that every deal in $D$ belongs either to $E$ or $F$; $D$ is a cotrump if and only if it is a cotrump for both $\varphi$ and $\psi$.
- $D$ is a trump for $(\forall y/W.\, \psi)[\vec{x}]$ if and only if the set $\{\, \vec{a}b \mid \vec{a} \in D, b \in A \,\}$ is a trump for $\psi[\vec{x}, y]$. $D$ is a cotrump if and only if it is non-empty and there is a cotrump $E$ for $\psi[\vec{x}, y]$ such that for every $\simeq_W$-set $F \subseteq D$ there is a $b$ such that $\{\, \vec{a}b \mid \vec{a} \in F \,\} \subseteq E$.
- $D$ is a trump for $(\exists y/W.\, \psi)[\vec{x}]$ if and only if there is a trump $E$ for $\psi[\vec{x}, y]$ such that for every $\simeq_W$-set $F \subseteq D$ there is a $b$ such that $\{\, \vec{a}b \mid \vec{a} \in F \,\} \subseteq E$; $D$ is a cotrump if and only if the set $\{\, \vec{a}b \mid \vec{a} \in D, b \in A \,\}$ is a cotrump for $\psi[\vec{x}, y]$.
- $D$ is a trump for $\sim\varphi$ if and only if $D$ is a cotrump for $\varphi$; $D$ is a cotrump for $\sim\varphi$ if and only if it is a trump for $\varphi$.

- $D$ is a trump (cotrump) for $\downarrow\varphi$ if and only if $D$ is a non-empty set of members (non-members) of trumps of $\varphi$.

A trump for $\varphi$ is essentially a set of winning positions for the model-checking game for $\varphi$, for a given *uniform* strategy, that is, a strategy where choices are uniform in the 'hidden' variables. The most intricate part of the above definition is the clause for $\exists y/W.\,\psi$: it says that a trump for $\exists y/W.\,\psi$ is got by adding a witness for $y$, uniform in the $W$-variables, to trumps for $\psi$.

In the absence of flattening, a cotrump is simply a trump for the game negation of a formula, in other words a set of winning positions for Abelard in the model-checking game. If one ignores flattening, it is not necessary to maintain cotrumps in the semantics, and so we shall often elide them.

It is easy to see that any subset of a trump is a trump. In the case of an ordinary first-order $\varphi(\vec{x})$, the set of trumps of $\varphi$ is just the power set of the set of tuples satisfying $\varphi$. To see how a more complex set of trumps emerges, consider the following formula, which has $x$ free: $\exists y/\{x\}.\,x = y$. Any singleton set of deals is a trump, but no other set of deals is a trump. Thus we obtain that $\forall x.\,\exists y/\{x\}.\,x = y$ has no trumps (unless the domain has only one element).

The strangeness of the trump definitions is partly to do with some more subtle features of IF logics, that we do not here have space to discuss, but which are considered in detail in Ahti-Veikko Pietarinen's thesis [Pie01]. However, to take one good example, raised by a referee, consider $\varphi = \exists x.\,\exists y/\{x\}.\,x = y$. What are its trumps? As above, the trumps of $\exists y/\{x\}.\,x = y$ are singleton sets of deals. The only potential trump for $\varphi$ is the set containing the empty deal $D = \{\langle\rangle\}$. Applying the definition, $D$ is a trump for $\varphi$ if and only if there is a singleton deal set $\{a\}$ for $x$ such that there is a $b$ such that $\{b\} \subseteq \{a\}$. The right hand side is true – take $b = a$ – so $D$ is a trump. How come, if there is more than one element in $A$? Surely we must choose $y$ independently of $x$, and therefore $\varphi$ can't be true? Not so: because the choices are both made by the same player (Eloise), she can, as it were, make a uniform choice of $y$ that, by 'good luck' agrees with her previous choice of $x$. Since she is not in the business of making herself lose, she will always do so. In game-theoretic terms, this is the difference between requiring a strategy to make uniform moves, and requiring a player to choose a strategy uniformly. In fact Hintikka and Sandu tried to hide this issue by only allowing the syntax to express quantifications independent in the other player's variables, which is in practice all one wishes to use in any case; but they also incorporated it by asserting that a player's choices are always independent of their own earlier choices, which is intuitively bizarre.

Hodges removed the syntactic restriction to make his semantics cleaner, exposing the issue more obviously.

A sentence is said to be true if $\{\langle\rangle\} \in T$ (the empty deal is a trump set), and false if $\{\langle\rangle\} \in C$; this corresponds to Eloise or Abelard having a uniform winning strategy. Otherwise, it is undetermined. Note that 'false' is reserved for a strong sense of falsehood – undetermined sentences are also not true, and in the simple cases where negation and flattening are not employed, an undetermined sentence is as good as false. Note also that the game negation $\sim$ provides the usual de Morgan dualities, but that it cannot be pushed through flattening.

### 2.4 IF-LFP

We now describe the addition of fixpoint operators to IF-FOL. This is slightly intricate, although the normal intuitions for understanding fixpoint logics still apply.

**Definition 4.** IF-LFP extends the syntax of IF-FOL as follows:

- There is a set $\mathrm{Var} = \{X, Y, \ldots\}$ of fixpoint variables. Each variable $X$ has an arity $\mathrm{ar}(X)$.
- If $X$ is a fixpoint variable, and $\vec{t}$ an $\mathrm{ar}(X)$-vector of terms then $X(\vec{t})$ is a formula.
- Let $\varphi$ be a formula with free fixpoint variable $X$. $\varphi$ has free individual variables $\vec{x} = \langle x_1, \ldots, x_{\mathrm{ar}(X)}\rangle$ for the elements of $X$, together with other free individual variables $\vec{z}$; let $\mathrm{fv}_\varphi(X)$ be the length of $\vec{z}$. Now if $\vec{t}$ is a sequence of $\mathrm{ar}(X)$ terms with free variables $\vec{y}$, then $(\mu X(\vec{x}).\varphi)(\vec{t})[\vec{z}, \vec{y}]$ is a formula; **provided that** $\varphi$ is IF-FOL$^+$. In this context, we write just $\mathrm{fv}(X)$ for $\mathrm{fv}_\varphi(X)$.
- Similarly for $\nu X(\vec{x}).\varphi$.

To give the semantics of IF-LFP, we first define valuations for free fixpoint variables, in the context of some IF-LFP formula.

**Definition 5.** A fixpoint valuation $\mathcal{V}$ maps each fixpoint variable $X$ to a pair

$$(\mathcal{V}_T(X), \mathcal{V}_C(X)) \in (\wp(\wp(A^{\mathrm{ar}(X)+\mathrm{fv}(X)})))^2.$$

Let $D$ be a non-empty set of deals for $X(\vec{t})[\vec{x}, \vec{z}, \vec{y}]$, where $\vec{y}$ are the free variables of $\vec{t}$ not already among $\vec{x}, \vec{z}$. A deal $d = \vec{a}\,\vec{c}\,\vec{b} \in D$, where $\vec{a}, \vec{c}, \vec{b}$ are the deals for $\vec{x}, \vec{z}, \vec{y}$ respectively, determines a deal $d' = \vec{t}(d)\vec{c}$ for $X[\vec{x}, \vec{z}]$. Let $D' = \{d' \mid d \in D\}$. The set $D$ is a trump for $X(\vec{t})$ if and only if $D' \in \mathcal{V}_T(X)$; it is a cotrump if and only if $D' \in \mathcal{V}_C X$.

The intuition here is that a fixpoint variable needs to carry the trumps and cotrumps both for the elements of the fixpoint and for any free variables, as we shall see below. Then we define a suitable complete partial order on the range of valuations, which will also be the range of denotations for formulae; it is simply the inclusion order on trump sets and the reverse inclusion order on cotrump sets:

**Definition 6.** If $(T_1, C_1)$ and $(T_2, C_2)$ are elements of $(\wp(\wp(A^n)))^2$, define $(T_1, C_1) \preceq (T_2, C_2)$ if and only if $T_1 \subseteq T_2$ and $C_1 \supseteq C_2$.

This order gives the standard basic lemma for fixpoint logics:

**Lemma 7.** If $\varphi(X)[\vec{x}, \vec{z}]$ is an IF-FOL$^+$ formula and $\mathcal{V}$ is a fixpoint valuation, the map on $(\wp(\wp(A^{\mathrm{ar}(X)+\mathrm{fv}(X)}))^2$ given by

$$(T, C) \mapsto [\![\varphi]\!]_{\mathcal{V}[X:=(T,C)]}$$

is monotone with respect to $\preceq$; hence it has least and greatest fixpoints, constructible by iteration from the bottom and top elements of the set of denotations.

Thus we have the familiar definition of the $\mu$ operator:

**Definition 8.** $[\![\mu X(x).\varphi(X)[\vec{x}, \vec{z}]]\!]$ is the least fixpoint of the map on $(\wp(\wp(A^{\mathrm{ar}(X)+\mathrm{fv}(X)}))^2$ given by

$$(T, C) \mapsto [\![\varphi]\!]_{\mathcal{V}[X:=(T,C)]};$$

and $[\![\nu X(x).\varphi(x)[\vec{x}, \vec{z}]]\!]$ is the greatest fixpoint. $\mu^{\zeta}_{X,x}\varphi$ means the $\zeta$th approximant of $\mu X(x).\varphi$, defined recursively by $\mu^{\zeta}_{X,x}\varphi = \varphi(\bigcup_{\xi<\zeta} \mu^{\xi}_{X,x}\varphi)$.

A distinctive feature of the definition, compared to the normal LFP definition, is the way that free variables are explicitly mentioned. Normally, one can fix values for the free variables, and then compute the fixpoint, but because of independent quantification this is not possible in the IF setting. For example, consider the formula fragment

$$\forall z. \ldots \mu X(x). \ldots \vee \exists y/\{z\}. X(y)$$

The independent choice of $y$ means that the trumps for the fixpoint depend on the possible deals for $z$, not just a single deal.

## 2.5 Examples of IF-LFP

In order to give some human-readable examples of IF-LFP, we here reproduce a section from [Bra03]. For convenience, we introduce the abbreviation $\varphi \Rightarrow \psi$ for $\psi \vee {\sim}\varphi$ provided that $\varphi$ is atomic.

Let $G = (V, E)$ be a directed graph. The usual LFP formula $R(y,z) := (\mu(X,x).z = x \vee \exists w.\, E(x,w) \wedge X(w))(y)$ asserts that the vertex $z$ is reachable from $y$. Hence the formula $\forall y.\, \forall z.\, R(y,z)$ asserts that $G$ is strongly connected. Now consider the IF-LFP formula

$$\forall y.\, \forall z.\, (\mu(X,x).z = x \vee \exists w/\{y,z\}.\, E(x,w) \wedge X(w))(y).$$

At first sight, one might think this asserts not only that every $z$ is reachable from every $y$, but that the path taken is independent of the choice of $y$ and $z$. This is true exactly if $G$ has a directed Hamiltonian cycle, a much harder property than being strongly connected.

Of course, the formula does not mean this, because the variable $w$ is fresh each time the fixpoint is unfolded. In the trump semantics, the denotation of the fixpoint will include all the possible choice functions at each step, and hence all possible combinations of choice functions. Thus the formula reduces to strong connectivity.

It may be useful to look at the approximants of this formula in a little more detail, to get some intuitions about the trump semantics. Considering just

$$H := (\mu(X,x).z = x \vee \exists w/\{y,z\}.\, E(x,w) \wedge X(w))[x,y,z],$$

we see that in computing each approximant, the calculation of $[\![\exists w/\{y,z\}.\, \ldots]\!]$ involves generating a trump for every possible value of a choice function $f\colon x \mapsto w$. This is a feature of the original trump semantics, and can be understood by viewing it as a second-order semantics: just as the compositional Tarskian semantics of $\exists x.\, \varphi(x)$ involves computing all the witnesses for $\varphi(x)$, so computing the trumps of $\exists x/\{y\}.\, \varphi$ involves computing all the Skolem functions; and unlike the first-order case, it is necessary to work with functions (as IF can express existential second-order logic). Consequently, the $n$th approximant includes all states such that $x \to f_1(x) \to f_2 f_1(x) \to \ldots \to f_n \ldots f_1(x) = z$ for any sequence of successor-choosing functions $f_i$. Thus we see that the cumulative effect is the same as for a normal $\exists w$, and the independent choice has indeed not bought us anything.

It is, however, possible to produce a slightly more involved formula expressing the Hamiltonian cycle property in this inductively defined way, by using the standard trick for expressing functions in Henkin quantifier logics. We replace the formula $H$ by

$$\begin{aligned}&\forall s.\, \exists t/\{y,z\}.\, E(s,t) \wedge (\mu X(x).x = z \vee \\ &\forall u.\, \exists v/\{x,y,z,s,t\}.\, (s = u \Rightarrow t = v) \wedge (x = u \Rightarrow X(v)))(y).\end{aligned}$$

This works because the actual function $f$ selecting a successor for every node is made outside the fixpoint by $\forall s.\, \exists t/\{y,z\}.\, E(s,t) \wedge \ldots$; then inside the fixpoint, a new choice function $g$ is made so that $X(g(x))$, and $g$ is constrained to be the same as $f$ by the clause $(s = u \Rightarrow t = v)$. (The reader who is not familiar with the IF/Henkin to existential second-order translation might wish to ponder why $\forall s.\, \exists t/\{y,z\}.\, E(s,t) \wedge \mu(X,x).x = z \vee (x = s \Rightarrow X(t))$ does not work.)

### 2.6 The issue of negation

As we remarked above, negation is a somewhat problematic concept in IF logics. In the game-theoretic presentation, IF sentences may be true, false, or undetermined. When we say that an IF logic is equivalent to a classical logic, such as IF-FOL $= \Sigma^1_1$, we mean that the IF formula is true if and only if its classical equivalent is true. Hence if an IF sentence $\varphi$ is undetermined, its translation $\hat{\varphi}$ is false; and so $\sim\varphi$ is certainly not equivalent to $\neg\hat{\varphi}$. Hodges' introduction of the $\downarrow$ operator provides a way to get classical negation into IF sentences: if $\varphi$ is an IF sentence that is undetermined, it has neither trumps nor cotrumps; hence $\downarrow\varphi$ has the empty deal as a co-trump, since $\langle\rangle$ is a non-member of the co-trumps of $\varphi$; and so $\sim\downarrow\varphi$ is true. The intuitive interpretation of $\downarrow$ on formulae with free variables is unclear, and its combination with game negation more so. Indeed, there is not a simple game account of the flattening operator.

The arguments we apply in the following results rely largely on the ability to simulate operations on functions by operations on trumps. It is unclear to us how to combine boolean negation on the functional side with game/flattening negation on the trump side. The model-checking upper bound holds also for the full IF logic with negation and flattening; and the complexity lower bounds hold *a fortiori* for the full logic. However, our lower bounds, obtaining by translating from classical logic to IF logic, use various techniques to avoid having to translate classical negation to IF negation, and it would be useful to know whether additional power is obtained from IF negation.

## 3 Second-order inductions and independence-friendly logics

It has been known from the early studies of Henkin quantifiers [Wal70, End70] that existential second-order sentences can be transformed into sentences with the Henkin quantifier, and thus into IF-FOL. A technique frequently used in our results is the translation of existential second order inductions into IF-LFP. For this we show that the translation of existential second-order logic into independence-friendly logic can be extended to a translation of positive existential second-order inductions into independence-friendly fixpoint logic. Throughout this paper we only consider finite structures. Therefore we only give the translation for finite structures here. We first give a formal definition of positive $\Sigma^1_1$-inductive formulae.

**Definition 9.** An $(n, k)$-ary third-order variable $\mathcal{R}$ is a variable interpreted by a set whose members are $n$-tuples of $k$-ary functions. Let, for some $k, n < \omega$, $\mathcal{R}$ be a $(n, k)$-ary third-order variable. A formula $\varphi(\mathcal{R}, f_1, \ldots, f_n)$ is $\Sigma^1_1$-*inductive* if it is built up by the usual formula building rules for $\Sigma^1_1$ augmented by a rule that allows the use of atoms $\mathcal{R}f_1 \ldots f_n$, where the $f_i$ are $k$-ary function symbols, provided that the variable $\mathcal{R}$ is only used positively in $\varphi$.

$\Sigma^1_1$-inductive formulae $\varphi$ can be used to define least fixpoint inductions in the same way as first-order formulae with a free relation variable in which they are positive are used to define fixpoint inductions. So we can define the stages $\mathcal{R}^\alpha$, $\alpha < \omega$, of the fixpoint induction in $\varphi$ which ultimately lead to the least fixpoint of the operator defined by the formula $\varphi$. We call a relation that is obtained as the least fixpoint of a $\Sigma^1_1$-inductive formula $\Sigma^1_1$-*inductive.* Note, that the $\Sigma^1_1$-inductive relations are third-order objects, *i.e.*, sets of functions.

We show next that any $\Sigma^1_1$-inductive third-order relation $\mathcal{R}$ can be defined by an IF-LFP-formula in the sense that there is a formula $\varphi(R, \vec{x}, y)$, positive in the second-order variable $R$, such that the maximal trumps in the least fixpoint of the operator defined by $\varphi$ are precisely the graphs of the functions in $\mathcal{R}$. For the sake of simplicity, we only consider the case of $(1, k)$-ary inductions, *i.e.*, where the fixpoint is a set of functions.

An important concept used in the following proofs is the notion of *functional trumps*; and a technically useful concept is that of *maximal trumps.*

**Definition 10.** Let $\varphi(\vec{x}, y)$ be a formula. A trump $T$ for $\varphi$ is *functional in* $\vec{x}$ *and* $y$, if for all pairs $(\vec{a}, b), (\vec{a}', b')$ of deals in $T$ we have $b = b'$ whenever $\vec{a} = \vec{a}'$. $T$ is *maximal* if there is no $T' \supsetneq T$ that is a trump for $\varphi$.

Note that because any subset of a trump is a trump, the trumps of a formula are determined by its maximal trumps. Of course, any subset of a functional trump is functional.

**Notation.** In the following proofs we will frequently use a construction like

$$\forall \vec{x}/\{\vec{x}_1, y_1, \ldots, \vec{x}_n, y_n\} \exists y/\{\vec{x}_1, y_1, \ldots, \vec{x}_n, y_n\} \big((\vec{x} = \vec{x}_1 \rightarrow y = y_1) \wedge \varphi\big)$$

for some formula $\varphi$. We will abbreviate this by

$$\forall \vec{x} \exists y \, \mathrm{clone}(\vec{x}_1, y_1; \vec{x}_2, y_2 \ldots, \vec{x}_n, y_n)\varphi.$$

and we will usually omit the list $(\vec{x}_2, y_2 \ldots, \vec{x}_n, y_n)$ of other variables which appear in the independence sets of the quantifiers, assuming that all other variables than the clones and originals are in that list. Essentially, this formula says that the Skolem functions $f_y$ and $f_{y_1}$ chosen for $y$ and $y_1$, respectively, are the same. The next lemma makes this precise and establishes some useful properties of the clone construction.

**Lemma 11.** Let $\mathbf{A}$ be a structure and let $\vec{x}$ be a $k$-tuple of variables.

(i) Let $\psi$ be a formula defined as $\psi(\vec{x}, y) := \forall \vec{x}' \exists y' \text{clone}(\vec{x}, y)\psi'$. Then the trumps for $\psi$ are precisely the sets of deals $(\ldots, \vec{a}, f(\vec{a}))$ for $\vec{a} \in A^{|\vec{x}|}$ functional in $\vec{x}$ and $y$ with some Skolem function $f$, such that the deals $(\ldots, \vec{a}, f(\vec{a}), \vec{a}', f(\vec{a}'))$ for $\vec{a} \in A^{|\vec{x}|}$ form a trump for $\psi'$. In particular, if $\psi'$ is the formula **true**, then the trumps of $\psi$ are exactly the sets of deals functional in $\vec{x}$ and $y$.

(ii) Let $\varphi(\vec{x}', y')$ be a formula with only functional trumps and let $\psi$ be defined as $\psi(\vec{x}, y) := \forall \vec{x}' \exists y' \text{clone}(\vec{x}, y)\ \varphi$. Then the trumps for $\psi$ and the trumps for $\varphi$ are the same, in the sense that for every trump $T' \subseteq A^{k+1}$ of $\varphi$ there is a trump $T \subseteq A^{k+1}$ of $\psi$ such that an assignment of elements $\vec{a}$ to the variables $\vec{x}'$ and $b$ to $y'$ is a deal in $T'$ if, and only if, the corresponding assignment of $\vec{a}$ to $\vec{x}$ and $b$ to $y$ is a deal in $T$ and, conversely, for every trump $T$ of $\psi$ there is a corresponding trump $T'$ for $\varphi$.

*Proof.* We first prove Part (i) of the lemma. Following our notation, the formula $\psi$ is an abbreviation for

$$\forall \vec{x}'/\{\vec{x}, y\} \exists y'/\{\vec{x}, y\}(\vec{x} = \vec{x}' \rightarrow y = y') \wedge \psi'.$$

Towards a contradiction, suppose there was a non-functional trump $T$ for $\psi$, *i.e.*, $T$ contains deals $(\vec{a}, b)$ and $(\vec{a}, b')$ for some $\vec{a}$ and $b \neq b'$. By the semantics of universal quantifiers, this implies that there must be a trump for $\exists y_1/\{\vec{x}, y\}(\vec{x} = \vec{x}' \rightarrow y = y') \wedge \psi'$ containing $(\vec{a}, b, \vec{a})$ and $(\vec{a}, b', \vec{a})$. But then, the set $\{(\vec{a}, b, \vec{a}), (\vec{a}, b', \vec{a})\}$ is a $\{\vec{x}, y\}$-set (recall Definition 3). Hence there must be a trump $T'$ for $(\vec{x} = \vec{x}' \rightarrow y = y') \wedge \psi'$ and an element $c$ so that $T'$ contains the deals $(\vec{a}, b, \vec{a}, c)$ and $(\vec{a}, b', \vec{a}, c)$. But this is impossible as not both $b = c$ and $b' = c$ can be true but obviously every deal $(\vec{d}, e, \vec{d}', e')$ in a trump for $(\vec{x} = \vec{x}' \rightarrow y = y')$ satisfies the condition that if $\vec{d} = \vec{d}'$ then also $e = e'$. Finally, if $T$ is a functional trump, then the corresponding $T'$ must be a trump for $\psi'$, and so the deals $(\vec{a}, b, \vec{a}, b)$ must be a trump for $\psi'$. Part (ii) of the lemma follows analogously. q.e.d.

The next lemma shows that every formula in $\Sigma^1_1$ is equivalent to a formula in IF-LFP. The proof of the lemma follows easily from the work on Henkin-quantifiers. However, as we need the lemma also for formulae with free function variables we give an explicit translation of $\Sigma^1_1$-formulae into IF-FOL.

**Lemma 12.** Let $\varphi(f_1, \ldots, f_n)$ be a $\Sigma^1_1$-formula with free function variables $f_1, \ldots, f_k$. Then there is a formula $\hat{\varphi}(\vec{x}_{f_1}, y_{f_1}, \ldots, \vec{x}_{f_k}, y_{f_k}) \in$ IF-FOL such that for every structure $\mathbf{A}$ a set $T$ is a maximal trump for $\hat{\varphi}$ if, and only if, there are functions $F_1, \ldots, F_k$ such that $\mathbf{A} \models \varphi(F_1, \ldots, F_k)$ and

$$T = \{(\vec{a}_1, b_1, \ldots, \vec{a}_k, b_k) : F_i(\vec{a}_i) = b_i \text{ for all } 1 \leq i \leq k\}.$$

*Proof.* We first present the standard translation of $\Sigma^1_1$ into independence-friendly first-order logic (in a less efficient but more transparent form than normal). Let $\psi$ be a first-order formula containing a free function variable $g$. For the variable $g$, introduce variables $\vec{x}_0, y_0$, and for each of the $i = 1, \ldots, n$ applications $g(\vec{t}_i)$ of $g$ introduce variables $\vec{x}_i, y_i$. Then the $\Sigma^1_1$ sentence $\exists g.\, \psi$ is translated to $\hat{\psi}_0 = \forall \vec{x}_0.\, \exists y_0.\, \hat{\psi}_1$, where

$$\hat{\psi}_i = \forall \vec{x}_i.\, \exists y_i.\, \mathrm{clone}(\vec{x}_{i-1}, y_{i-1}) \hat{\psi}_{i+1}$$

for $i = 1, \ldots, n$, and $\hat{\psi}_{n+1}$ is $\psi$ with each $g$-containing atom $Q(g(\vec{t}_i))$ replaced by $\vec{x}_i = \vec{t}_i \Rightarrow Q(y_i)$.

We can stop short of the final $\exists f$, and translate $\psi$ itself to $\hat{\psi}_1$, giving the lemma for one function symbol. By Lemma 11(i), the trumps for $\hat{\psi}_n$ are the sets of deals functional in $\vec{x}_{n-1}$ and $y_{n-1}$ such that when extended by a Skolem function they are trumps for $\hat{\psi}_{n+1}$. Now $\hat{\psi}_{n+1}$ is classical, so its trumps are just the sets of deals satisfying it classically.

Now repeated application of Lemma 11 to $\hat{\psi}_{n-1}, \ldots, \hat{\psi}_1$ that the trumps for $\hat{\psi}_1$ are functional in $\vec{x}_0$ and $y_0$, with a Skolem function $g$ such that $\hat{\psi}_{n+1}$ is true when every $y_i$ is replaced by $g(\vec{x}_i)$, as required.

It remains to deal with multiple function variables, and to allow some of them to be explicitly closed by existential quantification. To manage multiple function symbols, it is easiest to process them in parallel: if we are dealing with $g$ and $h$, then extend the above translation to work simultaneously with $\vec{x}, y$ for $g$ and $\vec{u}, v$ for $h$, in the obvious way: put both sets of function-constraining conjuncts in, and both sets of $\forall\exists$ quantifiers. The two sets of variables should made independent of each other. (This would look much more obvious in Henkin quantifier notation.) If it is desired to quantify some of the function variables by $\exists$, then simply apply the final stage of the translation to those variables. q.e.d.

We are now ready to prove the main theorem of this section.

**Theorem 13.** Let $\mathcal{R}$ be a $(1,k)$-ary third-order variable and let $\varphi(\mathcal{R}, f)$ be a $\Sigma_1^1$-inductive formula where $f$ is a $k$-ary function symbol. Then there is a formula $\varphi^*(R, \vec{x}, y) \in$ IF-LFP, where $R$ is a $k+1$-ary second-order variable that only occurs positively in $\varphi$ and $\vec{x}$ is a $k$-tuple of variables, such that the least fixpoint $R^\infty$ of $\varphi$ satisfies the following properties.

1. Every trump $T$ in $R^\infty$ is functional.
2. Every maximal trump encodes the graph of a function in $\mathcal{R}^\infty$ and, conversely,
3. for every function $f \in \mathcal{R}^\infty$ there is a trump $T$ in $R^\infty$ encoding the graph of $f$.

*Proof.* Let $\varphi(\mathcal{R}, f)$ be as in the statement of the theorem. Without loss of generality we assume that $\varphi$ has the form

$$\varphi(\mathcal{R}, f_0) := \varphi_0(f_0) \vee \exists f_1 \ldots \exists f_n ((\bigwedge_{i=1}^{n} \mathcal{R} f_i) \wedge \varphi_1)$$

so that $\mathcal{R}$ does not occur in $\varphi_0$ or $\varphi_1$. (See [EbbFlu99] for a proof of this normal form for existential first-order inductions. The proof for this case is analogous.) The formula $\varphi$ is translated into a formula $\hat{\varphi}(R, \vec{x}, y) \in$ IF-LFP defined as follows:

$$\hat{\varphi}(R, \vec{x}, y) \quad := \quad \forall \vec{x}_1 . \exists y_1 . \mathrm{clone}(\vec{x}, y) \big( \psi_0(\vec{x}, y) \ \vee \ \psi_1(R, \vec{x}_1, y_1) \big)$$

where

$$\psi_0(\vec{x}, y) \quad := \quad \forall \vec{x}_{f_0} \exists y_{f_0} \, \mathrm{clone}(\vec{x}, y) \ \hat{\varphi}_0(\vec{x}_{f_0}, y_{f_0})$$

and

$$\psi_1(R, \vec{x}_1, y_1) \quad := \quad \forall \vec{x}_{f_0} \exists y_{f_0} \, \mathrm{clone}(\vec{x}_1, y_1) \ \psi_1'(\vec{x}_{f_0}, y_{f_0})$$

and

$$\begin{aligned} \psi_1'(R, \vec{x}_{f_0}, y_{f_0}) \quad := \quad & \forall \vec{x}_{f_1} \exists y_{f_1} \ldots \forall \vec{x}_{f_n} \exists y_{f_n} \\ & \textstyle\bigwedge_{i=1}^{n} (\forall \vec{x}' \exists y' \mathrm{clone}(\vec{x}_{f_i}, y_{f_i}) R \vec{x}' y') \wedge \\ & \hat{\varphi_1}(\vec{x}_{f_0}, y_{f_0}, \vec{x}_{f_1}, y_{f_1}, \ldots, \vec{x}_{f_n}, y_{f_n}). \end{aligned}$$

We claim that the formula $\hat{\varphi}$ satisfies the properties stated in the theorem. Let $\mathbf{A}$ be a structure with universe $A$. By Lemma 11(i), the trumps $T$ for are functional in $\vec{x}$ and $y$, with Skolem function $g$ such that $g$ satisfies $\psi_0$ or $\psi_1$.

We show by ordinal induction that every maximal trump in $R^\alpha$ is the graph of a function in $\mathcal{R}^\alpha$ and, conversely, the graph of every function in $\mathcal{R}^\alpha$

is a trump in $R^\alpha$. For limit ordinals (including 0) this follows immediately from the induction hypothesis.

Now let $\alpha = 1$. We show first that every maximal trump in $R^1$ is the graph of a function in $\mathcal{R}^1$. By definition, $R^0 = \varnothing$ and therefore the subformula $\psi_1$ can not be satisfied by any trump. Hence, the only trumps in $R^1$ are the functional trumps for $\psi_0$. By Lemma 12, a set $T \subseteq A^{k+1}$ is a trump for $\hat{\varphi}_0$ if, and only if, there is a function $f : A^k \to A$ such that $(\mathbf{A}, f) \models \varphi_0$ and for all deals $(\vec{a}, b) \in T$ we have $f(\vec{a}) = b$. Hence, the maximal trumps for $\hat{\varphi}_0$ are precisely the graphs of functions satisfying $\varphi_0$. Thus, by Part (*ii*) of Lemma 11, we get that the maximal trumps for $\forall \vec{x}_1 \exists y_1 \text{clone}(\vec{x}, y)\ \hat{\varphi}_0$ are the graphs of functions satisfying $\varphi_0$.

Conversely, let $f$ be a function satisfying $\varphi_0$. By Lemma 12 the trump $T := \{(\vec{a}, b) : f(\vec{a}) = b\}$ is a trump for $\hat{\varphi}_0$ and hence for $\forall \vec{x}_1 \exists y_1 \text{clone}(\vec{x}, y)\ \hat{\varphi}_0$.

Now let $\alpha > 1$ be a successor ordinal, *i.e.*, $\alpha = \beta+1$ for some $\beta > 0$. Again we first show that every maximal trump in $R^\alpha$ is the graph of a function $f \in \mathcal{R}^\alpha$. This is clear for all trumps of $\psi_0$ as they are already contained in $R^1$. Now consider the formula $\psi_1$. For simplicity we assume without loss of generality that all tuples $\vec{x}_{f_0}, \dots, \vec{x}_{f_n}$ are of arity $k$. By Lemma 12, the maximal trumps for $\hat{\varphi}_1$ are the sets $T \subseteq A^{(n+1)(k+1)}$ such that there are functions $f_0, \dots, f_n$ satisfying $\varphi_1$ and for all tuples $(\vec{a}_0, b_0, \dots, \vec{a}_n, b_n)$ we have $(\vec{a}_0, b_0, \dots, \vec{a}_n, b_n) \in T$ if, and only if, $f_i(\vec{a}_i) = b_i$ for all $i$. Further, applying the induction hypothesis and Lemma 11, we get that the maximal trumps for $\bigwedge_{i=1}^n (\forall \vec{x}' \exists y' \text{clone}(\vec{x}_{f_i}, y_{f_i})\ R\vec{x}'y')$ are sets $T$ of deals such that there are functional trumps $T_1, \dots, T_n \in R^\beta$ which, by induction hypothesis, are the graphs of functions $f_1, \dots, f_n \in \mathcal{R}^\beta$, and for all tuples $(\vec{a}_1, b_1, \dots, \vec{a}_n, b_n)$ we have $(\vec{a}_1, b_1, \dots, \vec{a}_n, b_n) \in T$ if, and only if, $f_i(\vec{a}_i) = b_i$ for all $i$. Thus, every maximal trump of $\bigwedge_{i=1}^n (\forall \vec{x}' \exists y' \text{clone}(\vec{x}_{f_i}, y_{f_i})\ R\vec{x}'y') \wedge \hat{\varphi}_1$ is functional in $\vec{x}_{f_0}$ and $y_{f_0}$ and this function $f$ satisfies $\varphi_1$ for some interpretation of the variables $f_1, \dots, f_n$ by functions from $\mathcal{R}^\beta$. Therefore $f$ is contained in $\mathcal{R}^\alpha$. Thus, by Lemma 11, every maximal trump for $\psi_1$ encodes the graph of such a function $f \in \mathcal{R}^\alpha$.

The converse is again easily seen. For every function $f \in \mathcal{R}^\alpha \backslash \mathcal{R}^\beta$ choose functions $f_1, \dots, f_n \in R^\beta$ so that $\mathbf{A} \models \varphi_1(f_0, \dots, f_n)$. By induction hypothesis, the graphs of these functions are trumps in $R^\beta$. Hence, we can choose these trumps in the conjunction $\bigwedge_{i=1}^n (\forall \vec{x}' \exists y' \text{clone}(\vec{x}_i, y_i)\ R\vec{x}'y')$. Lemma 12, then, establishes the claim. q.e.d.

## 4 Independence-Friendly vs. Partial Fixpoint Logic

By definition, independence-friendly fixpoint logic is a least fixpoint logic. However, contrary to the fixpoint logics usually considered in finite model theory, here the fixpoints are not sets of elements but sets of trumps and therefore essentially third-order objects. In particular, it is no longer guaranteed that any fixpoint induction closes in polynomially many steps in the size of the structure – to the contrary, it may take an exponential number of steps to close. We will see below, that this greatly increases the expressive power of IF-LFP compared to normal least fixpoint logics.

As a first step in this direction we relate independence-friendly fixpoint logic to partial fixpoint logic. Partial fixpoint logic is an important logic in finite model theory. Among the various fixpoint logics commonly considered in finite model theory, it is the most expressive subsuming logics such as LFP and IFP and, on ordered structures, even second-order logic SO.

**Definition 14 (Partial Fixpoint Logic).** *Partial fixpoint logic* (PFP) is defined as the extension of first-order logic by the following formula building rule. If $\varphi(R, \vec{x})$ is a formula with free first-order variables $\vec{x} := x_1, \ldots, x_k$ and a free second-order variable $R$ of arity $k$, then $\psi := [\mathbf{pfp}_{R,\vec{x}}\,\varphi](\vec{t})$ is also a formula, where $\vec{t}$ is a tuple of terms of the same length as $\vec{x}$. The free variables of $\psi$ are the variables occurring in $\vec{t}$ and the free variables of $\varphi$ other than $\vec{x}$.

Having defined the syntax, we now turn to the definition of the semantics. Let $\psi := [\mathbf{pfp}_{R,\vec{x}}\,\varphi](\vec{t})$ be a formula and let $\mathbf{A}$ be a finite structure with universe $A$ providing an interpretation of the free variables of $\varphi$ other than $\vec{x}$. Consider the following sequence of stages induced by $\varphi$ on $\mathbf{A}$, where $F_\varphi$ is the functional defined by $\varphi$.

$$\begin{aligned} R^0 &:= \varnothing \\ R^{\alpha+1} &:= F_\varphi(R^\alpha) \end{aligned}$$

As there are no restrictions on $\varphi$, this sequence need not reach a fixpoint. In this case, $\psi$ is equivalent on $\mathbf{A}$ to false. Otherwise, if the sequence becomes stationary and reaches a fixpoint $R^\infty$, then for any tuple $\vec{a} \in A$,

$$\mathbf{A} \models [\mathbf{pfp}_{R,\vec{x}}\,\varphi](\vec{a}) \text{ if, and only if, } \vec{a} \in R^\infty.$$

As mentioned above, PFP is among the fixpoint logics commonly considered in finite model theory the most expressive – especially on ordered structures. A central issue in finite model theory is to relate the expressive power of logics to the computational complexity of classes of structures definable in the logic. Of particular interest are so-called *capturing results.*

**Definition 15.** A logic $\mathcal{L}$ captures a complexity class $\mathfrak{C}$ if every class of finite structures definable in $\mathcal{L}$ can be decided in $\mathfrak{C}$ and conversely, for every class $\mathcal{C}$ of finite structures which can be decided in $\mathfrak{C}$ there is a sentence $\varphi \in \mathcal{L}$ such that for all structures $\mathbf{A}$, $\mathbf{A} \models \varphi$ if, and only if, $\mathbf{A} \in \mathcal{C}$.

Capturing results are important as in the case that a logic $\mathcal{L}$ captures a complexity class $\mathfrak{C}$, the logic provides a logical characterisation of the complexity class, *i.e.*, a characterisation independent of any machine models or time or space bounds. In particular, non-expressibility results on $\mathcal{L}$ transfer directly into non-definability results on $\mathfrak{C}$. As such results are notoriously hard to come by, capturing results provide an interesting alternative for proving non-definability of problems in a complexity class.

Much effort has been spent on capturing results and for all major complexity classes such results have been found (see [EbbFlu99] for a summary). However, in many cases it could only be shown that a logic captures a complexity class on the class of ordered structures.

**Theorem 16 (Abiteboul, Vianu, [AbiVia89]).** Partial fixpoint logic captures PSPACE on the class of finite ordered structures.

As every class of structures definable in second-order logic is decidable in the polynomial time hierarchy, it follows immediately that PFP contains SO on ordered structures. One feature that makes PFP so expressive is its ability to define fixpoint inductions of exponential length in the size of the structure. We show next that every formula of PFP is equivalent to one in IF-LFP.

**Theorem 17.** For every formula $\varphi \in$ PFP there is an equivalent formula $\psi \in$ IF-LFP.

*Proof.* It is known that every PFP formula is equivalent to one with a single fixpoint, so we need deal only with a PFP formula $\mathbf{pfp}_{R,\vec{x}}\varphi(R, \vec{x})$, where $\varphi$ is first order. We assume that the fixpoint of $\varphi$ always exists. See [EbbFlu99] for a proof that both assumptions can be made without loss of generality.

To calculate a partial fixpoint, one needs to check whether two consecutive stages of the inductive approximation are equal, and so one needs to use the stages both positively and negatively. We get round this by building up the relation and its complement simultaneously. Let $\varphi^p$ be the negation normal form of $\varphi$ and $\varphi^n$ be the negation normal form of $\neg\varphi$. Further, let $k$ be the arity of $\vec{x}$, *i.e.*, the number of free variables in $\varphi$. Consider the following $\Sigma^1_1$ formula $\psi(P, \mathtt{f})$, where $P$ is a third-order relation symbol and $\mathtt{f}$ is a $k$-ary function symbol:

$$\begin{aligned}\psi(P, \mathtt{f}) := \quad & \forall\vec{x}\big((\varphi^p(\varnothing, \vec{x}) \wedge \mathtt{f}(\vec{x}) = 1) \vee (\varphi^n(\varnothing, \vec{x}) \wedge \mathtt{f}(\vec{x}) = 0)\big) \vee \\ & \exists \mathtt{f}' \in P \forall\vec{x}\big((\varphi^p(\vec{x}, R\vec{u}/\mathtt{f}'(\vec{u}) = 1, \neg R\vec{u}/\mathtt{f}'(\vec{u}) = 0) \wedge \mathtt{f}(\vec{x}) = 1) \\ & \vee(\varphi^n(\vec{x}, R\vec{u}/\mathtt{f}'(\vec{u}) = 1, \neg R\vec{u}/\mathtt{f}'(\vec{u}) = 0) \wedge \mathtt{f}(\vec{x}) = 0)\big)\end{aligned}$$

By $\varphi^p(\vec{x}, R\vec{u}/\mathtt{f}'(\vec{u}) = 1, \neg R\vec{u}/\mathtt{f}'(\vec{u}) = 0)$ we mean the formula obtained from $\varphi^p$ by replacing every positive occurrence of an atom $R\vec{u}$, for some tuple

$\vec{u}$ of terms, by $\texttt{f}'(\vec{u}) = 1$ and every negative occurrence of an atom $R\vec{u}$, for some tuple $\vec{u}$ of terms, by $\texttt{f}'(\vec{u}) = 0$. Thus the formula $\psi$ obtained in this way is positive in $P$ and its least fixpoint exists. We claim that for all functions $\texttt{f} \in R^\infty$ the set $\{\vec{a} : \texttt{f}(\vec{a}) = 1\}$ is a stage in the induction on $\varphi$. This is clear for the function $\texttt{f}$ satisfying $((\varphi^p(\emptyset, \vec{x}) \wedge \texttt{f}(\vec{x}) = 1) \vee (\varphi^n(\emptyset, \vec{x}) \wedge \texttt{f}(\vec{x}) = 0))$ as $\texttt{f}$ encodes the first stage of the induction on $\varphi$. Further, if $\texttt{f}' \in P^\infty$ encodes a stage $R^\alpha$ of the induction on $\varphi$ in the sense described above, then the function $\texttt{f}$ satisfying

$$(\varphi^p(\vec{x}, R\vec{u}/\texttt{f}'(\vec{u}) = 1, \neg R\vec{u}/\texttt{f}'(\vec{u}) = 0) \wedge \texttt{f}(\vec{x}) = 1) \vee$$
$$(\varphi^n(\vec{x}, R\vec{u}/\texttt{f}'(\vec{u}) = 1, \neg R\vec{u}/\texttt{f}'(\vec{u}) = 0) \wedge \texttt{f}(\vec{x}) = 0)$$

encodes the next stage $R^{\alpha+1}$.

Thus the formula $[\mathbf{pfp}_{R,\vec{x}}\varphi](\vec{x})$ is equivalent to the formula

$$\vartheta(\vec{x}) := \exists \texttt{f}\ \texttt{f}(\vec{x}) = 1 \wedge [\mu P(\texttt{f}).\psi](\texttt{f}) \wedge \forall \vec{x}(\varphi(\vec{x}, R\vec{u}/\texttt{f}(\vec{u}) = 1) \leftrightarrow \texttt{f}(\vec{x}) = 1)$$

stating that there is a function $\texttt{f}$ in the fixpoint of the $\Sigma^1_1$-formula $\psi$, $\texttt{f}$ is the partial fixpoint of $\varphi$, and $\vec{x}$ occurs in this partial fixpoint, *i.e.*, $\texttt{f}(\vec{x}) = 1$. Now, the theorem follows from Lemma 12. q.e.d.

We have already mentioned that pure independence-friendly logic is equivalent to $\Sigma^1_1$ and therefore an ordering on the universe of a structure can be defined in IF-LFP even on classes of otherwise unordered structures. Thus the theorem above implies that IF-LFP contains SO on all rather than just ordered structures.

**Corollary 18.** On finite structures, every formula of SO is equivalent to a formula in IF-LFP.

In the next section we will derive some further corollaries of this theorem concerning the model-checking complexity of IF-LFP.

# 5 Complexity of Independence-Friendly Fixpoint Logic

In this section we analyse the complexity of IF-LFP on finite structures, both with respect to data and model-checking complexity. By data-complexity we understand the complexity of deciding for a fixed formula $\varphi \in$ IF-LFP and a given structure $\mathbf{A}$ whether $\mathbf{A} \models \varphi$. In particular, the input only consists of the structure $\mathbf{A}$. By model-checking we mean the problem of deciding for a given finite structure $\mathbf{A}$ and formula $\varphi \in$ IF-LFP whether $\mathbf{A} \models \varphi$. Here, both $\varphi$ and $\mathbf{A}$ are part of input.

We begin our analysis with data-complexity. In [Bra03], the first author already noticed that any given formula of IF-LFP can be evaluated in time exponential in the size of the structure. For, every fixpoint $\mu R(\vec{x}).\varphi$ (or $\nu R(\vec{x}).\varphi$) can be evaluated in time linear in the number of trumps for $\varphi$ and therefore exponential in the size of the structure.

**Proposition 19.** IF-LFP has exponential time data-complexity.

We aim at a much stronger result. Not only will we show that IF-LFP is EXPTIME-complete with respect to data-complexity but we will prove that it actually captures EXPTIME, *i.e.*, every class of structures decidable by an exponential time Turing-machine can be defined in IF-LFP and vice versa every class of structures definable in IF-LFP can be decided in deterministic exponential time.

**Theorem 20.** IF-LFP captures EXPTIME.

*Proof.* We follow the usual approach to show capturing results in finite model theory by simulating the run of a Turing-machine by a fixpoint induction in IF-LFP. Let $M$ be an exponentially time-bounded Turing-machine over the alphabet $\{0, 1\}$. On any input of size $n$, $M$ can make at most $2^{n^k}$ steps, for some constant $k$ independent of the input. We first show how to simulate the run of $M$ on any input structure $\mathbf{A}$ by an third-order induction on a $\Sigma^1_1$-formula, *i.e.*, by a formula of the form $\mu R(\mathtt{s}, \mathtt{p}, c).\varphi(R, \mathtt{s}, \mathtt{p}, c)$, where $\varphi \in \Sigma^1_1$, $\mathtt{s}$ and $\mathtt{p}$ are $k$-ary function symbols, $c$ is an individual variable and $R$ is a third-order relation symbol. (Strictly speaking, individual variables $c$ are not allowed in $\Sigma^1_1$-inductions as defined in Definition 9. However, these can easily be replaced by nullary function variables.) Throughout this proof we use typewriter font for function variables and italics font for individual variables.

As an ordering on the universe $A$ of the structure $\mathbf{A}$ is definable in $\Sigma^1_1$ we assume without loss of generality that $\mathbf{A}$ is ordered and that there are two constants 0 and 1 interpreted by distinct elements. Further, we assume that we are given an ordering on the space of functions from $A^k$ to $\{0, 1\}$. Again such an ordering can easily be defined in $\Sigma^1_1$.

Given this ordering on the function space, we can code the content of the Turing-tape in a relation $P(\mathtt{p}, c)$ such that $(\mathtt{p}, c) \in P$ if, and only if, $\mathtt{p}$ is the $i$th function with respect to the ordering on the space of $k$-ary functions, $c \in \{0, 1\}$, and the $i$th cell on the Turing-tape contains $c$. With this, we can encode the evolution of the Turing-tape during a run of $M$ in a relation $R(\mathtt{s}, \mathtt{p}, c)$, such that if $\mathtt{s}$ is the $i$th function with respect to the ordering on the function space, then the set $\{(\mathtt{p}, c) : (\mathtt{s}, \mathtt{p}, c) \in R\}$ encodes the Turing-tape after $M$ has made $i$ steps. To define this relation inductively, we need a formula $\mathtt{init}(\mathtt{p}, c)$ which defines the encoding of the input structure $\mathbf{A}$ on the Turing-tape with the head reading position 0 and a formula $\mathtt{next}(R, \mathtt{s}, \mathtt{p}, c)$ which defines for any given $\mathtt{s}$ in $(\mathtt{p}, c)$ the successor configuration of the configuration stored in $R$ for time step $\mathtt{s}$. Finally, we need a formula $\mathtt{accept}(R, \mathtt{s})$ which is true for $\mathtt{s}$ and $R$ if the configuration coded in $R$ for time step $\mathtt{s}$ is accepting.

Due to space restrictions we refrain from giving the formulae here. Similar formulae are widely used in the finite model theory community to encode the run of polynomial time Turing-machines in LFP. See [Grä07] or [EbbFlu99]. The $\Sigma^1_1$-formulae needed here can be obtained from these via trivial modifications implementing the above encoding of Turing-tapes and time steps. Now the run of $M$ on input $\mathbf{A}$ is accepting if, and only if, the formula

$$\exists \mathtt{s}, \mathtt{p}\ [\mu R(\mathtt{s}, \mathtt{p}, c).\ \begin{array}{l} (\mathtt{s} = \textsc{min} \wedge \mathtt{init}(\mathtt{p}, c)) \vee \\ (\exists \mathtt{s}'\ (\mathtt{s} = \mathtt{s}' + 1 \wedge \mathtt{next}(\mathtt{s}', \mathtt{p}, c))) \vee \\ (\exists \mathtt{s} \mathtt{accept}(R, \mathtt{s}) \wedge c = \textsc{acc}) \end{array} ](\mathtt{s}, \mathtt{p}, \textsc{acc})$$

is true in $\mathbf{A}$, where MIN denotes the minimal element in the ordering on the function space, $+1$ refers to the successor relation with respect to this ordering, and ACC is an arbitrary element distinct from 0 and 1 used to mark an accepting configuration. q.e.d.

Clearly, if a logic $\mathcal{L}$ captures a complexity class $\mathfrak{C}$, then the evaluation problem of $\mathcal{L}$ must be $\mathfrak{C}$-complete with respect to data complexity. Thus we get the following simple corollary.

**Corollary 21.** IF-LFP has EXPTIME-complete data-complexity.

Capturing results relate the expressive power of a given logic to the computational complexity of the classes of structures definable in the logic. The study of data complexity corresponds to the study of the computational complexity of a problem, where the size of the program or algorithm used to solve a problem is ignored.

However, when actually evaluating a formula in a structure this approach is not satisfactory. For instance, monadic DATALOG and monadic second-order logic (MSO) have the same expressive power on trees and therefore the same data-complexity on trees, but whereas monadic DATALOG programs can be evaluated in time linear both in the size of the DATALOG program and the input tree, the evaluation of MSO-formulae is PSPACE-complete on trees. In fact, it was shown in [GroSch03] that any translation of a MSO-formula on trees to an equivalent monadic DATALOG program necessarily increases the formula size non-elementary.

Thus, data-complexity only gives limited information about the complexity of actually evaluating a formula in a structure. We therefore continue our complexity analysis of IF-LFP with the study of its model-checking complexity. In particular, we will prove that model-checking for IF-LFP is hard for exponential space. For an upper bound, it is easily seen that for any given structure $\mathbf{A}$ and formula $\varphi$ the formula can be evaluated in $\mathbf{A}$ using space doubly exponential in $|\varphi|$ and exponential in $|\mathbf{A}|$. For, every evaluation of a (least or greatest) fixpoint only needs enough space to store all possible trumps, and the number of trumps is bounded by $O(2^{\mathbf{A}^{|\varphi|}})$.

**Theorem 22.** Every formula $\varphi \in$ IF-LFP can be evaluated in a structure $\mathbf{A}$ in space doubly exponential in $|\varphi|$ and exponential in $|\mathbf{A}|$.

The theorem gives an upper bound on the model checking complexity of IF-LFP. We have seen in Section 4 above that every formula of PFP is equivalent to one of IF-LFP. Further, the translation is polynomial in the size of the PFP-formula. Consequently, model-checking for IF-LFP is at least as complex as it is for PFP. As model-checking for PFP is known to be hard for exponential space – in fact even complete for exponential space – we get the following theorem.

**Theorem 23.** The model-checking problem for IF-LFP is hard for exponential space.

## 6 Conclusion

In this paper we studied the computational complexity of various problems related to IF-LFP. As we have seen, adding independence to least fixpoint logic increases the expressive power and complexity significantly. Another indicator for this is the translation of formulae of PFP to formulae of IF-LFP. This showed that IF-LFP is even more expressive than second-order logic – unless, of course, PSPACE = EXPTIME.

Looking at the various proofs given for the results, it becomes clear that the common technique used in all proofs was to use independent quantification to define functions and then show that these functions can be passed through the fixpoint induction. This suggests that there might be a more general relation between independence-friendly logic and second-order logic, namely that the two logics are actually equivalent. Showing this, however, requires a careful analysis of the rôle of negation in independence friendly logics and is far from obvious. This is part of ongoing work.

It is also notable that all our hardness results involve constructions requiring only least fixed points. For LFP, it is well-known that all properties can be expressed with a single least fixed point. An analogous theorem of IF-LFP has not been shown.

# References.


[AbiVia89] Serge **Abiteboul** and Victor **Vianu**, Fixpoint Extensions of First-Order Logic and Datalog-like Languages, *in:* [Mey89, p. 71–79]

[BaaMak03] Matthias **Baaz** and Johann A. **Makowsky** (*eds.*), Computer Science Logic, Proceedings of the 17th International Workshop, CSL 2003, 12th Annual Conference of the EACSL, and 8th Kurt Güdel Colloquium, KGC 2003, Vienna, Austria, August 25-30, 2003, Springer 2003 [Lecture Notes in Computer Science 2803]

[Bra99] Julian C. **Bradfield**, Fixpoints in Arithmetic, Transition systems and Trees, **Theoretical Informatics and Applications** 33 (1999), p. 341–356

[Bra00] Julian C. **Bradfield**, Independence: Logics and Concurrency, *in:* [CloSch00, p. 247–261]

[Bra03] Julian C. **Bradfield**, Parity of Imperfection, *in:* [BaaMak03, p. 72–85]

[BraFrö02] Julian C. **Bradfield** and Sibylle B. **Fröschle**, Independence-Friendly Modal Logic and True Concurrency, **Nordic Journal of Computing** 9 (2002), p. 102–117

[CloSch00] Peter **Clote** and Helmut **Schwichtenberg** (*eds.*), Computer Science Logic, Proceedings of the 14th Annual Conference of the EACSL, Fischbachau, Germany, August 21-26, 2000, Springer 2000 [Lecture Notes in Computer Science 1862]

[EbbFlu99] Heinz-Dieter **Ebbinghaus** and Jörg **Flum**, Finite Model Theory, 2nd edition, Springer 1999

[End70] Herbert B. **Enderton**, Finite Partially Ordered Quantifiers, **Zeitschrift für Mathematische Logik und Grundlagen der Mathematik** 16 (1970), p. 393–397

[Grä07] Erich **Grädel**, Finite Model Theory and Descriptive Complexity, *in:* [Grä+07, p. 125–230],

[Grä+07] Erich **Grädel**, Phokion G. **Kolaitis**, Leonid **Libkin**, Maarten **Marx**, Joel **Spencer**, Moshe Y. **Vardi**, Yde **Venema**, and Scott **Weinstein**, Finite Model Theory and Its Applications, Springer 2007 [EATCS Series Texts in Theoretical Computer Science]

[GroSch03] Martin **Grohe** and Nicole **Schweikardt**, Comparing the Succinctness of Monadic Query Languages over Finite Trees, *in:* [BaaMak03, p. 226–240]

[HinSan96] Jaakko **Hintikka** and Gabriel **Sandu**, A Revolution in Logic?, **Nordic Journal of Philosophical Logic** 1 (1996), p. 169–183

[Hod97] Wilfried **Hodges**, Compositional Semantics for a Language of Imperfect Information, **Logic Journal of the IGPL** 5 (1997), p. 539–563

[Mey89] Albert **Meyer** (*ed.*), Proceedings of the 4th Annual IEEE Symposium on Logic in Computer Science, Asilomar, California, June 5-8, 1989, IEEE Computer Society Press 1989

[Pie01] Ahti-Veikko **Pietarinen**, Semantic Games in Logic and Language, *PhD thesis*, University of Helsinki 2001

[Wal70] Wilbur J. **Walkoe**, Finite Partially-Ordered Quantification, **The Journal of Symbolic Logic** 35 (1970), p. 535–555

**Received**: April 2nd, 2005;
**In revised version**: September 16th, 2005;
**Accepted by the editors**: October 31st, 2005.

Stefan **Bold**, Benedikt **Löwe**,
Thoralf **Räsch**, Johan **van Benthem** (*eds.*)
**Foundations of the Formal Sciences V**
Infinite Games

# An Infinite Game over $\omega$-Semigroups

JÉRÉMIE CABESSA AND JACQUES DUPARC

Centre romand de Logique, Histoire et Philosophie des Sciences
Université de Lausanne
Chemin de la Colline 12
1015 Lausanne, Switzerland
{jeremie.cabessa,jacques.duparc}@unil.ch

ABSTRACT. Jean-Éric Pin introduced the structure of an $\omega$-semigroup as an algebraic counterpart to the concept of automaton reading infinite words. It has been well studied since, specially by Carton, Perrin, and Wilke. We introduce a reduction relation on subsets of $\omega$-semigroups defined by way of an infinite two-player game. Both Wadge hierarchy and Wagner hierarchy become special cases of the hierarchy induced by this reduction relation. But on the other hand, set theoretical properties that occur naturally when studying these hierarchies, happen to have a decisive algebraic counterpart. A game theoretical characterization of basic algebraic concepts follows.

## 1 Introduction

This work comes from an interaction between classical game theory, and the algebra of automata theory, which rests on the following main facts. In case of finite words, a well-known correspondence between an automaton and a finite semigroup exists: from any finite automaton $\mathcal{A}$ recognizing a regular language $L$, one can build a finite semigroup $S_{\mathcal{A}}$ recognizing (in an algebraic way) the same language, and vice-versa [PerPin04]. Moreover, this correspondence generalizes in case of infinite words. Indeed, for that purpose, Jean-Éric Pin introduced the structure of $\omega$-semigroup [PerPin04] as an algebraic counterpart to the concept of an automaton on infinite words. More precisely, he proved *the equivalence* between a finite Büchi automaton and a finite $\omega$-semigroup.

This paper presents a game theoretical study of the structure of $\omega$-semigroup, leading to an expected new foundation of the Wagner hierarchy, but also to promising general set theoretical, and algebraic results.

## 2 Preliminaries

We recall that a relation $R$ is a *preorder* if it is reflexive, and transitive. It is a *partial order* if it is reflexive, transitive, and antisymmetric. And it is an *equivalence relation* if it is reflexive, transitive, and symmetric.

Given a set $A$ (called the alphabet), we respectively denote by $A^*$, $A^+$, $A^\omega$, the sets of finite words over $A$, non empty finite words over $A$, and infinite words over $A$. We set $A^\infty := A^* \cup A^\omega$, and the empty word is denoted by $\varepsilon$. Given two words $u$ and $v$ ($u$ finite), we write $uv$ for the concatenation of $u$ and $v$, $u \subseteq v$ for "$u$ is an initial segment of $v$", $v \upharpoonright n$ for the restriction of $v$ to its $n$ first letters. Given $X \subseteq A^*$, and $Y \subseteq A^\infty$, we set: $XY := \{xy : x \in X \wedge y \in Y\}$, $X^* := \{x_1 \cdots x_n : n \geq 0 \wedge x_1, \ldots, x_n \in X\}$, $X^+ := \{x_1 \cdots x_n : n > 0 \wedge x_1, \ldots, x_n \in X\}$, and $X^\omega := \{x_0 x_1 x_2 \cdots : \forall\, n \geq 0, x_n \in X\}$. The class of *$\omega$-rational* subsets of $A^\infty$ is the smallest class of subsets of $A^\infty$ containing the finite subsets of $A^\infty$, and closed under finite union, finite product, and both operations $X \to X^*$, and $X \to X^\omega$.

A *semigroup* $(S, \cdot)$ is a set $S$ equipped with an associative operation from $S \times S$ into $S$. A *morphism of semigroups* is a map $\varphi$ from a semigroup $S$ into a semigroup $T$ such that $\forall\, s_1, s_2 \in S, \varphi(s_1 s_2) = \varphi(s_1)\varphi(s_2)$ holds. A *monoid* is a set equipped with an associative operation, and an identity element. If $S$ is a semigroup, $S^1$ denotes $S$ if $S$ is a monoid, and $S \cup \{\mathbf{1}\}$ otherwise (with the operation of $S$ completed as follows: $\mathbf{1} \cdot s = s \cdot \mathbf{1} = s$ for all $s \in S$). A *group* $G$ is a monoid such that every element has an inverse, *i.e.*, $\forall\, s \in G\ \exists\, s^{-1} \in G$ such that $s^{-1} \cdot s = s \cdot s^{-1} = \mathbf{1}$.

For any set $A$, the set $A^\omega$ is a topological space equipped with the product topology of the discrete topology on $A$. The basic open sets of $A^\omega$ are of the form $WA^\omega$, where $W \subseteq A^*$. Given a topological space $E$, the class of *Borel* subsets of $E$ is the smallest class containing the open sets, and closed under countable union, and complementation. Let $F \subseteq 2^\omega$, F is a *flip set* if and only if $\forall\, x, y \in 2^\omega (\exists !\ k \in \omega\ (x(k) \neq y(k))) \to (x \in F \leftrightarrow y \notin F)$. We use the fact that a flip set cannot be Borel (as it doesn't satisfy the Baire property).

Let $\Sigma$ be a set, and $A \subseteq \Sigma^\omega$. The *Gale-Stewart game* $\mathsf{G}(A)$ [GalSte53] is a two-player infinite game with perfect information where players take turn playing letters from $\Sigma$. Player I begins. After $\omega$ moves, they produce an infinite word $\alpha \in \Sigma^\omega$. Player I wins if and only if $\alpha \in A$. A play of this game is illustrated below.

$$
\begin{array}{lccccccccccc}
\mathrm{I}: & x_0 & & & & x_2 & & \cdots & x_n & & & & x_{n+2} & \cdots \\
 & & \searrow & & \nearrow & & \searrow & & & \searrow & & \nearrow & & \\
\mathrm{II}: & & & x_1 & & & & \cdots & & & x_{n+1} & & & \cdots
\end{array}
$$

Let $\Sigma_A$, $\Sigma_B$ be two sets, and $A \subseteq {\Sigma_A}^\omega$, $B \subseteq {\Sigma_B}^\omega$. The *Wadge game*

$\mathsf{W}(A, B)$ [Wad72] is a two-player infinite game with perfect information, where player I is in charge of subset $A$, and player II is in charge of subset $B$. Players take turn playing letters from $\Sigma_A$ and $\Sigma_B$, respectively. Player I begins. Player II is allowed to skip provided he plays infinitely many letters; player I is not. After $\omega$ moves, player I and II have respectively produced two infinite words $\alpha \in {\Sigma_A}^\omega$, and $\beta \in {\Sigma_B}^\omega$. Player II wins in $\mathsf{W}(A, B)$ if and only if $(\alpha \in A \leftrightarrow \beta \in B)$. A play of this game is illustrated below.

$$\begin{array}{lccccccc}
(A)\ \mathrm{I}: & a_0 & & & a_1 & & \cdots \quad \xrightarrow{\text{after } \omega \text{ moves}} & \alpha = a_0 a_1 a_2 \cdots \\
 & & \searrow & \nearrow & & \searrow & & \\
(B)\ \mathrm{II}: & & b_0 & & & b_1 & \xrightarrow{\text{after } \omega \text{ moves}} & \beta = b_0 b_1 b_2 \cdots
\end{array}$$

## 3 $\omega$-semigroups

Jean-Éric Pin introduced the structure of an $\omega$-semigroup [PerPin04] in order to give an algebraic counterpart to the notion of automaton reading infinite words. He showed the equivalence between a finite Büchi automaton and a finite $\omega$-semigroup in the following sense:

- For any finite Büchi automaton $\mathcal{A}$ recognizing the language $L(\mathcal{A})$, one can build a finite $\omega$-semigroup $S_{\mathcal{A}}$ recognizing (in an algebraic sense) the same language $L(\mathcal{A})$.
- For any finite $\omega$-semigroup $S$ recognizing the language $L(S)$, one can build a finite Büchi automaton recognizing the same language $L(S)$.

**Definition 1 (Perrin, Pin, [PerPin04]).** An *$\omega$-semigroup* is an algebra consisting in two components, $S = (S_+, S_\omega)$, and equipped with the following operations:

- A binary operation defined on $S_+$ and denoted multiplicatively.
- A mapping $S_+ \times S_\omega \to S_\omega$ called mixed product, that associates with each pair $(s, t) \in S_+ \times S_\omega$ an element $st$ of $S_\omega$.
- A surjective mapping $\pi_S : {S_+}^\omega \to S_\omega$ called infinite product.

Moreover, these three operations must satisfy the following properties:

1. $S_+$ equipped with the binary operation is a semigroup,
2. $\forall\ s, t \in S_+\ \forall\ u \in S_\omega\ s(tu) = (st)u$,
3. the infinite product $\pi_S$ is $\omega$-associative, meaning that for every strictly increasing sequence of integers $(k_n)_{n>0}$, and for every sequence $(s_n)_{n\in\omega} \in {S_+}^\omega$, we have

$$\pi_S(s_0 s_1 \cdots s_{k_1 - 1}, s_{k_1} \cdots s_{k_2 - 1}, \ldots) = \pi_S(s_0, s_1, s_2, \ldots),$$

4. $\forall\ s \in S_+\ \forall\ (s_n)_{n\in\omega} \in {S_+}^{\omega}$

$$s\pi_S(s_0, s_1, s_2, \ldots) = \pi_S(s, s_0, s_1, s_2, \ldots).$$

Intuitively, an $\omega$-semigroup is just a semigroup equipped with a suitable infinite product. It is *finite* precisely when $S_+$ is finite. Otherwise it is *infinite.* A subset $X \subseteq S_\omega$ is called an *$\omega$-subset.* We focus on those subsets in the sequel.

**Definition 2.** Let $S = (S_+, S_\omega)$, $T = (T_+, T_\omega)$ be two $\omega$-semigroups. A *morphism of $\omega$-semigroups* from $S$ into $T$ is a pair $\varphi = (\varphi_+, \varphi_\omega)$, where $\varphi_+ : S_+ \longrightarrow T_+$ is a morphism of semigroups, and $\varphi_\omega : S_\omega \longrightarrow T_\omega$ is a mapping preserving the infinite product, *i.e.*, for every sequence $(s_n)_{n\in\omega}$ of elements of $S_+$, one has

$$\varphi_\omega\big(\pi_S(s_0, s_1, s_2, \ldots)\big) = \pi_T\big(\varphi_+(s_0), \varphi_+(s_1), \varphi_+(s_2), \ldots\big).$$

**Example 3.** Let $A$ be an alphabet. The $\omega$-semigroup

$$A^\infty := (A^+, A^\omega)$$

equipped with the usual concatenation is the *free $\omega$-semigroup* over alphabet $A$. It is free in the sense that, for any $\omega$-semigroup $S = (S_+, S_\omega)$, any function $f$ from $A$ into $S_+$ can uniquely be extended to a morphism of $\omega$-semigroups $\bar{f} = (f_+, f_\omega)$ from $A^\infty$ into $S$ [CarPer97]. We do this by setting $f_+ : A^+ \longrightarrow S_+$ defined by

$$f_+(a_0a_1\cdots a_n) = f(a_0)f(a_1)\cdots f(a_n)\ ,\ \text{with } a_i \in A\ (\forall i \leq n),$$

and $f_\omega : A^\omega \longrightarrow S_\omega$ defined by

$$f_\omega(a_oa_1a_2\cdots) = \pi_S(f(a_0), f(a_1), f(a_2), \ldots)\ ,\ \text{with } a_i \in A\ (\forall i).$$

So, sets of $\omega$-words, in other words sets of reals, are the less constraint ones with regard to the algebraic structure.

In order to state further results, we put the following topology on $\omega$-subsets:

**Definition 4.** Let $S = (S_+, S_\omega)$ be any $\omega$-semigroup, and $X \subseteq S_\omega$, we set:

$$X \text{ is a } \textit{basic open} \text{ if and only if } \pi_S^{-1}(X) \text{ is an open of } {S_+}^{\omega}$$

where ${S_+}^{\omega}$ is equipped with the product topology of the discrete topology on $S_+$.

**Remark 5.** For any $\omega$-semigroup $S = (S_+, S_\omega)$, the infinite product $\pi_S$ is a continuous function by definition of the previous topology.

**Remark 6.** At first glance, the topology defined by taking $sS_\omega := \{st : t \in S_\omega\}$ as a basic open set (for any $s \in S_+$) would look much nicer. Unfortunately, this topology is much too weak for our purpose. Indeed, with this particular topology, in case $S_+$ is a group, Borel subsets of $S_\omega$ come down to the empty set and the whole space; the reason being that, given $sS_\omega$ any basic open set, then $S_\omega = ss^{-1}S_\omega \subseteq sS_\omega$, meaning that $sS_\omega = S_\omega$. We certainly need much more than that as we'll see in the last section.

## 4 An infinite game over $\omega$-semigroups

In this section, we define a reduction relation between $\omega$-subsets by use of an infinite two-player game over $\omega$-semigroups. We then state some general properties of this reduction relation in order to characterize the set hierarchy that it generates.

### 4.1 Definitions

**Definition 7.** Let $S = (S_+, S_\omega)$, $T = (T_+, T_\omega)$ be two $\omega$-semigroups, and $X$, $Y$ be two $\omega$-subsets of $S_\omega$ and $T_\omega$, respectively. The infinite two-player game $\mathsf{SG}(X, Y)$ is defined as follows: player I is in charge of subset $X$, player II is in charge of subset $Y$. Players I and II alternately play elements of $S_+$ and $T_+ \cup \{\varepsilon\}$, respectively. Player I begins, player II is allowed to skip its turn (by playing $\varepsilon$) provided he plays infinitely many moves, otherwise he loses the play. Player I cannot skip its turn. After $\omega$ moves, players I and II have respectively produced two infinite sequences $\langle s_0, s_1, \ldots\rangle$, and $\langle t_0, t_1, \ldots\rangle$. A play of this game is illustrated below.

$$\begin{array}{llccccccc}
(X)\ \mathrm{I}: & s_0 & & & s_1 & & \cdots & \xrightarrow{\text{after } \omega \text{ moves}} & \langle s_0, s_1, s_2, \ldots\rangle \\
 & & \searrow & \nearrow & & \searrow & & & \\
(Y)\ \mathrm{II}: & & t_0 & & & & t_1 & \xrightarrow{\text{after } \omega \text{ moves}} & \langle t_0, t_1, t_2, \ldots\rangle
\end{array}$$

The winning condition is the following: player II wins in $\mathsf{SG}(X, Y)$ if and only if

$$\pi_S(s_0, s_1, \ldots) \in X \Leftrightarrow \pi_T(t_0, t_1, \ldots) \in Y$$

where $\pi_S$ and $\pi_T$ are the infinite products of $S$ and $T$ respectively, and $\pi_T(t_0, \ldots, t_{n-1}, \varepsilon, t_n, \ldots) := \pi_T(t_0, \ldots, t_{n-1}, t_n, \ldots)$, meaning that the skipping moves of II are not considered in the infinite product.

*A strategy* for player II is a mapping $\sigma : {S_+}^+ \to T_+ \cup \{\varepsilon\}$. A strategy for player I is defined similarly. A *winning strategy* for a player is a strategy such that the player always wins when using it. We can now define the following reduction relation:

$$X \leq_{\mathsf{SG}} Y :\Leftrightarrow \text{ II has a winning strategy in } \mathsf{SG}(X,Y)$$

and of course

$$X <_{\mathsf{SG}} Y :\Leftrightarrow X \leq_{\mathsf{SG}} Y \text{ but } Y \not\leq_{\mathsf{SG}} X$$

$$X \equiv_{\mathsf{SG}} Y :\Leftrightarrow X \leq_{\mathsf{SG}} Y \text{ and } Y \leq_{\mathsf{SG}} X$$

Following the terminology of Wadge games, we set that:

- an $\omega$-subset $X$ is *self-dual* if and only if

  $$X \equiv_{\mathsf{SG}} X^{\mathsf{C}}$$

  where $X^{\mathsf{C}}$ stands for the complement of $X$. Otherwise, we say that $X$ is *non-self-dual*;

- an $\omega$-subset $X$ is *initializable* if and only if there exists $Y$ such that

  $$X \equiv_{\mathsf{SG}} Y \text{ and } Y \equiv_{\mathsf{SG}} s^{-1}Y \ , \ \forall\, s \in S_+$$

  where $s^{-1}Y = \{x \in S_\omega : x = \pi_S(u_1, u_2, \ldots) \wedge \pi_S(s, u_1, u_2, \ldots) \in Y\}$. From a playful point of view, a player in charge of a initializable set $X$ in the SG-game never loses his playful strength during the play. Indeed, for any position $s \in S_+$ that he reaches, he remains as strong as at the beginning, when being in charge of the whole subset $X$.

**Example 8.** Let $S = (S_+, S_\omega)$ be any $\omega$-semigroup, and $X \subseteq S_\omega$, with $X \neq \varnothing, S_\omega$.

- The relation $\varnothing \leq_{\mathsf{SG}} X$ holds. Indeed, we give a winning strategy for player II in the game $\mathsf{SG}(\varnothing, X)$. At the end of the play, the infinite product of any infinite sequence played by I obviously doesn't belong to $\varnothing$. So the winning strategy for II simply consists in playing in order to be outside $X$ at the end of the play (possible, as $X \neq S_\omega$).
- Similarly, the relation $S_\omega \leq_{\mathsf{SG}} X$ holds. The winning strategy for II in the game $\mathsf{SG}(X, S_\omega)$ consists in in playing in order to be inside $X$ at the end of the play (possible, as $X \neq \varnothing$).
- The relation $\varnothing \not\leq_{\mathsf{SG}} S_\omega$ holds. Indeed, at the end of the play, the infinite product of any infinite sequence played by I doesn't belong to $\varnothing$, and the infinite product of any infinite sequence played by II belongs to $S_\omega$, so that II cannot win against I in any case.

- Similarly, the relation $S_\omega \not\leq_{\mathsf{SG}} \varnothing$ holds, as there is no possible winning strategy for II in the game $\mathsf{SG}(S_\omega, \varnothing)$.

This shows that the empty set and the whole space are non-self-dual sets, since no one is equivalent to its complement. Moreover, any other set reduces to both of them.

## 4.2 Properties of the SG-relation

Not using yet any determinacy principle for this game, one cannot say much of the SG-relation, except that it is a partial ordering with no particular interesting properties. However, Martin's Borel Determinacy result [Mar75] easily induces Borel Determinacy for SG-games. As it is the case with the Wadge ordering, this property turns the SG-relation into a much more interesting one.

**Theorem 9 (Martin).** Let $\Sigma$ be a set. If $A$ is a Borel subset of $\Sigma^\omega$, then $\mathsf{G}(A)$ is determined.

**Corollary 10 (SG-Borel Determinacy).** Let $S = (S_+, S_\omega), T = (T_+, T_\omega)$ be two $\omega$-semigroups, and $X \subseteq S_\omega$, $Y \subseteq T_\omega$ be two Borel $\omega$-subsets. Then $\mathsf{SG}(X, Y)$ is determined.

*Proof.* We define a Borel subset $Z \subseteq ({S_+}^\omega \cup {T_+}^\omega \cup \{\varepsilon\})^\omega$ such that a player $P$ has a winning strategy in $\mathsf{G}(Z)$ if and only if the same player $P$ has a winning strategy in $\mathsf{SG}(X, Y)$. Let $p_1$ and $p_2$ be the following continuous projections from $(S_+ \cup T_+ \cup \{\varepsilon\})^\omega$ into $(S_+ \cup T_+ \cup \{\varepsilon\})^\omega$ defined by $p_1(u_0u_1u_2u_3\ldots) := u_0u_2u_4\ldots$, and $p_2(u_0u_1u_2u_3\ldots) := u_1u_3u_5\ldots$. Let $X', X'', Y', Y'' \subseteq (S_+ \cup T_+ \cup \{\varepsilon\})^\omega$ be defined by

$$\begin{aligned}
X' &:= \{\alpha = u_0u_1u_2\ldots : \pi_S(u_0, u_2, u_4, \ldots) \in X\} = p_1^{-1}(\pi_S^{-1}(X)) \\
X'' &:= \{\alpha = u_0u_1u_2\ldots : \pi_S(u_0, u_2, u_4, \ldots) \in X^{\mathsf{C}}\} = p_1^{-1}(\pi_S^{-1}(X^{\mathsf{C}})) \\
Y' &:= \{\alpha = u_0u_1u_2\ldots : \pi_T(u_1, u_3, u_5, \ldots) \in Y\} = p_2^{-1}(\pi_T^{-1}(Y)) \\
Y'' &:= \{\alpha = u_0u_1u_2\ldots : \pi_T(u_1, u_3, u_5, \ldots) \in Y^{\mathsf{C}}\} = p_2^{-1}(\pi_T^{-1}(Y^{\mathsf{C}}))
\end{aligned}$$

By continuity of the functions $p_1, p_2, \pi_S, \pi_T$, these sets are all Borel, and we conclude by taking $Z := (X' \cap Y') \cup (X'' \cap Y'')$. q.e.d.

The following interesting results are similar to the results on the Wadge ordering, and they are a consequence of Borel determinacy for these games. The first one is an immediate consequence of determinacy. The second one is a corollary of the first one: it states that, for this partial ordering $\leq_{\mathsf{SG}}$, the antichains have length at most two. The third one is a result from Martin and Monk establishing the wellfoundedness of this $\leq_{\mathsf{SG}}$-relation on Borel $\omega$-subsets.

**Corollary 11.** Let $S = (S_+, S_\omega)$, $T = (T_+, T_\omega)$ be two $\omega$-semigroups, and $X \subseteq S_\omega$, $Y \subseteq T_\omega$ be two Borel $\omega$-subsets. Then

$$X \not\leq_{\mathsf{SG}} Y \Rightarrow Y \leq_{\mathsf{SG}} X^{\mathsf{C}}.$$

*Proof.* The relation $X \not\leq_{\mathsf{SG}} Y$ means that player II doesn't have a winning strategy in $\mathsf{SG}(X, Y)$. Hence, by determinacy, player I has a winning strategy $\sigma$ in this game. So Player II has the following winning strategy in $\mathsf{SG}(Y, X^{\mathsf{C}})$: he copies the first move of player I in $\mathsf{SG}(X, Y)$, and then, at each step $n$, he plays $\sigma(x_0 \cdots x_n)$, where $x_0, \ldots, x_n$ are the moves already played by I in $\mathsf{SG}(Y, X^{\mathsf{C}})$. q.e.d.

**Corollary 12 (Wadge's Lemma).** Let $S = (S_+, S_\omega)$, $T = (T_+, T_\omega)$ be two $\omega$-semigroups, and $X \subseteq S_\omega$, $Y \subseteq T_\omega$ be two Borel $\omega$-subsets. Then only one of these possibilities occurs:

- $X \leq_{\mathsf{SG}} Y$ and $Y \not\leq_{\mathsf{SG}} X$, which implies $X <_{\mathsf{SG}} Y$.
- $X \leq_{\mathsf{SG}} Y$ and $Y \leq_{\mathsf{SG}} X$, which implies $X \equiv_{\mathsf{SG}} Y$.
- $X \not\leq_{\mathsf{SG}} Y$ and $Y \not\leq_{\mathsf{SG}} X$, which implies $X \equiv_{\mathsf{SG}} Y^{\mathsf{C}}$.
- $X \not\leq_{\mathsf{SG}} Y$ and $Y \leq_{\mathsf{SG}} X$, which implies $Y <_{\mathsf{SG}} X$.

*Proof.* The first, second and fourth cases come from the very definition. The third case comes by the previous proposition, and by the obvious fact that $A \leq_{\mathsf{SG}} B \Leftrightarrow A^{\mathsf{C}} \leq_{\mathsf{SG}} B^{\mathsf{C}}$ holds, for any $\omega$-subset $A$ and $B$. q.e.d.

**Proposition 13 (Martin, Monk).** The partial ordering $<_{\mathsf{SG}}$ is well-founded on Borel $\omega$-subsets, meaning that there is no infinite sequence of Borel $\omega$-subsets $(A_i)_{i\in\omega}$ such that

$$A_0 >_{\mathsf{SG}} A_1 >_{\mathsf{SG}} \cdots >_{\mathsf{SG}} A_n >_{\mathsf{SG}} A_{n+1} >_{\mathsf{SG}} \cdots$$

*Proof.* Towards contradiction, assume that there exists an infinite sequence of $\omega$-semigroups $\{S_i = (S_{i,+}, S_{i,\omega})\}_{i\in\omega}$, and an infinite strictly $<_{\mathsf{SG}}$-descending sequence of Borel $\omega$-subsets $(A_n)_{n\in\omega}$, where $A_i \subseteq S_{i,\omega}$, (any $i \in \omega$). For all $n \geq 0$, the relation $A_n >_{\mathsf{SG}} A_{n+1}$ implies that both $A_n \not\leq_{\mathsf{SG}} A_{n+1}$ and $A_n^{\mathsf{C}} \not\leq_{\mathsf{SG}} A_{n+1}$ hold, meaning that player I has winning strategy $\sigma_n^0$ and $\sigma_n^1$ in both games $\mathsf{SG}(A_n, A_{n+1})$ and $\mathsf{SG}(A_n^{\mathsf{C}}, A_{n+1})$, respectively. Let $\alpha \in 2^\omega$ define the following sequence of strategies $(\sigma_k^{\alpha(k)})_{k\in\omega}$. We now consider $\omega$ many SG-games linked this way: in the first game, player I applies strategy $\sigma_0^{\alpha(0)}$ to II's play. Since it is a strategy for I, it gives the first letter $a_0^0$ before II has ever played anything, but then, applying $\sigma_0^{\alpha(0)}$ means to know II's first move $a_0^1$. Precisely, II copies I's moves in the second game, in which I applies the winning strategy $\sigma_1^{\alpha(1)}$. And so on for every game. This means, in game number $n$, player I applies strategy $\sigma_n^{\alpha(n)}$, and II scrupulously copies I's moves in the game number $n+1$. These $\omega$ many games chained together are illustrated below. Normal arrows denote the action of playing while dotted ones denote the action of copying.

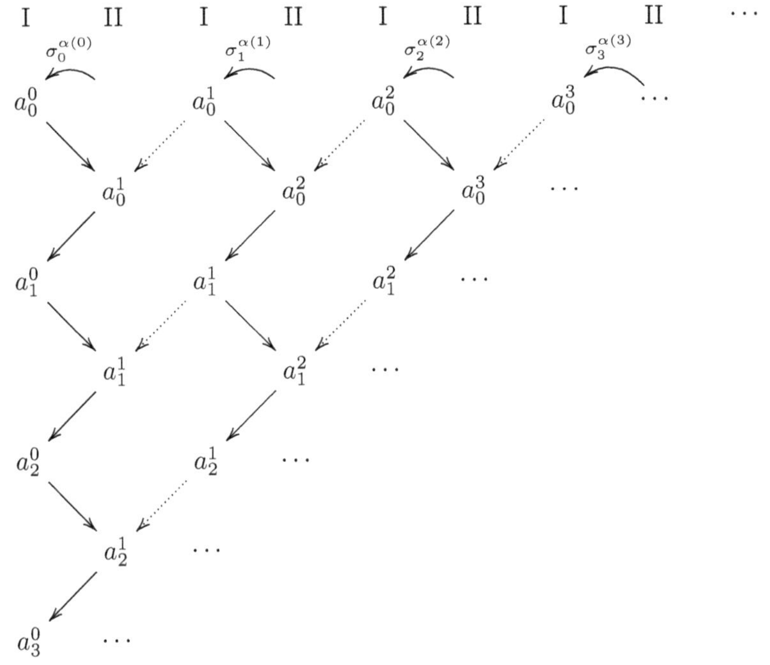


Let $x_\alpha = \prod_{k\in\omega} a_k^0$ be the infinite word played by player I in the first game, $\varphi : 2^\omega \to {S_{0,+}}^\omega$ defined by $\varphi(\alpha) = x_\alpha$, and $\psi = \pi_{S_0} \circ \varphi : 2^\omega \to S_{0,\omega}$ defined by $\psi(\alpha) = \pi_{S_0}(x_\alpha) = \pi_{S_0}(\prod_{k\in\omega} a_k^0)$. By definition of these chained games, $\varphi$ is continuous. Indeed, we remark that the $k$ first letters of $x_\alpha$ only depend on the $k$ first letters of $\alpha$, as we completely don't need games number $k+1, k+2, \ldots$ to determine $x_\alpha \restriction k$. So, for any $U \subseteq {S_{0,+}}^*$, $\varphi^{-1}(U{S_{0,+}}^\omega) = V2^\omega$, with $V \subseteq 2^*$, meaning that the pre-image by $\varphi$ of a basic open set is a basic open set. As $\varphi$ and $\pi_{S_0}$ are continuous, so is $\psi$. Consider $B = \psi^{-1}(A_0)$. By construction of these chained games, we notice that if $\alpha$ and $\alpha'$ only differ by one position (*i.e.*, there is exactly one $i$ such that $\alpha(i) \neq \alpha'(i)$), then $\alpha \in B \Leftrightarrow \alpha' \notin B$. This means that $B$ is a flip set, and it is Borel as $\psi$ is continuous, a contradiction. q.e.d.

**Remark 14.** Taking the $\equiv_{\mathsf{SG}}$-equivalence classes of Borel $\omega$-subsets leads to a hierarchy of classes of Borel $\omega$-subsets called the SG-*hierarchy*. As already mentioned, the previous results state the wellfoundedness of this hierarchy together with the fact that the antichains have length at most two. The SG-hierarchy has thus the same familiar "scaling shape" as the Borel hierarchy or the Wadge hierarchy: an increasing sequence of pairs of non-self-dual classes with single self-dual classes in between. This hierarchy is illustrated in Figure 14. Circles represent classes of Borel $\omega$-subsets, and arrows represent the fact of "being SG-smaller than".

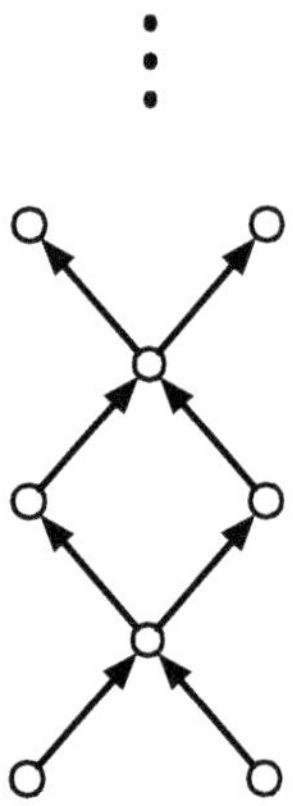

Figure 1. The SG-hierarchy.

**Definition 15.** The SG-*degree* of Borel $\omega$-subsets is defined by induction. At the bottom, we find $\varnothing$ and $\varnothing^{\mathsf{C}}$ since there is no non-empty set $A$ such that $A \leq_{\mathsf{SG}} \varnothing$ holds, and there is also no other smaller set than the whole space, which is incomparable to the empty set (see Example 8). So we set:

$$d^o_{\mathsf{SG}}(\varnothing) := d^o_{\mathsf{SG}}(\varnothing^{\mathsf{C}}) := 0,$$

and for any Borel $\omega$-subset $A >_{\mathsf{SG}} \varnothing$

$$d^o_{\mathsf{SG}}(A) := \sup\{d^o_{\mathsf{SG}}(B) + 1 : B <_{\mathsf{SG}} A\}.$$

## 5 Basic results about this game

In this section, we give some general results about both this infinite game over $\omega$-semigroups, and more precisely about the SG-hierarchy. We state that two important hierarchies become particular cases of the SG-hierarchy. But the most striking thing is that very essential algebraic notions turn out to correspond to very natural properties stated in a game theoretical way.

### 5.1 The Wadge hierarchy

In the late sixties, William W. Wadge introduced a very deep refinement of the Borel hierarchy of sets of the Baire space (or of the Cantor space as well) [Wad72]. The *Wadge hierarchy* is induced by the following relation on sets: $A \leq_{\mathsf{W}} B :\Leftrightarrow$ there is a continuous function $f$ such that $f^{-1}(B) = A$ $\Leftrightarrow$ Player II has a winning strategy in $\mathsf{W}(A, B)$ [Wad72].

**Proposition 16.** The SG-hierarchy restricted to Borel $\omega$-subsets of free $\omega$-semigroups corresponds exactly to the Wadge hierarchy of Borel subsets.

*Proof.* When restricted to free $\omega$-semigroups, the SG-game is exactly the same as the Wadge game. q.e.d.

**Remark 17.** As a matter of fact, the SG-hierarchy should be regarded as a widening of the Wadge hierarchy. Not only more sets are involved, but the algebraic structure of semigroups enriches the way one can describe or characterize Borel sets. For instance, some of them may "live" in an $\omega$-semigroup generated by a monoid, or even group, while most don't.

### 5.2 The Wagner hierarchy

In 1979, Klaus Wagner described a hierarchy among languages recognized by Muller automata called the *Wagner hierarchy* [Wag79]. This hierarchy has height $\omega^\omega$ and actually coincides with the restriction of the Wadge hierarchy to $\omega$-rational languages. In other words, it is the hierarchy induced by the following ordering on Muller automata: $\mathcal{A} \leq_{\mathsf{W}} \mathcal{B}$ if and only if the language recognized by $\mathcal{A}$ is the inverse image of the language recognized by $\mathcal{B}$ by a continuous function. This section shows that the Wagner hierarchy is a particular case of the SG-hierarchy.

**Proposition 18.** The SG-hierarchy restricted to subsets of finite $\omega$-semigroups is classwise isomorphic to the Wagner hierarchy.

*Proof.* In the forthcoming paper [CabDup$\infty$]. q.e.d.

The decidability of the Wagner hierarchy also holds in the following sense:

**Proposition 19.** Let $S = (S_+, S_\omega)$ be a finite $\omega$-semigroup, and $X \subseteq S_\omega$ be Borel. One can associate to $X$ an ordinal $\xi_X \in \omega^\omega$ being its degree in the Wagner hierarchy.

*Proof.* In the forthcoming paper [CabDup∞]. q.e.d.

### 5.3 Basic algebraic properties

Important algebraic notions can be expressed in a natural game theoretical way by use of the SG-game. These results militate in favor of developing the use of game theoretical tools in algebra. The two following propositions give a game theoretical approach of the algebraic concepts of monoid and group.

**Proposition 20.** Let $S = (S_+, S_\omega)$ be any $\omega$-semigroup, and $X \subseteq S_\omega$ be any Borel $\omega$-subset. The following conditions are equivalent:

**(1)** $X \not\leq_{\mathsf{SG}} X^{\mathsf{C}}$ (*i.e.*, $X$ is non-self-dual).

**(2)** Every player in charge of $X$ in the SG-game is allowed to skip his turn, provided he plays infinitely many letters, otherwise he loses.

**(3)** There exists an $\omega$-semigroup $T = (T_+, T_\omega)$ and a Borel $\omega$-subset $Y \subseteq T_\omega$ such that $T_+$ is a monoid, and $X \equiv_{\mathsf{SG}} Y$.

*Proof (Sketch).* "(1) ⇒ (2)": We show that we can assume without loss of generality that any player in charge of $X$ in the SG-game can skip his turn, provided he plays infinitely many letters. In other words, we show that a player in charge of $X$ that is allowed to skip is not stronger than (or can be beaten by) a player in charge of $X$ that is not allowed to. Let $\overline{\mathsf{SG}}(\_,\_)$ be the same infinite game as $\mathsf{SG}(\_,\_)$, instead that player I is allowed to skip — provided he plays infinitely often— while player II is not. By hypothesis, there exists a winning strategy $\sigma$ for I in the game $\mathsf{SG}(X, X^{\mathsf{C}})$. Then $\sigma$ is also a winning strategy for II in the game $\overline{\mathsf{SG}}(X, X)$.

"(2) ⇒ (1)": By hypothesis, every player in charge of $X$ is allowed to skip its turn, provided he plays infinitely letters. The winning strategy for player I in the game $\mathsf{SG}(X, X^{\mathsf{C}})$ consists in skipping the first move, and then copy player II.

"(3) ⇒ (1)": By hypothesis, $X \equiv_{\mathsf{SG}} Y$, with $Y \subseteq T_+$, and $T_+$ is a monoid. We thus show that $Y$ is non-self-dual by giving a winning strategy for I in the game $\mathsf{SG}(Y, Y^{\mathsf{C}})$: player I fist plays **1**; then when II doesn't skip, I copies II, and when II skips, I plays **1**. As $Y$ is non-self-dual, so is $X$.

"(1) $\Rightarrow$ (3)": A consequence of [Dup01] and [Dup$\infty$]. Basically, the idea is to consider the set $Z = \pi_S^{-1}(X)$. Viewed as a subset of the free $\omega$-semigroup $({S_+}^+, {S_+}^\omega)$ —with ${S_+}^\omega$ equipped with the usual topology (the product topology of the discrete topology over $S_+$)— it satisfies $Z \equiv_{\mathsf{SG}} X$. Since $Z$ is Borel and non-self-dual, it follows from both [Dup01], and [Dup$\infty$] that there exists some $\bar{Y} \subseteq {S_+}^{\leq\omega}$ verifying:

- $\bar{Y}^b \equiv_W Z$, where $\bar{Y}^b$ stands for all $\omega$-sequences $x$ built over the alphabet $S_+ \cup \{b\}$ —where $b$ stands for any new letter not in $S_+$— that verify: "$x$ in which every occurrence of the letter $b$ has been erased, belongs to $\bar{Y}$."
- $\bar{Y} \cap {S_+}^\omega = Z$

As $\bar{Y}^b \equiv_W Z$ holds, then $\bar{Y}^b \equiv_{\mathsf{SG}} Z$, when these sets are considered as subsets of the free $\omega$-semigroups $((S_+ \cup \{b\})^+, (S_+ \cup \{b\})^\omega)$, and $({S_+}^+, {S_+}^\omega)$, respectively. As $Z \equiv_{\mathsf{SG}} X$ also holds, then $\bar{Y}^b \equiv_{\mathsf{SG}} X$. By identifying $b$ and the identity element, *i.e.*, by setting the monoid $T_+ = (S_+ \cup \{b\})^+ = (S_+ \cup \{\mathbf{1}\})^+$, $T_\omega = (S_+ \cup \{b\})^\omega = (S_+ \cup \{\mathbf{1}\})^\omega$, and by taking $Y = \bar{Y}^b \subseteq T_\omega$, one gets the result. q.e.d.

**Proposition 21.** Let $S = (S_+, S_\omega)$ be any $\omega$-semigroup, and $X \subseteq S_\omega$ be any Borel subset. The following conditions are equivalent:

**(1)** $X \leq_{\mathsf{SG}} s^{-1}X, \forall\, s \in S_+$ (*i.e.*, X is initializable).

**(2)** Every player in charge of $X$ in the SG-game is allowed to erase his moves, provided he plays infinitely many letters, otherwise he loses.

**(3)** There exists an $\omega$-semigroup $T = (T_+, T_\omega)$ and a Borel $\omega$-subset $Y \subseteq T_\omega$ such that $T_+$ is a group, and $X \equiv_{\mathsf{SG}} Y$.

*Proof (Sketch).* "(3) $\Rightarrow$ (2)": We show that we can assume without loss of generality that any player in charge of $X$ in the SG-game can erase his moves, provided he plays infinitely many letters. In other words, we show that a player in charge of $X$ that is allowed to erase is not stronger than (or can be beaten by) a player in charge of $X$ that is not allowed to. Let $\tilde{\mathsf{SG}}(\_,\_)$ be the same infinite game as $\mathsf{SG}(\_,\_)$, instead that player I is allowed to erase his moves —provided he plays infinitely often— while player II is not. We first show that player II has a winning strategy in $\tilde{\mathsf{SG}}(X,Y)$. By hypothesis, II has a winning strategy $\sigma$ in the game $\mathsf{SG}(X,Y)$. This leads the following winning strategy for II in $\tilde{\mathsf{SG}}(X,Y)$: II copies I, and when I erases a part of his position, then II "cancels" a piece of his by playing the suitable inverse element, in order to come back to the expected situation. By hypothesis, II also has a winning strategy in the game $\mathsf{SG}(Y, X)$ (where no one can erase his moves). Then by composition of strategies, II has a winning strategy in the game $\tilde{\mathsf{SG}}(X, X)$.

"(2) $\Rightarrow$ (1)": Let $s \in S_+$. By hypothesis, we can give the following winning strategy $\sigma$ in the game $\mathsf{SG}(X, X)$, but where player II has already played the element $s$: player II erases $s$, and then copies player I. By the previous point, we can find a winning strategy $\sigma'$ in the game $\tilde{\mathsf{SG}}(X, X)$ (where I can erase, while II cannot). The composition of these strategies $\sigma'' = \sigma' \circ \sigma$ is winning in the game $\mathsf{SG}(X, s^{-1}X)$.

"(1) $\Rightarrow$ (3)": A consequence of [Dup01] and [Dup$\infty$]. First, since $X$ is clearly non-self-dual, one can assume without loss of generality that $S_+$ is a monoid with $\mathbf{1}$ as identity (otherwise, from previous proposition, one can get some $X'$ satisfying this property). Then, here also, the idea is to consider the set $Z = \pi_S^{-1}(X)$. Viewed as a subset of the free $\omega$-semigroup $({S_+}^+, {S_+}^\omega)$ —with ${S_+}^\omega$ equipped with the usual topology— one has $X \equiv_{\mathsf{SG}} Z$. Since $Z$ is Borel and initializable, from [Dup01] and [Dup$\infty$], we know that there exists some set $B \subseteq \{0,1\}^{\leq\omega}$ such that:

- $(B^\sim)^b \equiv_{\mathsf{W}} Z$, where $B^\sim$ is defined as $B$ plus an additional eraser, and $B^b$ is defined as $\bar{Y}^b$ was in last proposition ($b$ stands for "blank", it behaves just like a mute letter). In a few words, this means that a player in charge of $(B^\sim)^b$ in a Wadge game (either player I or player II) is like a player in charge of $B$ with two extra possibilities. This player can:

  – play $b$, which is just like skipping, except that here, one can decide to skip forever, which is materialized by playing infinitely many $b$'s;

  – erase all or part of his/her last moves ($b$ is just like a skip, it doesn't count as a true letter).

  After $\omega$ such moves, the resulting sequence played is the limit of what has been played, forgetting about the blanks. And $(B^\sim)^b$ is the set of all infinite sequences that can possibly be played such that their limits belong to $B$.

- Now, add two more letters $0^{-1}$, and $1^{-1}$ viewed as the inverse elements of respectively 0, and 1. Consider the free semigroup $\{0, 1, 0^{-1}, 1^{-1}, b\}^*$, where the concatenation operation moreover verifies $0^{-1}0 = 00^{-1} = b$, and $1^{-1}1 = 11^{-1} = b$. Take $b = \mathbf{1}$, and set $Y$ as the set of all infinite sequences over $\{0, 1, 0^{-1}, 1^{-1}, \mathbf{1}\}$, such that, once every possible "erasing" of the form $0^{-1}0 = 00^{-1} = \mathbf{1}$, or $1^{-1}1 = 11^{-1} = \mathbf{1}$ has been processed, yields an infinite sequence that belongs to $B^b$ (which is $B^{\mathbf{1}}$, since $b = \mathbf{1}$), if one forgets about the subscripts $^{-1}$ (*i.e.*, identifying $0^{-1}$ with 0 and $1^{-1}$ with 1). It is easy to see that $(B^{\sim})^b \equiv_{\mathsf{W}} Y$.

One gets the result by considering the $\omega$-semigroup

$$T = (T_+, T_\omega) = \left(\{0, 1, 0^{-1}, 1^{-1}, \mathbf{1}\}^+, \{0, 1, 0^{-1}, 1^{-1}, \mathbf{1}\}^\omega\right),$$

and the subset $Y \subseteq T_\omega$ as defined above. Indeed, one has $Z \equiv_{\mathsf{W}} (B^{\sim})^b \equiv_{\mathsf{W}} Y$, meaning that $Z \equiv_{\mathsf{SG}} Y$ (by treating $Z$, and $Y$ as subsets of the suitable $\omega$-semigroups). So, $X \equiv_{\mathsf{SG}} Z \equiv_{\mathsf{SG}} Y$. q.e.d.

## 6 Conclusion

The way we see it, further developments in the Wadge hierarchy, for instance, should be deeply related to the SG-hierarchy. It seems to be of a major interest to be able to characterize a Borel set of reals, by the type of $\omega$-semigroups where a complete set for the Wadge class it generates may "live". This should be a way of identifying the algebraic properties hidden behind various "Borel attitudes" of sets, *i.e.*, an algebraic way of classifying Borel sets: a very promising approach.

## References.

[CabDup∞] Jérémie **Cabessa** and Jacques **Duparc**, Constructing the Algebraic Counterpart of the Wagner Hierarchy by Way of Games, *to be submitted*

[CarPer97] Olivier **Carton** and Dominique **Perrin**, Chains and Superchains for $\omega$-Rational Sets, Automata and Semigroups, **International Journal of Algebra and Computation** 7 (1997), p. 673–695

[CarPer99] Olivier **Carton** and Dominique **Perrin**, The Wagner Hierarchy of $\omega$-Rational Sets, **International Journal of Algebra and Computation** 9 (1999), p. 697–620

[Dup01] Jacques **Duparc**, Wadge Hierarchy and Veblen Hierarchy, Part 1: Borel Sets of Finite Rank, **Journal of Symbolic Logic** 66 (2001), p. 56–86

[Dup∞] Jacques **Duparc**, Wadge Hierarchy and Veblen Hierarchy, Part II: Borel Sets of Infinite Rank, *submitted to:* **Journal of Symbolic Logic**

[DupRis06] Jacques **Duparc** and Mariane **Riss**, The Missing Link for $\omega$-Rational Sets, Automata, and Semigroups, **International Journal of Algebra and Computation** 16 (2006), p. 161–186

[GalSte53] David **Gale** and Frank M. **Stewart**, Infinite Games with Perfect Information, *in:* [KuhTuc53, p. 245–266]

[Jec02] Thomas **Jech**, Set Theory, the third millennium edition, revised and expanded, Springer 2002

[Kec94] Alexander S. **Kechris**, Classical Descriptive Set Theory, Springer 1994 [Graduate Texts in Mathematics 156]

[KuhTuc53] Harold W. **Kuhn** and Albert W. **Tucker** (*eds.*), Contributions to the Theory of Games, volume 2, Princeton University Press 1953 [Annals of Mathematics Studies 28]

[Mar75] Donald A. **Martin**, Borel Determinacy, **Annals of Mathematics** 102 (1975), p. 363–371

[PerPin04] Dominique **Perrin** and Jean-Éric **Pin**, Infinite Words: Automata, Semigroups, Logic and Games, Elsevier 2004 [Pure and Applied Mathematics Series 141]

[Wad72] William W. **Wadge**, Degrees of Complexity of Subsets of the Baire Space, **Notices of the American Mathematical Society** 19 (1972), p. 714–715

[Wad83] William W. **Wadge**, Reducibility and Determinateness on the Baire Space, *PhD thesis*, University of California at Berkeley, 1983

[Wag79] Klaus W. **Wagner**, On $\omega$-Regular Sets, **Information and Control** 43 (1979), p. 123–177

**Received**: February 28th, 2005;
**In revised version**: August 3rd, 2005;
**Accepted by the editors**: August 31st, 2005.

Stefan **Bold**, Benedikt **Löwe**,
Thoralf **Räsch**, Johan **van Benthem** (*eds.*)
**Foundations of the Formal Sciences V**
Infinite Games

# Nonmonotone Game Labellings

Tikitu de Jager and Benedikt Löwe*

Institute for Logic, Language and Computation
Universiteit van Amsterdam
Nieuwe Doelenstraat 15
1012 CP Amsterdam, The Netherlands
S.T.deJager@uva.nl

Institute for Logic, Language and Computation
Universiteit van Amsterdam
Plantage Muidergracht 24
1018 TV Amsterdam, The Netherlands
bloewe@science.uva.nl

ABSTRACT. We extend the traditional Gale-Stewart algorithm (backward induction) for the combinatorial graph game to an asymmetric variant; the extension makes the algorithm non-monotonic, but a linear-time formulation is still possible.

## 1 Introduction

If **G** is a graph, the (symmetric) combinatorial game on **G** is played by two players pushing a token on the graph. Whoever moves the token into a terminal node, wins. An example of a game of this type is the game of *Nim* (removing matches from a number of rows of matches until the game board is empty). If **G** was cyclic, then it is not guaranteed that one of the players will push the token into a terminal node. An infinite walk through **G** is considered a draw in the combinatorial game.[1]

Combinatorial games are perfect information games with simple payoff sets, and thus by the Gale-Stewart Theorem determined [GalSte53]. You can determine the winner of the game by unfolding the game into a game

*The authors would like to thank Philipp Rohde (Aachen) and Nick Bezhanishvili (Amsterdam) for fruitful discussions of topics connected to different aspects of this paper.

[1]More details can be found in the four-volume second edition of *Winning Ways* by Berlekamp, Conway and Guy.

tree $\mathbf{T_G}$, then labelling the tree via the Gale-Stewart labelling and read off the winner from the label of the node $s$.

However, analyzing combinatorial games via their game trees might not be optimal for several reasons:

Firstly, if $\mathbf{G}$ was cyclic, the game tree will be infinite and the labelling of the game tree will be an infinitary, possibly transfinite procedure.

Secondly, the Gale-Stewart procedure is not metamathematically parsimonious. There are computable trees with no computable winning strategy, and the Gale-Stewart theorem on determinacy of open games is equivalent to a nontrivial metamathematical statement of second-order arithmetic by a theorem of Steel.[2]

Thus, a labelling procedure directly on the graph that doesn't need unfolding into the game tree is a *desideratum*. The existence of such a procedure is well-known in automata theory (*cf.*, *e.g.*, [Maz02, Exercise 2.6]); in this paper we consider a variant of the standard combinatorial game, which gives rise to an interesting complication for the corresponding labelling procedure.

The asymmetric combinatorial game on $\mathbf{G}$ (starting at $s$) is a variant of the combinatorial games where one player has to play into terminal nodes and the other has to keep the game alive for an infinite number of steps. Again, this game is a perfect information game with simple payoff, and thus could be analyzed via the Gale-Stewart technique with similar drawbacks.

It is the goal of this paper to give a finitary algorithm for asymmetric combinatorial games directly on $\mathbf{G}$. The algorithm we give is strongly influenced by the non-monotonic Gale-Stewart technique developed in [Löw03].

In Section 2, we define some basic graph-theoretical notions used in Section 3 where we define our games and discuss labellings and their connections to games abstractly, understanding labellings on graphs as quotients of the labellings on their associated game trees.

Finally, following the ideas of quotient labellings, in Section 4 we develop two variant algorithms for the asymmetric combinatorial games (shown in Figures 3 and 4) and discuss their running times.

# 2 Graphs

## 2.1 Graphs & Bisimulations

Our graphs $\mathbf{G} = \langle V_{\mathbf{G}}, E_{\mathbf{G}} \rangle$ are directed graphs (digraphs), *i.e.*, $V_{\mathbf{G}}$ is a set of vertices and $E \subseteq V_{\mathbf{G}} \times V_{\mathbf{G}}$ is an arbitrary binary relation. If $\equiv$ is an

[2]The statement is $\mathsf{ATR}_0$, a set-theoretic existence statement for sets defined by transfinite recursion along an arithmetically defined wellorder. *Cf.* [Ste76, Tan90]; for a detailed overview in the context of Reverse Mathematics, *cf.* [Sim99, § V.8].

equivalence relation on $V_{\mathbf{G}}$, we can define a graph structure on the set of $\equiv$-equivalence classes $V_{\mathbf{G}/\equiv} := \{[v]_\equiv \,;\, v \in V_{\mathbf{G}}\}$ as follows:

$$\langle [v]_\equiv, [w]_\equiv \rangle \in E_{\mathbf{G}/\equiv} :\Longleftrightarrow \exists v', w'(v \equiv v' \;\&\; w \equiv w' \;\&\; \langle v, w \rangle \in E_{\mathbf{G}}).$$

We write $\mathbf{G}/\equiv := \langle V_{\mathbf{G}/\equiv}, E_{\mathbf{G}/\equiv} \rangle$ for the quotient graph.

If $s \in V_{\mathbf{G}}$, we call the pair $\langle \mathbf{G}, s \rangle$ a **pointed graph**. As usual, the natural numbers are identified with the sets of their predecessors, *i.e.*, $0 = \varnothing$ and $n + 1 = \{0, \ldots, n\}$. If $N \in \mathbb{N} \cup \{\mathbb{N}\}$, we call a function $W\colon N \to V$ a **walk through $\langle \mathbf{G}, s \rangle$ of length $N$** if

1. for each $n + 1 \in N$, we have $\langle W(n), W(n+1) \rangle \in E_{\mathbf{G}}$, and
2. $W(0) = s$.

A walk is called **finite** if $N \in \mathbb{N}$. It is called **maximal** if it is either infinite or finite of length $n + 1$ where $W(n)$ is a terminal node of $\mathbf{G}$. We define the **connected component of $v$ in $\mathbf{G}$** (in symbols: $\mathbf{C}^v_{\mathbf{G}}$) to be the set of vertices $w$ such that there is a walk $W$ of length $n + 1 \in \mathbb{N}$ through $\langle \mathbf{G}, v \rangle$ with $W(n) = w$. A pointed graph $\langle \mathbf{G}, v \rangle$ is called **connected** if $V_{\mathbf{G}} = \mathbf{C}^v_{\mathbf{G}}$.

If $\langle \mathbf{G}, s \rangle$ and $\langle \mathbf{H}, t \rangle$ are pointed graphs, then a function $Z\colon V_{\mathbf{G}} \to V_{\mathbf{H}}$ is called a **bounded epimorphism** if the following conditions hold:

1. $Z(s) = t$;
2. $Z$ is surjective;
3. if $v_0 \in V_{\mathbf{G}}$ and $\langle v_0, v_1 \rangle \in E_{\mathbf{G}}$, then $\langle Z(v_0), Z(v_1) \rangle \in E_{\mathbf{H}}$; and
4. if $w_0 \in V_{\mathbf{H}}$ and $\langle w_0, w_1 \rangle \in E_{\mathbf{H}}$ and $Z(v_0) = w_0$, then there is some $v_1 \in V_{\mathbf{G}}$ such that $Z(v_1) = w_1$ and $\langle v_0, v_1 \rangle \in E_{\mathbf{G}}$.

If $Z$ is a bounded epimorphism between $\mathbf{G}$ and $\mathbf{H}$, and we can define an equivalence relation $\equiv_Z$ on $V_{\mathbf{G}}$ by

$$v \equiv_Z w :\Longleftrightarrow Z(v) = Z(w).$$

**Proposition 1.** Let $\langle \mathbf{G}, s \rangle$ and $\langle \mathbf{H}, t \rangle$ be pointed graphs and $Z$ a bounded epimorphism between them. Let $\equiv_Z$ be the equivalence relation on $V_{\mathbf{G}}$ defined via $Z$. Then $\langle \mathbf{G}/\equiv_Z, [s]_{\equiv_Z} \rangle \cong \langle \mathbf{H}, t \rangle$.

*Proof.* Define $\widehat{Z}\colon V_{\mathbf{G}}/\equiv_Z \to V_{\mathbf{H}}$ by

$$\widehat{Z}([v]_{\equiv_Z}) := Z(v).$$

This function is clearly well-defined and a bijection. Using the fact that $Z$ is a bounded epimorphism, it is easy to see that $\widehat{Z}$ is structure preserving.

q.e.d.

### 2.2 The unravelled tree and the alternating graph

Let $\langle \mathbf{G}, s\rangle$ be a pointed graph. Define the set $V_{\mathbf{T}^s_{\mathbf{G}}}$ to be the set of finite walks through $\langle \mathbf{G}, s\rangle$ (*i.e.*, finite sequences of nodes in $\mathbf{C}^s_{\mathbf{G}}$ connected by $E_{\mathbf{G}}$ and starting with $s$). For walks $W_0$ of length $n$ and $W_1$ of length $n+1$, we let $\langle W_0, W_1\rangle \in E_{\mathbf{T}^s_{\mathbf{G}}}$ if and only if $W_0 = W_1{\restriction}n$. Furthermore, let $\mathrm{root}_{\mathbf{G},s} := \{\langle 0, s\rangle\}$ be the unique walk of length 1. Then

$$\mathbf{T}^s_{\mathbf{G}} := \langle V_{\mathbf{T}^s_{\mathbf{G}}}, E_{\mathbf{T}^s_{\mathbf{G}}}\rangle.$$

We call $\langle \mathbf{T}^s_{\mathbf{G}}, \mathrm{root}_{\mathbf{G},s}\rangle$ the **unravelled tree of** $\langle \mathbf{G}, s\rangle$.

In the following, we will use the parity function $\mathrm{par}\colon \mathbb{N} \to 2$ assigning to each natural number its parity. Let $V_{\mathbf{A}_{\mathbf{G}}} := 2 \times V_{\mathbf{G}}$,

$$\langle\langle e, v\rangle, \langle 1-e, w\rangle\rangle \in E_{\mathbf{A}_{\mathbf{G}}} :\Longleftrightarrow \langle v, w\rangle \in E_{\mathbf{G}},$$

and call $\mathbf{A}_{\mathbf{G}} := \langle V_{\mathbf{A}_{\mathbf{G}}}, E_{\mathbf{A}_{\mathbf{G}}}\rangle$ the **alternating graph of G**. If $s \in V_{\mathbf{G}}$, then we let $\mathbf{A}^s_{\mathbf{G}}$ be the connected component of $\langle 0, s\rangle$ in $\mathbf{A}_{\mathbf{G}}$.

**Proposition 2.** If $\langle \mathbf{G}, s\rangle$ is a pointed graph, there are bounded epimorphisms from $\langle \mathbf{T}^s_{\mathbf{G}}, \mathrm{root}_{\mathbf{G},s}\rangle$ to $\langle \mathbf{G}, s\rangle$, from $\langle \mathbf{A}^s_{\mathbf{G}}, \langle 0, s\rangle\rangle$ to $\langle \mathbf{G}, s\rangle$ and from $\langle \mathbf{T}^s_{\mathbf{G}}, \mathrm{root}_{\mathbf{G},s}\rangle$ to $\langle \mathbf{A}^s_{\mathbf{G}}, \langle 0, s\rangle\rangle$.

*Proof.* Let $e \in 2$, $v \in V_{\mathbf{G}}$, and $\mathrm{dom}(W) = n+1$ with $W(n) = v$. Then define

$$\begin{aligned} Z_{\mathbf{T}}(W) &:= v, \\ Z_{\mathbf{A}}(\langle e, v\rangle) &:= v, \text{ and} \\ Z_{\mathbf{T},\mathbf{A}}(W) &:= \langle \mathrm{par}(n), v\rangle. \end{aligned}$$

The functions $Z_{\mathbf{T}}$, $Z_{\mathbf{A}}$ and $Z_{\mathbf{T},\mathbf{A}}$ are bounded epimorphisms. q.e.d.

If $\langle \mathbf{G}, s\rangle$ is a pointed graph, $v \in V_{\mathbf{G}}$ and $W$ is a walk through $\langle \mathbf{G}, s\rangle$ of length $n+1$ such that $W(n) = v$, then the connected component of $W$ in $\mathbf{T}^s_{\mathbf{G}}$ and the graph $\mathbf{T}^v_{\mathbf{G}}$ are isomorphic as graphs. As a consequence, we get a slightly more general version of Proposition 2:

**Proposition 3.** Let $\langle \mathbf{G}, s\rangle$ be a pointed graph, and $W$ be a walk through $\langle \mathbf{G}, s\rangle$ of length $n+1$ such that $W(n) = v$. Then there are bounded epimorphisms from $\langle \mathbf{T}^s_{\mathbf{G}}, W\rangle$ to $\langle \mathbf{G}, v\rangle$ and from $\langle \mathbf{T}^s_{\mathbf{G}}, W\rangle$ to $\langle \mathbf{A}^v_{\mathbf{G}}, \langle 0, v\rangle\rangle$.

*Proof.* Compose the bounded epimorphisms $Z_{\mathbf{T}}$ between $\mathbf{T}^v_{\mathbf{G}}$ and $\mathbf{G}$ and $Z_{\mathbf{T},\mathbf{A}}$ between $\mathbf{T}^v_{\mathbf{G}}$ and $\mathbf{A}^v_{\mathbf{G}}$ with the mentioned graph isomorphism. q.e.d.

# 3 Games

## 3.1 Games and game equivalences

Given a graph $\mathbf{G} = \langle V_{\mathbf{G}}, E_{\mathbf{G}} \rangle$ and $s \in V_{\mathbf{G}}$, we define the **(symmetric) combinatorial game on** $\langle \mathbf{G}, s \rangle$ (in symbols: $\mathsf{S}(\mathbf{G}, s)$): at the beginning of the game, a token is positioned in the vertex $s$; players 0 and 1 move in turn with player 0 starting by pushing the token along the edges of $\mathbf{G}$; the player making the last move wins the game. If the game goes on for infinitely many steps, the outcome of the game is a draw. We define the **inverted symmetric combinatorial game** $\overline{\mathsf{S}}(\mathbf{G}, s)$ to be the game played like the symmetric combinatorial game, just with the rôles of the two players interchanged, *i.e.*, player 1 starts.

In the asymmetric version, the rôles of player 1 and the draw are interchanged: Given a graph $\mathbf{G} = \langle V_{\mathbf{G}}, E_{\mathbf{G}} \rangle$ and $s \in V_{\mathbf{G}}$, we define the **asymmetric combinatorial game on** $\langle \mathbf{G}, s \rangle$ (in symbols: $\mathsf{A}(\mathbf{G}, s)$): at the beginning of the game, a token is positioned in the vertex $s$; players 0 and 1 move in turn with player 0 starting by pushing the token along the edges of $\mathbf{G}$; if player 0 pushes the token into a terminal node, he wins; if player 1 pushes the token into a terminal node, the game is a draw. If the game goes on for infinitely many steps, player 1 wins. Again, we define the **inverted asymmetric game** $\overline{\mathsf{A}}(\mathbf{G}, s)$ to be the game with the players interchanged.

**Strategies** in these combinatorial games are simply functions that tell the players which edge $\langle v_0, v_1 \rangle$ to use if they are presented with the token in vertex $v_0$ (it is obvious that such memory-free strategies suffice for these games). A strategy is **winning** if the player following the strategy wins the game regardless of how the other player plays, and a strategy is called **nonlosing** if the game in which one player follows the strategy results in either a win for that player or a draw.

By the determinacy theorem of Gale and Stewart (for details, *cf.* Section 3.2) each of the games $\mathsf{X} \in \{\mathsf{A}, \overline{\mathsf{A}}, \mathsf{S}, \overline{\mathsf{S}}\}$ defined above will have one of the following three **values**, denoted by $\mathrm{val}(\mathsf{X})$:

- W. Player 0 has a winning strategy,
- D. both players have a nonlosing strategy,
- L. Player 1 has a winning strategy.

On the set $L = \{\mathsf{L}, \mathsf{D}, \mathsf{W}\}$ of these values, we define a lattice structure by

$\mathsf{L} \leq \mathsf{D} \leq \mathsf{W}$ and an inversion function $\mathrm{inv}\colon L \to L$ defined by

$$\begin{array}{ccc} \mathsf{W} & \mapsto & \mathsf{L} \\ \mathsf{D} & \mapsto & \mathsf{D} \\ \mathsf{L} & \mapsto & \mathsf{W}. \end{array}$$

Let $\mathsf{X}, \mathsf{Y} \in \{\mathsf{A}, \overline{\mathsf{A}}, \mathsf{S}, \overline{\mathsf{S}}\}$. We say that $\mathsf{X}$ and $\mathsf{Y}$ are **equivalent** if they have the same value. We say that they are **anti-equivalent** if $\mathrm{val}(\mathsf{X}) = \mathrm{inv}(\mathrm{val}(\mathsf{Y}))$.

We define the notion of a X-Y-(anti)-equivalence: Let $\mathbf{G}$ and $\mathbf{H}$ be graphs and let $f\colon V_{\mathbf{G}} \to V_{\mathbf{H}}$ be a function. Then $f$ is called a **X-Y-(anti)-equivalence** if for all $v \in V_{\mathbf{G}}$, we have that $\mathsf{X}(\mathbf{G}, v)$ and $\mathsf{Y}(\mathbf{H}, f(v))$ are (anti)-equivalent.

There are some obvious facts about equivalence of combinatorial games:

**Proposition 4.** For every pointed graph $\langle \mathbf{G}, v \rangle$, the games $\mathsf{S}(\mathbf{G}, v)$ and $\overline{\mathsf{S}}(\mathbf{G}, v)$ are anti-equivalent. In other words, $\mathrm{id}\colon V_{\mathbf{G}} \to V_{\mathbf{G}}$ is an $\mathsf{S}$-$\overline{\mathsf{S}}$-anti-equivalence.

*Proof.* Obvious. q.e.d.

**Proposition 5.** Let $\mathbf{G}$ and $\mathbf{H}$ be graphs, let $\mathsf{X} \in \{\mathsf{S}, \overline{\mathsf{S}}, \mathsf{A}, \overline{\mathsf{A}}\}$, and let $F\colon V_{\mathbf{G}} \to V_{\mathbf{H}}$ be a function. If $F$ is a bounded epimorphism, then it is a X-X-equivalence.

*Proof.* Obvious. q.e.d.

An immediate consequence of Propositions 2, 3 and 5 is that in order to analyze arbitrary combinatorial games, it is enough to analyze games on trees:

**Corollary 6.** Let $\mathsf{X} \in \{\mathsf{S}, \overline{\mathsf{S}}, \mathsf{A}, \overline{\mathsf{A}}\}$ and let $\langle \mathbf{G}, s \rangle$ be a pointed graph. Then the games $\mathsf{X}(\mathbf{G}, s)$ and $\mathsf{X}(\mathbf{T}^s_{\mathbf{G}}, \mathrm{root}_{\mathbf{G},s})$ are equivalent. Also, for any walk $W$ through $\langle \mathbf{G}, s \rangle$ with length $n+1$ and $W(n) = v$, the games $\mathsf{X}(\mathbf{G}, v)$ and $\mathsf{X}(\mathbf{T}^s_{\mathbf{G}}, W)$ are equivalent.

### 3.2 A translation into the usual Gale-Stewart theory of infinite games

Corollary 6 is the underlying methodology of the Gale-Stewart theory [GalSte53]. Instead of looking at the (possibly cyclic) graph, we look at the unravelled tree and analyze the game on the tree via backwards induction with a (possibly transfinite) labelling construction.

We shall translate our tree games into the usual topological notation of Gale-Stewart theory: Look at the space ${V_{\mathbf{G}}}^{\mathbb{N}}$ of functions from $\mathbb{N}$ into $V_{\mathbf{G}}$, endowed with the product topology of the discrete topology on $V_{\mathbf{G}}$.

We define three infinite games $\mathsf{G}_0(\mathbf{G}, v)$, $\mathsf{G}_1(\mathbf{G}, v)$, and $\mathsf{H}_1(\mathbf{G}, v)$. In all of the games, players 0 and 1 play elements of $V_{\mathbf{G}}$ in turn and produce an element of ${V_{\mathbf{G}}}^{\mathbb{N}}$, let's call it $X$. We assume that $X(0) = v$ and that player 0 plays the odd digits and player 1 plays the even digits. The payoff sets of the games are defined as follows:

- In $\mathsf{G}_0(\mathbf{G}, v)$, player 0 wins if the least $n+1$ such that $X{\restriction}n+1$ is not a walk through $\langle\mathbf{G}, v\rangle$ exists and is odd. Otherwise, player 1 wins.
- In $\mathsf{G}_1(\mathbf{G}, v)$, player 0 wins if either $X$ is an infinite walk through $\langle\mathbf{G}, v\rangle$, or the least $n+1$ such that $X{\restriction}n+1$ is not a walk through $\langle\mathbf{G}, v\rangle$ is odd. Otherwise, player 1 wins.
- In $\mathsf{H}_1(\mathbf{G}, v)$, player 0 wins if there is a least $n+1$ such that $X{\restriction}n+1$ is not a walk through $\langle\mathbf{G}, v\rangle$ and either $n+1$ is odd or $X(n)$ is a terminal node of $\mathbf{G}$. Otherwise, player 1 wins.

The payoff sets for player 0 in the defined three games are either open ($\mathsf{G}_0$ and $\mathsf{H}_1$) or closed ($\mathsf{G}_1$) in the topology defined on ${V_{\mathbf{G}}}^{\mathbb{N}}$, and by the usual Gale-Stewart theorem for open and closed games without draw, one of the two players has a winning strategy, *i.e.*, the values are either W or L.

It is easy to see that these infinite Gale-Stewart games correspond to the combinatorial games as follows:

**Proposition 7.** For every pointed graph $\langle\mathbf{G}, v\rangle$, the following equivalences hold:

$$
\begin{aligned}
\mathrm{val}(\mathsf{G}_0(\mathbf{G}, v)) = \mathsf{W} &\iff \mathrm{val}(\mathsf{S}(\mathbf{T}^v_{\mathbf{G}}, \mathrm{root}_{\mathbf{G},v})) = \mathsf{W}\\
&\iff \mathrm{val}(\mathsf{A}(\mathbf{T}^v_{\mathbf{G}}, \mathrm{root}_{\mathbf{G},v})) = \mathsf{W}\\
\mathrm{val}(\mathsf{G}_0(\mathbf{G}, v)) = \mathsf{L} &\iff \text{player 1 has a nonlosing strategy for } \mathsf{S}(\mathbf{T}^v_{\mathbf{G}}, \mathrm{root}_{\mathbf{G},v})\\
&\iff \text{player 1 has a nonlosing strategy for } \mathsf{A}(\mathbf{T}^v_{\mathbf{G}}, \mathrm{root}_{\mathbf{G},v})\\
\mathrm{val}(\mathsf{G}_1(\mathbf{G}, v)) = \mathsf{W} &\iff \text{player 0 has a nonlosing strategy for } \mathsf{S}(\mathbf{T}^v_{\mathbf{G}}, \mathrm{root}_{\mathbf{G},v})\\
\mathrm{val}(\mathsf{G}_1(\mathbf{G}, v)) = \mathsf{L} &\iff \mathrm{val}(\mathsf{S}(\mathbf{T}^v_{\mathbf{G}}, \mathrm{root}_{\mathbf{G},v})) = \mathsf{L}\\
\mathrm{val}(\mathsf{H}_1(\mathbf{G}, v)) = \mathsf{W} &\iff \text{player 0 has a nonlosing strategy for } \mathsf{A}(\mathbf{T}^v_{\mathbf{G}}, \mathrm{root}_{\mathbf{G},v})\\
\mathrm{val}(\mathsf{H}_1(\mathbf{G}, v)) = \mathsf{L} &\iff \mathrm{val}(\mathsf{A}(\mathbf{T}^v_{\mathbf{G}}, \mathrm{root}_{\mathbf{G},v})) = \mathsf{L}
\end{aligned}
$$

As a consequence, we get a proof of the claim (see above) that the values W, L and D are the only possible values for our games.

### 3.3 Sound labellings

Let $L := \{\mathsf{L}, \mathsf{D}, \mathsf{W}\}$, $\mathbf{G}$ be a graph and $\mathsf{X} \in \{\mathsf{S}, \overline{\mathsf{S}}, \mathsf{A}, \overline{\mathsf{A}}\}$. An $L$-labelling $\ell\colon V_{\mathbf{G}} \to L$ is called **X-sound** if it is a total function and if for each vertex $v \in V_{\mathbf{G}}$, we have

$$\ell(v) = \mathrm{val}(\mathsf{X}(\mathbf{G}, v)).$$

Because of Proposition 4, the notions of S-soundness and $\overline{\mathsf{S}}$-soundness are closely connected:

**Corollary 8.** Let $\mathbf{G}$ be a graph and $\ell\colon V_{\mathbf{G}} \to L$ be S-sound. Then $\overline{\ell}$ defined by $\overline{\ell}(v) := \mathrm{inv}(\ell(v))$ is $\overline{\mathsf{S}}$-sound.

The Gale-Stewart analysis for games on trees gives a (possibly transfinite) recursive procedure to actually compute an S-sound labelling:

**Theorem 9 (Gale & Stewart; 1953).** If $\mathbf{T}$ is a tree, then there is recursive procedure that computes (in less than $\mathrm{Card}(V_{\mathbf{T}})^+$ steps) the S-sound labelling $\ell$.

**Proposition 10.** Let $\langle \mathbf{G}, s\rangle$ be a connected pointed graph. If $\ell$ is the S-sound labelling on $\mathbf{T}^s_{\mathbf{G}}$, then the quotient labelling $\ell/{\equiv_{Z_{\mathbf{T}}}}$ on $\mathbf{G}$ defined by

$$\ell/{\equiv_{Z_{\mathbf{T}}}}(v) := \ell(W)$$

(where $W$ is any walk through $\langle \mathbf{G}, s\rangle$ with length $n+1$ such that $W(n) = v$) is well-defined and is the S-sound labelling for $\mathbf{G}$.

*Proof.* Let us show that $\equiv_{Z_{\mathbf{T}}}$ respects $\ell_{\mathbf{T}}$:

Suppose $W \equiv_{Z_{\mathbf{T}}} W'$, *i.e.*, $Z_{\mathbf{T}}(W) = Z_{\mathbf{T}}(W') = v$ for some $v$. Corollary 6 tells us that $\mathsf{S}(\mathbf{T}^s_{\mathbf{G}}, W)$ and $\mathsf{S}(\mathbf{T}^s_{\mathbf{G}}, W')$ are both equivalent to $\mathsf{S}(\mathbf{G}, v)$, so in particular, $\ell(W) = \ell(W')$, and the quotient labelling is sound. q.e.d.

For asymmetric combinatorial games, we don't have the symmetry of Corollary 8:

**Proposition 11.** If $\mathrm{val}(\mathsf{A}(\mathbf{G}, v)) = \mathsf{W}$, then $\mathrm{val}(\overline{\mathsf{A}}(\mathbf{G}, v)) \neq \mathsf{L}$. All other combinations are possible.

*Proof.* For player 0, a winning strategy is a strategy that forces the token into a terminal node in an odd number of moves. Such a strategy is a non-losing strategy for player 1 in the inverted game.

In Figure 1, examples for all eight combinatorially possible situations are given. q.e.d.

Because of the asymmetry indicated by Proposition 11, we define the following notion: A (partial) function $\ell\colon V_{\mathbf{G}} \to L^2$ is called a **(partial)**

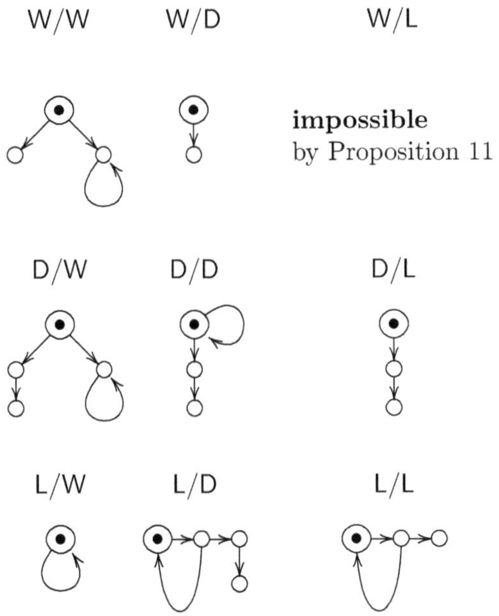


Figure 1. Examples for the eight possible value combinations in the asymmetric and the inverted asymmetric combinatorial game.

**$L$-bilabelling on G.** We write $\ell(v) = \langle \ell_0(v), \ell_1(v) \rangle$. An $L$-bilabelling $\ell\colon V_{\mathbf{G}} \to L^2$ is called **A-sound** if it's total and for each vertex $v \in V$, we have

$$\ell_0(v) = \mathrm{val}(\mathsf{A}(\mathbf{G}, v)) \text{ and } \ell_1(v) = \mathrm{val}(\overline{\mathsf{A}}(\mathbf{G}, v)).$$

In analogy to Theorem 9, the Gale-Stewart analysis gives us the existence of A-sound labellings for trees **T**. Among several ways to produce such a labelling, there is one procedure that was the motivation for the algorithm described in Section 4: in [Löw03], the author gives a non-monotone variant of the Gale-Stewart procedure which can be used to construct the A-sound labelling for trees.

**Proposition 12.** Let $\langle \mathbf{G}, s \rangle$ be a connected pointed graph. If $\ell$ is the A-sound labelling on $\mathbf{T}^s_{\mathbf{G}}$, then the quotient labelling $\ell/{\equiv_{Z_{\mathbf{T},\mathbf{A}}}}$ on $\mathbf{A}^s_{\mathbf{G}}$ defined by

$$\ell/{\equiv_{Z_{\mathbf{T},\mathbf{A}}}}(\langle e, v \rangle) := \ell(W)$$

(where $W$ is any walk through $\langle \mathbf{G}, s \rangle$ with length $2n + e + 1$ such that $W(2n + e) = v$) is well-defined and is the A-sound labelling for $\mathbf{A}^s_{\mathbf{G}}$.

We would like to extend this to an A-sound bilabelling, but not all nodes in $\mathbf{G}$ are necessarily reachable by both players. The following simple construction helps:

If $\langle \mathbf{G}, s\rangle$ is a connected pointed graph, let $\mathbf{G}^*_s$ be defined by

$$\begin{aligned} V_{\mathbf{G}^*_s} &:= V_{\mathbf{G}} \cup \{x\}, \\ E_{\mathbf{G}^*_s} &:= E_{\mathbf{G}} \cup \{\langle x, s\rangle\} \end{aligned}$$

(where $x \notin V_{\mathbf{G}}$). The connectedness of $\langle \mathbf{G}, s\rangle$ implies that

$$\mathbf{C}^{\langle 0,s\rangle}_{\mathbf{A}_{\mathbf{G}}} \cup \mathbf{C}^{\langle 0,x\rangle}_{\mathbf{A}_{\mathbf{G}^*_s}} = 2 \times V_{\mathbf{G}};$$

moreover, if $W$ is a walk through $\langle \mathbf{T}^s_{\mathbf{G}}, \mathrm{root}_{\mathbf{G},s}\rangle$ of length $n+1$ and $W'$ is a walk through $\langle \mathbf{T}^x_{\mathbf{G}^*_s}, \mathrm{root}_{\mathbf{G}^*_s,x}\rangle$ of length $m+1$ with $W(n) = W'(m)$ and $\mathrm{par}(n) = \mathrm{par}(m)$, then they represent exactly the same position in the game on $\mathbf{G}$ (albeit with different game histories), so if $\ell$ is $\mathsf{A}$-sound on $\mathbf{T}^s_{\mathbf{G}}$ and $\ell^*$ is $\mathsf{A}$-sound on $\mathbf{T}^x_{\mathbf{G}^*_s}$, then $\ell(W) = \ell^*(W')$. As a consequence, we get:

**Proposition 13.** If $\langle \mathbf{G}, s\rangle$ is a connected pointed graph, $\ell$ is an $\mathsf{A}$-sound labelling on $\mathbf{T}^s_{\mathbf{G}}$ and $\ell^*$ is an $\mathsf{A}$-sound labelling on $\mathbf{T}^x_{\mathbf{G}^*_s}$, then the bilabelling $\ell^\dagger$ defined by

$$\ell^\dagger_e(v) := \begin{cases} \ell/{\equiv_{Z_{\mathbf{T},\mathbf{A}}}}(\langle e, v\rangle) & \text{if } \langle e, v\rangle \in \mathbf{C}^{\langle 0,s\rangle}_{\mathbf{A}_{\mathbf{G}}}, \text{ or} \\ \ell^*/{\equiv_{Z_{\mathbf{T},\mathbf{A}}}}(\langle e, v\rangle) & \text{otherwise} \end{cases}$$

is welldefined, total and an $\mathsf{A}$-sound bilabelling on $\mathbf{G}$.

In the following section, we shall give an algorithm to compute $\ell^\dagger$ directly without going through the tree unravelling.

## 4 The algorithm

We fix a graph $\mathbf{G}$. For the purpose of this section, we assume that $V_{\mathbf{G}}$ is finite. Using the lattice structure on $L$, we can define an ordering $\leq^*$ on $L^2$ as the product ordering of $\langle L, \overline{\leq}\rangle$ with $\langle L, \leq\rangle$ as depicted in Figure 2.[3]

As mentioned, a labelling procedure for the graph (instead of the unravelled tree) is a "folk result" in automata theory (see for instance [Fra97, p. 15], albeit in a more graph-theoretic context). We call this procedure **Backward Induction (Graph)** to distinguish it from the Gale-Stewart tree analysis.

We let $\mathrm{BIG}^0(v) := \mathsf{L}$ for all terminal nodes $v$. After that, we define

[3]Here $\overline{\leq}$ denotes the inverse ordering of $\leq$.

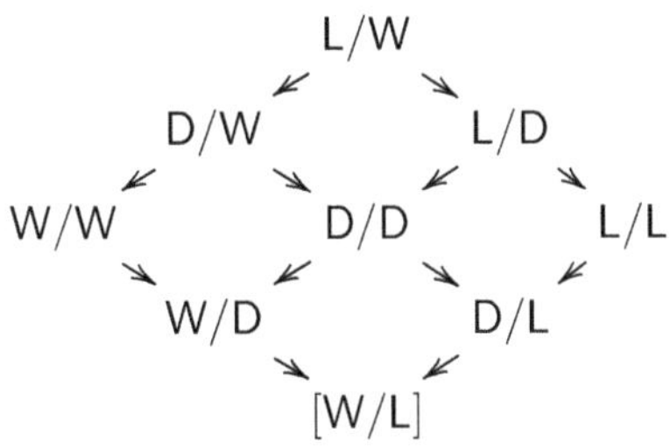


Figure 2. The ordering $\leq^*$ on $L^2$.

$$\mathrm{BIG}^{n+1}(v) := \begin{cases} \mathrm{BIG}^n(v) & \text{if } v \in \mathrm{dom}(\mathrm{BIG}^n), \\ \mathsf{W} & \text{if there is } \langle v, w\rangle \in E \text{ and } \mathrm{BIG}^n(w) = \mathsf{L}, \text{ or} \\ \mathsf{L} & \text{if for all } \langle v, w\rangle \in E, \text{ we have } \mathrm{BIG}^n(w) = \mathsf{W}. \end{cases}$$

For some $N$, we have $\mathrm{BIG}^N = \mathrm{BIG}^{N+1}$, then we let

$$\mathrm{BIG}(v) := \begin{cases} \mathrm{BIG}^N(v) & \text{if } v \in \mathrm{dom}(\mathrm{BIG}^N), \text{ or} \\ \mathsf{D} & \text{otherwise.} \end{cases}$$

This algorithm produces an S-sound labelling in $O(|V_{\mathbf{G}}| + |E_{\mathbf{G}}|)$ steps, and is essentially the quotient labelling of the Gale-Stewart labelling on $\mathbf{T_G}$: we label a vertex $v \in V_{\mathbf{G}}$ as soon as some $W$ with $Z_{\mathbf{T}}(W) = v$ is labelled in the Gale-Stewart construction. As the Gale-Stewart procedure, this labelling is monotonic in the sense that whenever a vertex is labelled, it will retain that label for ever.

Following this idea and injecting non-monotonicity in the spirit of [Löw03, § 5] into the procedure, we shall now give an algorithm `NMBIG` (for "non-monotonic backward induction (graph)") that produces the A-sound bilabelling.

We give the algorithm in pseudocode in Figure 3. Let us explain the two special datatypes `label` and `graph` used in the pseudocode:

Variables of type `label` can take the values `W`, `D`, and `L` representing W, D, and L. We have a binary relation < defined for variables of type `label`, and `X` < `Y` is `TRUE` if and only if X < Y. In addition, there is a unary function `INV` defined on `label` corresponding to the inversion function inv.

The datatype `graph` encodes the bilabelled graph structure. Let $\langle \mathbf{G}, \ell\rangle$ be a bilabelled graph with $V_{\mathbf{G}} = \{v_i\,;\, 0 \leq i \leq N\}$. If `G` is a variable of type `graph` representing $\langle \mathbf{G}, \ell\rangle$, then the following objects are defined:

```
Procedure:MAIN(G:graph)
  for i ← 1 to Nvert[G] do
3     Ell[G,i,0] ← L;
4     Ell[G,i,1] ← W;
  for i ← 1 to Nvert[G] do
      if Outbound[G,i] == ∅ then
7         Ell[G,i,0] ← D;
8         Ell[G,i,1] ← L;
9         foreach j ∈ Inbound[G,i] do
              Label(G,j,1);
              Label(G,j,0);

Procedure:Label(G:graph;i:integer;e:binary)
13 aux ← L;
14 foreach j ∈ Outbound[G,i] do
15     aux ← aux + INV[Ell[G,j,1-e]];
       if Ell[G,i,e] ≠ aux then
17         Ell[G,i,e] ← aux;
18         foreach j ∈ Inbound[G,i] do
               Label(G,j,1-e);
```

Figure 3. The algorithm NMBIG for non-monotonic backward induction graph labelling.

- Nvert[G], the number of vertices of the graph, *i.e.*, $N+1$;
- for each $i \leq N$, the objects Outbound[G, i] and Inbound[G, i] representing the sets $\{v_j\,;\,\langle v_i, v_j\rangle \in E_{\mathbf{G}}\}$ and $\{v_j\,;\,\langle v_j, v_i\rangle \in E_{\mathbf{G}}\}$, respectively;
- for each $i \leq N$ and $e \in 2$, an object Ell[G, i, e] of type label, representing $\ell_e(v_i)$.

If the graph structure is stored using an adjacency matrix, then iterating over the outbound or inbound set for a particular vertex can be done in linear time with respect to the size of the vertex set.

We run the algorithm NMBIG on (a representation of) $\mathbf{G}$. For any $t \in \mathbb{N}$, we let $\ell_e^t(v_i)$ be the value of Ell[G, i, e] at time $t$ of the algorithm,[4] and

$$\ell^t(v_i) := \langle \ell_0^t(v_i), \ell_1^t(v_i)\rangle.$$

[4]Code lines 3 and 4 make sure that the values $\ell_e(t)(v_i)$ are defined early on in the algorithm (in step $e \cdot N + i$), so from step $2 \cdot N$ onwards, $\ell^t$ is a total function. In the following, we shall mostly ignore these first $2 \cdot N$ steps of the algorithm.

**Proposition 14.** For each $i \leq N$, the sequence $\langle \ell^t(v_i)\,;\, t \in \mathbb{N}\rangle$ is $\leq^*$-decreasing in $L^2$. Equivalently, the sequence $\langle \ell_0^t(v_i)\,;\, t \in \mathbb{N}\rangle$ is $\overline{\leq}$-decreasing and the sequence $\langle \ell_1^t(v_i)\,;\, t \in \mathbb{N}\rangle$ is $\leq$-decreasing.

*Proof.* Let $t+1$ be the least number such that $\ell^{t+1}(v_i) <^* \ell^t(v_i)$ for some $i$. Clearly, the value of the bilabeling can only be changed by code lines 7, 8 and 17 of the algorithm.

Since the procedure `MAIN` can only change the values at terminal nodes from L/W to D/W and then to D/L, the lines 7 and 8 cannot create a decrease in $\leq^*$. Note that this means that $v_i$ is not a terminal node since `Label(G, i, e)` is only called if there is an edge from $v_i$ to somewhere.

Also, we cannot be in the first call of `Label(G, i, e)` at time $t$ since the bilabelling is initialized with L/W which is the top element of $\langle L^2, \leq^*\rangle$. Consequently, there are some $s_0 < s_1 < t$ such that at both $s_0$ and $s_1$, the procedure `Label(G, i, e)` is called. Let $s_0$ be largest with that property. By our assumption, we have that $\ell^{t+1}(v_i) <^* \ell^t(v_i) = \ell^{s_1}(v_i)$. By code lines 13 to 15, we have

$$\ell_e^{t+1}(v_i) := \sup_{\leq}\{\mathrm{inv}(\ell_{1-e}^{s_1}(w))\,;\, \langle v_i, w\rangle \in E_{\mathbf{G}}\}, \text{ and}$$

$$\ell_e^{s_1}(v_i) := \sup_{\leq}\{\mathrm{inv}(\ell_{1-e}^{s_0}(w))\,;\, \langle v_i, w\rangle \in E_{\mathbf{G}}\}.$$

But this means that there is some $w$ such that $\ell^{s_0}(w) <^* \ell^{s_1}(w)$, contradicting the choice of $t+1$ as minimal. q.e.d.

**Proposition 15.** The procedure `Label` is called at most $4 \cdot |E_{\mathbf{G}}|$ times.

*Proof.* By the loops 9 and 18, each call of `Label`$(G, i, e)$ is associated with an edge $\langle v_i, w\rangle$, and by code lines 7, 8 and 17, preceded by a change of $\ell(w)$. By Proposition 14, this means that each such an edge can be used for calls of the procedure `Label` at most four times. q.e.d.

**Theorem 16.** The running time of `NMBIG` is $O(|V_{\mathbf{G}}| + |E_{\mathbf{G}}|^2)$. If $\mathbf{G}$ is connected, this is $O(|E_{\mathbf{G}}|^2)$.

*Proof.* The procedure `NMBIG` itself (without the recursive calls of `Label`) takes at most $4 \cdot |V_{\mathbf{G}}| + 2 \cdot |E_{\mathbf{G}}|$ steps. Each call of `Label` has running time $O(|E_{\mathbf{G}}|)$, so by Proposition 15, the entire running time is $O(|V_{\mathbf{G}}| + |E_{\mathbf{G}}|^2)$. q.e.d.

The critical lines in the algorithm that push the running time from linear to quadratic are the loop from line 14: every time we run `Label` for $v_i$, we have to check the current values of all its successors. Let $\mathbf{G}_n$ be the graph with a root $v_0$ and $n$ immediate successors of the root which are terminal nodes, *i.e.*, $|E_{\mathbf{G}_n}| = n$. Then $\mathtt{Label}(\mathtt{G}, \mathtt{1}, \mathtt{e})$ is called $n$ times (once for each terminal node) and each time its running time is at least $n$ because it has to check each of the terminal nodes, so the total running time is at least $|E_{\mathbf{G}_n}|^2$.

The running time can be pushed to linear in the size of the edge set, at the cost of making the algorithm significantly less readable. In Section 4.1 we give the refined algorithm, and show that it computes the same labelling as `NMBIG`. The proof of the following theorem is however less tedious on the simpler algorithm:

**Theorem 17.** If run on the graph $\mathbf{G}$, the algorithm `NMBIG` computes the A-sound bilabelling on $\mathbf{G}$.

*Proof.* Again, let $\ell_e^t(v_i)$ be the value of $\mathtt{Ell}[\mathtt{G}, \mathtt{i}, \mathtt{e}]$ at time $t$. By Theorem 16, these values stabilize at some finite time $N$. Let $\ell_e(v_i) := \ell_e^N(v_i)$ be the eventual value.

This proof follows essentially the idea of the Gale-Stewart proof (a.k.a. "backwards induction"). The main ingredient of that idea is that players have strategies that force the following to be true: if $W$ is a run of the game according to the strategy, then the sequence of time indices of the label assignments of $W(n)$ during the algorithm is a decreasing sequence of integers. By wellfoundedness of $\mathbb{N}$, it can be deduced that we hit one of the basic cases eventually. Unfortunately, in our case, the nonmonotonicity of the algorithm causes some problems: it is possible that vertex $v$ is labelled at time $t$, but some successors receive their label later. In order to deal with this, we have to go through the different cases in detail.

For any $v \in V_{\mathbf{G}}$, let

$$\mathrm{ind}_e(v) := \min\{t\,;\, \ell_e^t(v) = \ell_e(v)\}$$

be the $e$-**index** of $v_i$. (Note that by Proposition 14, for all $\mathrm{ind}_e(v) \leq t \leq N$, we have that $\ell_e^t(v) = \ell_e(t)$.)

We shall discuss the properties of the six possible cases:

**Case 1.** If $\ell_0(v_i) = \mathsf{W}$, then there is some $w$ with $\langle v_i, w\rangle \in E_{\mathbf{G}}$, $\ell_1(w) = \mathsf{L}$, and $\mathrm{ind}_1(w) < \mathrm{ind}_0(v_i)$.

[The vertex $v_i$ has been labelled by code line 15, and this means that in the preceding call of code line 15 some successor was 1-labelled $\mathsf{L}$. By Proposition 14, a 1-label $\mathsf{L}$ can never be changed anymore.]

**Case 2.** If $\ell_0(v_i) = \mathsf{D}$, then there is no $w$ with $\langle v_i, w\rangle \in E_{\mathbf{G}}$ and $\ell_1(w) = \mathsf{L}$. Also, either $v_i$ is terminal or there is some $w$ with $\langle v_i, w\rangle \in E_{\mathbf{G}}$ and $\ell_1(w) = \mathsf{D}$, and for all such $w$, $\mathrm{ind}_1(w) < \mathrm{ind}_0(v_i)$.

[If $v_i$ is terminal, the claim is trivial, so let $v_i$ be nonterminal. Let $t := \mathrm{ind}_0(v_i)$ which is the time of a call of code line 15. Therefore, at time $t$, we have

$$(\star_t) \quad \forall w\,(\langle v_i, w\rangle \in E_{\mathbf{G}} \rightarrow \ell_1^t(w) \neq \mathsf{L}) \quad \& \\ \exists w\,(\langle v_i, w\rangle \in E_{\mathbf{G}} \;\&\; \ell_1^t(w) = \mathsf{D}).$$

By Proposition 14, none of the vertices that are 1-labelled $\mathsf{L}$ can change their labelling anymore and the vertices 1-labelled $\mathsf{D}$ can only change their label to $\mathsf{L}$. Suppose that there is some $t < s \leq N$ such that $(\star_s)$ is not true anymore. Then at time $s$, we have a call of code line 15 and all successors of $v_i$ are 1-labelled $\mathsf{L}$. In the subsequent call of `Label[G, i, 0]`, the label of $v_i$ will be changed to $\mathsf{W}$ in contradiction to the assumption.]

**Case 3.** If $\ell_0(v_i) = \mathsf{L}$, then for all $w$ with $\langle v_i, w\rangle \in E_{\mathbf{G}}$, we have $\ell_1(w) = \mathsf{W}$. Moreover, both $v_i$ and all of its successors have received that label at the beginning of the algorithm (code lines 3 and 4).

[Obvious from Proposition 14 and code lines 15 and 17.]

**Case 4.** If $\ell_1(v_i) = \mathsf{W}$, then there is some $w$ with $\langle v_i, w\rangle \in E_{\mathbf{G}}$ such that $\ell_0(w) = \mathsf{L}$ and both $v_i$ and $w$ have been labelled at the beginning of the algorithm (code lines 3 and 4).

[This is dual to Case 3.]

**Case 5.** If $\ell_1(v_i) = \mathsf{D}$, then there is no $w$ with $\langle v_i, w\rangle \in E_{\mathbf{G}}$ and $\ell_0(w) = \mathsf{L}$. Also, there is some $w$ with $\langle v_i, w\rangle \in E_{\mathbf{G}}$ and $\ell_0(w) = \mathsf{D}$, and for all such $w$, $\mathrm{ind}_0(w) < \mathrm{ind}_1(v_i)$.

[This is dual to Case 2, except that terminal nodes cannot be 1-labelled $\mathsf{D}$.]

**Case 6.** If $\ell_1(v_i) = \mathsf{L}$, then for all $w$ with $\langle v_i, w\rangle \in E_{\mathbf{G}}$, we have $\ell_0(w) = \mathsf{W}$ and $\mathrm{ind}_0(w) < \mathrm{ind}_1(v_i)$.

[This is dual to Case 1.]

With our six cases in mind, we can now define strategies for player $e \in 2$ as follows:

The strategy $\sigma_e$ plays from $v$ into some successor $w$ such that

$$\ell_{1-e}(w) = \mathrm{inv}(\ell_e(v)),$$

and –whenever possible– such that

$$\mathrm{ind}_{1-e}(w) < \mathrm{ind}_e(v).$$

We shall show that $\sigma_0$ and $\sigma_1$ are witnesses for value $\ell_0(v)$ in $\mathsf{A}(\mathbf{G}, v)$ and value $\ell_1(v)$ in $\overline{\mathsf{A}}(\mathbf{G}, v)$, respectively. This will finish the proof of Theorem 17.

**Case A:** $\ell_0(v) = \mathsf{W}$.

Let $W$ be a maximal walk through $\langle \mathbf{G}, v\rangle$ where player 0 follows $\sigma_0$ (*i.e.*, $W(2n + 2) = \sigma_0(W(2n + 1)))$. By Cases 1 and 6, we have $\ell_0(W(2n)) = \mathsf{W}$ and $\ell_1(W(2n + 1)) = \mathsf{L}$. If $W$ is finite (say, of length $n + 1$), then $W(n)$ is terminal, so $\ell(W(n)) = \mathsf{D}/\mathsf{L}$, hence $n$ is odd and player 0 has won the game with run $W$.

If $W$ is infinite, then define

$$i_k := \mathrm{ind}_{\mathrm{par}(k)}(W(k)). \qquad (\dagger)$$

By definition of $\sigma_0$ and by Cases 1 and 6, this is a strictly decreasing sequence of natural numbers which is absurd.

**Case B:** $\ell_1(v) = \mathsf{W}$.

Let $W$ be a maximal walk through $\langle \mathbf{G}, v\rangle$ where player 1 follows $\sigma_1$. By Cases 3 and 4, we have $\ell_0(W(2n)) = \mathsf{L}$ and $\ell_1(W(2n+1)) = \mathsf{W}$. This implies that none of the vertices in $W$ can be terminal, and thus $W$ is infinite and player 1 wins the game with run $W$.

**Case C:** $\ell_0(v) = \mathsf{D}$.

Let $W$ be a maximal walk through $\langle \mathbf{G}, v\rangle$ where player 0 follows $\sigma_0$. By Cases 2 and 5, the following three subcases cover all possibilities:

**Subcase C1.** There is some $n$ such that $\ell_0(W(2n)) = \mathsf{W}$. By Case A, player 0 wins.

**Subcase C2.** For all $k$, $\ell_{\mathrm{par}(k)}(W(k)) = \mathsf{D}$ and $W$ is finite (say, of length $n+1$). Then $W(n)$ is a terminal node, and since $\ell_{\mathrm{par}(n)}(W(n)) = \mathsf{D}$, we have that $n$ is even, so the game is a draw.

**Subcase C3.** For all $k$, $\ell_{\mathrm{par}(k)}(W(k)) = \mathsf{D}$ and $W$ is infinite. Now by Cases 2 and 5 and the definition of $\sigma_0$, the sequence $i_k$ as defined in (†) is a strictly descending sequence of natural numbers, yielding a contradiction.

Similarly (using Case B instead of Case A), we can show that $\sigma_1$ is a nonlosing strategy for player 1. Together, $\sigma_0$ and $\sigma_1$ witness that $\mathrm{val}(\mathsf{A}(\mathbf{G}, v)) = \mathsf{D}$.

**Case D:** $\ell_0(v) = \mathsf{L}$.

By Case 3, this means that player 0 is forced into a position $w$ with $\ell_1(w) = \mathsf{W}$. Now apply Case B.

**Case E:** $\ell_1(v) = \mathsf{D}$.

This is dual to Case B.

**Case F:** $\ell_1(v) = \mathsf{L}$.

By **Case 6.**, this means that player 1 is forced into a position $w$ with $\ell_0(w) = \mathsf{W}$. Now apply Case A. q.e.d.

### 4.1 True linear time

The 'linear-time' variant of this algorithm derives from the observation that only the *number* of labelled successors of a given vertex matters to the algorithm, and not (directly) their identities. Thus if we can by bookkeeping accurately track this number without storing and modifying sets, we avoid the additional cost of the inner loop.[5]

We define the auxiliary bookkeeping variable `Succ[G,j,e,a]`. This will store the number of successors of $v_j$ labelled with `a` for player `e`. The complete algorithm is given in Procedures `MAIN` and `Label`, figure 4. The proof of linear time shows that the same operations are performed as in the original algorithm, but for the replacement of an inner loop by a constant-time lookup.

The additional data access required by this algorithm means that we cannot ensure that the data structure is notionally consistent at every timestep of the algorithm. For instance, during the loop at line 28, the values stored in `Succ[G,k,e,oldLbl]` will be incorrect for some `k`.

[5] A related approach was used in [Gij+76] to efficiently solve Edward de Bono's "L-Game"; here instead of tracking the number of non-lost successor positions the authors keep a 'safe' move from each non-lost position and update it efficiently if the 'safe' position becomes lost. The present authors are grateful to Peter van Emde Boas for bringing this report to their attention.

```
Procedure:MAIN(G:graph)
initialise loop:
for i ← 1 to Nvert[G] do
    Ell[G,i,0] ← L;
    Ell[G,i,1] ← W;
    foreach j ∈ InBound[G,i] do
        Succ[G,j,0,L] ← Succ[G,j,0,L] + 1;
        Succ[G,j,1,W] ← Succ[G,j,1,W] + 1;
main loop:
for i ← 1 to Nvert[G] do
    Label(G,i,0);
    Label(G,i,1);

Procedure:Label(G:graph;i:integer;e:binary)
oldLbl ← Ell[G,j,e];
if Succ[G,j,1-e,L] > 0 then
    newLbl ← W;
else if Succ[G,j,1-e,D] > 0 then
    newLbl ← D;
else if Succ[G,j,1-e,W] > 0 then
    newLbl ← L;
else if e == 0 then
    newLbl ← D;
else
    newLbl ← L;
update loop:
if newLbl ≠ oldLbl then
    Ell[G,j,e] ← newLbl;
    foreach k ∈ Inbound[G,j] do
        Succ[G,k,e,oldLbl] ← Succ[G,k,e,oldLbl] - 1;
        Succ[G,k,e,newLbl] ← Succ[G,k,e,newLbl] + 1;
    recursive call loop:
    foreach k ∈ Inbound[G,j] do
        Label[G,k,1-e];
```

Figure 4. The algorithm `NMBIG-Lin` for non-monotonic backward induction labelling in linear time.

**Definition 18.** We define the **checkpoints** of the algorithm as all calls to `Label`, and number them $t \in \mathbb{N}$ in computation order. The labelling function $\ell_e^t(v_i)$ is the value of `Ell[G,i,e]` at *checkpoint* $t$, rather than computation step $t$.

**Lemma 19.** At each checkpoint $t \in \mathbb{N}$,

$$\texttt{Succ[G,j,e,a]} = \left|\{k \in \texttt{Outbound[G,j]} \; ; \texttt{Ell[G,k,e] == a}\}\right|.$$

*Proof.* This value is initialised in lines 7 and 8, and is by inspection correct when the main loop is entered at line 9. The value is also altered at lines 29 and 30, and here lines 14 and 27 ensure that the alterations reflect the changes being made to `Ell[G,j,e]`, by the time the recursive call at line 33 is reached. q.e.d.

**Proposition 20.** For each $i \leq N$, the sequence $\langle \ell^t(v_i) \; ; \; t \in \mathbb{N}\rangle$ is $\leq^*$-decreasing in $L^2$. In other words, the sequence $\langle \ell^t_0(v_i) \; ; \; t \in \mathbb{N}\rangle$ is $\leq$-decreasing and the sequence $\langle \ell^t_1(v_i) \; ; t \in \mathbb{N}\rangle$ is $\leq$-decreasing.

*Proof.* Let $t+1$ be the least number such that

$$\ell^t(v_i) <^* \ell^{t+1}(v_i) \qquad (*)$$

for some $i$. The value of the bilabelling can be changed at code lines 4 and 5 of `MAIN`, and line 27 of `Label`. As in the original algorithm, the first two of these are simply initialisation to the top value of the lattice and need not concern us.

The new value at line 27 is the result of the if-cascade from line 15. We can show that a situation as in $(*)$ can only occur at time $t+1$ if it has already occurred at some $s \leq t$, thus that no such least time exists. The proof is by cases.

Suppose $\ell^t_1(v_i) = \mathsf{D}$ and $\ell^{t+1}_1(v_i) = \mathsf{W}$. Since `Ell[G,i,1]` is initialised to `W`, there must be some $s \leq t$ least such that $\ell^s_1(v_i) \neq \mathsf{D}$ and $\ell^{s+1}_1(v_i) = \mathsf{D}$ (that is, between $s$ and $s+1$, $v_i$ gets labelled by line 27). By the if-cascade from line 15, at checkpoint $s$ we have `Succ[G,i,0,L]` $= 0$, but at $t$ `Succ[G,i,0,L]` $> 0$. By Lemma 19 this means that for some $j \in$ `OutBound[G,i]`, $\ell^s_0(v_j) \in \{\mathsf{W},\mathsf{D}\} \mathrel{\overline{<}} \ell^t_0(v_j) = \mathsf{L}$. But then $s < t$ and $\ell^s(v_j) <^* \ell^t(v_j)$, contradicting choice of $t+1$ least.

The cases for $\ell^t_1(v_i) = \mathsf{L}$ and $\ell^{t+1}_1(v_i) \in \{\mathsf{W},\mathsf{D}\}$ are similar. Here the if-cascade guarantees for some $s \leq t$ and $j \in$ `OutBound[G,i]` that $\ell^s_0(v_j) = \mathsf{W} \mathrel{\overline{<}} \ell^t_0(v_j) \in \{\mathsf{L},\mathsf{D}\}$. Again this gives us $s < t$ such that $\ell^s_0(v_j) <^* \ell^t_0(v_j)$, contradicting choice of $t+1$ least.

The three cases for player 0 are symmetrical. First, suppose $\ell^t_0(v_i) = \mathsf{D}$ and $\ell^{t+1}_0(v_i) = \mathsf{L}$. Initialisation of `Ell[G,i,0]` to `L` means that there is least $s \leq t$ such that $\ell^s_0(v_i) \neq \mathsf{D}$ and $\ell^{s+1}_0(v_i) = \mathsf{D}$. This means the labelling at $s+1$ was produced by line 27. By the if-cascade, at $s$ no successor of $v_i$ is 1-labelled with $\mathsf{L}$, and for some $j \in$ `OutBound[G,i]`, $\ell_1^s(v_j) = \mathsf{D}$. But at checkpoint $t$ the if-cascade guarantees that $\ell_1^t(v_j) = \mathsf{W}$, and thus $\ell^s(v_j) <^* \ell^t(v_j)$, contradicting choice of $t+1$ least.

For the next two cases, suppose $\ell_0^t(v_i) = \mathsf{W}$ and $\ell_0^t(v_i) \in \{\mathsf{D}, \mathsf{L}\}$. Initialisation and the if-cascade give us $s \leq t$ and $j \in$ `OutBound[G,i]` such that $\ell_1^s(v_j) = \mathsf{L}$, and $\ell_1^t(v_j) \in \{\mathsf{D}, \mathsf{W}\}$. But then $s < t$ and $\ell^s(v_j) <^* \ell^t(v_j)$, contradicting choice of $t+1$ least. q.e.d.

**Proposition 21.** The procedure `Label` is called at most $2 \cdot |V_G| + 4 \cdot |E_G|$ times.

*Proof.* The main loop at line 9 calls `Label` twice for every vertex. Lines 26 and 31 associate with each recursive call an edge and a relabelling. By Proposition 20 each edge can be used for no more than four relabellings. q.e.d.

**Theorem 22.** The running time of `NMBIG-Lin` is $O(|V_G| + |E_G|)$. If $G$ is connected, this is $O(|E_G|)$.

*Proof.* Apart from the calls to `Label`, `MAIN` walks every vertex and every edge in $G$. We can associate each pass through the loop at line 28 with a recursive call to `Label`. Apart from these recursive calls, `Label` is constant-time. Since `Label` is called $O(|E_G|)$ times, the total running time is $O(|V_G| + |E_G|)$. q.e.d.

**Theorem 23.** If run on the graph $G$, the algorithm `NMBIG-Lin` computes the $\mathsf{A}$-sound bilabelling on $G$.

*Proof.* It is sufficient to show that this algorithm performs the same labellings (in the same order) as `NMBIG`.

First, note that the loop from line 9 only directly relabels vertices with no successors. Such vertices are labelled $\mathsf{D}/\mathsf{L}$, as in `NMBIG`, by the if-cascade. This means that calls to `Label` from `MAIN` have exactly the same effect as in `NMBIG`.

Next, given Lemma 19 it is easy to see that the if-cascade assigns to `newLbl` the same value that line 15 of the `NMBIG` algorithm assigns to `aux`. This ensures that the same recursive calls are made as in `NMBIG`, and since the update loop of line 28 finishes before the recursive labelling loop starts, the results of these calls will also be the same. q.e.d.

# References.


[Fra97] Aviezri S. **Fraenkel**, Combinatorial Game Theory Foundations Applied to Digraph Kernels, **Electronic Journal of Combinatorics** 4 (1997), R10 [The Wilf Festschrift]

[GalSte53] David **Gale** and Frank M. **Stewart**, Infinite Games with Perfect Information, *in:* [KuhTuc53, p. 245–266]

[Gij+76] Vincent W. **Gijlswijk**, Gerard A. P. **Kindervater**, Koos G. J. **van Tubergen**, and Jan J. O. O. **Wiegerinck**, Computer Analysis of E. de Bono's L-Game, Technical Report # 76-18, Department of Mathematics, University of Amsterdam 1976

[GräThoWil02] Erich **Grädel**, Wolfgang **Thomas**, and Thomas **Wilke** (*eds.*), Automata, Logics, and Infinite Games, Springer 2002 [Lecture Notes in Computer Science 2500]

[KuhTuc53] Harold W. **Kuhn** and Albert W. **Tucker** (*eds.*), Contributions to the Theory of Games, volume 2, Princeton University Press 1953 [Annals of Mathematics Studies 28]

[Löw03] Benedikt **Löwe**, Determinacy for Infinite Games with More Than Two Players with Preferences, ILLC Publications PP-2003-19

[Maz02] René **Mazala**, Infinite Games, *in:* [GräThoWil02, p. 23–42]

[Sim99] Stephen G. **Simpson**, Subsystems of Second Order Arithmetic, Springer 1999 [Perspectives in Mathematical Logic]

[Ste76] John R. **Steel**, Determinateness and Subsystems of Analysis, *PhD thesis*, University of California at Berkeley, 1976

[Tan90] Kazuyuki **Tanaka**, Weak Axioms of Determinacy and Subsystems of Analysis I: $\Delta^0_2$ Games, **Zeitschrift für Mathematische Logik und Grundlagen der Mathematik** 36 (1990), p. 481–491



**Received:** October 19th, 2005;
**In revised version:** December 11th, 2006;
**Accepted by the editors:** December 11th, 2006.

FotFS V

Stefan **Bold**, Benedikt **Löwe**,
Thoralf **Räsch**, Johan **van Benthem** (*eds.*)
**Foundations of the Formal Sciences V**
Infinite Games

# Cooperative Games with Infinite Number of Players, Projective Limits and Cores

DIETER DENNEBERG AND GLEB KOSHEVOY*

Fachbereich Mathematik und Informatik
Universität Bremen
Bibliothekstraße 1
28359 Bremen, Germany
denneberg@math.uni-bremen.de

Central Institute of Economics and Mathematics
Russian Academy of Sciences
Nahimovskii prospekt 47
117418 Moscow, Russia
koshevoy@cemi.rssi.ru

ABSTRACT. Let $\mathcal{A}$ be an algebra, not necessary finite, consider the set $\widetilde{\mathcal{A}}$ of all finite subalgebras of $\mathcal{A}$ which is a partially ordered set with an order $\mathcal{A}_1 < \mathcal{A}_2$ if $\mathcal{A}_2$ is a subalgebra of $\mathcal{A}_1$. This allows to consider games on $\mathcal{A}$, and similarly cores of the games or, equivalently, finitely-additive measures, as elements of projective limits. On this way we obtain decomposition theorems for cores of the form of commuting diagrams of some canonical isomorphisms (analogues of Möbius inversion).

## 1 Introduction

Constructing solutions of a game is a central problem in the theory of cooperative and non-cooperative games. The core is a kind of set-valued solution of a game. A drawback of this solution is that the core can be empty. However, a better understanding of the core, even in the case of its emptiness, is of importance for constructing solutions based on the core concept, for example the Shapley value. For a cooperative game with a finite number of

*We thank the referee for remarks and suggestions.

players [DanKos00] obtained a decomposition of the core as the Minkowski difference of the cores of two totally monotone games. The core of a totally monotone game is the Minkowski sum of the cores of simple games (simplices). Thus, in view of this decomposition theorem, even an empty core is represented as Minkowski difference of nonempty cores.

In this paper we obtain a similar decomposition for cooperative games with an infinite number of players. Viewing cooperative games as non-additive measures, our decomposition result is a kind of generalization of Jordan decomposition of charges [Yos65].

In [GilSch95, Mar96] such a problem has been considered for cores consisting of $\sigma$-additive measures. However a nice answer was not obtained since for defining the core the choice of topology is of importance in such a case. Our approach is different. We define cores in the space of finitely-additive measures. Namely, the set of cooperative games with a finite set of players is a finite-dimensional vector space. In case of infinitely many players, we define the corresponding space as a projective limit. Similarly, we define the core as a subset of the projective limit of the space of measures. On this way we consider finitely additive measures as elements of the projective limit of measures on finite subalgebras. Using this approach, we obtain the decomposition theorem in the form of an isomorphism between the corresponding subsets of the projective limits.

## 2 Projective limits

Let $I$ be a partially ordered set and let $\mathcal{C}$ be a category. If we are given an object $X_i$ of the category $\mathcal{C}$ for each $i \in I$ and a morphism $\varphi_{ij} : X_i \to X_j$ of $\mathcal{C}$ for each pair of elements of $I$ with $j \leq i$, and if the conditions

1. $\varphi_{ii} = \mathrm{id}$,
2. $\varphi_{ik} = \varphi_{jk} \circ \varphi_{ij}$, $i \geq j \geq k$,

are satisfied, then we call the system $(X_i, \varphi_{ij})$ a *projective system*. Assume that the Cartesian product $\prod_I X_i$ is an object of the category $\mathcal{C}$. Let $P$ be the subset of the Cartesian product $\prod_I X_i$ defined by

$$P = \{(x_i) \,|\, \varphi_{ij}(x_i) = x_j \; (i \geq j)\}, \tag{1}$$

and let $p_i : P \to X_i$ be the canonical mapping. We call $(P, p_i)$ the *projective limit* of the projective system $(X_i, \varphi_{ij})$ over $I$ and denote it by proj lim $X_i$. In a general category, the projective limit is defined as an object $P \in \mathcal{C}$ and morphisms $p_i : P \to X_i$ satisfying

**P1** $\varphi_{ij} \circ p_i = p_j$, $i \geq j$;

**P2** for any set $X$, and for any system of mappings $q_i : X \to X_i$ satisfying $\varphi_{ij} \circ q_i = q_j$, $i \geq j$, there exists a unique mapping $p : X \to P$ such that $p_i \circ p = q_i$, $i \in I$.

Let $(X_i, \varphi_{ij})$ and $(X'_i, \varphi'_{ij})$ be two projective systems over the same index set $I$, and let $\varphi_i : X_i \to X'_i$, $i \in I$, be morphisms satisfying $\varphi'_{ij} \circ \varphi_i = \varphi_j \circ \varphi_{ij}$, $i \geq j$. Then the system $(\varphi_i)$ is called a *morphism* between the projective systems. If proj lim $X_i$ and proj lim $X'_i$ exist, then $(\varphi_i)$ induces a morphism proj lim $X_i \to$ proj lim $X'_i$ in a natural way. Observe that, if the morphism $(\varphi_i)$ is an isomorphism, then it induces an isomorphism of projective limits.

# 3 Games with an infinite population

Let $(\Omega, \mathcal{A})$ be a *measurable space*, *i.e.*, $\Omega$ is a non-void set and $\mathcal{A}$ is an algebra of subsets of $\Omega$. Recall that a system $\mathcal{A}$ of sets is an *algebra* if $\varnothing \in \mathcal{A}$ and it is closed under complementation and finite union. Then a game is a function $v : \mathcal{A} \to \mathbb{R}$ with $v(\varnothing) = 0$. We understand elements of $\Omega$ as players, elements of $\mathcal{A}$ as admissible coalitions and $v(A)$, $A \in \mathcal{A}$, as the value of coalition $A$.

## 3.1 Games as projective limits

Let us first consider the case of a finite algebra $\mathcal{A}$, *i.e.*, it consists of a finite number of subsets of $\Omega$. In this case, the set of games is the finite dimensional vector space $\mathbb{R}^{\mathcal{A}'}$, $\mathcal{A}' := \mathcal{A} \setminus \{\varnothing\}$ (for simplicity we suppress the prime $'$ in the sequel).

Let $\mathcal{A}$ be an algebra, not necessarily finite. Let us consider the set of all finite subalgebras of $\mathcal{A}$, which will be denoted $\widetilde{\mathcal{A}}$. The set $\widetilde{\mathcal{A}}$ is a partially ordered set with the order $\mathcal{A}_1 \geq \mathcal{A}_2$ if $\mathcal{A}_2$ is a subalgebra of $\mathcal{A}_1$. Letting

$$\varphi_{\mathcal{A}_1,\mathcal{A}_2} : \mathbb{R}^{\mathcal{A}_1} \to \mathbb{R}^{\mathcal{A}_2}, \quad v \mapsto v|_{\mathcal{A}_2},$$

be the natural projection, which restricts a game on $\mathcal{A}_1$ to $\mathcal{A}_2$, then it is easy to check that $(\mathbb{R}^{\mathcal{A}_1}, \varphi_{\mathcal{A}_1,\mathcal{A}_2})$ is a projective system.

A *game* with the sets of players and coalitions $(\Omega, \mathcal{A})$ is an element of proj $\lim_{\widetilde{\mathcal{A}}}(\mathbb{R}^{\mathcal{A}_1}, \varphi_{\mathcal{A}_1,\mathcal{A}_2})$. Note, that if $\mathcal{A}$ is a finite algebra, we have $\mathbb{R}^{\mathcal{A}} \cong$ proj $\lim_{\widetilde{\mathcal{A}}}(\mathbb{R}^{\mathcal{A}_1}, \varphi_{\mathcal{A}_1,\mathcal{A}_2})$.

For establishing the core concept for such games we need an appropriate space of measures. For a finite subalgebra $\mathcal{A}_1$, we can identify measures on $(\Omega, \mathcal{A}_1)$ with additive games. Let $\mathrm{M}(\mathcal{A}_1) \subseteq \mathbb{R}^{\mathcal{A}_1}$ denote this subset of measures. Then we get the set of finitely additive measures on $(\Omega, \mathcal{A})$ as the projective limit proj $\lim_{\widetilde{\mathcal{A}}}(\mathrm{M}(\mathcal{A}_1), \varphi^*_{\mathcal{A}_1,\mathcal{A}_2})$ (any finite decomposition lives in all "sufficiently fine" elements of the projective system). Here $\varphi^*_{\mathcal{A}_1,\mathcal{A}_2}$ : $\mathrm{M}(\mathcal{A}_1) \to \mathrm{M}(\mathcal{A}_2)$ is the restriction $m \mapsto m|_{\mathcal{A}_2} = \varphi_{\mathcal{A}_1,\mathcal{A}_2}(m)$.

### 3.2 Möbius transform

In this subsection the Möbius transform is introduced for our setup. In case of a finite algebra $\mathcal{A}$ there is the well-known Möbius transform $\mu_{\mathcal{A}} : \mathbb{R}^{\mathcal{A}} \to \mathbb{R}^{\mathcal{A}}$, $x \mapsto \mu_{\mathcal{A}}(x)$, defined as

$$\mu_{\mathcal{A}}(x)_A = \sum_{B \subseteq A,\, B \in \mathcal{A}} (-1)^{|A \setminus B|} x_B, \tag{2}$$

where $x_A$ denotes the $A$th coordinate of $x \in \mathbb{R}^{\mathcal{A}}$, $A \in \mathcal{A}$. Observe that the function of two variables $(B, A)$ given by $(-1)^{|A \setminus B|}$ for $B \subseteq A$ and 0 else is the Möbius function of the poset $\mathcal{A}$ endowed with the order given by set inclusion. The Möbius transform is an isomorphism of $\mathbb{R}^{\mathcal{A}}$ onto itself with the zeta transform as inverse,

$$\mu_{\mathcal{A}}^{-1}(y)_A = \sum_{B \subseteq A,\, B \in \mathcal{A}} y_B. \tag{3}$$

Define a mapping $\mu_{\mathcal{A}_1,\mathcal{A}_2} : \mathbb{R}^{\mathcal{A}_1} \to \mathbb{R}^{\mathcal{A}_2}$, $\mathcal{A}_1 \geq \mathcal{A}_2$, by

$$\mu_{\mathcal{A}_1,\mathcal{A}_2} = \mu_{\mathcal{A}_2} \circ \varphi_{\mathcal{A}_1,\mathcal{A}_2} \circ \mu_{\mathcal{A}_1}^{-1}.$$

Because $\mu_{\mathcal{A}_i}$ are isomorphisms, mappings $\mu_{\mathcal{A}_1,\mathcal{A}_2}$ satisfy conditions 1 and 2 in Section 2. Hence, $(\mathbb{R}^{\mathcal{A}_1}, \mu_{\mathcal{A}_1,\mathcal{A}_2})$ comes out as a projective system. Define the *set of Möbius transforms of games* as the projective limit

$$\text{proj lim}_{\tilde{\mathcal{A}}}(\mathbb{R}^{\mathcal{A}_1}, \mu_{\mathcal{A}_1,\mathcal{A}_2}).$$

Because the $\mu_{\mathcal{A}_i}$ are isomorphisms, there is a natural isomorphism

$$\text{proj lim}_{\tilde{\mathcal{A}}}(\mathbb{R}^{\mathcal{A}_1}, \varphi_{\mathcal{A}_1,\mathcal{A}_2}) \cong \text{proj lim}_{\tilde{\mathcal{A}}}(\mathbb{R}^{\mathcal{A}_1}, \mu_{\mathcal{A}_1,\mathcal{A}_2}), \quad v \mapsto \mu(v). \tag{4}$$

This isomorphism generalises the Möbius transform to infinite algebras as one can check that, in case of a finite algebra $\mathcal{A}$, the above construction provides $\text{proj lim}_{\tilde{\mathcal{A}}}(\mathbb{R}^{\mathcal{A}_1}, \varphi_{\mathcal{A}_1,\mathcal{A}_2}) \cong \mathbb{R}^{\mathcal{A}}$, $\text{proj lim}_{\tilde{\mathcal{A}}}(\mathbb{R}^{\mathcal{A}_1}, \mu_{\mathcal{A}_1,\mathcal{A}_2}) \cong \mathbb{R}^{\mathcal{A}}$, and the isomorphism (4) is nothing but (2).

### 3.3 Cores of games

Recall, that for a finite algebra $\mathcal{A}$, the core $C(v)$ of a game $v : \mathcal{A} \to \mathbb{R}$, $v(\emptyset) = 0$, is the set of all $m \in \mathrm{M}(\mathcal{A})$ such that $m(A) \geq v(A)$ for all $A \in \mathcal{A}$ and $m(\Omega) = v(\Omega)$.

Let $\mathcal{A}$ be an arbitrary algebra and pick $v \in \text{proj lim}_{\tilde{\mathcal{A}}}(\mathbb{R}^{\mathcal{A}_1}, \varphi_{\mathcal{A}_1,\mathcal{A}_2})$. Given $\mathcal{A}_1 \geq \mathcal{A}_2$, consider the mapping $\xi_{\mathcal{A}_1,\mathcal{A}_2}(v) : C(p_{\mathcal{A}_1}(v)) \to C(p_{\mathcal{A}_2}(v))$

which is the restriction $m \mapsto m|_{\mathcal{A}_2} = \varphi^*_{\mathcal{A}_1,\mathcal{A}_2}(m)$. Obviously, conditions P1 and P2 are satisfied for $\xi$. Therefore, given $v$, the system

$$(C(p_{\mathcal{A}_1}(v)), \xi_{\mathcal{A}_1,\mathcal{A}_2})(v))$$

is a projective system and we can define

$$C(v) := \text{proj lim}_{\tilde{\mathcal{A}}}(C(p_{\mathcal{A}_1}(v)), \xi_{\mathcal{A}_1,\mathcal{A}_2}(v)).$$

By construction, the set $C(v)$ is contained in the set of finitely additive measures proj $\lim_{\tilde{\mathcal{A}}}(\mathrm{M}(\mathcal{A}_1), \varphi^*_{\mathcal{A}_1,\mathcal{A}_2})$. We call this set $C(v)$ the *core* of the game $v$.

### 3.4 The core of totally monotone games

A game $v$ on $\mathcal{A}$ with non-negative Möbius transform $\mu = \mu_{\mathcal{A}}(v)$ (which is defined for any game via the construction above) is said to be a *totally monotone game*. Suppose in this section that $v$ is totally monotone.

For finite $\mathcal{A}$, denote with $\mathbb{A}$ the set of atoms of the algebra $\mathcal{A}$. We have the following decomposition of the core [DanKos00]

$$C(v) = \bigoplus_{A \in \mathcal{A}} \mu_{\mathcal{A}}(v)_A \, \Delta_A, \tag{5}$$

where $\oplus$ denotes the Minkowski sum of sets, $\Delta_A$ is the face of the unit simplex in $\mathbb{R}^{\mathbb{A}}$ with the set of vertices being the atoms of $\mathcal{A}$ which belong to the set $A$. Compare (5) with (*cf.* (3))

$$v = \sum_{A \in \mathcal{A}} \mu_{\mathcal{A}}(v)_A \, u_A$$

and observe that $\Delta_A$ is the core of the simple game $u_A(B) = 1$ if $B \supset A$ and 0 else on $\mathcal{A}$.

Given a pair $\mathcal{A}_1 \geq \mathcal{A}_2$ of subalgebras, define the projection of the atom spaces $a_{\mathcal{A}_1,\mathcal{A}_2} : \mathbb{R}^{\mathbb{A}_1} \to \mathbb{R}^{\mathbb{A}_2}$ as follows

$$(a_{\mathcal{A}_1,\mathcal{A}_2}(x))_{\mathbf{a}} = \sum_{\mathbf{b} \subseteq \mathbf{a},\, \mathbf{b} \in \mathbb{A}_1} x_{\mathbf{b}}, \quad \mathbf{a} \in \mathbb{A}_2. \tag{6}$$

**Lemma 1.** The mappings $a_{\mathcal{A}_1,\mathcal{A}_2} : \mathbb{R}^{\mathbb{A}_1} \to \mathbb{R}^{\mathbb{A}_2}$, $\mathcal{A}_1 \geq \mathcal{A}_2$, satisfy conditions P1 and P2 in Section 2 and, given $v \in \mathbb{R}^{\mathcal{A}_1}$, apply the core of $v$ to the core of $v|_{\mathcal{A}_2}$,

$$a_{\mathcal{A}_1,\mathcal{A}_2}\left( \bigoplus_{A \in \mathcal{A}_1} \mu_{\mathcal{A}_1}(p_{\mathcal{A}_1}(v))_A \, \Delta_A \right) = \bigoplus_{A \in \mathcal{A}_2} \mu_{\mathcal{A}_2}(p_{\mathcal{A}_2}(v))_A \, \Delta_A. \tag{7}$$

*Proof.* Observe that the mapping $a_{\mathcal{A}_1,\mathcal{A}_2}$ makes commutative the following diagram

$$\begin{array}{ccc} \mathbb{R}^{\mathcal{A}_1} & \overset{\varphi_{\mathcal{A}_1,\mathcal{A}_2}}{\longrightarrow} & \mathbb{R}^{\mathcal{A}_2} \\ \cup\!| & & \cup\!| \\ \mathrm{M}(\mathcal{A}_1) & \overset{\varphi_{\mathcal{A}_1,\mathcal{A}_2}}{\longrightarrow} & \mathrm{M}(\mathcal{A}_2) \\ \updownarrow & & \updownarrow \\ \mathbb{R}^{\mathbb{A}_1} & \overset{a_{\mathcal{A}_1,\mathcal{A}_2}}{\longrightarrow} & \mathbb{R}^{\mathbb{A}_2} \end{array}$$

Hence the map $a_{\mathcal{A}_1,\mathcal{A}_2}$ satisfies conditions P1 and P2. Because $\mathrm{M}(\mathcal{A}_i)$ is isomorphic to $\mathbb{R}^{\mathbb{A}_i}$ the map $\varphi_{\mathcal{A}_1,\mathcal{A}_2}$ provides the required mapping of cores.

q.e.d.

By these properties of the map $a_{\mathcal{A}_1,\mathcal{A}_2}$, given a game $v$, we have a new projective system $\left(\bigoplus_{A\in\mathcal{A}_1} \mu_{\mathcal{A}_1}(p_{\mathcal{A}_1}(v))_A \Delta_A, a_{\mathcal{A}_1,\mathcal{A}_2}\right)$. Now (7) gives us the following isomorphism

$$C(v) \cong \text{proj lim}_{\tilde{\mathcal{A}}} \left( \bigoplus_{A\in\mathcal{A}_1} \mu_{\mathcal{A}_1}(p_{\mathcal{A}_1}(v))_A \, \Delta_A \, , \, a_{\mathcal{A}_1,\mathcal{A}_2} \right). \tag{8}$$

The isomorphism (8) is said to be an *integral decomposition of the core.* This is an analogue of the decomposition (5) of the core via cores of simple games. It could be written

$$C(v) = \int \Delta_A \, d\mu(A),$$

where $\mu = \mu(v)$ is the Möbius transform of the game $v$ defined via the isomorphism (4) and $\int$ denotes the "Minkowski integral". For interpreting $\mu$ as a finitely additive measure, *cf.* [Den97].

### 3.5 Decomposition of the core of arbitrary games

Here we obtain a decomposition of an arbitrary game similarly to (8) for totally monotone games.

Recall, that, for a case of a finite algebra $\mathcal{A}$, there is the following decomposition of the core of a game via cores of simple games [DanKos00]

$$C(v) = \left( \bigoplus_{A\in\mathcal{A},\mu(v)_A>0} \mu(v)_A \Delta_A \right) \ominus \left( \bigoplus_{A\in\mathcal{A},\mu(v)_A<0} (-\mu(v)_A) \, \Delta_A \right),$$

where $S \ominus T$ denotes the Minkowski difference of sets, that is a maximal closed set $W$ such that $T + W \subseteq S$ (note that equality often does not hold).

Because projective limits commute with Minkowski sum and difference and due to the Lemma above, we get the following decomposition

$$C(v) \cong \text{proj}\lim_{\widetilde{\mathcal{A}}} \left( \left( \bigoplus_{\substack{A \in \mathcal{A}_1 \\ \mu_{\mathcal{A}_1}(p_{\mathcal{A}_1}(v))_A > 0}} \mu_{\mathcal{A}_1}(p_{\mathcal{A}_1}(v))_A \, \Delta_A \right) \ominus \left( \bigoplus_{\substack{A \in \mathcal{A}_1 \\ \mu_{\mathcal{A}_1}(p_{\mathcal{A}_1}(v))_A < 0}} (-\mu_{\mathcal{A}_1}(p_{\mathcal{A}_1}(v))_A) \, \Delta_A \right), a_{\mathcal{A}_1,\mathcal{A}_2} \right).$$

# References.


[Cho53] Gustave **Choquet**, Theory of Capacities, **Annales de l'Institute Fourier** 5 (1953), p. 131–295

[DanKos00] Vladimir I. **Danilov** and Gleb A. **Koshevoy**, Cores of Cooperative Games, Superdifferentials of Functions and Minkowski Difference of Sets, **Journal of Mathematical Analysis and Applications** 247 (2000), p. 1–14

[Den97] Dieter **Denneberg**, Representation of the Choquet Integral with the $\sigma$-Additive Möbius Transform, **Fuzzy Sets and Systems** 92 (1997), p. 139–156

[GilSch95] Itzhak **Gilboa** and David **Schmeidler**, Canonical Representation of Set Functions, **Mathematics of Operations Research** 20 (1995), p. 197–212

[Mar96] Massimo **Marinacci**, Decomposition and Representation of Coalitional Games, **Mathematics of Operations Research** 21 (1996), p. 1000–1015

[Yos65] Kōsaku **Yosida**, Functional Analysis, Springer 1965 [Grundlehren der Mathematischen Wissenschaften 123]



**Received**: March 3rd, 2005;
**In revised version**: September 22nd, 2007;
**Accepted by the editors**: September 27th, 2007.

Stefan **Bold**, Benedikt **Löwe**,
Thoralf **Räsch**, Johan **van Benthem** (*eds.*)
**Foundations of the Formal Sciences V**
Infinite Games

# An $\omega$-Power of a Context-Free Language Which is Borel Above $\mathbf{\Delta}^0_\omega$

JACQUES DUPARC AND OLIVIER FINKEL*

Centre romand de Logique, Histoire et Philosophie des Sciences
Université de Lausanne
Chemin de la Colline 12
1015 Lausanne, Switzerland
jacques.duparc@unil.ch

CNRS et École Normale Supérieure de Lyon
Laboratoire de l'Informatique du Parallélisme
Equipe Modèles de Calcul et Complexité
46, Allée d'Italie
69364 Lyon Cedex 07, France
Olivier.Finkel@ens-lyon.fr

ABSTRACT. We use eraser-like basic operations on words to construct a set that is both Borel and above $\mathbf{\Delta}^0_\omega$, built as a set $V^\omega$ where $V$ is a language of finite words accepted by a pushdown automaton. In particular, this gives a first example of an $\omega$-power of a context free language which is a Borel set of infinite rank.

## 1 Preliminaries

Given a set $A$ (called the alphabet) we write $A^*$, and $A^\omega$, for the sets of finite, and infinite words over $A$. We denote the empty word by $\varepsilon$. In order to facilitate the reading, we use $u, v, w$ for finite words, and $x, y, z$ for infinite words. Given two words $u$ and $v$ (respectively, $u$ and $y$), we write $uv$ (respectively,$uy$) for the concatenation of $u$ and $v$ (respectively, of $u$ and $y$). Let $U \subseteq A^*$ and $Y \subseteq A^* \cup A^\omega$, we set $UY := \{uv, uy : u \in X \wedge v, y \in Y\}$.

We recall that, given a language $V \subseteq A^*$, the $\omega$-*power* of this language is

$$V^\omega = \{x = u_1 u_2 \dots u_n \dots \in A^\omega : \ \forall n < \omega \ u_n \in V \backslash \{\varepsilon\}\}.$$

*We wish to thank an anonymous referee for useful comments on a previous version of this paper.

The set $A^\omega$ is equipped with the usual topology, *i.e.*, the product of the discrete topology on the alphabet $A$. So that every open set is of the form $WA^\omega$ for any $W \subseteq A^*$. Or, to say it differently, every closed set is defined as the set of all infinite branches of a tree over $A$. We work within the Borel hierarchy of sets which is the strictly increasing (for inclusion) sequence of classes of sets $(\mathbf{\Sigma}^0_\xi)_{\xi<\omega_1}$, together with the dual classes $(\mathbf{\Pi}^0_\xi)_{\xi<\omega_1}$ and the ambiguous ones $(\mathbf{\Delta}^0_\xi)_{\xi<\omega_1}$, which reports how many operations of countable unions and intersections are necessary to produce a Borel set on the basis of the open ones.

A reduction relation between sets $X$, $Y$ is a partial ordering $X \leq Y$ which expresses that the problem of knowing whether any element $x$ belongs to $X$ is at most as complicated as deciding whether $f(x)$ belongs to $Y$, for some given *simple* function $f$. A very natural reduction relation between sets of infinite words (closely related to reals), has been thoroughly studied by Wadge in the seventies. From the topological point of view, *simple* means continuous, therefore the Wadge ordering compares sets of infinite sequences with respect to their fine topological complexity. Associated with determinacy, this partial ordering becomes a pre-wellordering with anti-chains of length at most two. The so called Wadge Hierarchy it induces incredibly refines the old Borel Hierarchy. Determinacy makes it way through a representation of continuous functions in terms of strategies for player II in a suitable two-player game: the Wadge game $\mathsf{W}(X, Y)$. In this game, players I and II, take turn playing letters of the alphabet corresponding to $X$ for I, and letters of the alphabet corresponding to $Y$ for II. In order to get the right correspondence between a strategy for player II and a continuous function, player II is allowed to skip, whereas I is not. However, II must play infinitely many letters.

As usual, reduction relations induce the notion of a complete set: a set that both belongs to some class, whose members it also reduces. In the context of Wadge reducibility, a set is complete if it belongs to some class closed by inverse image of continuous functions, and reduces everyone of its members. A class which admits a complete set is called a Wadge class. As a matter of fact, all $\mathbf{\Sigma}^0_\xi$ and $\mathbf{\Pi}^0_\xi$ are Wadge classes, whereas $\mathbf{\Delta}^0_\xi$ (for $\xi > 1$) are not.

For instance, the set of all infinite sequences that contains a 1 is $\mathbf{\Sigma}^0_1$-complete, the one that contains infinitely many 1's is $\mathbf{\Pi}^0_2$-complete. As a matter of fact, reaching complete sets for upper levels of the Borel hierarchy, requires other means which we introduce in next sections.

## 2 Erasers

For climbing up along the finite levels of the Borel hierarchy, we use eraser-like moves, *cf.* [Dup01]. For simplicity, imagine a player (either I or II) playing a Wadge game, in charge of a set $X \subseteq A^\omega$, with the extra possibility to delete any terminal part of her last moves.

We recall the definition of the operation $X \mapsto X^\approx$ over sets of infinite words. It was first introduced in [Fin01] by the second author, and is a simple variant of the first author's operation of exponentiation $X \mapsto X^\sim$ which first appeared in [Dup01].

Given a finite alphabet $A$, we write $A^{\leq\omega}$ for $A^* \cup A^\omega$. We take $\twoheadleftarrow \notin A$ and $B = A \cup \{\twoheadleftarrow\}$. For $x \in B^{\leq\omega}$, $x^{\twoheadleftarrow}$ denotes the string $x$, once every $\twoheadleftarrow$ occurring in $x$ has been "evaluated" to the backspace operation (the one familiar to your computer!), proceeding from left to right inside $x$. In other words $x^{\twoheadleftarrow}$ is $x$ after removing all intervals of the form "$a\twoheadleftarrow$" ($a \in A$). By convention, we assume $(u\twoheadleftarrow)^{\twoheadleftarrow}$ is undefined when $u^{\twoheadleftarrow}$ is the empty sequence, *i.e.*, when the last letter $\twoheadleftarrow$ cannot be used as an eraser (because every letter of $A$ in $u$ has already been erased by some eraser $\twoheadleftarrow$ placed in $u$). We remark that the resulting word $x^{\twoheadleftarrow}$ may be finite or infinite. We denote $|v|$ the length of any finite word $v$. If $|v| = 0$ , $v$ is the empty word. If $v = v_1 v_2 \ldots v_k$ where $k \geq 1$ and each $v_i$ is in $A$, then $|v| = k$ and we write $v(i) = v_i$ and $v[i] = v(1)\ldots v(i)$ for $i \leq k$ ; so $v[0] = \varepsilon$. The prefix relation is denoted $\sqsubseteq$: the finite word $u$ is a prefix of the finite word $v$ (denoted $u \sqsubseteq v$) if and only if there exists a (finite) word $w$ such that $v = uw$. the finite word $u$ is a prefix of the $\omega$-word $x$ (denoted $u \sqsubseteq x$) if and only if there exists an $\omega$-word $y$ such that $x = uy$.

**Definition 1.** Let $A$ be any finite alphabet, $\twoheadleftarrow \notin A$, $B = A \cup \{\twoheadleftarrow\}$, and $x \in B^{\leq\omega}$, then $x^{\twoheadleftarrow}$ is inductively defined by:

1. $\varepsilon^{\twoheadleftarrow} = \varepsilon$,
2. $(ua)^{\twoheadleftarrow} = u^{\twoheadleftarrow} a$, if $a \in A$ and $u \in (A \cup \{\twoheadleftarrow\})^*$ is finite,
3. $(u\twoheadleftarrow)^{\twoheadleftarrow} = u^{\twoheadleftarrow}$ with its last letter removed if $|u^{\twoheadleftarrow}| > 0$ and $u \in (A \cup \{\twoheadleftarrow\})^*$ is finite,
4. $(u\twoheadleftarrow)^{\twoheadleftarrow}$ is undefined if $|u^{\twoheadleftarrow}| = 0$ and $u \in (A \cup \{\twoheadleftarrow\})^*$ is finite,
5. $(u)^{\twoheadleftarrow} = \lim_{n\in\omega}(u[n])^{\twoheadleftarrow}$ if $u$ infinite, where, given $\beta_n$ and $v$ in $A^*$, $v \sqsubseteq \lim_{n\in\omega} \beta_n \leftrightarrow \exists n \forall p \geq n \quad \beta_p[|v|] = v$.

For instance, if $n \geq 1$ and $u = (a\twoheadleftarrow)^n$ or $u = (a\twoheadleftarrow)^\omega$, then $(u)^{\twoheadleftarrow} = \varepsilon$; if $u = (ab\twoheadleftarrow)^\omega$ then $(u)^{\twoheadleftarrow} = a^\omega$; if $u = bb(\twoheadleftarrow a)^\omega$ then $(u)^{\twoheadleftarrow} = b$; if $u = \twoheadleftarrow(a\twoheadleftarrow)^\omega$ or $u = a\twoheadleftarrow\twoheadleftarrow a^\omega$ or $u = (a\twoheadleftarrow\twoheadleftarrow)^\omega$ then $(u)^{\twoheadleftarrow}$ is undefined.

**Definition 2.** For $X \subseteq A^\omega$,

$$X^\approx = \{x \in (A \cup \{\twoheadleftarrow\})^\omega : x^{\twoheadleftarrow} \in X\}.$$

The following result easily follows from [Dup01] and was applied in [Fin01, Fin04] to study the $\omega$-powers of finitary context free languages.

**Theorem 3.** Let $n$ be an integer $\geq 2$ and $X \subseteq A^\omega$ be a $\mathbf{\Pi}^0_n$-complete set. Then $X^\approx$ is a $\mathbf{\Pi}^0_{n+1}$-complete subset of $(A \cup \{\twoheadleftarrow\})^\omega$.

Consider the following function

$$f: \ x \in (A \cup \{\twoheadleftarrow\})^\omega \mapsto y \in A^\omega$$

defined by:

- $y = 0^\omega$ if $x^\twoheadleftarrow$ is finite or undefined,
- $y = x^\twoheadleftarrow$ otherwise.

The function $f$ is clearly Borel. In fact, a quick computation shows that the inverse image of any basic clopen set is Borel of low finite rank.

Let $X$ be any subset of the Cantor space $\{0,1\}^\omega$, and $f$ as above. If $0^\omega \notin X$, then for any $x \in \{0, 1, \twoheadleftarrow\}^\omega$

$$x \in X^\approx \iff f(x) \in X.$$

In other words, $X^\approx = f^{-1}X$. In particular, if $X$ is Borel, so is $X^\approx$.

## 3 Increasing sequences of erasers

The following construction has been partly used by the second author in [Fin04] to construct a Borel set of infinite rank which is an $\omega$-power, *i.e.*, in the form $V^\omega$, where $V$ is a set of finite words over a finite alphabet $\Sigma$. We iterate the operation $X \mapsto X^\approx$ finitely many times, and take the limit. The following definition makes this more precise:

**Definition 4.** Given any set $X \subseteq A^\omega$:

- $X_k^{\approx 0} = X$, $X_k^{\approx 1} = X^\approx$, $X_k^{\approx 2} = (X_k^{\approx 1})^\approx$,
- $X_k^{\approx (k)} = (X_k^{\approx (k-1)})^\approx$, where we apply $k$ times the operation $X \mapsto X^\approx$ with different new letters $\twoheadleftarrow_k$, $\twoheadleftarrow_{k-1}$, $\ldots$, $\twoheadleftarrow_2$, $\twoheadleftarrow_1$, in such a way that we have $X_k^{\approx 0} = X \subseteq A^\omega$, $X_k^{\approx 1} \subseteq (A \cup \{\twoheadleftarrow_k\})^\omega$, $X_k^{\approx 2} \subseteq (A \cup \{\twoheadleftarrow_k, \twoheadleftarrow_{k-1}\})^\omega$, and $X_k^{\approx (k)} \subseteq (A \cup \{\twoheadleftarrow_1, \twoheadleftarrow_2, \ldots, \twoheadleftarrow_k\})^\omega$.
- We set $X^{\approx (k)} = X_k^{\approx (k)}$.

We define $X^{\approx \infty} \subseteq (A \cup \{\twoheadleftarrow_n : \ 0 < n < \omega\})^\omega$ by saying that $x \in X$ if and only if

1. for each integer $n$, $x_n = x^{\twoheadleftarrow_1 \cdots \twoheadleftarrow_{n-1} \twoheadleftarrow_n}$ is defined and infinite, and

2. $x_\infty = \lim_{n<\omega} x_n$ is defined, infinite, and belongs to $X$.

Consider the following sequence of functions:

- $f_0(x) = x$ ($f_0$ is the identity),
- $f_{k+1} : (A \cup \{\twoheadleftarrow_n : k < n < \omega\})^\omega \longmapsto (A \cup \{\twoheadleftarrow_n : k+1 < n < \omega\})^\omega$ defined by:
  - $f_{k+1}(x) = x^{\twoheadleftarrow_{k+1}}$ if $x^{\twoheadleftarrow_{k+1}}$ is infinite,
  - $f_{k+1}(x) = 0^\omega$ if $x^{\twoheadleftarrow_{k+1}}$ is finite or undefined,

By induction on $k$, one shows that every function $f_k$ is Borel — and even Borel of finite rank. Moreover, since Borel functions are closed under taking the limits [Kur61], the function

$$f_\infty : (A \cup \{\twoheadleftarrow_n : 0 < n < \omega\})^\omega \longmapsto A^\omega$$

is Borel which is defined by

- $f_\infty(x) = \lim_{n<\omega} f_n(x)$ if $\lim_{n<\omega} f_n(x)$ is defined and infinite,
- $f_\infty(x) = 0^\omega$ otherwise.

**Remark 5.** Let $X \subset \{0,1\}^\omega$ with $0^\omega \notin X$, then we have for any element $x \in (\{0,1\} \cup \{\twoheadleftarrow_n : 0 < n < \omega\})^\omega$

$$x \in X^{\approx\infty} \Longleftrightarrow f_\infty(x) \in X.$$

In other words, $X^{\approx\infty} = {f_\infty}^{-1}(X)$, which shows that whenever $X$ is Borel, $X^{\approx\infty}$ is Borel too.

In fact, with tools described in [Dup01], and [Dup$\infty$], it is possible to show that given any $\mathbf{\Pi}^0_1$-complete set $Y$, the set $Y^{\approx\infty}$ belongs to $\mathbf{\Pi}^0_{\omega+2}$. If $X$ is the set of infinite words over the alphabet $\{0,1\}$ which contains an infinite number of 1s, then it is also possible to show that $X^{\approx\infty}$ is Borel by completely different methods involving decompositions of $\omega$-powers [FinSim03, Fin04].

**Proposition 6.** Let $X$ be the set of infinite words over $\{0,1\}$ that contain infinitely many 1's,

$$X^{\approx\infty} \in \mathbf{\Delta}^1_1 \setminus \mathbf{\Delta}^0_\omega.$$

*Proof.* The fact $X^{\approx\infty}$ is Borel is Remark 5. As for $X^{\approx\infty} \notin \mathbf{\Delta}^0_\omega$, it is a consequence of the fact that the operation $Y \longmapsto Y^{\approx}$ is strictly increasing (for the Wadge ordering) inside $\mathbf{\Delta}^0_\omega$ (*cf.* [Dup01, Dup$\infty$]). In other words, for any $Y \in \mathbf{\Delta}^0_\omega$ the relation $Y <_W Y^{\approx}$ holds ($<_W$ stands for the strict Wadge ordering). But, as a matter of fact, $(X^{\approx\infty})^{\approx} \leq_W X^{\approx\infty}$ holds which forbids $X^{\approx\infty}$ to belong to $\mathbf{\Delta}^0_\omega$.

Indeed, to see that $(X^{\approx\infty})^{\approx} \leq_W X^{\approx\infty}$ holds, it is enough to describe a winning strategy for player II in the Wadge game $W\left((X^{\approx\infty})^{\approx}, X^{\approx\infty}\right)$. In this game, player II uses $\omega$ many different erasers: $\twoheadleftarrow_1, \twoheadleftarrow_2, \twoheadleftarrow_3, \ldots$ whose strength is opposite to their indices ($\twoheadleftarrow_k$ erases all erasers $\twoheadleftarrow_j$ for any $j > k$ but no $\twoheadleftarrow_i$ for $i \leq k$). While player I uses the same erasers as player II does, plus an extra one ($\twoheadleftarrow$) which is stronger than all the other ones.

The winning strategy for II derives from ordinal arithmetic: $1 + \omega = \omega$. It consists in copying I's run with a shift on the indices of erasers:

- if I plays a letter 0 or 1, then II plays the same letter,
- if I plays an eraser $\twoheadleftarrow_n$, II plays the eraser $\twoheadleftarrow_{n+1}$.
- if I plays the eraser $\twoheadleftarrow$ (the first one that will be taken into account when the erasing process starts), then II plays $\twoheadleftarrow_0$.

This strategy is clearly winning. q.e.d.

## 4 Simulating $X^{\approx\infty}$ by the $\omega$-power of a context-free language

It was already known that there exists an $\omega$-power of a finitary language which is Borel of infinite rank [Fin04]. But the question was left open whether such a finitary language could be *context free*.

This article provides effectively a context free language $V$ such that $V^\omega$ is a Borel set of infinite rank, and uses infinite Wadge games to show that this $\omega$-power $V^\omega$ is located above $\mathbf{\Delta}^0_\omega$ in the Borel hierarchy.

The idea is to have $X^{\approx\infty}$, where $X$ stands for the set of all infinite words over $\{0,1\}$ that contain infinitely many 1s to be of the form $V^\omega$ for some language $V$ recognized by a (non-deterministic) Pushdown Automaton. We first recall the notion of pushdown automaton [Ber79, AutBerBoa97].

**Definition 7.** A pushdown automaton is a 7-tuple

$$M = (Q, A, \Gamma, \delta, q_0, Z_0, F)$$

where $Q$ is a finite set of states, $A$ is a finite input alphabet, $\Gamma$ is a finite pushdown alphabet, $q_0 \in Q$ is the initial state, $Z_0 \in \Gamma$ is the start symbol, $\delta$ is a mapping from $Q \times (A \cup \{\varepsilon\}) \times \Gamma$ to finite subsets of $Q \times \Gamma^*$, and $F \subseteq Q$ is the set of final states.

If $\gamma \in \Gamma^+$ describes the pushdown store content, the leftmost symbol of $\gamma$ will be assumed to be on "top" of the store. A configuration of a pushdown automaton is a pair $(q, \gamma)$ where $q \in Q$ and $\gamma \in \Gamma^*$.

For $a \in A \cup \{\varepsilon\}$, $\gamma, \beta \in \Gamma^*$ and $Z \in \Gamma$, if $(p, \beta)$ is in $\delta(q, a, Z)$, then we write $a : (q, Z\gamma) \mapsto_M (p, \beta\gamma)$. We denote by $\mapsto^*_M$ the transitive and reflexive closure of $\mapsto_M$.

Let $u = a_1 a_2 \ldots a_n$ be a finite word over $A$. A finite sequence of configurations $r = (q_i, \gamma_i)_{1 \leq i \leq p}$ is called a run of $M$ on $u$, starting in configuration $(p, \gamma)$, if and only if

1. $(q_1, \gamma_1) = (p, \gamma)$, and
2. for each $i$, $1 \leq i \leq p - 1$, there exists $b_i \in A \cup \{\varepsilon\}$ satisfying $b_i : (q_i, \gamma_i) \mapsto_M (q_{i+1}, \gamma_{i+1})$ such that $a_1 a_2 \ldots a_n = b_1 b_2 \ldots b_{p-1}$.

This run is simply called a run of $M$ on $u$ if it starts from configuration $(q_0, Z_0)$.

Finally, the language accepted by a pushdown automaton $M$ is the set of words $L(M) = \{u \in A^*$: there is a run $r$ of $M$ on $u$ ending in a final state$\}$.

For instance, the set $0^*1 \subset \{0, 1\}^*$ is trivially context-free.

**Proposition 8 (Finkel).** Let $L_n$ be the maximal subset of the set of finite sequences with $n$ different erasers $\{0, 1, \twoheadleftarrow_1, \twoheadleftarrow_2, \ldots, \twoheadleftarrow_n\}^*$ such that

$$L_n^{\twoheadleftarrow_1 {}^{\twoheadleftarrow_2 {}^{\cdots \twoheadleftarrow_n}}} = 0^*1.$$

Then $L_n$ is context-free.

This was first noticed by the second author in [Fin01]. To be more precise, by $u \in L_n$ we mean: we start with some $u$, then we evaluate $\twoheadleftarrow_1$ as an eraser, and obtain $u_1$ (providing that we must never use $\twoheadleftarrow_1$ to erase the empty sequence, *i.e.*, every occurrence of a $\twoheadleftarrow_1$ symbol does erase a letter 0 or 1 or an eraser $\twoheadleftarrow_i$ for $i > 1$). Then we start again with $u_1$, this time we evaluate $\twoheadleftarrow_2$ as an eraser, which yields $u_2$, and so on. When there is no more symbol $\twoheadleftarrow_i$ to be evaluated, we are left with $u_n \in \{0, 1\}^*$. We define $u \in L_n$ if and only if $u_n \in 0^*1$.

To make a pushdown automaton recognize $L_n$, the idea is to have it guess (non-deterministically), for each single letter that it reads, whether this letter will be erased later or not. Moreover, the pushdown automaton should also guess for each eraser it encounters, whether this eraser should be used as an eraser or whether it should not — for the only reason that it will be erased later on by a *stronger* eraser. During the reading, the stack should be used to accumulate all pendant guesses, in order to verify later on that they are fulfilled.

We would very much like to prove that $L_\infty = \bigcup_{n<\omega} L_n$ is context-free. Unfortunately, we cannot get such a result. However, we are able to show that a slightly more complicated set (strictly containing $L_\infty$) is indeed context-free.

Of course, the first problem that comes to mind when working with $L_\infty$, is to handle $\omega$ many different erasers with a finite alphabet. This implies that erasers must be coded by finite words. This was done by the second author in [Fin03b]. Roughly speaking, the eraser $\twoheadleftarrow_n$ is coded by the word $\alpha B^n C^n D^n E^n \beta$ with new letters $\alpha, B, C, D, E, \beta$. It is a little bit tricky, but the pushdown automaton must really be able to read the number $n$ identifying the eraser four times.

The very definition of the sets $L_n$, requires the erasing operations to be executed in an increasing order: in a word that contains only the erasers $\twoheadleftarrow_1, \ldots, \twoheadleftarrow_n$, one must consider first the eraser $\twoheadleftarrow_1$, then $\twoheadleftarrow_2$, and so on. Therefore this erasing process satisfy the following properties:

(a) An eraser $\twoheadleftarrow_j$ may only erase letters $c \in \{0, 1\}$ or erasers $\twoheadleftarrow_k$ with $k > j$.

(b) Assume that in a word $u \in L_n$, there is a sequence $cvw$ where $c$ is either in $\{0, 1\}$ or in the set $\{\twoheadleftarrow_1, \ldots, \twoheadleftarrow_{n-1}\}$, and $w$ is (the code of) an eraser $\twoheadleftarrow_k$ which erases $c$ once the erasing process is achieved. If there is in $v$ (the code of) an eraser $\twoheadleftarrow_j$ which erases $e$, where $e \in \{0, 1\}$ or $e$ is (the code of) another eraser, then $e$ must belong to $v$ (it is between $c$ and $w$ in the word $u$) ; moreover the erasing by the eraser $\twoheadleftarrow_j$ has been achieved before the other one with the eraser $\twoheadleftarrow_k$. This implies $j \leq k$ and hence $k \geq \max\{j : \text{an eraser } \twoheadleftarrow_j \text{ was used inside } v\}$.

The essential difference with the case studied in [Fin03b] is that here an eraser $\twoheadleftarrow_j$ may only erase letters 0 or 1 or erasers $\twoheadleftarrow_k$ for $k > j$, while in [Fin03b] an eraser $\twoheadleftarrow_j$ was assumed to be only able to erase letters 0 or 1 or erasers $\twoheadleftarrow_k$ for $k < j$. So the above inequality was replaced by $k \leq \min\{j :$ an eraser $\twoheadleftarrow_j$ was used inside $v\}$.

However, with a slight modification, we can construct a pushdown automaton $\mathcal{B}$ which, among words where letters $\alpha, \beta, B, C, D, E$ are only used to code erasers of the form $\twoheadleftarrow_j$, accepts exactly the words which belong to the language $L_\infty$. We now explain the behavior of this pushdown automaton. (For simplicity, we sometimes talk about the eraser $\twoheadleftarrow_j$ instead of its code $\alpha B^j C^j D^j E^j \beta$.)

Assume that $\mathcal{A}$ is a finite automaton accepting (by final state) the finitary language $0^*1$ over the alphabet $A = \{0, 1\}$. We can informally describe the behavior of the pushdown automaton $\mathcal{B}$ when reading a word $u$ such that the letters $\alpha, B, C, D, E, \beta$ are only used in $u$ to code the erasers $\twoheadleftarrow_j$ for $1 \leq j$.

The automaton $\mathcal{B}$ simulates the automaton $\mathcal{A}$ until it guesses (non-deterministically) that it begins to read a segment $w$ which contains erasers which really erase and some letters of $A$ or some other erasers which are erased when the operations of erasing are achieved in $u$.

Then, still non-deterministically, when $\mathcal{B}$ reads a letter $c \in A$ it may guess that this letter will be erased and push it in the pushdown store, keeping in memory the current state of the automaton $\mathcal{A}$.

In a similar manner when $\mathcal{B}$ reads the code $\twoheadleftarrow_j = \alpha B^j C^j D^j E^j \beta$, it may guess that this eraser will be erased (by another eraser $\twoheadleftarrow_k$ with $k < j$) and then may push in the store the finite word $\gamma E^j \nu$, where $\gamma$, $E$, $\nu$ are in the pushdown alphabet of $\mathcal{B}$.

But $\mathcal{B}$ may also guess that the eraser $\twoheadleftarrow_j = \alpha B^j C^j D^j E^j \beta$ will really be used as an eraser. If it guesses that the code of $\twoheadleftarrow_j$ will be used as an eraser, $\mathcal{B}$ has to pop from the top of the pushdown store either a letter $c \in A$ or the code $\gamma E^i.\nu$ of another eraser $\twoheadleftarrow_i$, with $i > j$, which is erased by $\twoheadleftarrow_j$.

In this case, it is easy for $\mathcal{B}$ to check whether $i > j$ when reading the initial segment $\alpha B^j$ of $\twoheadleftarrow_j$.

But as we remarked in $(b)$, the pushdown automaton $\mathcal{B}$ must also check that the integer $j$ is greater than or equal to every integer $p$ such that an eraser $\twoheadleftarrow_p$ has been used since the letter $c \in A$ or the code $\gamma E^i.\nu$ was pushed in the store. Then, after having pushed some letter $t \in A$ or the code $t = \gamma E^i.\nu$ of an eraser in the pushdown store, and before popping it from the top of the stack, $\mathcal{B}$ must keep track of the integer $k = \max\{p :$ some eraser $\twoheadleftarrow_p$ has been used since $t$ was pushed in the stack$\}$ in the memory stack.
For that purpose $\mathcal{B}$ pushes the finite word $L_2 S^k L_1$ in the pushdown store ($L_1$ is pushed first, then $S^k$ and the letter $L_2$), with $L_1, L_2$ and $S$ are new letters added to the pushdown alphabet.

So, when $\mathcal{B}$ guesses that $\twoheadleftarrow_j = \alpha B^j C^j D^j E^j \beta$ will be really used as an eraser, there is on top of the stack either a letter $c \in A$ or a code $\gamma E^i.\nu$ of an eraser which will be erased or a code $L_2 S^k L_1$. The behavior of $\mathcal{B}$ is then as follows.

Assume first there is a code $L_2 S^k L_1$ on top of the stack. Then $\mathcal{B}$ firstly checks that $j \geq k$ holds by reading the segment $\alpha B^j C$ of the eraser $\alpha B^j C^j D^j E^j \beta$.

If $j \geq k$ holds, then using $\varepsilon$-transitions, $\mathcal{B}$ completely pops the word $L_2 S^k L_1$ from the top of the stack. ($\mathcal{B}$ has already checked it is allowed to use the eraser $\twoheadleftarrow_j$).

Then, in each case, the top of the stack contains either a letter $c \in A$, or the code $\gamma E^i \nu$ of an eraser which should be erased later. $\mathcal{B}$ pops this letter $c$ or the code $\gamma E^i.\nu$ ( having checked that $j < i$ after reading the segment

$\alpha B^j C^j$ of the eraser $\alpha B^j C^j D^j E^j \beta$ ).

At this point, we must have a look at the top stack symbols. There are three cases:

1. The top stack symbol is the bottom symbol $Z_0$. In which case, the pushdown automaton $\mathcal{B}$, after having completely read the eraser $\twoheadleftarrow_j$, may pursue the simulation of the automaton $\mathcal{A}$ or guess that it begins to read another segment $v$ which will be erased. Hence the next letter $c \in A$ or the next code $\alpha B^m.C^m.D^m.E^m.\beta$ of the word will be erased. Then $\mathcal{B}$ pushes the letter $c \in A$ or the code $\gamma E^m.\nu$ of $\twoheadleftarrow_m$ in the pushdown store.
2. If the top stack symbol is either a letter $c' \in A$ or a code $\gamma E^m.\nu$, then $\mathcal{B}$ pushes the code $L_2 S^j L_1$ in the pushdown store ( $j$ is then the maximum of the set of integers $p$ such that an eraser $\twoheadleftarrow_p$ has been used since the letter $c'$ or the code $\gamma E^m.\nu$ has been pushed into the stack).
3. If the top stack symbols are a code $L_2 S^l L_1$, then the pushdown automaton $\mathcal{B}$ must compare the integers $j$ and $l$, and replace $L_2 S^l L_1$ by $L_2 S^j L_1$ in case $j > l$. $\mathcal{B}$ achieves this task while reading the segment $D^j E^j \beta$ of the eraser $\alpha B^j C^j D^j E^j \beta$.
   The pushdown automaton $\mathcal{B}$ pops a letter $S$ for each letter $D$ it reads. Then it checks whether $j \geq l$ is satisfied.
   If $j \geq l$ then it pushes $L_2 S^j L_1$ while it reads the segment $E^j \beta$ of the eraser $\twoheadleftarrow_j$.
   In case $j < l$, after it reads $D^j$, the part $S^{l-j} L_1$ of the code $L_2 S^l L_1$ remains in the stack. The pushdown automaton then pushes again $j$ letters $S$ and a letter $L_2$ while reading $E^j \beta$.

When again the stack only contains $Z_0$ (the initial stack symbol), $\mathcal{B}$ resumes the simulation of the automaton $\mathcal{A}$ or it guesses that it begins to read a new segment which will be erased later.

We are confronted with the fact $\mathcal{B}$ will also accept some words where the letters $\alpha, \beta, B, C, D, E$ are not used to code erasers. How can we make sure that this pushdown automaton is not misled by such wrong codes of erasers?

## 5 Wrong codes of erasers and the right $\omega$-power

In fact, one cannot make sure that a pushdown automaton notices the discrepancy between right codes of the form $\alpha B^j C^j D^j E^j \beta$ and wrong ones (of the form $\alpha B^b C^c D^d E^e \beta$ where $b, c, d, e$ are not all the same integer for instance). However, there is a satisfactory solution: instead of having a pushdown automaton reject these wrong codes, simply let it accept all of them. Accepting a word if it contains a wrong code of an eraser is trivial for a non-deterministic pushdown automaton. So instead of a pushdown automaton $\mathcal{B}$ that accepts precisely $L_\infty$ (up to the coding of erasers), we let $W$ stands for the set of all finite words which host a wrong code. In the following, $L_\infty$ really is $L_\infty$ where erasers are replaced by their correct codes, and $\mathcal{L}(\mathcal{B})$ is the language recognized by $\mathcal{B}$.

**Proposition 9.** There exists a pushdown automaton $\mathcal{B}$ such that $\mathcal{L}(\mathcal{B}) = L_\infty \cup W$.

Now everything is ready for the main result.

**Theorem 10.** The $\omega$-power $Y = \mathcal{L}(\mathcal{B})^\omega$ of the context-free language $L(\mathcal{B})$ described above satisfies $Y \in \mathbf{\Delta}^1_1 \backslash \mathbf{\Delta}^0_\omega$.

*Proof.* To begin with, the set $Y$ is the disjoint union of three different sets: $Y = Y_0 \cup Y_\infty \cup Y_*$, where $Y_0$ is the set of all infinite sequences in $Y$ with no wrong code in them, $Y_\infty$ the set of all infinite sequences with infinitely many wrong codes, and $Y_*$ the set of infinite sequences with finitely many wrong codes (at least one). We remark that:

- $Y_0$ is Wadge equivalent to the set $X^{\approx\infty}$ as defined in 6, *i.e.*, the set of all $\omega$-words that, after taking care of the erasing process, ultimately reduce to words with infinitely many 1s. To be more precise, it is this very same set up to a renaming of the erasers. So $Y_0$ belongs to $\mathbf{\Delta}^1_1$.
- $Y_\infty$ is Wadge equivalent to $X$, so it is $\mathbf{\Pi}^0_2$-complete.
- $Y_*$ is more complicated. However, it is of the form $Y_* = WY_0$, where $W$ is the set of all finite words with at least an occurrence of a wrong code. So $Y_*$ is a countable union of sets, each of which is Wadge equivalent to $Y_0$. Hence, $Y_*$ is a countable union of Borel sets, therefore $Y_*$ is Borel too.

All three cases put together show that $Y = Y_0 \cup Y_\infty \cup Y_*$, is a finite union of Borel sets, hence it Borel too.

It remains to prove that $Y \notin \mathbf{\Delta}^0_\omega$. This, in fact, is immediate from Proposition 6 which stated that $X^{\approx\infty} \notin \mathbf{\Delta}^0_\omega$. Because there is an obvious winning strategy for player II in the Wadge game $\mathsf{W}(X^{\approx\infty}, Y)$. It consists in never playing a wrong code, and copying I's run up to the renaming of the erasers. Since $X^{\approx\infty}$ is clearly Wadge equivalent to $Y_0$ this strategy works perfectly well and shows that $Y \notin \mathbf{\Delta}^0_\omega$. q.e.d.

This quick study gives an example of how an infinite game theoretical approach leads to intriguing results in Theoretical Computer Science. On one hand, the notion of erasers is highly related to the dynamic behavior of players in games. And, on the other hand, non-determinism provides very effective ways to deal with the erasing process. So, all together, they afford a method for describing (topological) complexity of very effective sets of reals.

## References.


[AutBerBoa97] Jean-Michel **Autebert**, Jean **Berstel**, and Luc **Boasson**, Context Free Languages and Pushdown Automata, *in:* [RozSal97a, p. 111–174]

[Ber79] Jean **Berstel**, Transductions and Context Free Languages, Teubner 1979 [Studienbücher Informatik]

[dBadRoRoz94] Jaco W. **de Bakker**, Willem P. **de Roever**, and Grzegorz **Rozenberg**, A Decade of Concurrency, Springer 1994 [Lecture Notes In Computer Science 803]

[Dup01] Jacques **Duparc**, Wadge Hierarchy and Veblen Hierarchy, Part 1: Borel Sets of Finite Rank, **Journal of Symbolic Logic** 66 (2001), p. 56–86

[Dup$\infty$] Jacques **Duparc**, Wadge Hierarchy and Veblen Hierarchy, Part 2: Borel Sets of Infinite Rank, *submitted to:* **Journal of Symbolic Logic**

[Fin01] Olivier **Finkel**, Topological Properties of Omega Context Free Languages, **Theoretical Computer Science** 262 (2001), p. 669–697

[Fin03a] Olivier **Finkel**, Borel Hierarchy and Omega Context Free Languages, **Theoretical Computer Science** 290 (2003), p. 1385–1405

[Fin03b] Olivier **Finkel**, On Omega Context Free Languages Which are Borel Sets of Infinite Rank, **Theoretical Computer Science** 299 (2003), p. 327–346

[Fin04] Olivier **Finkel**, An $\omega$-Power of a Finitary Language Which is a Borel Set of Infinite Rank, **Fundamenta Informaticae** 62 (2004), p. 333–342

[FinSim03] Olivier **Finkel** and Pierre **Simonnet**, Topology and Ambiguity in Omega Context Free Languages, **Bulletin of the Belgian Mathematical Society** 10 (2003), p. 707–722

[HopUll69] John E. **Hopcroft** and Jeffrey D. **Ullman**, Formal Languages and Their Relation to Automata, Addison-Wesley 1969

[Kec94] Alexander S. **Kechris**, Classical Descriptive Set Theory, Springer 1994 [Graduate Texts in Mathematics 156]

[KecMarMos83] Alexander S. **Kechris**, Donald A. **Martin**, and Yiannis N. **Moschovakis** (*eds.*), Cabal Seminar 79-81, Proceedings of the Caltech-UCLA Logic Seminar 19769-1981, Springer 1983 [Lecture Notes in Mathematics 1019]

[KecMos78] Alexander S. **Kechris** and Yiannis N. **Moschovakis** (*eds.*), Cabal Seminar 76-77, Proceedings of the Caltech-UCLA Logic Seminar 1976-1977, Springer 1978 [Lecture Notes in Mathematics 689]

[Kur61] Casimir **Kuratowski**, Topologie I et II, tome I, 4e édition, 1958 et tome II, 3e édition, 1961, reprint, Editions Jacques Gabay 1992

[Lec01] Dominique **Lecomte**, Sur les ensembles de phrases infinies constructibles á partir d'un dictionnaire sur un alphabet fini, **Séminaire d'Initiation a l'Analyse** 1 (2001)

[Lec05] Dominique **Lecomte**, Omega-Powers and Descriptive Set Theory, **Journal of Symbolic Logic** 70 (2005), p. 1210–1232

[LesTho94] Helmut **Lescow** and Wolfgang **Thomas**, Logical Specifications of Infinite Computations, *in:* [dBadRoRoz94, p. 583–621]

[Lou83] Alain **Louveau**, Some Results in the Wadge Hierarchy of Borel Sets, *in:* [KecMarMos83, p. 28–55]

[Mar75] Donald A. **Martin**, Borel Determinacy, **Annals of Mathematics** 102 (1975), p. 363–371

[Mos80] Yiannis N. **Moschovakis**, Descriptive Set Theory, North Holland 1980 [Studies in Logic and the Foundations of Mathematics 100]

[Niw90] Damian **Niwinski**, Problem on $\omega$-Powers, *in:* [Tho92, p. 24]

[PauSal97] Gheorghe **Paun** and Arto **Salomaa** (*eds.*), New Trends in Formal Languages: Control, Coperation, and Combinatorics, Springer 1997 [Lecture Notes in Computer Science 1218]

[PerPin04] Dominique **Perrin** and Jean-Éric **Pin** (*eds.*), Infinite Words: Automata, Semigroups, Logic and Games, Elsevier 2004

[RozSal97a] Grzegorz **Rozenberg** and Arto **Salomaa** (*eds.*), Handbook of Formal Languages, volume 1, Word Language Grammar, Springer 1997

[RozSal97b] Grzegorz **Rozenberg** and Arto **Salomaa** (*eds.*), Handbook of Formal Languages, volume 3, Beyond Words, Springer 1997

[Sim92] Pierre **Simonnet**, Automates et Théorie Descriptive, *PhD thesis*, Université Paris 7, 1992

[Sta86] Ludwig **Staiger**, Hierarchies of Recursive $\omega$-Languages, **Journal of Information Processing and Cybernetics EIK** 22 (1986), p. 219–241

[Sta97a] Ludwig **Staiger**, $\omega$-Languages, *in:* [RozSal97b, p. 339–388]

[Sta97b] Ludwig **Staiger**, On $\omega$-Power Languages, *in:* [PauSal97, p. 377–393]

[Tho90] Wolfgang **Thomas**, Automata on Infinite Objects, *in:* [vLe90, p. 133–191]

[Tho92] Wolfgang **Thomas** (*ed.*), Proceedings of the ASMICS-Workshop "Logics and Recognizable Sets", Dersau, October 8-10, 1990, Bericht 9104, Institut für Informatik und Praktische Mathematik, Christian-Albrechts-Universität zu Kiel, Germany, 1992

[vLe90] Jan **van Leeuwen** (*ed.*), Handbook of Theoretical Computer Science, volume B, Elsevier 1990

[Van78] Robert A. **Van Wesep**, Wadge Degrees and Descriptive Set Theory, *in:* [KecMos78, p. 151-170]

[Veb08] Oswald **Veblen**, Continuous Increasing Functions of Finite and Transfinite Ordinals, **Transactions of the American Mathematical Society** 9 (1908), p. 280–292

[Wad72] William W. **Wadge**, Degrees of Complexity of Subsets of the Baire Space, **Notices of the American Mathematical Society** 19 (1972), p. 714–715

[Wad83] William W. **Wadge**, Reducibility and Determinateness on the Baire Space, *PhD thesis*, University of California at Berkeley, 1983

**Received**: March 5th, 2005;
**In revised version**: November 17th, 2006;
**Accepted by the editors**: June 7th, 2007.

Stefan **Bold**, Benedikt **Löwe**,
Thoralf **Räsch**, Johan **van Benthem** (*eds.*)
**Foundations of the Formal Sciences V**
Infinite Games

# A Playful Approach to Silver and Mathias Forcings

LORENZ HALBEISEN*

Institut für Informatik und angewandte Mathematik
Universität Bern
Neubrückstrasse 10
3012 Bern, Switzerland
halbeis@iam.unibe.ch

ABSTRACT. Forcing is a method to extend models of Set Theory in order to get independence or at least consistency results. For generalized Silver and Mathias forcings it is shown how infinite games between two players, say Death and the Maiden, and in particular the absence of a winning strategy for the Maiden, can be used to predict combinatorial properties of the extended model. For example it is shown that Mathias forcing restricted to certain game families adds dominating reals, has pure decision, and does not add Cohen reals, and that Silver forcing restricted to some weaker game families does not add unbounded reals, adds splitting reals, and is minimal.

## Outline

The aim of this paper is to show how infinite games can be used in the investigation of forcing extensions of models of Set Theory. In particular two types of forcing notions are considered, namely Mathias and Silver forcings, and it is shown how infinite two-player games, especially the absence of a winning strategy for one of the players, can be used to predict combinatorial properties of the corresponding extended models.

Our system of set-theoretic axioms includes the axioms of Zermelo and Fraenkel as well as the Axiom of Choice. This system is usually denoted ZFC. All our set-theoretic notations and definitions are standard and can be found in textbooks such as [Jec03], [Kun83] or [BarJud95]. A brief

*I would like to thank the referee for valuable comments on a former version of this paper (which led to the discussion of happy families in Section 6).

introduction to the forcing technique can be found for example in [Jec86] and [She98, Chapter I, § 1]. However, to make this paper self-contained, we also provide a short introduction to forcing here.

The paper is organized as follows: In the first section, a brief introduction to forcing is given and some combinatorial properties of forcing extensions are discussed. Then in Section 2, two types of forcing notions are introduced which are investigated in the last two sections. In Section 3, families defined by the absence of a winning strategy for player I are introduced. These families play the key rôle in the investigation of Mathias and Silver forcings in Section 4 and Section 5 respectively.

# 1 The Notion of Forcing

In modern set theory, one usually gets consistency results by a forcing construction. Forcing was invented by Paul Cohen in the early 1960s to show that the Axiom of Choice AC as well as the Continuum Hypothesis CH are not provable in ZF (which is Zermelo-Fraenkel Set Theory without AC). In fact he showed that ¬AC is relatively consistent with ZF and that ¬CH is relatively consistent with ZFC. Forcing is a technique to extend models of set theory in such a way that certain statements become true in the extension, no matter if they were true or false in the ground model. In other words, forcing adds new sets to some ground model and by choosing the right forcing notion, which is essentially a partial ordering, we can make sure that the new sets have some desired properties. So, the main ingredients of a forcing construction are a model of ZFC, usually denoted by $\mathbf{V}$, and a partial ordering $\mathbb{P} = (P, \leq)$.

## 1.1 Partial orderings, generic filters, and names

Let $\mathbf{V}$ be a model of Set Theory and let $\mathbb{P} = (P, \leq)$ be a partial ordering defined in this model. The elements of $P$ are usually called **conditions**. Two conditions $p_1$ and $p_2$ of $P$ are called **incompatible**, denoted $p_1 \perp p_2$, if there is no $q \in P$ such that $p_1 \leq q \geq p_2$. A set $D \subseteq P$ is called **dense** if for every condition $p \in P$ there is a $q \in D$ such that $p \leq q$. A set $D \subseteq P$ is called **open** (or upwards closed) if $p \in D$ and $q \geq p$ implies $q \in D$. A set $F \subseteq P$ is called a **filter** if it is directed (*i.e.*, for all $p_1, p_2 \in F$ there is a $q \in F$ such that $p_1 \leq q \geq p_2$) and downwards closed (*i.e.*, if $p \in F$ and $q \leq p$ implies $q \in F$). A filter $G$ is called a **generic filter** if for each dense open set $D \subseteq P$ which belongs to $\mathbf{V}$ we have $G \cap D \neq \varnothing$. If $G \subseteq P$ is a generic filter, then we say that $G$ is **$\mathbb{P}$-generic over $\mathbf{V}$**. Notice that if the partial ordering $\mathbb{P}$ has no trivial branch in the sense that for every condition $p \in P$ there are $q_1, q_2 \in P$ such that $q_1 \geq p \leq q_2$ and $q_1 \perp q_2$, then a generic filter cannot belong to $\mathbf{V}$ (otherwise, the set $P \setminus G$ would be a dense open subset of $P$ belonging to $\mathbf{V}$).

**Theorem 1 (The Generic Model Theorem).** Let $\mathbf{V}$ be a model of ZFC, called the **ground model**, and let $\mathbb{P} = (P, \leq)$ be a partial ordering defined in $\mathbf{V}$. If $G$ is $\mathbb{P}$-generic over $\mathbf{V}$, then there is a model $\mathbf{V}[G]$ of ZFC, called the **generic extension** of $\mathbf{V}$, such that $\mathbf{V} \subseteq \mathbf{V}[G]$ and $G \in \mathbf{V}[G]$, and every model of ZFC containing $\mathbf{V}$ and $G$ contains also $\mathbf{V}[G]$.

In the sequel, let $\mathbf{V}$ be a model of ZFC, let $\mathbb{P}$ be a partial ordering defined in $\mathbf{V}$, and let $G$ be $\mathbb{P}$-generic over $\mathbf{V}$. In general, $G \notin \mathbf{V}$ and therefore people living in $\mathbf{V}$ do not have knowledge of all the sets in $\mathbf{V}[G]$. On the other hand, people living in $\mathbf{V}$ have so-called "names" for each member of $\mathbf{V}[G]$, but before we introduce the notion of names and their interpretation in $\mathbf{V}[G]$, let us recall the definition of the rank of a set: The **rank** of a set $x \in \mathbf{V}$, denoted $\mathrm{rk}(x)$, is $\bigcup\{\mathrm{rk}(y)+1 : y \in x\}$ where $\mathrm{rk}(\varnothing) = 0$. Notice that since the union of a set of ordinals is an ordinal, $\mathrm{rk}(x)$ is an ordinal if defined, and by the Axiom of Foundation $\mathrm{rk}(x)$ is defined for every set $x \in \mathbf{V}$.

Now we can define by induction on $\alpha$ what is a $\mathbb{P}$-name $\underset{\sim}{\tau}$ of rank less than or equal to $\alpha$ as well as its interpretation $\underset{\sim}{\tau}[G]$ in $\mathbf{V}[G]$: $\underset{\sim}{\tau}$ is a $\mathbb{P}$-name with $\mathrm{rk}(\underset{\sim}{\tau}) \leq \alpha$ if it has the form

$$\underset{\sim}{\tau} = \{(p_\iota, \underset{\sim}{\tau_\iota}) : \iota \in I\}$$

where $I$ is some set and for each $\iota \in I$ we have $p_\iota \in P$ and $\mathrm{rk}(\underset{\sim}{\tau_\iota}) < \alpha$. The interpretation $\underset{\sim}{\tau}[G]$ of $\underset{\sim}{\tau}$ in $\mathbf{V}[G]$ is

$$\{\underset{\sim}{\tau_\iota}[G] : (p_\iota, \underset{\sim}{\tau_\iota}) \in \underset{\sim}{\tau} \text{ and } p_\iota \in G\}.$$

Since $G \in \mathbf{V}[G]$ and $V \subseteq \mathbf{V}[G]$, there is a $\mathbb{P}$-name for $G$ as well as for each set in $\mathbf{V}$. For a set $x \in \mathbf{V}$, $\dot{x}$ is a $\mathbb{P}$-name for $x$ defined by induction on $\mathrm{rk}(x)$ as follows:

$$\dot{x} = \{(p, \dot{y}) : p \in P \text{ and } y \in x\}.$$

Further, the $\mathbb{P}$-name for $G$, denoted by $\underset{\sim}{G}$, is defined as follows:

$$\underset{\sim}{G} = \{(p, \dot{p}) : p \in P\}.$$

Notice that for every $\mathbb{P}$-generic filter $G$ we have $\dot{x}[G] = x$ and $\underset{\sim}{G}[G] = G$. In the sequel we identify the names for sets in the ground model with the corresponding sets and omit the dots.

Let $\underset{\sim}{\tau_1}, \ldots, \underset{\sim}{\tau_n}$ be $\mathbb{P}$-names and $\varphi(x_1, \ldots, x_n)$ be a first-order formula of the language of Set Theory. For a condition $p \in P$ we write

$$p \Vdash_{\mathbb{P}} \text{“}\varphi(\underset{\sim}{\tau_1}, \ldots, \underset{\sim}{\tau_n})\text{”}$$

and say $p$ **forces** $\varphi(\underset{\sim}{\tau_1},\ldots,\underset{\sim}{\tau_n})$, if for every $\mathbb{P}$-generic filter $G$ containing $p$ we have $\varphi(\underset{\sim}{\tau_1}[G],\ldots,\underset{\sim}{\tau_n}[G])$ is true in $\mathbf{V}[G]$, in symbols:

$$\mathbf{V}[G] \models \varphi\big(\underset{\sim}{\tau_1}[G],\ldots,\underset{\sim}{\tau_n}[G]\big).$$

Notice that for any conditions $p, q \in \mathbb{P}$ and any first-order sentence of the forcing language $\varphi$, if $p \Vdash_{\mathbb{P}} \varphi$ and $q \geq p$, then also $q \Vdash_{\mathbb{P}} \varphi$.

**Theorem 2 (The Forcing Theorem).** If $\varphi(\underset{\sim}{\tau_1},\ldots,\underset{\sim}{\tau_n})$ is a first-order sentence of the forcing language, then for every $\mathbb{P}$-generic filter $G$ we have

$$\mathbf{V}[G] \models \varphi\big(\underset{\sim}{\tau_1}[G],\ldots,\underset{\sim}{\tau_n}[G]\big) \iff \exists p \in G\big(p \Vdash_{\mathbb{P}} \text{“}\varphi(\underset{\sim}{\tau_1},\ldots,\underset{\sim}{\tau_n})\text{”}\big).$$

## 1.2 Dominating, unbounded, and splitting reals

In the following we characterize a few real numbers which might appear in a generic extension, but first we have to introduce some notations.

The set $\{0,1,2,\ldots\}$ of **natural numbers** is denoted by $\omega$ and we usually consider a natural number $n$ as the set of all numbers smaller than $n$, so, $n = \{k \in \omega : k < n\}$ and $n+1 = n \cup \{n\}$, which is also denoted $n^+$. The **cardinality** of a set $x$ is denoted by $|x|$. In particular, for every natural number $n$ we have $|n| = n$.

The set of all infinite subsets of $\omega$ is denoted by $[\omega]^\omega$, the set of all functions from $\omega$ to $\omega$ is denoted by ${}^\omega\omega$, and the set of all functions from $\omega$ to $\{0,1\} = 2$ is denoted by ${}^\omega 2$. Each of the sets $[\omega]^\omega$, ${}^\omega\omega$, and ${}^\omega 2$ can be identified with the set of real numbers and in the sequel we usually call their members just “reals”.

For two functions $f, g \in {}^\omega\omega$ we say that $g$ is **dominated** by $f$, denoted $g <^* f$, if there is an $n \in \omega$ such that for all $k \geq n$ we have $g(k) < f(k)$. For two sets $x, y \in [\omega]^\omega$ we say that $x$ **splits** $y$ if both sets $y \cap x$ and $y \setminus x$ are infinite.

Now let $\mathbf{V}$ be any model of ZFC and let $\mathbf{V}[G]$ be a generic extension (with respect to some forcing notion $\mathbb{P}$). A function $f \in {}^\omega\omega$ in $\mathbf{V}[G]$ is called a **dominating real** if each function $g \in {}^\omega\omega \cap \mathbf{V}$ is dominated by $f$, and $f$ is called an **unbounded real** if it is not dominated by any function $g \in {}^\omega\omega \cap \mathbf{V}$. Further, a set $x \in [\omega]^\omega$ in $\mathbf{V}[G]$ is called a **splitting real** if it splits each set $y \in [\omega]^\omega \cap \mathbf{V}$.

**Proposition 3.** If $\mathbf{V}[G]$ contains a dominating real, then it also contains a splitting real.

*Proof.* We can just follow the proof of [vDo84, Theorem 3.1 (a)]: Since $\mathbf{V}[G]$ is a model of ZFC we have that if a function $f \in {}^\omega\omega$ belongs $\mathbf{V}[G]$, then also the set

$$\sigma_f = \bigcup\Big\{\big[f^{2n}(0), f^{2n+1}(0)\big) : n \in \omega\Big\}$$

belongs to $\mathbf{V}[G]$, where $[a,b) = \{k \in \omega : a \leq k < b\}$ and $f^{n+1}(0) = f\big(f^n(0)\big)$ with $f^0(0) := 0$. Now let $f \in {}^\omega\omega$ be a dominating real. Without loss of generality we may assume that $f$ is strictly increasing and that $f(0) > 0$. Fix any $x \in [\omega]^\omega \cap \mathbf{V}$ and let $g_x$ be the (unique) strictly increasing bijection $\omega \to x$. Since $f$ is dominating we have $g_x <^* f$, which implies that there is an $n \in \omega$ such that for all $k \geq n$ we have $g_x(k) < f(k)$. Because $k \leq g_x(k)$ for all $k \in \omega$, we get that if $k \geq n$ then

$$f^n(0) \leq g_x\big(f^n(0)\big) < f\big(f^n(0)\big) = f^{n+1}(0)\,.$$

Hence $g_x\big(f^n(0)\big) \in \sigma_f$ if $n$ is even and $g_x\big(f^n(0)\big) \notin \sigma_f$ if $n$ is odd, which shows that $\sigma_f$ is splitting. q.e.d.

A forcing notion is called ${}^\omega\omega$**-bounding** if there are no unbounded reals in the generic extension, or in other words, if every function is dominated by some function in the ground model. Obviously, a forcing notion which adds a dominating real also adds unbounded reals and therefore cannot be ${}^\omega\omega$-bounding, and by Proposition 3, such a forcing notion also adds splitting reals. On the other hand, none of these implications is reversible. For example a forcing notion which is ${}^\omega\omega$-bounding but adds splitting reals is Silver forcing (investigated in Section 5), and Cohen forcing, discussed below, is an example of a forcing notion which adds unbounded and splitting reals but does not add dominating reals.

### 1.3 Cohen forcing

The Cohen partial ordering is certainly one of the simplest non-trivial forcing notions. Cohen forcing is denoted by $\mathbb{C} = (C, \leq)$ and defined as follows: The set of conditions $C$ consists of all functions from some $n \in \omega$ to $\{0,1\}$, and for two conditions $p, q \in C$ we define

$$p \leq q \iff q|_{\mathrm{dom}(p)} \equiv p\,,$$

or in other words, $p \leq q$ if $q$ extends $p$. If $G$ is $\mathbb{C}$-generic over $\mathbf{V}$, then $G$ generates a function $c \in {}^\omega 2$. To see this, notice that for every $n \in \omega$, the set $D_n = \{p \in C : n \in \mathrm{dom}(p)\}$ is dense open, and therefore, for any $n \in \omega$ there is a $p \in G$ such that $n \in \mathrm{dom}(p)$. Further, since any two members of $G$ are compatible, all conditions $p \in G$ which are defined on $n$ must agree at this point. Thus,

$$c(n) = \begin{cases} 0 & \text{if } \exists p \in G\,\big(p(n) = 0\big), \\ 1 & \text{if } \exists p \in G\,\big(p(n) = 1\big), \end{cases}$$

is a well-defined function from $\omega$ to $\{0,1\}$. The real $c \in {}^{\omega}2$ is called a **Cohen real** (over $\mathbf{V}$). Thus, a $\mathbb{C}$-generic filter generates a Cohen real, and vice versa, the $\mathbb{C}$-generic filter can be reconstructed from the corresponding Cohen real.

The following proposition gives some basic properties of Cohen forcing and Cohen reals respectively. Even though the proofs are straightforward, they involve some standard techniques which will be also used later in the investigation of Mathias and Silver forcings.

**Proposition 4.** Cohen forcing does not add dominating reals, but every Cohen real is unbounded and splitting.

*Proof.* First we show that a Cohen real is always unbounded: Let $\underset{\sim}{c}$ be a name for a Cohen real, then for any condition $p \in C$ we have

$$p \Vdash_{\mathbb{C}} \text{“}\underset{\sim}{c}|_{\mathrm{dom}(p)} \equiv p\text{”} .$$

Let $g \in {}^{\omega}2 \cap \mathbf{V}$ and $n \in \omega$, then there exists a $k \geq n$ and a condition $q \geq p$ such that $k \in \mathrm{dom}(q)$ and $q(n) > g(n)$, which shows that for every $n \in \omega$ the set of conditions $q \in C$ such that

$$q \Vdash_{\mathbb{C}} \text{“}\exists k \geq n \left(g(k) < \underset{\sim}{c}(k)\right)\text{”}$$

is dense open in $C$, which implies that no condition forces that $c$ is dominated by $g$, and since $g$ was arbitrary, $c$ is not dominated by any real in the ground model.

In a similar way one can show that a Cohen real is always splitting: Let $c$ be a Cohen real and let $\sigma_c := \{k \in \omega : c(k) = 1\}$, then for any infinite set $x \in [\omega]^{\omega} \cap \mathbf{V}$ and any $n \in \omega$, the set of conditions $p \in C$ such that

$$p \Vdash_{\mathbb{C}} \text{“}|x \cap \underset{\sim}{\sigma_c}| > n \text{ and } |x \setminus \underset{\sim}{\sigma_c}| > n\text{”}$$

is dense open, and therefore, $\sigma_c$ splits every real in the ground model.

Now let $f \in {}^{\omega}\omega$ be a function in $\mathbf{V}[c]$ and let $\underset{\sim}{f}$ be a $\mathbb{C}$-name for $f$. In order to show that $f$ is not dominating we have to find a function $g \in {}^{\omega}\omega \cap \mathbf{V}$ such that for every $n \in \omega$ there is a $k \geq n$ such that $g(k) \not< f(k)$. We can just follow the proof of [BarJud95, Lemma 3.1.2 (2)]: Let $\{p_k : k \in \omega\}$ be a countable dense subset of $C$. For every $k \in \omega$ define

$$g(k) = \min\left\{n : \exists q \geq p_k \left(q \Vdash_{\mathbb{C}} \underset{\sim}{f}(k) = n\right)\right\}.$$

For every condition $p \in C$ and every $n \in \omega$ there is a $k \geq n$ such that $p_k \geq p$, and we find a $q \geq p_k$ such that $q \Vdash_{\mathbb{C}} \underset{\sim}{f}(k) = g(k)$. Consequently, for every $n \in \omega$, the set

$$D_n = \left\{q \in C : q \Vdash_{\mathbb{C}} \text{“}\underset{\sim}{f}(k) = g(k) \text{ for some } k \geq n\text{”}\right\}$$

is dense open in $C$. Hence, by the Forcing Theorem 2 and since the Cohen real $c$ meets every dense open subset of $C$ we have

$$\mathbf{V}[c] \models \forall n \in \omega\, \exists k \geq n\, \big(g(k) \not< f(k)\big)$$

which shows that $g$ is not dominated by $f$. q.e.d.

## 2 Mathias and Silver Forcings

### 2.1 Free families

Before we introduce the forcing notions of Mathias and Silver, we consider certain families of subsets of $\omega$.

In the following the "ground set" will be $\omega$ and consequently for $x \subseteq \omega$ we define $x^{c} := \omega \setminus x$. Now a family $\mathscr{F} \subseteq [\omega]^{\omega}$ is called a **filter** if it is closed under intersections and supersets, or in other words, if for any $x, y \in [\omega]^{\omega}$ we have

1. if $x \in \mathscr{F}$ and $y \in \mathscr{F}$, then $x \cap y \in \mathscr{F}$,
2. if $x \in \mathscr{F}$ and $x \subseteq y$, then $y \in \mathscr{F}$.

The **Fréchet filter** is the filter consisting of all co-finite subsets of $\omega$, *i.e.*, all $x \in [\omega]^{\omega}$ such that $x^{c}$ is finite, and a filter $\mathscr{F} \subseteq [\omega]^{\omega}$ is called a **free filter** if it contains the Fréchet filter. For a filter $\mathscr{F} \subseteq [\omega]^{\omega}$, $\mathscr{F}^{+}$ denotes the collection of all subsets $x \subseteq \omega$ such $x^{c} \notin \mathscr{F}$. It is useful to notice that for a free filter $\mathscr{F}$, $\mathscr{F}^{+} = \{x \subseteq \omega : \forall z \in \mathscr{F}\, (|x \cap z| = \omega\}$ (*cf.* [Laf96, p. 52]). Further, a family $\mathscr{E}$ of subsets of $\omega$ is called a **free family** if there is a free filter $\mathscr{F} \subseteq [\omega]^{\omega}$ such that $\mathscr{E} = \mathscr{F}^{+}$. In particular, $[\omega]^{\omega}$ and all ultrafilters are free families.

Notice that a free family does not contain any finite sets and is closed under supersets. A filter $\mathscr{F} \subseteq [\omega]^{\omega}$ is called an **ultrafilter** if for all $x \subseteq [\omega]^{\omega}$ either $x$ or $x^{c}$ belongs to $\mathscr{F}$. It is easy to see that for a filter $\mathscr{F} \subseteq [\omega]^{\omega}$, $\mathscr{F} = \mathscr{F}^{+}$ if and only if $\mathscr{F}$ is an ultrafilter. Hence, a free family $\mathscr{E}$ is closed under intersections if and only if $\mathscr{E}$ is an ultrafilter. However, all free families have the following slightly weaker property.

**Lemma 5.** If $\mathscr{E}$ is a free family, $x \in \mathscr{E}$, $y \subseteq x$, and $y \notin \mathscr{E}$, then $x \setminus y$ belongs to $\mathscr{E}$.

*Proof.* Let $\mathscr{E} = \mathscr{F}^{+}$ where $\mathscr{F}$ is some free filter, let $x \in \mathscr{E}$, and let $y \subseteq x$ be such that $y \notin \mathscr{E}$. By definition, $x^{c} \notin \mathscr{F}$ and $y^{c} \in \mathscr{F}$. Since $x \in \mathscr{E}$, $x \cap z$ is infinite for all $z \in \mathscr{F}$. In particular, since $y^{c} \in \mathscr{F}$ and $\mathscr{F}$ is a free filter, $x \cap y^{c}$ as well as $x \cap (y^{c} \cap z) = (x \cap y^{c}) \cap z$ is infinite for all $z \in \mathscr{F}$, which implies that $x \cap y^{c} = x \setminus y$ belongs to $\mathscr{E}$. q.e.d.

## 2.2 Mathias forcing restricted to free families

In the sequel let $\mathscr{E}$ be an arbitrary free family. **Mathias forcing** restricted to $\mathscr{E}$, denoted $\mathbb{M}_{\mathscr{E}} = (M_{\mathscr{E}}, \leq)$, is defined as follows:

$$M_{\mathscr{E}} = \{(s,x) : s \subseteq \omega \text{ is finite}, x \in \mathscr{E}, \max(s) < \min(x)\}$$

and

$$(s,x) \leq (t,y) \iff s \subseteq t,\ y \subseteq x,\ t \setminus s \subseteq x\,.$$

The finite set $s$ of a Mathias condition $(s,x)$ is called the **stem** of the condition. Similar to Cohen forcing we can identify every $\mathbb{M}_{\mathscr{E}}$-generic filter with a real number, called **Mathias real**, which is in fact just the union of the stems of the conditions which belong to the generic filter.

In [Mat77], Mathias introduced and investigated rigorously his forcing notion in the case when $\mathscr{E}$ is a so-called happy family (defined and discussed in Section 6). Special cases of happy families are when $\mathscr{E} = [\omega]^{\omega}$ (in which case $\mathbb{M}_{\mathscr{E}}$ is known as unrestricted Mathias forcing) and when $\mathscr{E}$ is a Ramsey ultrafilter. Mathias showed that if $\mathscr{E}$ is a happy family, then $\mathbb{M}_{\mathscr{E}}$ has many interesting combinatorial properties. In the next section, so-called Ramsey families, defined in terms of infinite games, will be introduced and in Section 4 it will be shown that Mathias forcing restricted to such families has essentially the same combinatorial features as for example unrestricted Mathias forcing.

## 2.3 Silver forcing restricted to free families

In the sequel let again $\mathscr{E}$ be an arbitrary free family. For a set $x \subseteq \omega$, let ${}^{x}2$ denote the set of all functions $f : x \to \{0,1\}$. **Silver forcing** restricted to $\mathscr{E}$, denoted $\mathbb{S}_{\mathscr{E}} = (S_{\mathscr{E}}, \leq)$, is defined as follows:

$$S_{\mathscr{E}} = \bigcup \{ {}^{x}2 : x^{c} \in \mathscr{E} \}$$

and for $p, q \in S_{\mathscr{E}}$ we stipulate

$$p \leq q \iff q|_{\operatorname{dom}(p)} \equiv p\,.$$

Again we can identify every $\mathbb{S}_{\mathscr{E}}$-generic filter with a real number, called **Silver real**, which is in fact just the union of the functions which belong to the generic filter.

The original (or unrestricted) Silver forcing we get when $\mathscr{E} = [\omega]^{\omega}$ (*cf.* [Mat79, p. 112] or [Jec86, Part I, 3.10]). For $\mathscr{E}$ a P-point (defined below), restricted Silver forcing, also known as **Grigorieff forcing**, was introduced and investigated in depth by Grigorieff in [Gri71] (see also [Jec86, Part I, 3.22]). Unrestricted Silver forcing has essentially the same combinatorial properties as Grigorieff forcing. Moreover, there are families between

$[\omega]^\omega$ and P-points (introduced in the next section) such that Silver forcing restricted to such families has still the same combinatorial features as unrestricted Silver forcing or as Grigorieff forcing.

## 3 Infinite Games

Let $\mathscr{E}$ be an arbitrary free family. Associated with $\mathscr{E}$ we define two quite similar games between two players, say DEATH and the MAIDEN.

$\mathcal{G}_{\mathscr{E}}$ :

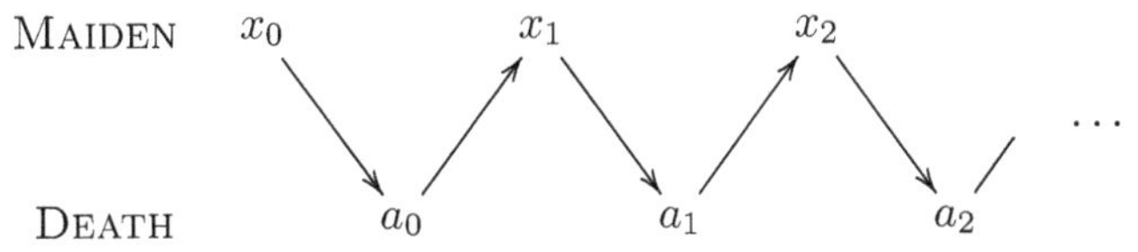


The rules for the game $\mathcal{G}_{\mathscr{E}}$ are as follows: For each $i \in \omega$, $x_i \in \mathscr{E}$ and $a_i \in x_i$, and further we require that $x_{i+1} \subseteq x_i$ and $a_i < a_{i+1}$. Finally, DEATH wins the game $\mathcal{G}_{\mathscr{E}}$ if and only if the sequence $\{a_i : i \in \omega\}$ belongs to the family $\mathscr{E}$.

A free family $\mathscr{E}$ is called a **Ramsey family** if the MAIDEN has no winning strategy for the game $\mathcal{G}_{\mathscr{E}}$. In other words, if $\mathscr{E}$ is a Ramsey family then DEATH can always defeat any given strategy of the MAIDEN, no matter how sophisticated her strategy is. (The only possibility for the MAIDEN to win against DEATH is to play randomly, *i.e.*, not according to any strategy.) Notice that this does not imply that DEATH has a winning strategy. Obviously, $[\omega]^\omega$ is a Ramsey family. On the other hand, there are also ultrafilters which are Ramsey families, namely the so-called Ramsey ultrafilters (and vice versa, every Ramsey family which is an ultrafilter is a Ramsey ultrafilter). According to [BarJud95], this was shown by Galvin and Shelah (*cf.* [BarJud95, Theorem 4.5.3]). So, Ramsey families are a kind of generalized Ramsey ultrafilters. These families were first introduced and studied by Laflamme in [Laf96] (where the filters associated to a Ramsey family are called +-Ramsey filters). As we will see in Section 6, every Ramsey family is a happy family (in the sense of [Mat77]), but not vice versa.

In the game $\mathcal{G}^*_{\mathscr{E}}$, DEATH has slightly more freedom, since he can play now finite sequences instead of just singletons.

$\mathcal{G}^*_{\mathscr{E}}$ :

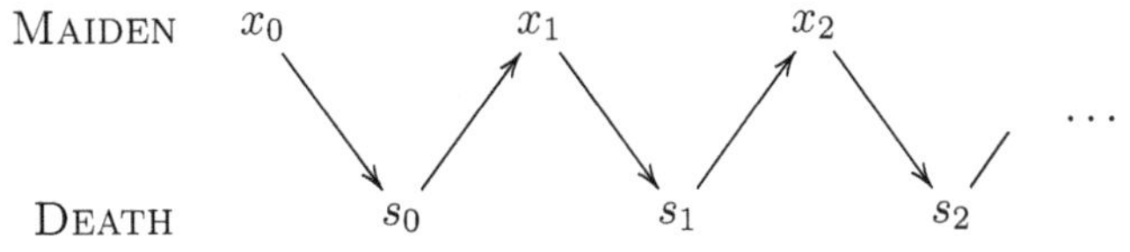

Again, the sets $x_i$ played by the MAIDEN must belong to the free family $\mathscr{E}$ and each finite set $s_i$ played by DEATH must be a subset of the corresponding $x_i$. Further, for each $i \in \omega$ we require that $x_{i+1} \subseteq x_i$ and $\max(s_i) < \min(s_{i+1})$. Finally, DEATH wins the game $\mathcal{G}^*_{\mathscr{E}}$ if and only if $\bigcup\{s_i : i \in \omega\}$ belongs to the family $\mathscr{E}$.

A free family $\mathscr{E}$ is called a **P-family** if the MAIDEN has no winning strategy for the game $\mathcal{G}^*_{\mathscr{E}}$. Obviously, $[\omega]^\omega$ is a P-family. On the other hand, there are also P-families which are ultrafilters, namely the so-called P-points (*cf.* [BarJud95, Theorem 4.4.4]). So, P-families can be considered as a generalization of P-points, which are a weaker form of Ramsey ultrafilters. P-families were first introduced and studied by Laflamme in [Laf96] (where the filters associated to a P-family are called P+-filters).

# 4 Properties of Mathias Forcing Notions

Throughout this section let $\mathscr{E}$ be an arbitrary but fixed Ramsey family. It will be shown that the forcing notion $\mathbb{M}_{\mathscr{E}}$ adds dominating reals, does not add Cohen reals, and that every infinite subset of a Mathias real is also a Mathias real.

Since every Ramsey family is happy (*cf.* Fact 19), the main results of this section follow from Mathias' investigations [Mat77, Section 4]. However, the technique used here provides a new and uniform approach to Mathias forcing notions and may be applied also in more general contexts.

## 4.1 $\mathbb{M}_{\mathscr{E}}$ adds dominating reals

**Theorem 6.** The forcing notion $\mathbb{M}_{\mathscr{E}}$ adds dominating reals.

*Proof.* We show that a Mathias real is always dominating: Let $m$ be $\mathbb{M}_{\mathscr{E}}$-generic over $\mathbf{V}$, let $p = (s, x)$ be an arbitrary $\mathbb{M}_{\mathscr{E}}$-condition, and let $g \in {}^\omega\omega \cap \mathbf{V}$ be any function in the ground model. It is enough to show that there exists a condition $q \geq p$ such that $q \Vdash_{\mathbb{M}_{\mathscr{E}}}$ "$\underset{\sim}{m}$ dominates $g$". In order to construct the condition $q$ we run the game $\mathcal{G}_{\mathscr{E}}$ where the MAIDEN plays according to the following strategy: The MAIDEN's first move is $x_0 := x \setminus \big(g(n_0)^+\big)$, where $n_0 = |s|$, and for $i \in \omega$ she plays $x_{i+1} := x_i \setminus \max\big\{g(n_0 + i)^+, a_i^+\big\}$. Since this strategy is not a winning strategy for the MAIDEN, DEATH can play such that $y := \{a_i : i \in \omega\} \in \mathscr{E}$. Now by construction we get that $(s, y) \geq p$ and

$$(s, y) \Vdash_{\mathbb{M}_{\mathscr{E}}} \forall k \geq n_0 \big(\underset{\sim}{m}(k) > g(k)\big)$$

which shows that $m$ is a dominating real. q.e.d.

As a consequence of Proposition 3 we get

**Corollary 7.** The forcing notion $\mathbb{M}_{\mathscr{E}}$ adds splitting reals.

### 4.2 $\mathbb{M}_{\mathscr{E}}$ has pure decision

The following property of Mathias forcing is known as **pure decision** (*cf.* [Mat77, Proposition 4.12 (2.9)]):

**Theorem 8.** *For every $\mathbb{M}_{\mathscr{E}}$-condition $(s,x)$ and for every sentence of the forcing language $\varphi$, there exists an $\mathbb{M}_{\mathscr{E}}$-condition $(s,y) \geq (s,x)$ with the same stem as $(s,x)$ such that $(s,y)$ decides $\varphi$, i.e.,*

$$(s,y) \Vdash_{\mathbb{M}_{\mathscr{E}}} \varphi \quad or \quad (s,y) \Vdash_{\mathbb{M}_{\mathscr{E}}} \neg\varphi .$$

Pure decision is one of the main features of Mathias forcing, but before we can prove the theorem, we have to introduce some terminology and prove some auxiliary results: For $\mathbb{M}_{\mathscr{E}}$-conditions $(s,x)$ let

$$[s,x] := \{z \in [\omega]^{\omega} : s \subseteq z \subseteq s \cup x\}.$$

For a (fixed) open set $\mathcal{O} \subseteq M_{\mathscr{E}}$ let $\bar{\mathcal{O}} := \bigcup\{[s,x] : (s,x) \in \mathcal{O}\}$. An $\mathbb{M}_{\mathscr{E}}$-condition $(s,x)$ is called **good** (with respect to $\mathcal{O}$), if there is a condition $(s,y) \geq (s,x)$ such that $[s,y] \subseteq \bar{\mathcal{O}}$; otherwise it is called **bad**. Further, the condition $(s,x)$ is called **ugly** if $(s \cup \{a\}, x \setminus a^+)$ is bad for all $a \in x$. Notice that if $(s,x)$ is ugly, then $(s,x)$ is bad, too. Finally, $(s,x)$ is called **completely ugly** if $(s \cup \{a_0,\ldots,a_n\}, x \setminus a_n^+)$ is bad for all $\{a_0,\ldots,a_n\} \subseteq x$ with $a_0 < \ldots < a_n$.

**Lemma 9.** *If an $\mathbb{M}_{\mathscr{E}}$-condition $(s,x)$ is bad, then there is a condition $(s,y) \geq (s,x)$ which is ugly.*

*Proof.* We run the game $\mathcal{G}_{\mathscr{E}}$ where the MAIDEN plays according to the following strategy: She starts the game by playing $x_0 := x$, and then, for $i \in \omega$, she plays $x_{i+1} \subseteq (x_i \setminus a_i^+)$ such that $[s \cup \{a_i\}, x_{i+1}] \subseteq \bar{\mathcal{O}}$ if possible, and $x_{i+1} = (x_i \setminus a_i^+)$ otherwise. Strictly speaking we assume that $\mathscr{E}$ is well-ordered and that $x_{i+1}$ is the first element of $\mathscr{E}$ with the required properties. However, since this strategy is not a winning strategy for the MAIDEN, DEATH can play such that $z := \{a_i : i \in \omega\} \in \mathscr{E}$ and let $y = \{a_i \in z : [s \cup \{a_i\}, x_{i+1}] \subseteq \bar{\mathcal{O}}\}$. Because $\mathscr{E}$ is a free family, by Lemma 5 we get that $y$ or $z \setminus y$ belongs to $\mathscr{E}$. If $y \in \mathscr{E}$, then $[s,y] \subseteq \bar{\mathcal{O}}$ which would imply that $(s,x)$ is good, but this contradicts the premise of the lemma. Hence, $z \setminus y \in \mathscr{E}$, which implies that $(s, z \setminus y)$ is ugly. q.e.d.

**Lemma 10.** *If an $\mathbb{M}_{\mathscr{E}}$-condition $(s,x)$ is bad, then there is a condition $(s,y) \geq (s,x)$ such that $(s,y)$ is completely ugly.*

*Proof.* This follows by an iterative application of Lemma 9. In fact, for every $i \in \omega$, the MAIDEN can play a set $x_i \in \mathscr{E}$ such that for each $t \subseteq \{a_0, \ldots, a_{i-1}\}$, either the condition $(s \cup t, x_i)$ is ugly or $[s \cup t, x_i] \subseteq \bar{\mathcal{O}}$. Now DEATH can play such that $y := \{a_i : i \in \omega\} \in \mathscr{E}$. Assume that there exists a finite set $t \subseteq y$ such that $(s \cup t, y \setminus \max(t)^+)$ is good. Such a set cannot be empty, since $(s, x)$ was assumed to be bad. Now let $t_0$ be a smallest finite subset of $y$ such that $q_0 = (s \cup t_0, y \setminus \max(t_0)^+)$ is good and let $t_0^- = t_0 \setminus \{\max(t_0)\}$. Then by definition of $t_0$, the condition $q_0^- = \big(s \cup t_0^-, y \setminus \max(t_0)\big)$ is not good, and hence, by the strategy of the MAIDEN, it must be ugly, but if $q_0^-$ is ugly, then $q_0$ is bad, which is a contradiction to our assumption. Thus, there is no finite set $t \subseteq y$ such that $(s \cup t, y \setminus \max(t)^+)$ is good, which implies that all these conditions are ugly, and therefore $(s, y)$ is completely ugly. q.e.d.

Now we are ready to proof Theorem 8 (*i.e.*, that the forcing notion $\mathbb{M}_{\mathscr{E}}$ has pure decision):

*Proof.* Let $(s, x)$ be an $\mathbb{M}_{\mathscr{E}}$-condition and let $\varphi$ be a sentence of the forcing language. With respect to $\varphi$ we define $\mathcal{O}_1 := \{q \in M_{\mathscr{E}} : q \Vdash_{\mathbb{M}_{\mathscr{E}}} \varphi\}$ and $\mathcal{O}_2 := \{q \in M_{\mathscr{E}} : q \Vdash_{\mathbb{M}_{\mathscr{E}}} \neg\varphi\}$. Clearly $\mathcal{O}_1$ and $\mathcal{O}_2$ are both open and $\mathcal{O}_1 \cup \mathcal{O}_2$ is even dense in $M_{\mathscr{E}}$. By Lemma 10 we know that for any $(s, x)$ there exists $(s, y) \geq (s, x)$ such that either $[s, y] \subseteq \bar{\mathcal{O}}_1$ or $[s, y] \cap \bar{\mathcal{O}}_1 = \varnothing$. In the former case we have $(s, y) \Vdash_{\mathbb{M}_{\mathscr{E}}} \varphi$ and we are done. In the latter case we find $(s, y') \geq (s, y)$ such that $[s, y'] \subseteq \bar{\mathcal{O}}_2$. (Otherwise we would have $[s, y] \cap (\bar{\mathcal{O}}_1 \cup \bar{\mathcal{O}}_2) = \varnothing$, which is impossible by the density of $\mathcal{O}_1 \cup \mathcal{O}_2$.) Hence, $(s, y') \Vdash_{\mathbb{M}_{\mathscr{E}}} \neg\varphi$. q.e.d.

For the following result, which is again a consequence of Lemma 10, see also [Mat77, Corollary 4.10 (ii)] (and for a kind of reverse implication see [Mat77, Theorem 2.10]).

**Proposition 11.** Every infinite subset of an $\mathbb{M}_{\mathscr{E}}$-generic real is also $\mathbb{M}_{\mathscr{E}}$-generic.

*Proof.* Let $\mathcal{D} \subseteq M_{\mathscr{E}}$ be an arbitrary dense open subset of $M_{\mathscr{E}}$ and let $\mathcal{D}'$ be the set of all conditions $(s, z) \in M_{\mathscr{E}}$ such that for all $t \subseteq s$, $[t, z] \subseteq \bar{\mathcal{D}}$.

First we show that $\mathcal{D}'$ is dense (and by definition also open) in $M_{\mathscr{E}}$: For this take an arbitrary condition $(s, x) \in \mathcal{D}$ and let $\{t_i : 0 \leq i \leq m\}$ be an enumeration of all subsets of $s$. Because $\mathcal{D}$ is dense open in $M_{\mathscr{E}}$, by Lemma 10 we find a condition $(s, y) \geq (s, x)$ such that $(s, y) \in \mathcal{D}'$, which implies that $\mathcal{D}'$ is dense in $M_{\mathscr{E}}$.

Let $m \in [\omega]^\omega$ be $\mathbb{M}_{\mathscr{E}}$-generic and let $m'$ be an infinite subset of $m$. Since $\mathcal{D}'$ is dense open and $m$ is $\mathbb{M}_{\mathscr{E}}$-generic, there exists a condition $(s, x) \in \mathcal{D}'$ such that $s \subseteq m \subseteq s \cup x$. Let $t = m' \cap s$, then $t \subseteq m' \subseteq t \cup x$ and by definition of $\mathcal{D}'$ we have $[t, x] \subseteq \bar{\mathcal{D}}$. Thus, $m'$ meets the dense open set $\mathcal{D}$, and since $\mathcal{D}$ was arbitrary, this completes the proof. q.e.d.

## 4.3 $\mathbb{M}_{\mathscr{E}}$ does not add Cohen reals

In Section 1 we have seen that Cohen forcing adds unbounded but not dominating reals. Now we will see that the forcing notion $\mathbb{M}_{\mathscr{E}}$, even though it adds dominating reals (*cf.* Theorem 6), it does not add Cohen reals:

**Theorem 12.** The forcing notion $\mathbb{M}_{\mathscr{E}}$ does not add Cohen reals.

*Proof.* Let $\underset{\sim}{f}$ be an $\mathbb{M}_{\mathscr{E}}$-name for a function in ${}^{\omega}2$ and let $m$ be $\mathbb{M}_{\mathscr{E}}$ generic over $\mathbf{V}$. We have to show that $\underset{\sim}{f}[m]$ is not $\mathbb{C}$-generic over $\mathbf{V}$, *i.e.*, there is a dense open set $\mathcal{D}_f \subseteq C$ in $\mathbf{V}$ such that for all $p \in \mathcal{D}_f$ we have $f[m]|_{\mathrm{dom}(p)} \not\equiv p$.

Notice that by Theorem 8, for every $\mathbb{M}_{\mathscr{E}}$-condition $(s, x)$ and for every $k \in \omega$ there exists a condition $(s, y)$ which decides $\underset{\sim}{f}(k)$, *i.e.*, $(s, y) \Vdash_{\mathbb{M}_{\mathscr{E}}} \underset{\sim}{f}(k) = 0$ or $(s, y) \Vdash_{\mathbb{M}_{\mathscr{E}}} \underset{\sim}{f}(k) = 1$. Consequently, for every $(s, x)$ and every $n \in \omega$ there exists a condition $(s, y)$ which decides $\underset{\sim}{f}(k)$ for all $k < n$.

In order to construct $\mathcal{D}_f$ we run the game $\mathcal{G}_{\mathscr{E}}$ where the MAIDEN plays according to the following strategy: For $i \in \omega$ she plays $x_i$ such that for all $t \subseteq \{a_0, \ldots, a_{i-1}\}$, the condition $(t, x_i)$ decides $\underset{\sim}{f}(k)$ for all $k < 2^{2^i}$. Further, for $t \subseteq \{a_0, \ldots, a_{i-1}\}$ let $p_t^i \in C$ be such that $\mathrm{dom}(p_t^i) = 2^{2^i}$ and $(t, x_i) \Vdash_{\mathbb{M}_{\mathscr{E}}} \underset{\sim}{f}|_{\mathrm{dom}(p_t^i)} \equiv p_t^i$. Since this strategy is not a winning strategy for the MAIDEN, DEATH can play such that $x := \{a_i : i \in \omega\} \in \mathscr{E}$. Now, let

$$\mathcal{C}_f = \{q \in C : \exists i \in \omega\, \exists t \subseteq x\, (q \leq p_t^i)\}$$

and let $\mathcal{D}_f := C \setminus \mathcal{C}_f$.

By construction $\mathcal{D}_f$ is open in $C$ and it remains to show that $\mathcal{D}_f$ is also dense: Firstly notice that for all $n \in \omega$, $(\varnothing, x) \Vdash_{\mathbb{M}_{\mathscr{E}}} \underset{\sim}{f}|_n \in \mathcal{C}_f$. Secondly notice that for any finite $t \subseteq x$ and for any $i \geq |t|$,

$$\left|\{q \in C : q \geq p_t^i,\ \mathrm{dom}(q) = 2^{2^{i+1}}\}\right| = 2^{2^i}\left(2^{2^i} - 1\right)$$

whereas

$$\left|\{q \in \mathcal{C}_f : q \geq p_t^i,\ \mathrm{dom}(q) = 2^{2^{i+1}}\}\right| = 2^{i+1}$$

which implies that $\mathcal{D}_f$ is dense in $C$ and completes the proof. q.e.d.

# 5 Properties of Silver Forcing Notions

Throughout this section let $\mathscr{E}$ be an arbitrary but fixed P-family. It will be shown that the forcing notion $\mathbb{S}_{\mathscr{E}}$ is ${}^{\omega}\omega$-bounding, *i.e.*, adds no unbounded reals (and consequently no Cohen reals), but adds splitting reals and is minimal, *i.e.*, if $g$ is a Silver real, then every real $f$ in the extension which does not belong to the ground model reconstructs $g$.

## 5.1 $\mathbb{S}_{\mathscr{E}}$ is ${}^{\omega}\omega$-bounding

Recall that a forcing notion is ${}^{\omega}\omega$-bounding if no function $f \in {}^{\omega}\omega$ in the generic extension is unbounded, *i.e.*, every function $f \in {}^{\omega}\omega$ in the generic extension is dominated by some function in the ground model. Before we can show that the forcing notion $\mathbb{S}_{\mathscr{E}} = (S_{\mathscr{E}}, \leq)$ is ${}^{\omega}\omega$-bounding, we have to introduce the following notation: Remember that a condition $p \in S_{\mathscr{E}}$ is a function from some $x \subseteq \omega$ to $\{0,1\}$, where $x^{c} \in \mathscr{E}$. For a condition $p \in S_{\mathscr{E}}$ and a finite set $t \subseteq \operatorname{dom}(p)$ let

$$\overline{p \sim t} = \left\{q \in S_{\mathscr{E}} : \operatorname{dom}(q) = \operatorname{dom}(p),\ q|_{\operatorname{dom}(q)\setminus t} \equiv p|_{\operatorname{dom}(p)\setminus t}\right\}.$$

**Theorem 13.** The forcing notion $\mathbb{S}_{\mathscr{E}}$ is ${}^{\omega}\omega$-bounding.

*Proof.* Let $G$ be $\mathbb{S}_{\mathscr{E}}$-generic over $\mathbf{V}$, let $f \in {}^{\omega}\omega$ be a function in $\mathbf{V}[G]$, and let $\underset{\sim}{f}$ be an $\mathbb{S}_{\mathscr{E}}$-name for $f$. In order to show that $f$ is bounded by some function some function in the ground model it is enough to prove that for every condition $p \in S_{\mathscr{E}}$ there is a condition $q_0 \geq p$ and a function $g \in {}^{\omega}\omega \cap \mathbf{V}$ such $q_0 \Vdash_{\mathbb{S}_{\mathscr{E}}}$ "$g$ dominates $\underset{\sim}{f}$".

We construct the condition $q_0$ by running the game $\mathcal{G}^*_{\mathscr{E}}$ where the MAIDEN plays according to the following strategy: Let $m_0 \in \omega$ be the least number for which there exists a condition $p_0 \geq p$ such that $p_0 \Vdash_{\mathbb{S}_{\mathscr{E}}} \underset{\sim}{f}(0) < m_0$. Then the MAIDEN plays $x_0 = \operatorname{dom}(p_0)^{c}$. For positive integers $i \in \omega$ let $t_i = \bigcup_{k \in i} s_k$, where $s_0, \ldots, s_{i-1}$ are the moves of DEATH, and let $m_i \in \omega$ be the least number for which there exists a condition $p_i \geq p_{i-1}$ with $\operatorname{dom}(p_i) \supseteq \operatorname{dom}(p_{i-1}) \cup t_i$ such that for all $q \in \overline{p_i \sim t_i}$ we have $p_i \Vdash_{\mathbb{S}_{\mathscr{E}}} \underset{\sim}{f}(i) < m_i$. Then the MAIDEN plays $x_i = \operatorname{dom}(p_i)^{c}$.

Since this strategy of the MAIDEN is not a winning strategy, DEATH can play such that $\bigcup_{i\in\omega} s_i \in \mathscr{E}$. Let $h = \bigcup_{i \in \omega} p_i$, then $h \in {}^{x}2$ for some $x \subseteq \omega$ (but $h$ is not necessarily a $\mathbb{S}_{\mathscr{E}}$-condition). Now let $q_0 \in S_{\mathscr{E}}$ be such that $\operatorname{dom}(q_0) = \operatorname{dom}(h) \setminus \bigcup_{i\in\omega} s_i$ and $q_0 \equiv h|_{\operatorname{dom}(q_0)}$, and define the function $g \in {}^{\omega}\omega$ by stipulating $g(i) := m_i$. Then $g$ belongs to the ground model $\mathbf{V}$ and by construction we have

$$q_0 \Vdash_{\mathbb{S}_{\mathscr{E}}} \forall i \in \omega \left(g(i) > \underset{\sim}{f}(i)\right)$$

which shows that $f$ is dominated by $g$. q.e.d.

By Proposition 4, Theorem 13 implies that the forcing notion $\mathbb{S}_{\mathscr{E}}$ does not add Cohen reals. However, it adds splitting reals:

**Proposition 14.** The forcing notion $\mathbb{S}_{\mathscr{E}}$ adds splitting reals.

*Proof (Sketch).* Let $f \in {}^{\omega}2$ be $\mathbb{S}_{\mathscr{E}}$-generic over $\mathbf{V}$. We can identify $f$ with the function $\bar{f} \in {}^{\omega}\omega$ by stipulating $\bar{f}(n) = k$ *iff* $f(k) = 1$ and $\big|\{m < k : f(m) = 1\}\big| = n$. Then the set

$$\sigma_f = \bigcup\Big\{\big[\bar{f}(2n), \bar{f}(2n+1)\big) : n \in \omega\Big\}$$

splits every real in the ground model. To see this, notice that for each $x \in [\omega]^{\omega} \cap \mathbf{V}$ and for every $n \in \omega$, the set

$$D_{x,n} = \{p \in S_{\mathscr{E}} : p \Vdash_{\mathbb{S}_{\mathscr{E}}} \text{“}|x \cap \underset{\sim}{\sigma_f}| > n \text{ and } |x \setminus \underset{\sim}{\sigma_f}| > n|\text{”}\}$$

is dense open in $S_{\mathscr{E}}$. q.e.d.

### 5.2 $\mathbb{S}_{\mathscr{E}}$ is minimal

A real $g$ is **minimal** over $\mathbf{V}$ if $g$ is not in the ground model $\mathbf{V}$ and every real $f$ in $\mathbf{V}[g]$ is either in $\mathbf{V}$ or it reconstructs $g$, *i.e.*, $g$ belongs to $\mathbf{V}[f]$ (where $\mathbf{V}[f]$ is the smallest model of ZFC which contains all sets of $\mathbf{V}$ as well as the function $f$). Let $\mathbb{P}$ be a forcing notion and let $G$ be $\mathbb{P}$-generic over $\mathbf{V}$. If there is a real $g$ such that $\mathbf{V}[g] = \mathbf{V}[G]$ and $g$ is minimal over $\mathbf{V}$, then the forcing notion $\mathbb{P}$ is called **minimal**.

In the following we show that the forcing notion $\mathbb{S}_{\mathscr{E}}$ is minimal. The result will be a consequence of the following lemmas, but first we have to introduce some terminology (*cf.* [Gri71, p. 375 f.]): Let $G$ be $\mathbb{S}_{\mathscr{E}}$-generic over $\mathbf{V}$ and let $\underset{\sim}{f}$ be an $\mathbb{S}_{\mathscr{E}}$-name for a function $f \in {}^{\omega}2 \cap \mathbf{V}[G]$. Two $\mathbb{S}_{\mathscr{E}}$-conditions $p$ and $q$ are called **$\underset{\sim}{f}$-compatible** if for all $k \in \omega$ and $\varepsilon \in \{0,1\}$ we have:

$$p \Vdash_{\mathbb{S}_{\mathscr{E}}} \underset{\sim}{f}(k) = \varepsilon \iff q \Vdash_{\mathbb{S}_{\mathscr{E}}} \underset{\sim}{f}(k) = \varepsilon$$

For conditions $p$ and functions $h \in {}^{u}2$, where $u \subseteq \omega$ is finite and $u \cap \mathrm{dom}(p) = \varnothing$, we write $p \cup h$ for the extension of $p$ by $h$, *i.e.*, $(p \cup h)|_{\mathrm{dom}(p)} \equiv p$ and $(p \cup h)|_u \equiv h$. We say that $n \in \omega$ is **$\underset{\sim}{f}$-indifferent** to a condition $p$ if $n \notin \mathrm{dom}(p)$ and for any $q \geq p$ we have either $n \in \mathrm{dom}(q)$ or the conditions $q \cup \langle n, 0\rangle$ and $q \cup \langle n, 1\rangle$ are $\underset{\sim}{f}$-compatible. Roughly speaking, $n$ is $\underset{\sim}{f}$-indifferent to $p$ if above $p$, $n$ is of no use for the interpretation of $\underset{\sim}{f}$.

For any condition $p$, two mutually exclusive cases are possible:

(i) $\exists q \geq p\, \forall r \geq q\, \forall n \in \omega$ ($n$ is not $\underset{\sim}{f}$-indifferent to $r$)

(ii) $\forall q \geq p \, \exists r \geq q \, \exists n \in \omega \, (n \text{ is } \underset{\sim}{f}\text{-indifferent to } r)$

Firstly consider the case when $p$ satisfies (i) (*cf.* [Gri71, Lemma 4.6]):

**Lemma 15.** If $p$ satisfies (i), then there is a condition $q \geq p$ such that for every $k \in \omega$ and for any distinct functions $t_1, t_2 : \operatorname{dom}(q)^{\mathsf{c}} \cap k \to 2$, the conditions $q \cup t_1$ and $q \cup t_2$ are $\underset{\sim}{f}$-incompatible.

*Proof.* Since by assumption $p$ satisfies (i), there is a $q_0 \geq p$ such that for *all* $r \geq q_0$ and for *all* $n \in \operatorname{dom}(r)^{\mathsf{c}}$ we have:

$$\exists r' \geq r \left(r' \cup \langle n, 0\rangle \text{ and } r' \cup \langle n, 1\rangle \text{ are } \underset{\sim}{f}\text{-incompatible}\right) \qquad (\clubsuit)$$

In order to construct the condition $q$ we run the game $\mathcal{G}^*_{\mathscr{E}}$ where the MAIDEN plays according to the following strategy: She begins by playing $x_0 := \operatorname{dom}(q_0)^{\mathsf{c}}$. Then, for positive integers $i \in \omega$, she plays $x_i := \operatorname{dom}(q_i)^{\mathsf{c}}$ where the condition $q_i$ has the following properties: $q_i \geq q_{i-1}$ and any two different conditions which belong to the set $\overline{q_i \,\tilde{}\, t_i}$, where $t_i = \bigcup_{j \in i} s_j$ and the $s_j$'s are the moves of DEATH, are $\underset{\sim}{f}$-incompatible.

Notice that by ($\clubsuit$) and since $q_i \geq q_{i-1} \geq q_0$, such a condition exists. Since this strategy is not a winning strategy for the MAIDEN, DEATH can play such that $x := \{s_i : i \in \omega\} \in \mathscr{E}$. Define $q$ by stipulating $\operatorname{dom}(q) := x^{\mathsf{c}}$ and $q \equiv \bigcup_{i \in \omega} q_i|_{\operatorname{dom}(q)}$, then by construction, $q$ belongs to $S_{\mathscr{E}}$ and has the desired properties. q.e.d.

Secondly consider the case when $p$ satisfies (ii) (*cf.* [Gri71, Lemma 4.7]):

**Lemma 16.** If $p$ satisfies (ii), then there is a condition $q \geq p$ which decides $\underset{\sim}{f}(k)$ for each $k \in \omega$.

*Proof.* Since by assumption $p$ satisfies (ii), for each $p' \geq p$ there is an $r \geq p'$ such that we have:

$$\exists n \in \operatorname{dom}(r)^{\mathsf{c}} \left(n \text{ is } \underset{\sim}{f}\text{-indifferent to } r\right) \qquad (\spadesuit)$$

For conditions $p' \in S_{\mathscr{E}}$ let

$$I_{p'} = \left\{n \in \operatorname{dom}(p')^{\mathsf{c}} : n \text{ is } \underset{\sim}{f}\text{-indifferent to } p'\right\}.$$

**Claim 17.** For each condition $p' \geq p$, $I_{p'}$ belongs to $\mathscr{E}$.

*Proof.* Assume towards a contradiction that $I_{p'} \notin \mathscr{E}$, then, since $\operatorname{dom}(p')^{\mathsf{c}} \in \mathscr{E}$, by Lemma 5 we get $\operatorname{dom}(p')^{\mathsf{c}} \setminus I_{p'} \in \mathscr{E}$. Let $r \geq p'$ be any condition with $\operatorname{dom}(r) = \operatorname{dom}(p') \cup I_{p'}$, then there is no $n \in \omega$ such that $n$ is $\underset{\sim}{f}$-indifferent to $r$, since such an $n$ would also be $\underset{\sim}{f}$-indifferent to $p'$, but this is impossible by the definition of $I_{p'}$ and completes the proof of the claim.

q.e.d. (Claim 17)

In order to construct the condition $q$ we run the game $\mathcal{G}^*_{\mathscr{E}}$ where the MAIDEN plays according to the following strategy: She starts the game by playing $x_0 := I_{p_0}$, where $p_0 \geq p$ is such that $p_0$ decides $\underset{\sim}{f}(0)$. In addition, she plays a condition $q_0 \geq p_0$ such that $\operatorname{dom}(q_0) = x_0^{\mathrm{c}}$. In general, for a positive integer $i \in \omega$ let $t_i = \bigcup_{j \in i} s_j$, where the $s_j$'s are the moves of DEATH, and let $p_i \geq q_{i-1}$ be such that $\operatorname{dom}(p_i) \supseteq x_{i-1}^{\mathrm{c}} \cup t_i$ and every $p' \in \overline{p_i \,\tilde{}\, t_i}$ decides $\underset{\sim}{f}(i)$. Now the MAIDEN plays $x_i := I_{p_i}$ and a condition $q_i \geq p_i$ such that $\operatorname{dom}(q_i) = x_i^{\mathrm{c}}$.

Notice that by (♠) and by the claim, the strategy of the MAIDEN is well-defined. Since her strategy is not a winning strategy, DEATH can play such that $x := \{s_i : i \in \omega\} \in \mathscr{E}$. Define $q$ by stipulating $\operatorname{dom}(q) := x^{\mathrm{c}}$ and $q \equiv \bigcup_{i \in \omega} q_i|_{\operatorname{dom}(q)}$, then by construction, $q$ belongs to $S_{\mathscr{E}}$ and has the desired properties.

q.e.d. (Lemma 16)

By combining the previous two lemmas we are now able to prove that the forcing notion $\mathbb{S}_{\mathscr{E}}$ is minimal, or equivalently, that each Silver real is minimal (*cf.* [Gri71, Theorem 4.1]).

**Theorem 18.** Each real $g \in {}^{\omega}2$ which is $\mathbb{S}_{\mathscr{E}}$-generic over $\mathbf{V}$ is minimal over $\mathbf{V}$, *i.e.*, for every real $f \in {}^{\omega}2 \cap \mathbf{V}[g]$, either $f \in \mathbf{V}$ or $g \in \mathbf{V}[f]$.

*Proof.* Let $G$ be $\mathbb{S}_{\mathscr{E}}$-generic over $\mathbf{V}$, let $g \in {}^{\omega}2$ be the Silver real which corresponds to $G$, and let $\underset{\sim}{f}$ be an $\mathbb{S}_{\mathscr{E}}$-name for a function $f \in {}^{\omega}2 \cap \mathbf{V}[g]$. We have to show that for $f = \underset{\sim}{f}[g]$, either $f \in \mathbf{V}$ or $g \in \mathbf{V}[f]$.

Let $D = D_1 \cup D_2$ where $D_1$ and $D_2$ are defined as follows:

$$D_1 = \{q \in S_{\mathscr{E}} : q \text{ as in Lemma 15 with respect to some } p \leq q\}$$
$$D_2 = \{q \in S_{\mathscr{E}} : q \text{ as in Lemma 16 with respect to some } p \leq q\}$$

By definition, $D$ is obviously dense open in $S_{\mathscr{E}}$ which implies that there exists a $q_0 \in G \cap D$. We have to consider the following two cases.

"$q_0 \in D_1$": In $\mathbf{V}[f]$ define the function $g' \in {}^{\omega}2 \cap \mathbf{V}[f]$ as follows. Firstly, on $\operatorname{dom}(q_0)$ define $g'$ such that $g'|_{\operatorname{dom}(q_0)} \equiv q_0$. Secondly, on $\operatorname{dom}(q_0)^{\mathrm{c}}$ define $g'$ by the following induction: Suppose that the function $g'$ is already defined on some $k \in \omega$. Let $t_k \equiv g'|_{k \cap \operatorname{dom}(q_0)}$ and let $m = \min(\operatorname{dom}(q_0)^{\mathrm{c}} \setminus k)$. Then, by the definition of $q_0$, the conditions $p^0_m := q_0 \cup t_k \cup \langle m, 0\rangle$ and $p^1_m := q_0 \cup t_k \cup \langle m, 1\rangle$ are $\underset{\sim}{f}$-incompatible, *i.e.*, there is an $n \in \omega$ such that $p^0_m \Vdash_{\mathbb{S}_\mathscr{E}} \underset{\sim}{f}(n) = \varepsilon$ and $p^1_m \Vdash_{\mathbb{S}_\mathscr{E}} \underset{\sim}{f}(n) = 1 - \varepsilon$ (for some $\varepsilon \in \{0,1\}$). Take the least such $n$ and define $g'(m) := f(n)$. Notice that this can be done since we work in the model $\mathbf{V}[f]$ in which we know the value $f(n)$. Notice also that since $f = \underset{\sim}{f}[g]$, for $i \in \{0,1\}$ the condition $p^i_m$ belongs to $G$ *iff* $p^i_m \Vdash_{\mathbb{S}_\mathscr{E}} \underset{\sim}{f}(n) = g'(m)$. Thus, $g'(m)$ decides which of the two incompatible condition, $p^0_m$ or $p^1_m$, belongs to $G$. Now, because $q_0$ is in $G$, we see inductively how the function $g'$ reconstructs the $\mathbb{S}_\mathscr{E}$-generic filter $G$ or equivalently the function $g$, and since $g' \in \mathbf{V}[f]$ we consequently have $g \in \mathbf{V}[f]$.

"$q_0 \in D_2$": By definition, $q_0$ decides $\underset{\sim}{f}(k)$ for each $k \in \omega$, which shows that the function $f$ belongs to $\mathbf{V}$.

Hence, we have either $g \in \mathbf{V}[f]$ (if $q_0 \in D_1$) or $f \in \mathbf{V}$ (if $q_0 \in D_2$), which completes the proof. q.e.d.

## 6 Happy Families and Their Relatives

Firstly we recall Mathias' notion of a happy family (*cf.* [Mat77]): Let $[\omega]^{<\omega}$ be the set of all finite subsets of $\omega$, and for $s \in [\omega]^{<\omega}$, let $\bar{s}^+ := (\max s) + 1$. A set $x \subseteq \omega$ is said to **diagonalize** the set $\{x_s : s \in [\omega]^{<\omega}\} \subseteq [\omega]^\omega$, if $x \subseteq x_\varnothing$ and for all $s \in [\omega]^{<\omega}$, if $(\max s) \in x$, then $x \setminus \bar{s}^+ \subseteq x_s$. For $\mathscr{A} \subseteq [\omega]^\omega$ we write $\operatorname{fil}\mathscr{A}$ for the filter generated by the members of $\mathscr{A}$, *i.e.*, $\operatorname{fil}\mathscr{A}$ consists of all subsets of $\omega$ which are supersets of intersections of finitely many members of $\mathscr{A}$. A free family $\mathscr{E}$ is called a **happy family** if whenever $\operatorname{fil}\{x_s : s \in [\omega]^{<\omega}\} \subseteq \mathscr{E}$, then there is an $x \in \mathscr{E}$ which diagonalizes the set $\{x_s : s \in [\omega]^{<\omega}\}$.

An obvious example of a happy family is the set $[\omega]^\omega$, and it is not hard to see that all Ramsey ultrafilters are happy (*cf.* [Mat77, Section 0]). Other examples of happy families are Ramsey families:

**Fact 19.** Every Ramsey family is happy.

*Proof.* Let $\mathscr{E}$ be a free family which is not happy and let $\{x_s : s \in [\omega]^{<\omega}\} \subseteq \mathscr{E}$ be such that $\operatorname{fil}\{x_s : s \in [\omega]^{<\omega}\} \subseteq \mathscr{E}$ but there is no $x \in \mathscr{E}$ which diagonalizes the set $\{x_s : s \in [\omega]^{<\omega}\}$. We leave it as an exercise to the reader to construct with the set $\{x_s : s \in [\omega]^{<\omega}\}$ a winning strategy for the Maiden for the game $\mathcal{G}_\mathscr{E}$. q.e.d.

More examples of happy families we obtain by maximal almost disjoint families: An infinite family $\mathscr{A} \subseteq [\omega]^\omega$ is called **maximal almost disjoint**, abbreviated *m.a.d.*, if any two distinct sets of $\mathscr{A}$ have a finite intersection and for every $y \in [\omega]^\omega \setminus \mathscr{A}$ there is an $x \in \mathscr{A}$ such that $x \cap y$ is infinite. Now, let $\mathscr{A} \subseteq [\omega]^\omega$ be a *m.a.d.* family and let $\mathscr{F} = \text{fil}\{\omega \setminus x : x \in \mathscr{A}\}$, then $\mathscr{E}(\mathscr{A}) := \mathscr{F}^+$ is a happy family (*cf.* [Mat77, Proposition 0.7]). This example of a happy family leads to the following:

**Proposition 20.** Not every happy family is Ramsey.

*Proof.* Let $S = \{s_i : i \in \omega\}$ be the set of all finite sequences of $\omega$, which is partially ordered by the extension relation, denoted "$\prec$". For infinite sequences $f \in {}^\omega\omega$, let $x_f := \{i \in \omega : \exists n \in \omega(f|_n = s_i)\}$. Obviously, for any distinct sequences $f, g \in {}^\omega\omega$ we have that $x_f \cap x_g$ is finite. Now, let $\mathscr{A}_0 := \{x_f : f \in {}^\omega\omega\}$, then $\mathscr{A}_0 \subseteq [\omega]^\omega$ is a set of pairwise almost disjoint sets which can be extended to a *m.a.d.* family, say $\mathscr{A}$.

We show that $\mathscr{E}(\mathscr{A})$ is not a Ramsey family: Let $x_0 := \omega$ be the first move of the MAIDEN and let $a_0$ be DEATH' response. In general, for $i \in \omega$ she plays

$$x_{i+1} = \{i \in \omega : s_{a_n} \prec s_i\}.$$

For every $n \in \omega$, $x_n$ is the union of infinitely many members of $\mathscr{A}_0$, and therefore is an element of $\mathscr{E}(\mathscr{A})$. Now, by the MAIDEN's strategy, $s_{a_0} \prec s_{a_1} \prec \cdots$ corresponds to an infinite sequence $f \in {}^\omega\omega$ and therefore, $\{a_n : n \in \omega\} \subseteq x_f$ for some $x_f \in \mathscr{A}_0$, which implies that $\{a_n : n \in \omega\} \notin \mathscr{E}(\mathscr{A})$ and that DEATH loses the game. Thus, the MAIDEN has a winning strategy for the game $\mathcal{G}_{\mathscr{E}(\mathscr{A})}$, and hence, $\mathscr{E}(\mathscr{A})$ is not Ramsey. q.e.d.

A *m.a.d.* family $\mathscr{A}$ is called **strongly maximal almost disjoint** if given countably many members of $\mathscr{E}(\mathscr{A})$, then there is a member of $\mathscr{A}$ that meets each of them in an infinite set.

For a free family $\mathscr{E}$, consider the following game $\bar{\mathcal{G}}_{\mathscr{E}}$: The moves of the MAIDEN are members of $\mathscr{E}$ and DEATH responses like in the game $\mathcal{G}_{\mathscr{E}}$. Further, DEATH wins if and only if the set of integers played by DEATH belongs to $\mathscr{A}$, but has infinite intersection with each set played by the MAIDEN.

If $\mathscr{A}$ is a *m.a.d.* family, then obviously the MAIDEN has a winning strategy for $\bar{\mathcal{G}}_{\mathscr{E}(\mathscr{A})}$ if and only if $\mathscr{A}$ is not strongly *m.a.d.*, which motivates the following:

**Question 21.** Is it the case that for a *m.a.d.* family $\mathscr{A}$, $\mathscr{E}(\mathscr{A})$ is Ramsey if and only if $\mathscr{A}$ is strongly *m.a.d.*?

Related to happy families are the so-called moderately happy families introduced by Mathias in [Mat77, Section 9]: A free family $\mathscr{E}$ is **moderately happy** if whenever $\text{fil}\{x_n : n \in \omega\} \subseteq \mathscr{E}$, then there is an $x \in \mathscr{E}$ such that for all $n \in \omega$, $x \setminus x_n$ is finite.

On the one hand, it is not hard to verify that every P-family is moderately happy. On the other hand, by similar arguments as in the proof of Proposition 20, one can show that there exist moderately happy families which are not P-families.

Now, since the results of Section 4 are also valid for Mathias forcing restricted to happy families, it is natural to ask whether something similar holds for Silver forcing with respect to moderately happy families:

**Question 22.** Are the results of Section 5 also valid for Silver forcing restricted to moderately happy families?

## References.

[BarJud95] Tomek **Bartoszyński** and Haim **Judah**, Set Theory: On the Structure of the Real Line, AK Peters 1995

[Gri71] Serge **Grigorieff**, Combinatorics on Ideals and Forcing, **Annals of Mathematical Logic** 3 (1971), p. 363–394

[Jec86] Thomas **Jech**, Multiple Forcing, Cambridge University Press 1986 [Cambridge Tracts in Mathematics 88]

[Jec03] Thomas **Jech**, Set Theory, Springer 2003 [Springer Monographs in Mathematics]

[Kun83] Kenneth **Kunen**, Set Theory: An Introduction to Independence Proofs, Elsevier 1983 [Studies in Logic and the Foundations of Mathematics 102]

[KunVau84] Kenneth **Kunen** and Jerry E. **Vaughan** (*eds.*), Handbook of Set-Theoretic Topology, Elsevier 1984

[Laf96] Claude **Laflamme**, Filter Games and Combinatorial Properties of Strategies, **Contemporary Mathematics** 192 (1996), p. 51–67

[Mat77] Adrian R. D. **Mathias**, Happy Families, **Annals of Mathematical Logic** 12 (1977), p. 59–111

[Mat79] Adrian R. D. **Mathias**, Surrealist Landscape with Figures: A Survey of Recent Results in Set Theory, **Periodica Mathematica Hungarica** 10 (1979), p. 109–175

[She98] Saharon **Shelah**, Proper and Improper Forcing, Springer 1998 [Perspectives in Mathematical Logic]

[vDo84] Eric K. **van Douwen**, The Integers and Topology, *in:* [KunVau84, p. 111–167]

**Received**: February 24th, 2005;
**In revised version**: August 28th, 2005;
**Accepted by the editors**: October 16th, 2005.

Stefan **Bold**, Benedikt **Löwe**,
Thoralf **Räsch**, Johan **van Benthem** (*eds.*)
**Foundations of the Formal Sciences V**
Infinite Games

# Determinacy in Second Order Arithmetic

CHRISTOPH HEINATSCH AND MICHAEL MÖLLERFELD

Institut für Mathematische Logik und Grundlagenforschung
Westfälische Wilhelms-Universität Münster
Einsteinstraße 62
48149 Münster, Germany
heinatc@math.uni-muenster.de

ABSTRACT. Determinacy axioms state the existence of winning strategies for infinite two player games played on the natural numbers. We give an overview about characterizing some frequently used axiom systems in terms of determinacy and state that a base theory enriched by a certain scheme of determinacy axioms is $\Pi^1_1$-equivalent to a subsystem of second order arithmetic called $\Pi^1_2$-comprehension.

## 1 Introduction

The main question in the program of reverse mathematics, founded by Harvey Friedman, is to determine which axioms are necessary for proving theorems of ordinary mathematics. That should be done by not only proving the theorems from the axioms but also by deriving the axioms from the theorems. Many results of the reverse mathematics program are collected in [Sim99]. Historically, reverse mathematics arose from H. Friedman's attempt to demonstrate the necessary use of higher set theory in mathematical practice, *cf.*, *e.g.*, [Fri71]. One of the axioms that are of interest in reverse mathematics is the axiom of arithmetical comprehension, which ensures that for every arithmetical formula $\varphi(x)$ the set $\{x \mid \varphi(x)\}$ exists. The following theorems are all equivalent to this axiom over a weak base theory (we will give the exact definitions later): (a) Every countable field is isomorphic to a subfield of a countable algebraically closed field; (b) every countable vector space over a countable field has a basis; (c) every countable commutative ring has a maximal ideal. The proofs of these results can be found in [Sim99, Chapter III, Theorems 3.2, 4.3 and 5.5].

In this article we will examine a special kind of theorem which we introduce next. A game between two players is given by a set $A$ of infinite sequences of natural numbers, the payoff set. The two players alternately choose natural numbers, and so they produce an infinite sequence of natural numbers.

$$\begin{array}{lcccccccccc} \mathrm{I} & x_0 & & x_2 & & x_4 & & x_6 & & & \\ & & & & & & & & & & \dots \\ \mathrm{II} & & x_1 & & x_3 & & x_5 & & x_7 & & \end{array}$$

Player I wins if and only if this sequence is an element of the payoff set $A$, otherwise player II wins. A strategy for player I is a function that maps a natural number (the next move for player I) to each sequence of even length (the positions where player I has to make the next move). A strategy for player II is defined similarly with "odd" instead of "even". A strategy for player I (player II) is a winning strategy if and only if every run of a game consistent with the strategy is in $A$ (the complement of $A$). A game $A$ is determined if and only if either I or II has a winning strategy. In the following, we identify games with their payoff set, and say that any set of infinite sequences of natural numbers is a game. Let $\mathcal{A}$ be a class of games. The scheme of $\mathcal{A}$-determinacy says that every game in $\mathcal{A}$ is determined.

Determinacy is of great interest in descriptive set theory, and for many set theoretic axiom systems it is known for what complexity of the payoff set determinacy is provable in the axiom system. $\mathsf{ZFC}$ for example proves determinacy for Borel sets but not for analytic sets (*cf.*, *e.g.*, [Kec95, Theorem 20.5] for the first and [Kan97, p. 379] for the second claim).

In context of reverse mathematics mainly axiom systems much weaker than $\mathsf{ZFC}$ are considered. The reason for this is that the axiom "For each set there exists the power set" is an axiom of $\mathsf{ZFC}$ that is not really needed for proving most theorems of ordinary mathematics and that is conversely not derivable from these theorems. An axiom system much weaker than $\mathsf{ZFC}$ is second order arithmetic ($\mathsf{SOA}$) which we introduce next.

## 2 Second order arithmetic

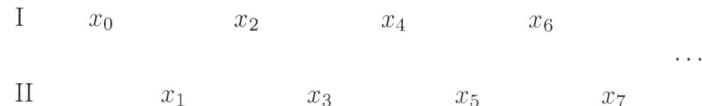

The language of second order arithmetic is a two sorted language, *i.e.*, we have variables for natural numbers and variables for sets of natural numbers. This language is sufficient to talk about real numbers because we can code a real number $r$ by the set of rational numbers which are less then $r$, and rational numbers are essentially pairs of natural numbers. Therefore we use set variables to represent real numbers. The language of second order arithmetic is strong enough to express wide parts of analysis.

**Definition 1 (The language of second order arithmetic $\mathcal{L}_2$).** The language of second order arithmetic $\mathcal{L}_2$ is a two sorted language and contains

- countably many number variables, denoted by $x_1$, $x_2$, ...,
- countably many variables for sets of natural numbers, denoted by $X_1$, $X_2$, ...,
- the constants 0 and 1,
- the binary function symbols $+$ and $\cdot$,
- the binary relation symbols $<$, $=$, $\in$,
- the logical symbols $\neg$, $\wedge$, $\vee$, $\exists$, $\forall$, $\rightarrow$, $\leftrightarrow$ and parentheses.

We define terms and formulas canonically. We allow quantification both for number and set variables.

**Definition 2 (The Lévy-hierarchy).** A formula is said to be $\Sigma^0_n$ ($\Pi^0_n$) for $n \in \mathbb{N}$ if and only if it is of the form $\exists x_1 \forall x_2 \exists x_3 \ldots x_n \varphi$ ($\forall x_1 \exists x_2 \forall x_3 \ldots x_n \varphi$) with $n$ alternating unbounded quantifiers in front and a formula $\varphi$ which contains no unbounded quantifiers (bounded quantifiers are $\forall x(x < m \rightarrow \ldots)$ and $\exists x(x < m \wedge \ldots)$). The formula $\varphi$ may contain further free number or set variables.

A formula is called arithmetical if and only if it contains no second order quantifiers. Analogously a formula is $\Sigma^1_n$ ($\Pi^1_n$) if and only if it is of the form

$$\exists X_1 \forall X_2 \exists X_3 \ldots X_n \varphi \ (\forall X_1 \exists X_2 \forall X_3 \ldots X_n \varphi)$$

for an arithmetical formula $\varphi$.

**Definition 3 ($\mathcal{L}_2$-structure).** An $\mathcal{L}_2$-structure is an ordered 7-tuple

$$M = (|M|, S_M, +_M, \cdot_M, 0_M, 1_M, <_M)$$

where $|M|$ is a set which serves as range of the number variables, $S_M$ is a set of subsets of $|M|$ serving as range of the set variables, $+_M$ and $\cdot_M$ are binary functions on $|M|$, $0_M$ and $1_M$ are distinguished elements of $|M|$, and $<_M$ is a binary relation on $|M|$. $=$ and $\in$ are interpreted as identity and membership relation.

The intended structure of $\mathcal{L}_2$ is $\mathcal{N} := (\mathbb{N}, \mathcal{P}(\mathbb{N}), +, \cdot, 0, 1, <)$ with $\mathcal{P}(A) := \{X | X \subseteq A\}$. A $\beta$-model is a model which satisfies all $\Sigma^1_1$-sentences which are true in $\mathcal{N}$; $\beta$-models are of importance because wellorderings are absolute for them. For more detailed information *cf.* [Sim99, Chapter VII].

Observe that according to our definition an $\mathcal{L}_2$-structure may be regarded as a first order structure. By interpreting $S_M$ as a predicate symbol, we can transform two sorted logic into first order logic; therefore there is a complete calculus by the completeness theorem for first order logic. Details can be found in [GlaPoh92, Chapter 5.1].

**Definition 4 (Second order arithmetic (SOA)).** The theory SOA is formulated in the language $\mathcal{L}_2$ and contains the universal closures of the following axioms and schemes:

- The base axioms

$$\neg(x+1=0);\ x+1=y+1 \to x=y;\ x+0=x;$$
$$x+(y+1)=(x+y)+1;\ x\cdot 0=0;\ x\cdot(y+1)=(x\cdot y)+x;$$
$$\neg(x<0);\ x<y+1 \leftrightarrow x<y \vee x=y.$$

- The induction axiom

$$\forall X\big((0\in X \wedge \forall x(x\in X \to x+1\in X)) \to \forall x(x\in X)\big).$$

- The comprehension scheme

  $\exists X \forall x(x\in X \leftrightarrow \varphi(x))$ for arbitrary formulas $\varphi$ in which $X$ does not occur freely.

  We allow parameters in $\varphi$, *i.e.*, for each $\varphi(x,\vec{y},\vec{Y})$

$$\forall\vec{y}\forall\vec{Y}\exists X(x\in X \leftrightarrow \varphi(x,\vec{y},\vec{Y}))$$

  is an instance of the comprehension scheme.

When we prove a theorem of ordinary mathematics in second order arithmetic, we mostly use comprehension and induction only for formulas with restricted complexity. Therefore we consider some subsystems of SOA.

**Definition 5 ($\mathsf{RCA}_0$).** The scheme of $\mathbf{\Delta}^0_1$-comprehension consists of all formulas

$$\forall x(\varphi(x) \leftrightarrow \psi(x)) \to \exists X\forall x(x\in X \leftrightarrow \varphi(x))$$

where $\varphi$ is $\Sigma^0_1$ and $\psi$ is $\Pi^0_1$. The scheme of $\mathbf{\Sigma}^0_1$-induction consists of all formulas

$$\varphi(0) \wedge \forall x(\varphi(x) \to \varphi(x+1))) \to \forall x\varphi(x)$$

where $\varphi$ is $\Sigma^0_1$. $\mathsf{RCA}_0$ is SOA with the induction axiom replaced by $\mathbf{\Sigma}^0_1$-induction and comprehension replaced by $\mathbf{\Delta}^0_1$-comprehension.

'RCA' stands for 'recursive comprehension axiom'. Roughly speaking, $\mathsf{RCA}_0$ is only strong enough to prove the existence of recursive sets of natural numbers. In reverse mathematics, it plays the rôle of a weak base theory which proves most of the equivalences between theorems of ordinary mathematics and axioms. Nevertheless $\mathsf{RCA}_0$ is strong enough to prove many theorems of analysis and algebra like the intermediate value theorem and the existence of an algebraic closure of a countable field (*cf.* [Sim99, Chapter II, Theorems 6.6 and 9.4]).

**Definition 6** ($\mathsf{ACA}_0$)**.** $\mathsf{ACA}_0$ is $\mathsf{SOA}$ with comprehension restricted to arithmetical formulas.

'ACA' stands for 'arithmetical comprehension axiom'. Examples for theorems of the same strength as $\mathsf{ACA}_0$ have been given in the introduction. These equivalences are provable in $\mathsf{RCA}_0$.

**Definition 7** ($\mathbf{\Pi}^1_{\mathsf{n}}$-$\mathsf{CA}_0$)**.** $\mathbf{\Pi}^1_{\mathsf{n}}$-$\mathsf{CA}_0$ is $\mathsf{SOA}$ with comprehension restricted to $\Pi^1_{\mathsf{n}}$-formulas.

There are other subsystems of second order arithmetic which are of interest in reverse mathematics; for example $\mathsf{ATR}_0$ which is $\mathsf{ACA}_0$ together with the scheme of transfinite arithmetical recursion which allows to iterate arithmetical comprehension along definable wellorderings; for an exact definition, *cf.* [Sim99, Theorem V.2.4].

## 3 Iterated inductive definitions

To examine for what complexity of payoff sets determinacy is provable in $\mathbf{\Pi}^1_2$-$\mathsf{CA}_0$ we will use that $\mathbf{\Pi}^1_2$-$\mathsf{CA}_0$ is strongly connected to a system of iterated monotone inductive definitions called $\mu$-calculus. For this we have to study monotone operators and their least fixed-points. An inductive definition is a monotone operator $\Gamma$, *i.e.*, a mapping from $\mathcal{P}(\mathbb{N})$ to $\mathcal{P}(\mathbb{N})$ with the property

$$X \subseteq Y \Rightarrow \Gamma(X) \subseteq \Gamma(Y).$$

A set $X$ with $\Gamma(X) \subseteq X$ is called $\Gamma$-closed. For each monotone operator there exists a (with respect to set inclusion) least fixed-point $I_\Gamma$ which is the intersection of all $\Gamma$-closed sets, *i.e.*,

$$I_\Gamma = \Gamma(I_\Gamma) = \bigcap\{X \subseteq \mathbb{N} \mid \Gamma(X) = X\}.$$

If $\alpha$ is an ordinal we define the $\alpha$th stage of a fixed-point by $I^\alpha_\Gamma := \Gamma(I^{<\alpha}_\Gamma)$ with $I^{<\alpha}_\Gamma := \bigcup_{\beta<\alpha} I^\beta_\Gamma$. Let

$$||\Gamma|| := \min \{\alpha \mid I^\alpha_\Gamma = I^{<\alpha}_\Gamma\}$$

be the closure ordinal of $\Gamma$. By cardinality reasons $||\Gamma||$ exists and is a countable ordinal, and we have $I_\Gamma^{||\Gamma||} = I_\Gamma$.

We can associate an operator $\Gamma_\varphi(X)$ to every formula $\varphi(x, X)$ by

$$\Gamma_\varphi(X) := \{x \mid \varphi(x, X)\}.$$

If $\varphi$ is $X$-positive (that means that $X$ occurs only within even numbers of negations) then $\Gamma_\varphi$ is monotone. If $\varphi$ contains further free variables $y$ and $Y$ then $I_\varphi$ also depends on $y$ and $Y$. If $Y$ occurs positively in $\varphi$ (negatively in $\psi$), then the operators $\Gamma_1(Y) := \{y \mid t \in I_\varphi(y, Y)\}$ ($\Gamma_2(Y) := \{y \mid t \notin I_\psi(y, Y)\}$) are also monotone (where $t$ denotes some first order term) and therefore have least fixed-points. The fixed-point of $\Gamma_1$ is arithmetical in the fixed-point of a positive arithmetical operator because the two inductive processes can be coded into one. This is in general not true for fixed-points of $\Gamma_2$. The idea of the $\mu$-calculus is to iterate that and to talk about complex nestings of fixed-points. For this we augment the language of second order arithmetic.

**Definition 8 (The language of the $\mu$-calculus $\mathcal{L}_\mu$).** We start with the language of second order arithmetic and add a set-constructor $\mu$. For each $X$-positive formula $\varphi(x, X)$ which contains no second order quantification we add a set term $\mu x X \varphi(x, X)$ which is intended to denote the least fixed-point of the monotone operator $\Gamma_\varphi(X)$. A free variable $Y$ occurs positively in $t \in \mu x X \varphi(x, X)$ ($t \notin \mu x X \varphi(x, X)$) if and only if $\varphi$ is $Y$-positive ($Y$-negative). The formula $\varphi$ may contain further $\mu$-terms such that nestings of fixed-points are possible.

**Definition 9 ($\mu$-calculus).** The $\mu$-calculus is formulated in $\mathcal{L}_\mu$ and comprises the following axioms:

- The axioms of $\mathsf{ACA}_0$ with comprehension for all $\mathcal{L}_\mu$-formulas without second order quantifiers.,
- $\forall x[x \in \mu x X \varphi(x, X) \leftrightarrow \varphi(x, \mu x X \varphi(x, X))]$ for each $X$-positive and first-order $\varphi$, which means that $\mu x X \varphi(x, X)$ is a fixed-point of the operator given by $\varphi$, and
- $\forall Y[\forall x(\varphi(x, Y) \to x \in Y) \to \mu x X \varphi(x, X) \subseteq Y]$ for each $X$-positive and first-order $\varphi$, which means that $\mu x X \varphi(x, X)$ is the least fixed-point.

We can study the $\mu$-calculus instead of $\mathbf{\Pi}^1_2$-$\mathsf{CA}_0$, because they are closely connected in the following way:

**Definition 10.** Two theories are $\Pi^1_1$-equivalent if and only if they prove the same $\Pi^1_1$-sentences.

**Theorem 11 (Möllerfeld).** The $\mu$-calculus and $\mathbf{\Pi}^1_2$-$\mathsf{CA}_0$ are $\Pi^1_1$-equivalent.

*Proof. Cf.* [Möl02]. q.e.d.

## 4 Some results about determinacy

To characterize the introduced theories in terms of determinacy we have to describe the complexity of payoff sets. We can view the natural numbers as a topological space with the discrete topology, *i.e.*, every subset is open. The Baire space is the product space of countable many copies of the natural numbers equipped with the product topology. Therefore the elements of the Baire space are infinite sequences of natural numbers, and a basis of this topology is formed by the sets $\mathcal{N}_s$, *i.e.*, the set of all infinite sequences which begin with the finite sequence $s$. The class of Borel sets (also called $\mathbf{\Delta}^1_1$-sets) is the least class which contains the open sets and is closed under complements and countable unions (and therefore also under countable intersections). In analogy to the Lévy hierarchy for formulas we define a hierarchy which orders the Borel sets.

- $\mathbf{\Sigma}^0_1 = \{X \mid X \text{ is open}\}$,
- $\mathbf{\Sigma}^0_\alpha = \{\bigcup_n X_n \mid \exists(\beta_n)_{n\in\omega}(1 \leq \beta_n < \alpha \wedge X_n \in \mathbf{\Pi}^0_{\beta_n})\}$ for $\alpha > 1$,
- $\mathbf{\Pi}^0_\alpha = \{X^{\mathsf{C}} \mid X \in \mathbf{\Sigma}^0_\alpha\}$ for $\alpha \geq 1$, where $X^{\mathsf{C}}$ denotes the complement of $X$.

Martin proved that in $\mathsf{ZFC}$ every game with a Borel payoff set is determined (*cf.* [Kec95, Theorem 20.5]). But this is far away from what is provable in $\mathsf{SOA}$. Let us for example have a look at the determinacy of open sets. Fix a payoff set $A = \bigcup_{s\in S}\mathcal{N}_s$ and suppose player I has no winning strategy. We have to produce a winning strategy for player II. First notice that in a position where player II has to make the next move and player I has no winning strategy, player II can choose her move in a way that after this move player I has still no winning strategy. On the other hand, if player I has no winning strategy in a position where she herself has to make the next move, she still has no winning strategy after her move, no matter what she played. Therefore player II can achieve that player I never gets into a position where she has a winning strategy. But that means that they never arrive at a position which has an initial sequence in $S$, for then player I would have a winning strategy (she can make arbitrary moves and always wins). Therefore the played sequence is not in $A$, and player II wins. This shows that in case of an open game, it is sufficient for player II to avoid losing positions to win the game. The determinacy of open games is known as Gale-Stewart Theorem (*cf.* [GalSte53]). In the spirit of Friedman's program of reverse mathematics we can ask now: Which axioms of SOA are necessary to prove the Gale-Stewart Theorem? The answer is the following theorem (*cf.* [Sim99, Theorem V.8.7]).

**Theorem 12.** $\mathsf{RCA}_0$ proves: The scheme of arithmetical transfinite recursion is equivalent to the scheme of $\mathbf{\Sigma}^0_1$-determinacy.

To characterize the complexity of the winning strategies we can ask where they appear in Gödel's constructible universe $\mathbf{L}$ (for $\mathbf{L}$, *cf.*, *e.g.*, [Jec97, Chapter 2.12]). If the game $A$ is recursively open (that means that $A = \bigcup_{s\in S}\mathcal{N}_s$ for recursive $S$), then there is either a winning strategy for player I in $\mathbf{L}_{\omega_1^{\mathrm{CK}}}$ or a winning strategy for player II in $\mathbf{L}_{\omega_1^{\mathrm{CK}}+1}$ ($\omega_1^{\mathrm{CK}}$ is the first ordinal which is not representable by a recursive wellordering).

Tanaka has proved some results about determinacy in some stronger subsystems of SOA. He formulated the theory $\mathbf{\Sigma}^1_1$-$\mathsf{MI}_0$, which is an extension of $\mathsf{ACA}_0$ that guarantees the existence of least fixed-points of monotone $\mathbf{\Sigma}^1_1$-definable operators. In this theory he could formalize Wolfe's proof of $\mathbf{\Sigma}^0_2$-determinacy (*cf.* [Mos80, Theorem 6A.3]).

**Theorem 13.**

1. $\mathsf{RCA}_0$ proves that $\mathbf{\Pi}^1_1$-comprehension and $\mathbf{\Sigma}^0_1 \wedge \mathbf{\Pi}^0_1$-determinacy are equivalent, where $\mathbf{\Sigma}^0_1 \wedge \mathbf{\Pi}^0_1$ is the class of all intersections of a $\mathbf{\Sigma}^0_1$ and a $\mathbf{\Pi}^0_1$-set (in terms of the difference hierarchy also $\mathrm{Diff}(2; \mathbf{\Sigma}^0_1)$, *cf.* Section 6).
2. $\mathsf{RCA}_0$ proves that $\mathbf{\Sigma}^1_1$-$\mathsf{MI}_0$ and $\mathbf{\Sigma}^0_2$-determinacy are equivalent.

*Proof. Cf.* [Sim99, Theorem VI.5.4] for a proof of the first and [Tan91] for the second claim. q.e.d.

Let $\sigma$ be the supremum of the ordinals $||\Gamma||$ for monotone, $\mathbf{\Sigma}^1_1$-definable operators $\Gamma$. Solovay showed that for recursively coded $\mathbf{\Sigma}^0_2$-games the winning strategies for player I occur in $\mathbf{L}_\sigma$, the winning strategies for player II lie in the next admissible set beyond $\mathbf{L}_\sigma$.[1] The following result of Philip Welch (*cf.* [Wel03]) implies that $\mathbf{\Pi}^1_3$-$\mathsf{CA}_0$ is much stronger than $\mathbf{\Sigma}^0_3$-determinacy.

**Theorem 14.** $\mathbf{\Pi}^1_3$-$\mathsf{CA}_0$ proves that there is a $\beta$-model of $\mathbf{\Delta}^1_3$-$\mathsf{CA}_0 + \mathbf{\Sigma}^0_3$-determinacy.

Let $\Game xX\varphi(x, X) := \{x \,|\, \text{Player I has a winning strategy in } \{X \,|\, \varphi(x, X)\}\}$. An operator is $\Game\mathbf{\Sigma}^0_3$-definable if and only if it is given by an arithmetical

[1] *Cf.* [Mos80, p. 414–415] and [Mar∞, Exercise 1.3.2]. For more on admissible sets, *cf.* [Bar75, Chapter I].

| Theory | Determinacy strength | Reference |
|---|---|---|
| ZFC | $\mathbf{\Delta}^1_1$ | [Kec95, Theorem 20.5]; [Kan97, p. 379] |
| SOA | $< \mathbf{\Sigma}^0_4$ | [Mar∞, Exercise 1.4.2] |
| $\mathbf{\Pi}^1_3$-$\mathsf{CA}_0$ | $> \mathbf{\Sigma}^0_3$ | [Wel03] |
| $\mathbf{\Pi}^1_2$-$\mathsf{CA}_0$ | ??? | |
| $\mathbf{\Sigma}^1_1$-$\mathsf{MI}_0$ | $\mathbf{\Sigma}^0_2$ | [Tan91] |
| $\mathbf{\Pi}^1_1$-$\mathsf{CA}_0$ | $\mathbf{\Sigma}^0_1 \wedge \mathbf{\Pi}^0_1$ | [Sim99, Theorem VI.5.4] |
| $\mathsf{ATR}_0$ | $\mathbf{\Sigma}^0_1$ | [Sim99, Theorem V.8.7] |

Table 1. Determinacy strength.

formula which may contain set terms of the form $\Game\varphi$ for a $\mathbf{\Sigma}^0_3$-formula $\varphi$. Let

$$\gamma := \sup\{||\Gamma|| \mid \Gamma \text{ is a monotone, } \Game\mathbf{\Sigma}^0_3\text{-definable operator}\}.$$

There is an unpublished result by Martin and Solovay that the winning strategies for player I in recursively coded $\mathbf{\Sigma}^0_3$-games lie in $\mathbf{L}_\gamma$, and John in [Joh86] proved something similar.

Adding stronger instances of the comprehension scheme does not yield much stronger determinacy results.

**Theorem 15.** SOA does not prove $\mathbf{\Sigma}^0_4$-determinacy.

In [Fri71], Harvey Friedman showed that SOA does not prove $\mathbf{\Sigma}^0_5$-determinacy; Martin later sharpened that result to $\mathbf{\Sigma}^0_4$.

We summarize these results in Table 1. By "determinacy strength" we mean an informal summary of the different ways of characterizing a theory in terms of determinacy, as seen in the previous theorems.

## 5 Generalized quantifiers

As pointed out in [Möl02], the $\mu$-calculus is strongly connected to a calculus with generalized quantifiers. A generalized quantifier $Q$ on the natural numbers is a subset of $\mathcal{P}(\mathbb{N})$ with the following properties:

- $\varnothing \notin \mathsf{Q}$,
- $\mathsf{Q} \neq \varnothing$, and
- $X \subseteq Y \wedge X \in \mathsf{Q} \Rightarrow Y \in \mathsf{Q}$.

The string $(\mathsf{Q}x)\varphi(x)$ is an abbreviation for $\{x \mid \varphi(x)\} \in \mathsf{Q}$. In this notation the quantifier $\forall$ is $\{\mathbb{N}\}$ and $\exists$ is $\{X \subseteq \mathbb{N} \mid X \neq \varnothing\}$. For each quantifier $\mathsf{Q}$ the inverse $\overline{\mathsf{Q}} := \{\{x \mid x \notin X\} \mid X \notin \mathsf{Q}\}$ is again a quantifier such that

$$\overline{\overline{\mathsf{Q}}} = \mathsf{Q} \text{ and } (\overline{\mathsf{Q}}x)\varphi(x) \leftrightarrow \neg\mathsf{Q}x\neg\varphi(x).$$

We fix a coding function $\langle\cdot\rangle$. Then $\langle\rangle$ is (the code of) the empty sequence. For each quantifier $\mathsf{Q}$ we define the next quantifier $\mathsf{Q}^{\vee}$:

$$(\mathsf{Q}^{\vee}x)\varphi(x) :\Leftrightarrow (\overline{\mathsf{Q}}x_0)(\overline{\mathsf{Q}}x_1)(\overline{\mathsf{Q}}x_2)\cdots \bigvee_{n\in\mathbb{N}} \varphi(\langle x_0,\ldots,x_n\rangle)$$

As an example we study the formula $\exists^{\vee}(x)\varphi(x)$. Since $\overline{\exists} = \forall$ we obtain that $\exists^{\vee}(x)\varphi(x)$ is equivalent to $\forall f \exists n \varphi(f[n])$, where $f[n]$ denotes the sequence $\langle f(0), f(1), \ldots, f(n)\rangle$ for a function $f$. The quantifier $\exists^{\vee}$ is commonly known as Suslin-quantifier. By iteration we obtain generalized quantifiers $\exists^n$ and $\forall^n$ for each $n \in \omega$ by $\exists^0 := \exists$, $\exists^{n+1} := (\exists^n)^{\vee}$ and $\forall^n := \overline{\exists^n}$.

We introduce a language and a theory where we can talk about $\exists^n$ and $\forall^n$.

**Definition 16 (The language $\mathcal{L}_{\mathfrak{D}}$ of the theory $\mathfrak{D}$ame).** We enlarge the language $\mathcal{L}_2$ by quantifiers $\forall^n$ and $\exists^n$ for each $n \in \omega$. These quantifiers can be used in formulas like the ordinary first order quantifiers $\forall$ and $\exists$ with the restriction that $\forall^n x\varphi(x)$ and $\exists^n\varphi(x)$ can only be built if $\varphi$ contains no second order quantifiers (but may contain further quantifiers $\exists^m$ or $\forall^m$).

**Definition 17 (The theory $\mathfrak{D}$ame).** The theory $\mathfrak{D}$ame is formulated in $\mathcal{L}_{\mathfrak{D}}$ and contains the following axioms:

- The axioms of $\mathsf{ACA}_0$, with comprehension for all $\mathcal{L}_{\mathfrak{D}}$-formulas without second order quantifiers,
- $\exists^0 x\varphi(x) \leftrightarrow \exists x\varphi(x)$,
- $\forall^n x\varphi(x) \leftrightarrow \neg\exists^n x\neg\varphi(x)$, and
- $\exists^{n+1} x\varphi(x,\vec{y},\vec{Y}) \leftrightarrow \forall X\big(\forall x(\varphi^{\exists^n}(x,\vec{y},X,\vec{Y}) \rightarrow x \in X) \rightarrow \langle\,\rangle \in X\big)$.

for each $\mathcal{L}_{\mathfrak{D}}$-formula $\varphi$ without second order quantifiers. The string $\varphi^{\mathsf{Q}}(s,\vec{y},X,\vec{Y})$ is an abbreviation for

$$\varphi(s,\vec{y},\vec{Y}) \vee (\overline{\mathsf{Q}}x)s^\frown\langle x\rangle \in X.$$

The formula $\varphi^{\mathsf{Q}}$ is $X$-positive, so

$$\forall X\big(\forall x(\varphi^{\exists^n}(x,\vec{y},X,\vec{Y}) \to x \in X) \to \langle\,\rangle \in X\big)$$

means " the least fixed-point of the operator given by $\varphi^{\exists^n}(x,X)$ contains $\langle\,\rangle$." So $\exists^{n+1}$ is the next quantifier of $\exists^n$.

It is immediate that the theory $\mathfrak{G}$ame is a subsystem of the $\mu$-calculus because of

$$\exists^{n+1}x\varphi(x,\vec{y},\vec{Y}) \leftrightarrow \langle\,\rangle \in \mu x X \varphi^{\exists^n}(x,\vec{y},X,\vec{Y}).$$

But the converse is also true: As shown in [Möl02] the theories $\mathfrak{G}$ame and the $\mu$-calculus are of equal strength in the following sense:

**Theorem 18 (Möllerfeld).** The $\mu$-calculus and $\mathfrak{G}$ame prove the same $\mathcal{L}_2$-sentences.

## 6 The main theorem

Let $\omega$ denote the natural numbers and $<\!\omega\text{-}\boldsymbol{\Sigma}^0_2$ the closure of $\boldsymbol{\Sigma}^0_2$ under complements and finite unions (and therefore also under finite intersections). We chose the notation $<\!\omega\text{-}\boldsymbol{\Sigma}^0_2$ to indicate that this is also the union of the finite levels of the difference hierarchy over $\boldsymbol{\Sigma}^0_2$ which are defined by

$$\mathrm{Diff}(n;\boldsymbol{\Sigma}^0_2) := \{A \mid \exists A_0,\ldots,A_{n-1} \in \boldsymbol{\Sigma}^0_2(A = \mathrm{Diff}_{i<n} A_i)\} \text{ for } n \in \omega$$

with

$$\mathrm{Diff}_{i<n} A_i := \{x \in A_0 \mid \text{the least } i \le n \text{ such that } x \notin A_i \text{ is odd}\}$$

where $A_n = \varnothing$. Then we obtain $<\!\omega\text{-}\boldsymbol{\Sigma}^0_2 = \bigcup_{n\in\omega} \mathrm{Diff}(n;\boldsymbol{\Sigma}^0_2)$. If $<\!\omega\text{-}\boldsymbol{\Sigma}^0_2\text{-}\mathsf{Det}_0$ is $\mathsf{ACA}_0$ together with determinacy for $<\!\omega\text{-}\boldsymbol{\Sigma}^0_2$-formulas, the following is the main theorem of this article:

**Theorem 19.** The $\mu$-calculus and $<\!\omega\text{-}\boldsymbol{\Sigma}^0_2\text{-}\mathsf{Det}_0$ prove the same $\mathcal{L}_2$-sentences.

Together with Theorem 11 we obtain

**Corollary 20.** $\boldsymbol{\Pi}^1_2\text{-}\mathsf{CA}_0$ and $<\!\omega\text{-}\boldsymbol{\Sigma}^0_2\text{-}\mathsf{Det}_0$ are $\Pi^1_1$-equivalent.

The starting point in the proof of the direction "The $\mu$-calculus proves $<\!\omega\text{-}\boldsymbol{\Sigma}^0_2\text{-}\mathsf{Det}$" is Wolfe's theorem (*cf.* [Mos80, Theorem 6A.3]) which says the every $\boldsymbol{\Sigma}^0_2$-set is determined. This was generalized by Bradfield to $<\!\omega\text{-}\boldsymbol{\Sigma}^0_2$-sets (*cf.* [Bra99]), and we formalized this proof in the $\mu$-calculus.

The other direction is more interesting in this context because it employs infinite games. We use the fact that the quantifiers $\exists^n$ and $\forall^n$ can be described by infinite games. Take for example the formula $\exists^1 x\varphi(x)$. We already mentioned that we can write this formula as $\forall f \exists n \varphi(f[n])$. Imagine a two player game where player I wants to show that the formula is true, player II wants to show that it is false. Player II has to play natural numbers, player I can decide before every move of player II if she wants to pass or play "stop".

| | | | | | | | | | |
|---|---|---|---|---|---|---|---|---|---|
| I | p | | p | | p | | p | | stop |
| II | | $x_0$ | | $x_1$ | | $x_2$ | | $x_3$ | |

If player I never plays "stop", she loses. Let $\langle x_1, \dots, x_i \rangle$ be the sequence played by player II until the first break. Then I wins if $\varphi(\langle x_1, \dots, x_i \rangle)$ is true, otherwise II wins. This game is open because if player I wins, she has won the game already after finitely many moves, namely after she played her "stop".

We can describe all quantifiers $\exists^n$ and $\forall^n$ in that way, and the complexity of the games is always in $<\omega$-$\mathbf{\Sigma}^0_2$. Using these games one can prove the remaining direction of Theorem 19. The proof is in [HeiMöl∞] and [Hei03].

## References.


[Bar75] Jon **Barwise**, Admissible Sets and Structures, Springer 1975

[Bra99] Julian C. **Bradfield**, Fixpoint Alteration and the Game Quantifier, *in:* [FluRod99, p. 350–361]

[FluRod99] Jörg **Flum** and Mario **Rodriguez-Artalejo** (*eds.*), Computer Science Logic, Proceedings of the 13th International Workshop, CSL'99, 8th Annual Conference of the EACSL, Madrid, Spain, September 20-25, 1999, Springer 1999 [Lecture Notes in Computer Science]

[Fri71] Harvey **Friedman**, Higher Set Theory and Mathematical Practice, **Annals of Mathematical Logic** 2 (1971), p. 326–357

[GalSte53] David **Gale** and Frank M. **Stewart**, Infinite Games with Perfect Information, *in:* [KuhTuc53, p. 245–266]

[GlaPoh92] Thomas **Glaß** and Wolfram **Pohlers**, An Introduction to Mathematical Logic, A Lecture by Wolfram Pohlers, *Lecture notes*, University of Münster 1992

[Hei03] Christoph **Heinatsch**, Zur Determiniertheitsstärke von $\Pi^1_2$-Komprehension, *Master thesis*, University of Münster 2003

[HeiMöl∞] Christoph **Heinatsch** and Michael **Möllerfeld**, The Determinacy Strength of $\Pi^1_2$-Comprehension, *in preparation*

[Jec97] Thomas **Jech**, Set Theory, 2nd edition, Springer 1997

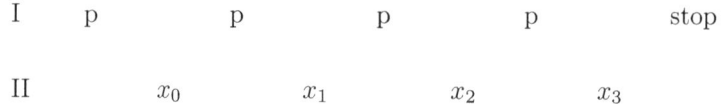

[Joh86] Thomas **John**, Recursion in Kolmogoroff's R-operator and the ordinal $\sigma_3$, **Journal of Symbolic Logic** 51 (1986), p. 1–11

[Kan97] Akihiro **Kanamori**, The Higher Infinite: Large Cardinals in Set Theory from Their Beginnings, second printing, Springer 1997 [Perspectives in Mathematical Logic]

[Kec95] Alexander S. **Kechris**, Classical Descriptive Set Theory, Springer 1995 [Graduate Texts in Mathematics 156]

[KuhTuc53] Harold W. **Kuhn** and Albert W. **Tucker** (*eds.*), Contributions to The Theory of Games, volume 2, Princeton University Press 1953 [Annals of Mathematics Studies 28]

[Mar∞] Donald A. **Martin** Proving Determinacy, *forthcoming*

[Möl02] Michael **Möllerfeld**, Generalized Inductive Definitions: The $\mu$-Calculus and $\Pi^1_2$-Comprehension, *PhD thesis*, University of Münster 2002

[Mos80] Yiannis N. **Moschovakis**, Descriptive Set Theory, North Holland 1980 [Studies in Logic and the Foundations of Mathematics 100]

[Sim99] Stephen G. **Simpson**, Subsystems of Second Order Arithmetic, Springer 1999

[Tan91] Kazuyuki **Tanaka**, Weak Axioms of Determinacy and Subsystems of Analysis II: $\Sigma^0_2$-Games, **Annals of Pure and Applied Logic** 52 (1991), p. 181–193

[Wel03] Philip **Welch**, Weak Systems of Determinacy and Arithmetical Quasi-Inductive Definitions, *preprint*, 2003

[Wol55] Philip **Wolfe**, The Strict Determinateness of Certain Infinite Games, **Pacific Journal of Mathematics** 5 (1955), p. 841–847


**Received**: March 1st, 2005;
**In revised version**: June 8th, 2005; November 15th, 2005;
**Accepted by the editors**: June 7th, 2007.

Stefan **Bold**, Benedikt **Löwe**,
Thoralf **Räsch**, Johan **van Benthem** (*eds.*)
**Foundations of the Formal Sciences V**
Infinite Games

# Games in Algebraic Logic: Axiomatisations and Beyond

ROBIN HIRSCH AND IAN HODKINSON*

Department of Computer Science
University College London
Gower Street
London WC1E 6BT, UK
R.Hirsch@cs.ucl.ac.uk

Department of Computing
Imperial College London
South Kensington Campus
London SW7 2AZ, UK
imh@doc.ic.ac.uk

ABSTRACT. A classical problem in algebraic logic is to characterise classes of representable algebras. Taking the example of the representable Tarskian relation algebras, we will discuss how games can help with such problems, and how they lead to a deeper study of representability.

## Introduction

A classical problem in algebraic logic is to characterise classes of representable algebras. Taking the example of the representable Tarskian relation algebras, we will discuss how games can help with such problems, and how they lead to a deeper study of representability. We will be able to use the games to help to explain some classical results in this area, and to discuss some more recent ones.

## 1 Algebras of relations

Algebraic formalisation of unary relations began with Boole in the 19th century. It was very successful. The boolean algebra axioms are sound and

*We would like to thank the referee for helpful comments on a draft of this paper. The second author thanks the organisers for inviting him to speak at the conference.

complete: every boolean algebra is isomorphic to a field of sets [Sto36].

De Morgan proposed considering *binary* (and higher-arity) relations. Peirce and Schröder developed the theory and established hundreds of laws of binary relations (see, *e.g.*, [Sch95]). [Mad91] has an interesting discussion of the history. But Pierce lamented:

> The logic of relatives is highly multiform; it is characterized by innumerable immediate conclusions from the same set of premises. ... The effect of these peculiarities is that this algebra cannot be subjected to hard and fast rules like those of the Boolian calculus; and all that can be done in this place is to give a general idea of the way of working with it. [Pei33, 3.342]

In the 1940s, Tarski and his collaborators began to investigate binary relations with modern algebra. Tarski laid down the notion of a *field of binary relations,* by which he meant a subalgebra of a product of algebras of the form

$$\mathfrak{Re}(X) = (\wp(X \times X), \cup, \setminus, \varnothing, X \times X, \mathrm{Id}_X, {}^{-1}, \mid),$$

for some set $X$, where

$$\begin{array}{rcl} \mathrm{Id}_X & = & \{(x,x) : x \in X\}, \\ R^{-1} & = & \{(y,x) : (x,y) \in R\}, \\ R|S & = & \{(x,y) : \exists z((x,z) \in R \wedge (z,y) \in S)\}. \end{array}$$

He wanted to characterise the algebras isomorphic to fields of binary relations. Such algebras are called *representable relation algebras,* the class of them is denoted **RRA**, and the isomorphism is called a *representation.*

It's easily seen why Tarski wanted to admit *subalgebras* of $\mathfrak{Re}(X)$. They are simply obtained by omitting some of the relations in $\mathfrak{Re}(X)$, but they still contain $\varnothing$, $X \times X$, and $\mathrm{Id}_X$, and are closed under the operations, so they can certainly be considered as algebras of binary relations.

But why *products?* One could argue that if $X_i$ $(i \in I)$ are pairwise disjoint and have union $X$, the product $\prod_{i \in I} \mathfrak{Re}(X_i)$ is isomorphic to the *relativisation* of $\mathfrak{Re}(X)$ to the equivalence relation $E = \bigcup_{i \in I}(X_i \times X_i)$ on $X$, defining 'being in the same $X_i$'. Relations not contained in $E$ are deleted, and the algebra operations are intersected with $E$: *e.g.*, $a\,;b$ in the relativisation is defined to be $c \cap E$, where $c$ is $a\,;b$ evaluated in $\mathfrak{Re}(X)$. Such a relativisation is some sort of algebra of binary relations, but maybe not the kind one would first think of considering. So perhaps a better answer is that under this 'subalgebras of products' definition, **RRA** is a *variety* — an equationally axiomatised class. This was proved by Tarski in [Tar55]. It follows from Birkhoff's theorem [Bir35] that **RRA** is closed under subalgebras, products, and homomorphic images.

An algebra is *simple* if it has no non-trivial proper homomorphic images. It can be shown that all simple representable relation algebras are isomorphic to subalgebras of $\mathfrak{Re}(X)$ for some $X$: there is no need to consider products. For simplicity of exposition, we will generally restrict our attention here to simple algebras; but most of what we say is either true for arbitrary ones, or can easily be generalised to them. We also generally consider only *non-degenerate* relation algebras, satisfying $0 \neq 1$. (When $0 = 1$, the algebra has only one element; it is isomorphic to $\mathfrak{Re}(\varnothing)$ and so is representable. This case is not interesting.)

**Relation algebras** In [Tar41], Tarski proposed axioms to capture **RRA**. These axioms defined the class **RA** of *'relation algebras'*.

**Definition 1.** A *relation algebra* is an algebra of the form $\mathcal{A} = (A, +, -, 0, 1, 1', \breve{}, ;)$ such that

- $(A, +, -, 0, 1)$ is a boolean algebra
- $(A, ;, 1')$ is a monoid
- 'Peircean law' (actually discovered by De Morgan): $(a ; b) \cdot c \neq 0 \iff (\breve{a} ; c) \cdot b \neq 0 \iff a \cdot (c ; \breve{b}) \neq 0$ for all $a, b, c \in A$.

As is standard, we use the notation $+, -, 0, 1, 1', \breve{}, ;$ for 'abstract' algebra operations corresponding to the 'concrete', set-theoretically defined operations $\cup, \setminus, \varnothing, X \times X, \mathrm{Id}_X, {}^{-1}, \mid$ (respectively) on algebras of binary relations. Considering *triangles* helps to make the point of the Peircean law clear:

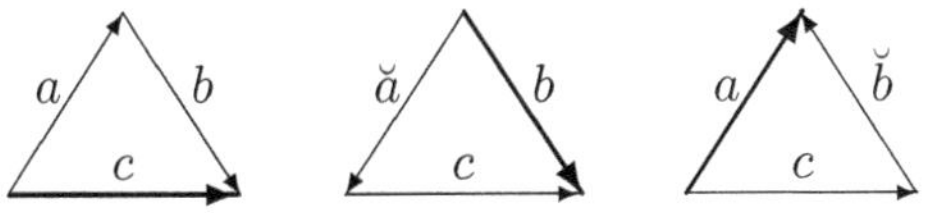


The axioms in Definition 1 are equivalent to Tarski's original ones, which were *equations*. We will not need it here, but it may be of interest to mention that the relation algebra axioms actually capture all equations valid in **RRA** that can be proved with 4 variables. The underlying proof system here can be (*e.g.*,) a sequent calculus or a Hilbert system for first-order logic, tuned to produce proofs using only 4 variables. For details, see [Mad83, TarGiv87, Mad89].

Did Tarski's axioms capture **RRA**? Well, soundness (**RRA** $\subseteq$ **RA**) is easily seen. But completeness failed. In a celebrated 1950 paper, Lyndon [Lyn50] gave an example of an algebra $\mathcal{A} \in$ **RA** $\setminus$ **RRA**. In 1964, Monk [Mon64], building on work of Lyndon [Lyn61] and Jónsson [Jón59], showed that **RRA** *is not finitely axiomatisable,* so proving the key 'negative' result in the field. Many other negative results about **RRA** are now known. One of the stronger ones is:

**Theorem 2 (Hirsch, Hodkinson, [HirHod02a, Theorem 18.13]).** There is no algorithm to tell whether an arbitrary finite relation algebra is representable.

The following *problem* was stated in [HenMonTar71] for 'cylindric algebras', but the version for relation algebras is just as pertinent: *find a simple intrinsic characterisation of (the algebras in)* **RRA**. In the next sections, we will look into this question using games.

# 2 Case study: atomic and finite relation algebras

First, we try to cast relation algebras and representations in a more manageable form. This is quite useful for *atomic* relation algebras, and for representations of *finite* relation algebras. We will consider the general case later.

## 2.1 Atomic relation algebras

An element $a$ of a relation algebra $\mathcal{A}$ is said to be an *atom* if $a$ is a minimal non-zero element with respect to the standard boolean algebra ordering '$\leq$', where $x \leq y \iff x + y = y$. $\mathcal{A}$ is said to be *atomic* if for every non-zero element $x$ of $\mathcal{A}$, there is an atom $a$ of $\mathcal{A}$ with $a \leq x$. All finite relation algebras are atomic, of course. We will say more about infinite atomic relation algebras in Sections 4 and 5.

Atomic relation algebras can be quite easily specified. One can prove from the **RA** axioms that $\breve{}$ and ; are *additive.* That is, $(a + b)\breve{} = \breve{a} + \breve{b}$, $(a + b) ; c = a ; c + b ; c$, and $a ;(b + c) = a ; b + a ; c$ are valid laws in relation algebras. We can even prove from the **RA** axioms that $\breve{}$ and ; are additive over infinite sums. It follows that in an atomic relation algebra $\mathcal{A}$, the operations $\breve{}$ and ; are determined by their values on atoms, and we can specify $\mathcal{A}$ by stating:

- the set $\mathrm{At}\,\mathcal{A}$ of atoms of $\mathcal{A}$, and which elements of $\mathcal{A}$ are the sum of which atoms (this pins down the boolean structure of $\mathcal{A}$),
- which atoms are $\leq 1$',
- $\breve{a}$, for each atom $a$ (it turns out that $\breve{a}$ is also an atom),
- for each $a, b, c \in \mathrm{At}\,\mathcal{A}$, whether $a ; b \geq c$ or not. In the case where $a ; b \geq c$, we say that $(a, b, c)$ is a *'consistent triple'.*

Remark: It follows from the Peircean law that $(a, b, c)$ is consistent if and only if its *Peircean transforms* $(a, b, c)$, $(\breve{a}, c, b)$, $(c, \breve{b}, a)$, $(b, \breve{c}, \breve{a})$, $(\breve{c}, a, \breve{b})$, $(\breve{b}, \breve{a}, \breve{c})$ are all consistent.

### 2.2 Ultrafilters

Given a relation algebra $\mathcal{A}$, we'll write $\mathcal{A}$ for its domain as well. An *ultrafilter* of $\mathcal{A}$ is a subset $\alpha \subseteq \mathcal{A}$ such that

1. $a, b \in \alpha \Rightarrow a \cdot b \in \alpha$,
2. $a \geq b \in \alpha \Rightarrow a \in \alpha$,
3. $\alpha$ contains precisely one of $a$, $-a$, for every $a \in \mathcal{A}$.

Examples of ultrafilters are sets $\alpha$ of the form $\{b \in \mathcal{A} : b \geq a\}$, for any $a \in \mathrm{At}\,\mathcal{A}$. Such 'atom-generated' ultrafilters are called *principal* ultrafilters. All ultrafilters $\alpha$ satisfy $\mathcal{A} \in \alpha$ and $0 \notin \alpha$.

Assume that $\mathcal{A}$ is simple, and suppose we are given a representation $h : \mathcal{A} \to \mathfrak{Re}(X)$ for some set $X$. For $x, y \in X$, let

$$h^{-1}(x, y) = \{a \in \mathcal{A} : (x, y) \in h(a)\}.$$

It is easy to check that

**Lemma 3.** $h^{-1}(x, y)$ is always an ultrafilter of $\mathcal{A}$.

### 2.3 Representations of finite simple relation algebras

The following is well known and easily proved:

**Lemma 4.** Any ultrafilter of a finite relation algebra is principal.

Hence, a representation $h : \mathcal{A} \to \mathfrak{Re}(X)$ of a finite (simple) relation algebra $\mathcal{A}$ can be viewed in a simple way as a complete labelled directed graph $M = (X, \lambda)$, where $X$ is a set and $\lambda : X \times X \to \mathrm{At}\,\mathcal{A}$ is a 'labelling function'. We just define $\lambda(x, y)$ to be the (unique) atom in $h^{-1}(x, y)$. It can be checked that for all $x, y, z \in X$,

- $\lambda(x, y) \leq 1' \iff x = y$.
- $\lambda(x, y) = \lambda(y, x)^{\smile}$.
- $\lambda(x, y) \leq \lambda(x, z) \,;\, \lambda(z, y)$. That is, *'all triangles are consistent'.*
- For all $a, b \in \mathrm{At}\,\mathcal{A}$, if $\lambda(x, y) \leq a \,;\, b$ then there is $w \in X$ with $\lambda(x, w) = a$ and $\lambda(w, y) = b$. *'All consistent triples are witnessed wherever possible.'*

Conversely, given a map $\lambda : X \times X \to \mathrm{At}\,\mathcal{A}$ satisfying these conditions, we can obtain a conventional representation $h : \mathcal{A} \to \mathfrak{Re}(X)$ by defining $h(a) = \{(x, y) \in X \times X : a \geq \lambda(x, y)\}$. The '$(X, \lambda)$' view of representations of finite relation algebras is very handy, as we will see.

### 2.4 Two finite relation algebras

1. *McKenzie's algebra $\mathcal{K}$.*

   Four atoms: $1', <, >, \sharp$ (so 16 elements altogether).

   $\breve{1'} = 1'$, $\quad \breve{<} = >$, $\quad \breve{>} = <$, $\quad \breve{\sharp} = \sharp$.

   All triples are consistent except Peircean transforms of: $(1', a, a')$ for $a \neq a'$, $(<, <, >)$, $(<, <, \sharp)$, and $(\sharp, \sharp, \sharp)$.
2. *The 'anti-Monk algebra' $\mathcal{M}$.* We use this name not out of lack of respect, but because $\mathcal{M}$ is in some way the opposite of what are known as 'Monk algebras'. We believe $\mathcal{M}$ was discovered by Maddux.

   Four atoms: $1', \mathsf{r}, \mathsf{b}, \mathsf{g}$.

   $\breve{a} = a$ for all atoms $a$. (So $\mathcal{M}$ is a relation algebra all of whose elements are self-converse. Such a relation algebra is said to be *symmetric*.)

   All triples are consistent except Peircean transforms of: $(1', a, a')$ for $a \neq a'$, and $(\mathsf{r}, \mathsf{b}, \mathsf{g})$.

These are both relation algebras. Can you tell if they are in **RRA** or not? Games will help to tell, as we will see.

## 3 Games and representability (finite relation algebras)

In [Lyn50], Lyndon characterised the *finite* representable relation algebras by a *'step by step' construction.* In a nutshell, his approach was this:

1. Try to build 'step by step' a representation of a given finite relation algebra.
2. Write first-order axioms expressing that you can succeed.

The resulting axioms will be true in a finite relation algebra $\mathcal{A}$ just when $\mathcal{A}$ has a representation.

It's interesting to compare Lyndon's method with the *Henkin construction* of a model of a consistent first-order theory $T$, as given in, *e.g.*, [Hod93, Theorem 6.1.1] or [ChaKei90, § 2.1]. In this construction, $T$ is extended, sentence by sentence, to a consistent theory $U$ in a larger signature with additional constants. These new constants are called 'witnesses', because the construction arranges that they witness truth of all existential statements in $U$. Together with other properties of $U$ enforced by the construction, this ensures that a model of $U$ (and hence of $T$) can be built easily from the witnesses.

The important point for us is that starting from an *inconsistent* $T$, the construction won't work, because it will get stuck somewhere. Consistency of the original $T$ is used to prove that the construction succeeds, never getting stuck. But this gives us *a test for consistency* of any theory $T$. We just try to do the construction, and see if it succeeds.

This is rather what Lyndon did. His construction of a representation of a finite relation algebra $\mathcal{A}$ succeeds precisely when $\mathcal{A}$ has a representation. The axioms he wrote expresses that the construction succeeds, and hence they characterise representability of $\mathcal{A}$.

We are now going to explain (a minor variant of) Lyndon's step by step characterisation in more detail, using a game.

### 3.1 Networks

The 'pieces' played during the game are called *networks*. A network is like a *piece of a representation* (though of course, the given algebra might not have a representation). It satisfies the *universal* conditions of 'representation'.

**Definition 5.** Let $\mathcal{A}$ be an atomic relation algebra. An $\mathcal{A}$-*network* is a complete labelled directed graph $N = (X, \lambda)$ where $X \neq \varnothing$ and $\lambda : X \times X \to \mathrm{At}\,\mathcal{A}$ is a labelling function satisfying, for all $x, y, z \in X$,

- $\lambda(x, y) \leq 1' \iff x = y$,
- $\lambda(x, y) = \lambda(y, x)^{\smile}$,
- $\lambda(x, y) \leq \lambda(x, z) ; \lambda(z, y)$ — *all triangles in $N$ are consistent.*

We write $N$ for any of $N, X, \lambda$. We rely on the context to tell which one is meant.

### 3.2 Games on $\mathcal{A}$-networks

Let $\mathcal{A}$ be a non-degenerate atomic relation algebra — so $\mathrm{At}\,\mathcal{A} \neq \varnothing$ — and let $n \leq \omega$. The game $G_n(\mathcal{A})$ has two players, $\forall$ (male) and $\exists$ (female), and $n$ rounds. If $n = 0$, there are no rounds and we declare $\exists$ the winner. Assume $n > 0$. *In round* 0, $\forall$ picks $a_0 \in \mathrm{At}\,\mathcal{A}$, and $\exists$ plays an $\mathcal{A}$-network $N_0$ with $a_0$ occurring as a label in it. *In round* $t$ $(1 \leq t < n)$, suppose that the current network at the start of the round is $N_{t-1}$. Play goes as follows. First, $\forall$ picks $x, y \in N_{t-1}$ and $a, b \in \mathrm{At}\mathcal{A}$ with $a ; b \geq N_{t-1}(x, y)$:

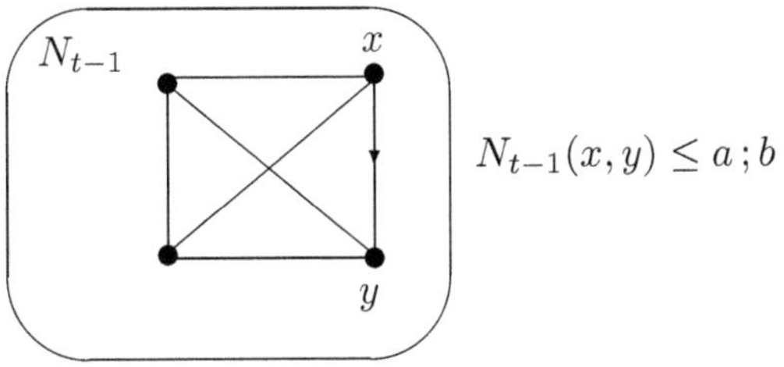

If there is already a node $z \in N_{t-1}$ such that $N_{t-1}(x, z) = a$ and $N_{t-1}(z, y) = b$, then $\exists$ simply sets $N_t = N_{t-1}$. If not, she has more work to do. She begins by adding a new node $z$ (say) to $N_{t-1}$, and labelling the edges $(x, z)$ with $a$ and $(z, y)$ with $b$. This forms the basis of the new network $N_t$:

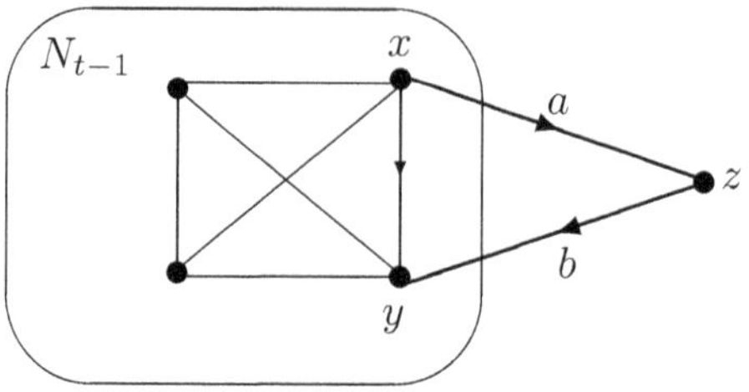


The player $\exists$ now has to *complete the labelling of* $N_t$, by defining $N_t(u, v)$ for all remaining pairs $(u, v)$ of nodes. These are the ones other than $(x, z)$, $(z, y)$, and pairs of nodes of $N_{t-1}$, whose labels are already fixed:

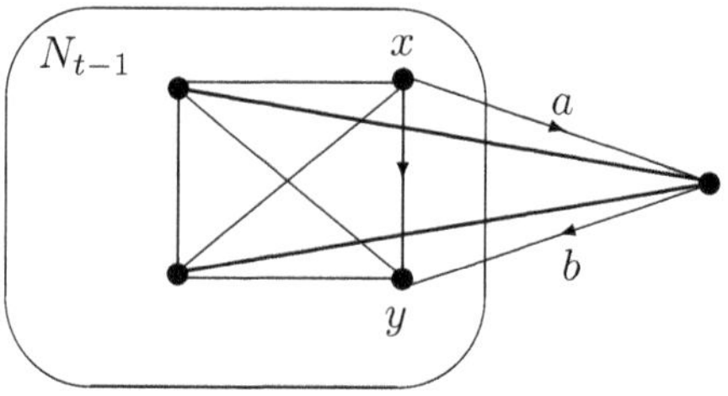


It can be very hard for $\exists$ to complete the labelling. $N_t$ must be a network, so all its triangles must be consistent. Worse still, $N_t$ is then passed on to the next round (if any), in which $\forall$ can make new choices. So even if $\exists$ succeeds in creating a *network* $N_t$, she may have left herself open to a lethal attack by $\forall$ in a later round. If in some round she cannot manage to complete the labelling and create a network, she loses. Thus, $\exists$ wins the play of $G_n(\mathcal{A})$ if she always responds legally to $\forall$'s moves.

Note that it is in $\exists$'s interests to play as small a network (with as few nodes) as possible. Although she is permitted, by the rules of the game, to make arbitrarily large extensions to the networks played in the game, she only needs to include the nodes shown in the diagrams above. Additional nodes are superfluous and will only make it easier for $\forall$ to win, by giving him more rope to hang her with. We will always assume that she plays this way, so that $N_0$ has at most two nodes, and for each $t$, $N_{t+1}$ has at most one more node than $N_t$.

The connection of the game to representability is given by the following theorem. It is more or less what Lyndon proved in [Lyn50] (but he didn't use games). The theorem is not restricted to simple relation algebras, but it only covers *finite* relation algebras; we will consider what to do about infinite relation algebras later.

**Theorem 6.** Let $\mathcal{A}$ be a finite relation algebra.

1. $\mathcal{A} \in \mathbf{RRA}$ if and only if $\exists$ has a winning strategy in $G_\omega(\mathcal{A})$.
2. The player $\exists$ has a winning strategy in $G_\omega(\mathcal{A})$ if and only if she has one in $G_n(\mathcal{A})$ for all finite $n$.
3. One can construct first-order sentences $\sigma_n$ for $n < \omega$ (independently of $\mathcal{A}$) such that $\mathcal{A} \models \sigma_n$ if and only if $\exists$ has a winning strategy in $G_n(\mathcal{A})$.

Hence, for a finite relation algebra $\mathcal{A}$, we have $\mathcal{A} \in \mathbf{RRA} \iff \mathcal{A} \models \{\sigma_n : n < \omega\}$.

*Proof.* We sketch the main ideas of the proof. For a more rigorous treatment, see [HirHod02a, Chapter 11].

1. If $\mathcal{A} \in \mathbf{RRA}$ then $\exists$ can use a representation as a guide in winning $G_\omega(\mathcal{A})$. Conversely, if she has a winning strategy in $G_\omega(\mathcal{A})$, then from plays of the game in which she uses her strategy and $\forall$ plays all possible moves at some stage, we can recover a representation of $\mathcal{A}$.
2. $\Rightarrow$ is clear. For the converse, we observe that because $\mathcal{A}$ is finite, $\exists$ has only finitely many possible responses to $\forall$'s move in any round. König's tree lemma can now be used to collimate her responses in the finite games into a single winning strategy in $G_\omega(\mathcal{A})$.
3. First, given an $\mathcal{A}$-network $N$, and $k < \omega$, we write an axiom $\tau_k(N)$ saying that $\exists$ *can win* $G_k(\mathcal{A})$ *starting from* $N$. We go by induction on $k$. The case $k = 0$ is easy: we need only say that $N$ is a network:

$$\begin{aligned}\tau_0(N) \quad = \quad & \bigwedge_{x \in N} \Big( N(x,x) \leq 1\text{'} \;\wedge \bigwedge_{y \in N \setminus \{x\}} N(x,y) \not\leq 1\text{'} \Big) \\ & \wedge \bigwedge_{x,y \in N} N(x,y) = N(y,x)^{\smile} \\ & \wedge \textstyle\bigwedge_{x,y,z \in N} N(x,y) \leq N(x,z) \,;\, N(z,y).\end{aligned}$$

The next formula $\tau_{k+1}(N)$ says that whatever move $\forall$ makes in the first round of the game, there is some $N'$ such that if $\exists$ responds to $\forall$'s move with $N'$ then she can win $G_k(\mathcal{A})$ starting from $N'$ — *i.e.*, such that $\tau_k(N')$ holds ($\tau_k(N')$ has been constructed inductively). Roughly, $\tau_{k+1}$ looks like this:

$$\tau_{k+1}(N) \quad = \quad \bigwedge_{x,y \in N} \forall a, b \Big( N(x,y) \leq a \,; b \rightarrow \exists N' \supseteq N \\ \big(\tau_k(N') \wedge \bigvee_{z \in N'} (N'(x,z) = a \wedge N'(z,y) = b)\big)\Big).$$

The formula uses *variables* to hold the labels on network edges. The expression $\exists N' \supseteq N$ in the middle is really shorthand for a string of quantifiers of the form $\exists v_1 \ldots \exists v_l$, relativised to atoms. The variables $v_i$ represent the atoms labelling the 'new' edges of $N'$ (if any) that are not already edges of $N$. We know how many there are, because $N'$ has at most one more node than $N$ does. The two possibilities — of $N'$ being the same as $N$ or bigger — mean that the rest of the formula is actually a disjunction to cope with these two cases. In the case $N' = N$, the variables $v_i$ are not used. For simplicity, this is not shown above. More details can be found in [HirHod97a, HirHod02a].

Finally, we let $\sigma_n = \forall a_0 \exists N (\tau_{n-1}(N) \wedge \bigvee_{x,y \in N} N(x,y) = a_0)$ for $n > 0$. Here, the $\exists N$ signifies $\exists v_{00} \exists v_{01} \exists v_{10} \exists v_{11}$. The variable $v_{ij}$ represents the atom $N(i,j)$. Again, these quantifiers are relativised to atoms of the algebra, and again, they are actually followed by a disjunction (not shown above) to allow for the possibility that in her first move, $\exists$ might pick a one-node network (in which case only $v_{00}$ is used) or a two-node network. We let $\sigma_0 = \top$.

q.e.d.

The axioms $\sigma_n$ (plus the **RA** axioms) seem to give an *intrinsic* characterisation of the finite algebras in **RRA**. But is it a *simple* one? Can you tell whether McKenzie's algebra and the anti-Monk algebra satisfy the $\sigma_n$ for all $n$?

It's easier to use the games $G_n$ directly.

**Example 7 (McKenzie's algebra $\mathcal{K}$).** Recall that this relation algebra has four atoms: 1', $<, >, \sharp$. We have $\breve{1'} = 1'$, $\breve{<} = >$, $\breve{>} = <$, $\breve{\sharp} = \sharp$. All triples of atoms are consistent except Peircean transforms of $(1', a, a')$ for $a \neq a'$, $(<, <, >)$, $(<, <, \sharp)$, and $(\sharp, \sharp, \sharp)$.

Consider the following play of $G_\omega(\mathcal{K})$. The player $\forall$ starts off by picking the atom $\sharp$. The player $\exists$ responds with the network $N_0$ as shown below.

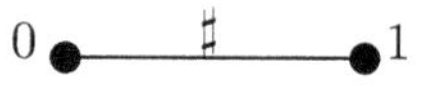

The edge $(0,1)$ is labelled by $\sharp$. We know that in any $\mathcal{K}$-network $N$ and nodes $x, y$ of $N$, we have $N(x,y) = 1$' if and only if $x = y$, and $N(y,x) = N(x,y)^{\smile}$. So $\exists$ has no choice over the labels of the remaining edges of $N_0$. We don't need an arrow on the edge in the diagram to indicate its direction, because $\breve{\sharp} = \sharp$, so the converse edge $(1,0)$ will also be labelled $\sharp$.

The player $\forall$ continues by choosing the two nodes $0, 1$ of $N_0$, and the atoms $>$ and $<$. The player $\exists$ has to add a new node, say 2, and label $(0,2)$ with $>$ and $(2,1)$ with $<$. She has no choice in labelling the remaining edges of her response, $N_1$:

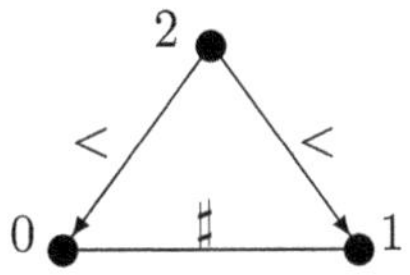


We prefer to show the edge $(2,0)$, which will be labelled $\breve{>} = <$.

The player $\forall$ now picks the nodes $0, 1$ again, and the atoms $<, >$. The player $\exists$ now has to add a node 3, with $(0,3)$ labelled $<$ and $(3,1)$ labelled $>$. She has no choice over the remaining edges: in particular, she must label the edge $(2,3)$ by $<$, since all other choices lead to inconsistency of the triangle $2, 0, 3$.

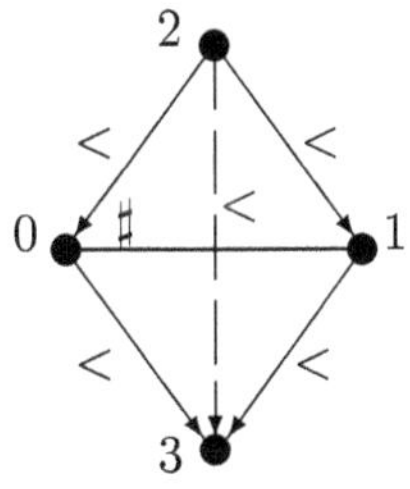


Now $\forall$ deals the killer blow, picking $2, 3$ and the atoms $\sharp, \sharp$. The player $\exists$ has to add a new node, say 4.

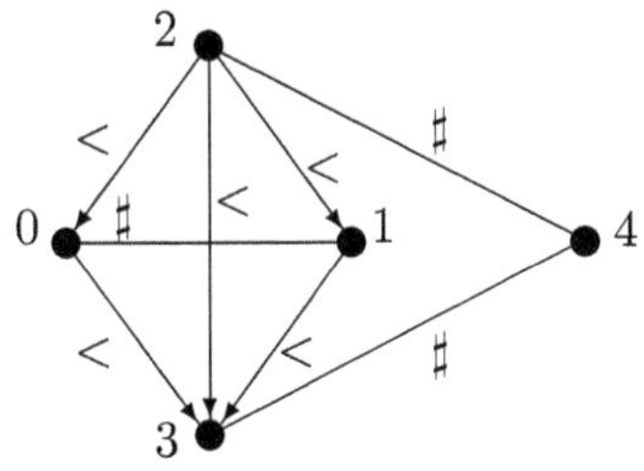

The player $\exists$ cannot consistently label the edge $(0, 4)$ by $<$ or 1' (because of the triangle $2, 0, 4$), nor by $>$ (because of the triangle $3, 0, 4$). She has to use $\sharp$. Similarly, she must label $(1, 4)$ with $\sharp$. But now, $0, 1, 4$ is an inconsistent triangle, and $\exists$ has lost. It is clear that she never had any real choice, so what we have described is a *winning strategy for* $\forall$ *in* $G_\omega(\mathcal{K})$ (and indeed in $G_4(\mathcal{K})$). The player $\exists$ has no winning strategy, so by Theorem 6, $\mathcal{K}$ is not representable.

**Example 8 (Anti-Monk algebra $\mathcal{M}$).** Recall that $\mathcal{M}$ has four atoms: 1', $\mathsf{r}, \mathsf{b}, \mathsf{g}$. We think of these as the colours red, blue, and green. $\mathcal{M}$ is symmetric: we have $\breve{x} = x$ for all atoms $x$. All triples of atoms are consistent except Peircean transforms of $(1', a, a')$ for $a \neq a'$, and $(\mathsf{r}, \mathsf{b}, \mathsf{g})$.

Consider a typical $\mathcal{M}$-network $N$ as shown below. Observe that all triangles involve at most two colours from $\mathsf{r}, \mathsf{b}, \mathsf{g}$, as required for consistency. We don't need any arrows at all on edges this time, since $\breve{a} = a$ for all atoms $a$, so the labels on an edge $(u, v)$ and the converse edge $(v, u)$ are always the same.

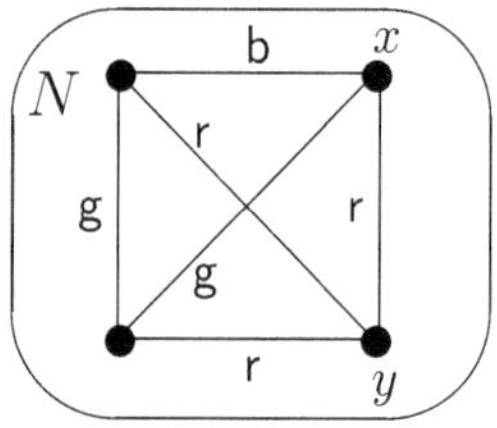


Suppose that $N$ is in play in some round of the game $G_\omega(\mathcal{M})$. A typical move of $\forall$ will be to pick two nodes and some atoms or other. We assume by way of example that he picks the two right-hand nodes $x, y$ in the diagram, and the atoms $p, q$, say. If there is a node $z$ in $N$ with $N(x, z) = p$ and $N(z, y) = q$, as in the game rules, then $\exists$ has an easy job. We'll assume there isn't; it follows that $p, q \neq 1'$. The player $\exists$ must now add a new node on the right as shown:

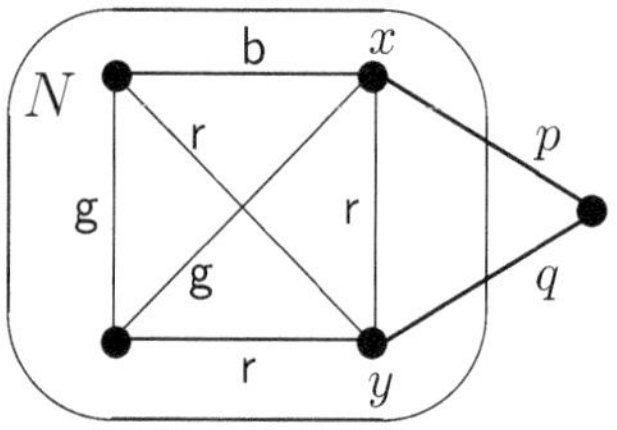

Then, she must fill in the remaining labels, to give a network $N'$, say:

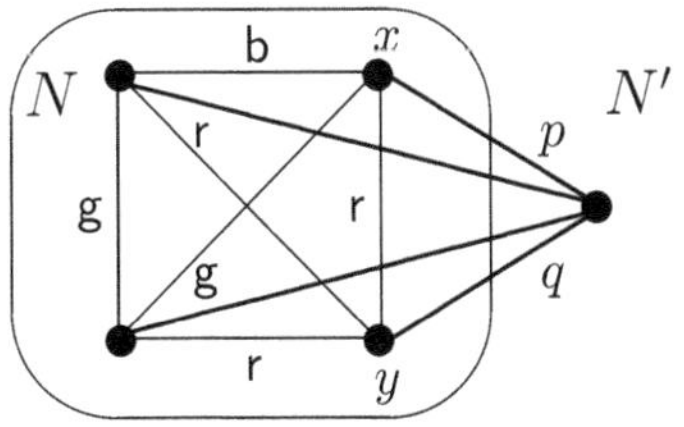


In this example, the edge $(x, y)$ that $\forall$ picked in $N$ is labelled $\mathsf{r}$. His chosen atoms $p, q$, combined with $\mathsf{r}$, must not all be different, or his choice would be illegal because $\mathsf{r} \not\leq p \,;q$. So two of $p, q, \mathsf{r}$ must be equal. There are two possibilities.

**Case 1**: $p = q$; so $N$ looks like:

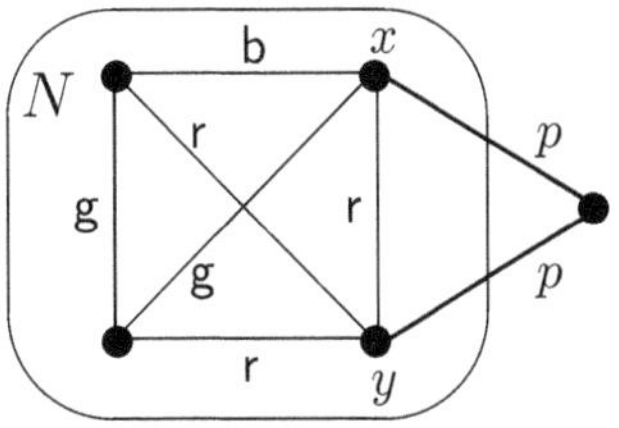


In this case, $\exists$ simply uses $p$ to label all remaining edges:

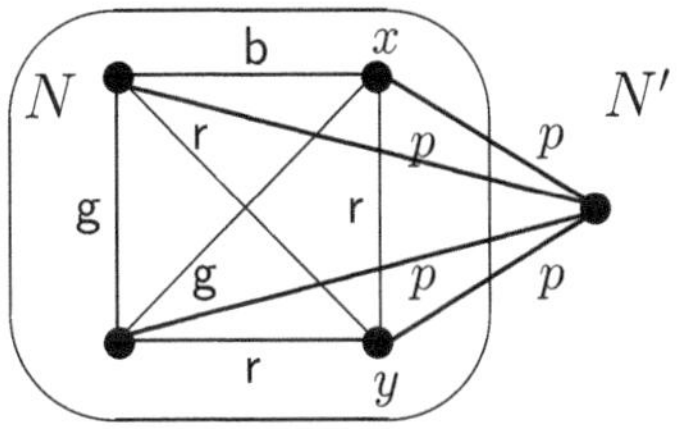


It is clear that all triangles have at least two edges of the same colour, so are consistent.

**Case 2**: $\mathsf{r} = p \neq q$ or $\mathsf{r} = q \neq p$. Let's suppose that $\mathsf{r} = q \neq p$ (the other case is similar), and that the new node is called $z$:

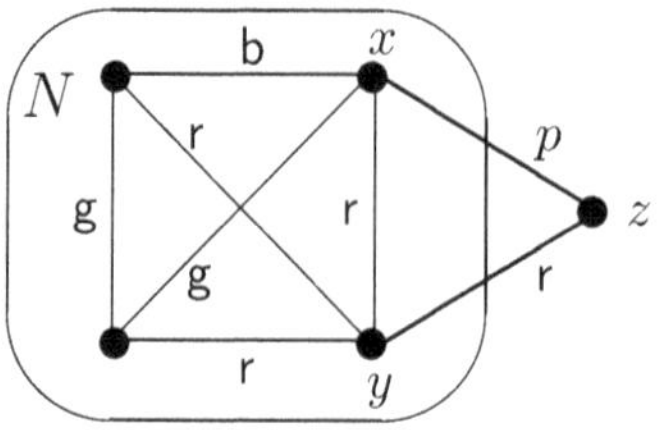


Observe that $x$ *and* $z$ *look the same as seen from* $y$*:* the labels on the edges $(y, x)$ and $(y, z)$ are the same. *The player* $\exists$ *tries to make this true for the other nodes, as well as* $y$. That is, she defines $N'(t, z) = N(t, x)$ for all nodes $t$ of $N$ other than $x, y$:

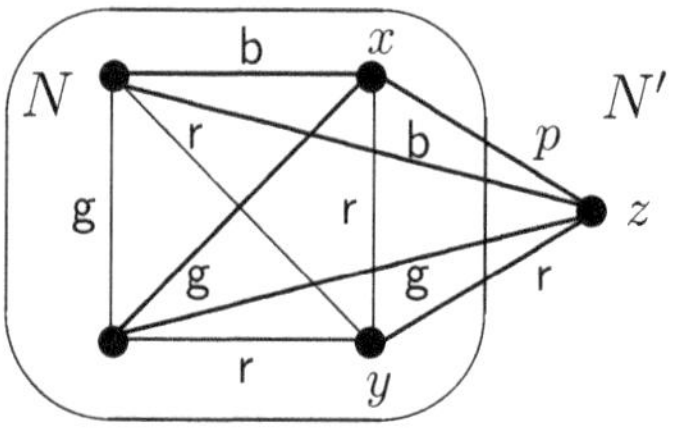


Now, there are three kinds of triangle in $N'$:

1. Triangles consisting of nodes of $N$. These are certainly consistent, because $N$ is a network.
2. Triangles of the form $t, x, z$, involving $x, z$. These have two edges with identical colours, because $N'(t, z) = N'(t, x)$. So they are consistent.
3. Triangles of the form $t, u, z$, involving $z$ but not $x$. The sides of such a triangle are coloured the same as in the triangle $t, u, x$ of $N$ (because $z$ looks the same as $x$ from $t$, and from $u$). But the triangle $t, u, x$ is consistent, by case 1, and hence, so is triangle $t, u, z$.

So all triangles of $N'$ are consistent, and $N'$ is a $\mathcal{M}$-network.

This can be elaborated into a winning strategy for $\exists$ in $G_\omega(\mathcal{M})$, showing that $\mathcal{M}$ is representable. This elegant strategy is due to Maddux (personal communication).

### 3.3 Summary

1. McKenzie's algebra $\mathcal{K} \notin \mathbf{RRA}$. So $\mathbf{RRA} \subset \mathbf{RA}$, as Lyndon (1950) showed. In fact, $\mathcal{K}$ is one of the smallest non-representable relation algebras. There are other 4-atom non-representable relation algebras, but all relation algebras with at most 3 atoms are representable.

2. The anti-Monk algebra $\mathcal{M} \in \mathbf{RRA}$.

   *Exercise:* Show that if $(X, \lambda)$ is any representation of $\mathcal{M}$, then $X$ is infinite. This is perhaps surprising. The obvious way of forcing a finite relation algebra to have only infinite representations is to include a relation $<$ like the one in $\mathcal{K}$, whose algebraic properties force it to be interpreted as a dense linear order. But $\mathcal{M}$ is finite and symmetric.

# 4 Infinite relation algebras

Games can still be used to characterise representability of infinite relation algebras. But there are some issues that need dealing with first.

## 4.1 Complete representations

Recall that a relation algebra is *atomic* if every non-zero element of it lies above an atom. All finite relation algebras are atomic, but not all infinite relation algebras are — indeed, some have no atoms at all. Even the atomic ones need care. Lemma 3 holds for infinite algebras, but Lemma 4 does not: not all ultrafilters of an infinite relation algebra, even an atomic one, are principal. So we cannot assume that in a representation of such an algebra, we can associate an atom with every edge in the representation.

Let us start by picking out the representations where we *can* associate atoms to edges.

**Definition 9.** A representation $h$ of a relation algebra $\mathcal{A}$ is said to be a *complete representation* if $h^{-1}(x, y)$ is a principal ultrafilter of $\mathcal{A}$ — it contains an atom of $\mathcal{A}$ — for every $x, y \in X$.

Complete representations are special kinds of representations. It is not hard to show that in the above notation,

**Theorem 10 (Hirsch, Hodkinson, [HirHod02a, Theorem 2.21]).** A representation $h$ is a complete representation just in case $h$ preserves all existing infima and suprema in $\mathcal{A}$: *i.e.*, if $S \subseteq \mathcal{A}$, and $S$ has a least upper bound $a \in \mathcal{A}$ (with respect to $\geq$), then

$$h(a) = \bigcup_{s \in S} h(s) \subseteq X \times X,$$

and similarly for greatest lower bounds.

This property gave rise to the name 'complete representation'. Any representation of a finite relation algebra is complete. A model-theoretic saturation argument will easily show that any infinite representable relation algebra has incomplete representations. So for infinite relation algebras, the question of interest is whether they have *any* complete representation at all.

**Definition 11.** A relation algebra is said to be *completely representable* if it has a complete representation. We write **CRA** for the class of completely representable relation algebras.

It is not hard to see that any completely representable relation algebra must be atomic. It's easy to find non-atomic representable relation algebras, and these cannot have any complete representation. But in fact, there are even *atomic* relation algebras that have a representation but don't have a complete representation. They are representable, but not completely representable. The first such relation algebra was given by Lyndon in [Lyn50], though it was not recognised as such at the time.

Games can help to analyse complete representations. We can generalise the game $G_n(\mathcal{A})$ seen earlier to a game $G_\kappa(\mathcal{A})$ with $\kappa$ rounds, where $\kappa$ is any cardinal. Then we can prove

**Theorem 12.** Let $\mathcal{A}$ be any atomic relation algebra. If $\mathcal{A}$ is completely representable, then $\exists$ has a winning strategy in $G_\kappa(\mathcal{A})$ for any $\kappa$. If $\exists$ has a winning strategy in $G_\kappa(\mathcal{A})$ for $\kappa = |\operatorname{At}\mathcal{A}| + \aleph_0$, then $\mathcal{A}$ is completely representable.

There is also an approximate characterisation of complete representability, generalising Theorem 6:

**Theorem 13 (Hirsch, Hodkinson, [HirHod97b, HirHod02a]).** For any atomic relation algebra $\mathcal{A}$, the following are equivalent:

1. The player $\exists$ has a winning strategy in $G_n(\mathcal{A})$ for all finite $n$,
2. $\mathcal{A}$ is elementarily equivalent to (*i.e.*, satisfies the same first-order sentences as) some completely representable relation algebra.

It is easily seen that the class **CRA** of completely representable relation algebras is pseudo-elementary (see [ChaKei90, Exercise 4.1.17] and [Hod93, § 5.2] for information about pseudo-elementary classes). However, there are many negative results about it. [HirHod97b] and [HirHod02a] used game-inspired relation algebras to show that **CRA** is not elementary (it is not definable by any set of first-order sentences). By Theorem 2, it is not definable by a second-order (or higher-order) sentence, or a sentence of fixed-point logic. The completely representable relation algebras with countably many atoms can be characterised using the infinitary logic $L_{\infty\omega}$, using Theorem 12 (this was observed by Väänänen at the meeting). But the countability assumption is necessary: there are atomic relation algebras $\mathcal{A}, \mathcal{B}$, the former with uncountably many atoms, that agree on all $L_{\infty\omega}$-sentences, with $\mathcal{B}$ completely representable and $\mathcal{A}$ not.[1] So **CRA** is not definable by a sentence of $L_{\infty\omega}$.

[1] In the notation of [HirHod02a, Theorem 17.25], take $\mathcal{A} = \mathcal{A}_{\mathsf{K}_{\omega_1},\mathsf{K}_\omega}$ and $\mathcal{B} = \mathcal{A}_{\mathsf{K}_\omega,\mathsf{K}_\omega}$.

## 4.2 Games and representations for infinite relation algebras

So much for *complete* representations. What about arbitrary ones? Can we use games to test whether an infinite relation algebra is representable?

Our game characterisation of the finite representable relation algebras in Theorem 6 relied on every edge in a representation being labelled by an atom — *i.e.*, on completeness of the representation. For infinite relation algebras, which may not have complete representations, this is not going to work.

There are two ways out of this difficulty. We can modify the games to handle arbitrary (complete or incomplete) representations. One of the changes is that player $\forall$ will choose arbitrary elements of the algebra, not just atoms. Then, we can use universal algebra to turn the $\sigma_n$ of Theorem 6 into equations. This gives an equational axiomatisation of **RRA**. The method is very close to one of Lyndon from [Lyn56]. For details, see [HirHod97a, HirHod02a].

Alternatively, we can take advantage of *canonical extensions*.

**Definition 14.** The *canonical extension* $\mathcal{A}^\sigma$ of a relation algebra $\mathcal{A}$ is a special relation algebra formed from the set $\mathrm{Uf}\,\mathcal{A}$ of all ultrafilters of $\mathcal{A}$. Its boolean part is just the full power set algebra $(\wp(\mathrm{Uf}\,\mathcal{A}), \cup, \setminus, \varnothing, \mathrm{Uf}\,\mathcal{A})$. So $\mathcal{A}^\sigma$ is atomic. We will identify an atom $\{\alpha\}$, consisting of a single ultrafilter $\alpha$, with the ultrafilter $\alpha$ itself. So the atoms of $\mathcal{A}^\sigma$ are essentially the ultrafilters of $\mathcal{A}$. Then:

- The atoms $\leq$ 1' (in the sense of $\mathcal{A}^\sigma$) are precisely the ultrafilters containing 1' (in the sense of $\mathcal{A}$).
- The converse of an atom (ultrafilter) $\alpha$ is the ultrafilter consisting of the converses of all the elements of $\alpha$: in symbols, $\breve{\alpha} = \{\breve{a} : a \in \alpha\}$. (The relation algebra axioms ensure that this is an ultrafilter.)
- A triple $(\alpha, \beta, \gamma)$ of ultrafilters is consistent just when every triple $(a, b, c)$ of elements of $\mathcal{A}$ taken from them (*i.e.*, $a \in \alpha$, $b \in \beta$, $c \in \gamma$) satisfies the consistency condition $(a\,;b) \cdot c \neq 0$. This generalises the consistency condition for atoms given in § 2. It is equivalent to say that $a\,;b \in \gamma$ whenever $a \in \alpha$ and $b \in \beta$.

Apart from some changes in notation, this definition is due to Jónsson and Tarski [JónTar51, JónTar52], and it generalises Stone's related construction for boolean algebras [Sto36]. Any relation algebra $\mathcal{A}$ has a canonical extension $\mathcal{A}^\sigma$, and $\mathcal{A}$ embeds in $\mathcal{A}^\sigma$ via $a \mapsto \{\alpha : \alpha \text{ an ultrafilter of } \mathcal{A},\ a \in \alpha\}$. For finite $\mathcal{A}$, we have $\mathcal{A} \cong \mathcal{A}^\sigma$. Thus, the following generalises Theorem 6:

**Theorem 15.** A relation algebra $\mathcal{A}$ is representable if and only if $\exists$ has a winning strategy in $G_n(\mathcal{A}^\sigma)$ for all finite $n$.

*Proof.* $\Rightarrow$: In an important result, Monk proved that if $\mathcal{A}$ is representable then $\mathcal{A}^\sigma$ is representable. (Monk did not publish it; his result is reported in his student McKenzie's Ph.D. dissertation [McK66].) In fact, it can even be shown that if $\mathcal{A}$ is representable then $\mathcal{A}^\sigma$ is *completely* representable [HirHod02a, Theorem 3.36]. So by Theorem 12, $\exists$ has a winning strategy in $G_n(\mathcal{A}^\sigma)$ for all finite $n$.

$\Leftarrow$: Assume that $\exists$ has a winning strategy in $G_n(\mathcal{A}^\sigma)$ for all finite $n$. By Theorem 13, $\mathcal{A}^\sigma$ is elementarily equivalent to some (completely) representable relation algebra $\mathcal{B}$. Up to isomorphism, $\mathcal{A}$ is a subalgebra of $\mathcal{A}^\sigma$. We saw in Section 1 that **RRA** is a variety, and so is closed under elementary equivalence and under taking isomorphic copies of subalgebras. So we obtain $\mathcal{A} \in \mathbf{RRA}$ as required. q.e.d.

This means that we can still use the games $G_n$ to characterise representability. We just need to play on the canonical extension, not the relation algebra itself. (For finite algebras $\mathcal{A}$, this makes no difference, since $\mathcal{A}^\sigma \cong \mathcal{A}$.) This characterisation of representability is perhaps not intrinsic, since it uses the canonical extension; but it is still useful.

## 5 Infinite atom structures

Recall from Section 2 that for an atomic relation algebra, if we know the value of the relation algebra operators applied to atoms, then we can determine these operators on arbitrary elements. For an atomic relation algebra $\mathcal{A}$, we call

$$\begin{aligned} \mathfrak{At}\,\mathcal{A} \;=\; \big(&\mathrm{At}\,\mathcal{A},\, \{a \in \mathrm{At}\,\mathcal{A} : a \leq 1'\}, \{(a, \breve{a}) : a \in \mathrm{At}\,\mathcal{A}\}, \\ &\{(a,b,c) : a,b,c \in \mathrm{At}\,\mathcal{A},\ a\,;b \geq c\}\big) \end{aligned}$$

the *atom structure* of $\mathcal{A}$. A tuple $\mathcal{S} = (S, I, f, C)$ is called an *atom structure* if it is the atom structure of some atomic relation algebra. It is not hard to derive from the relation algebra axioms a first-order sentence expressing that $\mathcal{S}$ is an atom structure. We used atom structures in Section 2 as a kind of notational device to allow us to present finite relation algebras more concisely. They certainly serve this function, but in some ways it is with infinite atomic relation algebras that connections between the representability of an algebra and the properties of its atom structure become most interesting.

Any atomic relation algebra uniquely determines its atom structure, but once we move away from finite relation algebras, we see that there can be many relation algebras possessing the same atom structure but with different (non-isomorphic) boolean structures. The boolean structure of $\mathcal{A}$ (*i.e.*, which suprema of sets of atoms exist in $\mathcal{A}$), together with the atom structure, determine $\mathcal{A}$ up to isomorphism. Informally, we have

atomic relation algebra = atomic boolean algebra + atom structure.

Now all boolean algebras are representable, but the representability problem for relation algebras is highly non-trivial. So we might surmise that the difficulties in representing an (atomic) relation algebra reside in its atom structure. More precisely, we might guess that whether an atomic relation algebra is representable or not is determined by its atom structure. For *complete* representations, in which all edges are labelled by atoms, this is of course true (though the 'completely representable atom structures' are at least as hard to characterise as the completely representable relation algebras). But for arbitrary representations, it is not so clear.

What are the possible atomic relation algebras with a given atom structure? At one end of the spectrum we can define the *complex algebra* $\mathrm{Cm}\,\mathcal{S}$ of an atom structure $\mathcal{S}$. This is the biggest atomic relation algebra whose atom structure is $\mathcal{S}$. Its domain is the full power set of the domain of $\mathcal{S}$, and the relation algebra operations are determined by the atom structure. If the cardinality of the atom structure $\mathcal{S}$ is $\lambda$ then $\mathrm{Cm}\,\mathcal{S}$ has cardinality $2^\lambda$. At the other end of the spectrum, the *term algebra* $\mathrm{Tm}\,\mathcal{S}$ is the smallest atomic relation algebra whose atom structure is $\mathcal{S}$. It is the subalgebra of $\mathrm{Cm}\,\mathcal{S}$ generated, using the relation algebra operations, by the atoms. The cardinality of the term algebra is $\lambda$, for infinite atom structures. It is easily seen that if $\mathcal{A}$ is an atomic relation algebra with $\mathfrak{At}\,\mathcal{A} = \mathcal{S}$, then up to isomorphism, $\mathcal{A}$ is a subalgebra of $\mathrm{Cm}\,\mathcal{S}$ and $\mathrm{Tm}\,\mathcal{S}$ is a subalgebra of $\mathcal{A}$.

So we may distinguish two types of representability for atom structures. An atom structure is *weakly representable* if it is the atom structure of *some* atomic representable relation algebra. An atom structure is *strongly representable* if *every* atomic relation algebra with that atom structure is representable. Since any subalgebra of a representable relation algebra is also representable, we can easily see that:

**Theorem 16.**

1. An atom structure is weakly representable if and only if its term algebra is representable.
2. An atom structure is strongly representable if and only if its complex algebra is representable.

For finite atom structures, the term algebra is the same as the complex algebra, so weak and strong representability coincide. Several questions immediately present themselves:

- Is representability of an atomic relation algebra determined by its atom structure? That is, could an (infinite) atom structure be weakly representable but not strongly representable?
- Is the class of weakly representable atom structures elementary?
- What about the class of strongly representable atom structures?
- Can we define either class with finitely many axioms?

The last question is easily dealt with: by Theorem 2, there can be no finite axiomatisation of either class. Also, since **RRA** is a variety, a result of [Ven97a] shows that the class of weakly representable atom structures is elementary.

The other questions are more tricky. To help us answer them, we look at a class of interesting atom structures obtained from graphs.

### 5.1 Graphs and relation algebras

By a *graph,* we mean an irreflexive symmetric 'edge' relation on a finite or infinite set of 'nodes'. A set $I$ of nodes of a graph is said to be *independent* if no two nodes in $I$ are connected by a graph edge. For finite $k$, a *$k$-colouring* of a graph is a partition of its nodes into at most $k$ independent sets. The *chromatic number* of a graph is the least finite $k$ for which it has a $k$-colouring, and if there is no such $k$ then the chromatic number is $\infty$.

Given a graph $\Gamma$, we can make an atom structure $\mathcal{S}(\Gamma) = (S, I, f, C)$ whose atoms are red, blue, and green copies of each node of $\Gamma$, plus 1' as an extra atom. That is, the set of atoms is

$$S = \{\mathsf{r}_x, \mathsf{g}_x, \mathsf{b}_x : x \in \Gamma\} \cup \{1'\}.$$

(Here and below, if $\Gamma$ is a graph, we also let $\Gamma$ denote its set of nodes.) The set $I$ of sub-identity atoms is just $\{1'\}$. The converse function $f$ leaves each atom fixed — $\mathcal{S}(\Gamma)$ is symmetric. To define $C$, we stipulate that all triples of atoms are consistent (included in $C$) except the following:

- Peircean transforms of $(1', a, a')$ for $a \neq a'$,
- monochromatic triples of nodes forming an independent set in $\Gamma$ — *i.e.*, triples $(\mathsf{r}_x, \mathsf{r}_y, \mathsf{r}_z)$ where $\{x, y, z\} \subseteq \Gamma$ is independent, and similarly for green and blue.

It turns out, for any graph $\Gamma$, that $\mathrm{Cm}(\mathcal{S}(\Gamma))$ is a simple relation algebra (to prove associativity of composition we need to take advantage of the three colours), and so $\mathcal{S}(\Gamma)$ is a genuine relation algebra atom structure. Surprisingly, perhaps, its strong representability is entirely determined by the chromatic number of $\Gamma$, in the case where $\Gamma$ is infinite:

**Theorem 17 (Hirsch, Hodkinson, [HirHod02b, HirHod02a]).** For any infinite graph $\Gamma$, the relation algebra $\mathrm{Cm}(\mathcal{S}(\Gamma))$ is representable if and only if $\Gamma$ has chromatic number $\infty$.

*Proof.* First, some notation: if $Z \subseteq \Gamma$, we let $\mathsf{r}_Z = \{\mathsf{r}_z : z \in Z\}$, and similarly we define $\mathsf{g}_Z, \mathsf{b}_Z$. Note that these are all in $\mathrm{Cm}\,\mathcal{S}(\Gamma)$, since the domain of the complex algebra is the full power set of the set of atoms.

$\Rightarrow$: Suppose that $h : \mathrm{Cm}\,\mathcal{S}(\Gamma) \to \mathfrak{Re}(X)$ is a representation. As usual, we write $+, \cdot, ;$ for the operations of $\mathrm{Cm}\,\mathcal{S}(\Gamma)$, and $\cup, \cap, |$ for those of $\mathfrak{Re}(X)$.

Supposing, for contradiction, that $\Gamma$ has finite chromatic number, its set of nodes can be partitioned into independent sets $I_0, \ldots, I_{n-1}$ for some finite $n$. Clearly, in $\mathrm{Cm}\,\mathcal{S}(\Gamma)$ we have

$$1' + \mathsf{r}_{I_0} + \mathsf{g}_{I_0} + \mathsf{b}_{I_0} + \cdots + \mathsf{r}_{I_{n-1}} + \mathsf{g}_{I_{n-1}} + \mathsf{b}_{I_{n-1}} = 1.$$

Now $h$ respects $+$: we have $h(a+b) = h(a) \cup h(b)$, for any $a, b \in \mathrm{Cm}\,\mathcal{S}(\Gamma)$, and this extends by induction to sums of any finite length. So for any distinct $x, y \in X$, since $(x, y) \notin h(1')$, we know that $(x, y) \in h(\mathsf{c}_{I_k})$ for some $k < n$ and some colour $\mathsf{c} \in \{\mathsf{r}, \mathsf{g}, \mathsf{b}\}$.

Observe that $X$ must be infinite (since $\mathcal{S}(\Gamma)$ is). So it follows from Ramsey's Theorem [Ram30] that there are distinct $x_i \in X$ $(i < \omega)$ and some element $a \in \mathrm{Cm}\,\mathcal{S}(\Gamma)$ of the form $\mathsf{c}_{I_k}$ for some colour $\mathsf{c}$ and $k < n$, such that $(x_i, x_j) \in h(a)$ for all $i < j < \omega$. So $(x_0, x_2) \in h(a)$. Also, $(x_0, x_1), (x_1, x_2) \in h(a)$, so that $(x_0, x_2) \in h(a)|h(a)$. Now $h$ is a representation, so it respects all the algebra operations. We deduce that

$$(x_0, x_2) \in h(a) \cap (h(a)|h(a)) = h(a \cdot (a\,;a)).$$

But for any nodes $p, q, s \in I_k$, we know that $\{p, q, s\}$ is an independent subset of $\Gamma$ (since $I_k$ is), and so $(\mathsf{c}_p, \mathsf{c}_q, \mathsf{c}_s)$ is not a consistent triple of atoms in $\mathcal{S}(\Gamma)$. Because ';' in $\mathrm{Cm}\,\mathcal{S}(\Gamma)$ is defined additively from the atoms, we have

$$a\,;a = \sum_{p,q \in I_k} \mathsf{c}_p\,;\mathsf{c}_q = \big\{s \in \mathcal{S}(\Gamma) : \exists p, q \in I_k((\mathsf{c}_p, \mathsf{c}_q, s) \text{ is consistent})\big\}.$$

It is clear that the '$s$' here cannot lie in $a$, so $a \cdot (a\,;a) = 0$, and $(x_0, x_2) \in h(0) = \varnothing$. This is impossible.

$\Leftarrow$: Assume $\Gamma$ has infinite chromatic number. We'll show that $\exists$ has a winning strategy in the game $G_\omega((\mathrm{Cm}\,\mathcal{S}(\Gamma))^\sigma)$ played on the canonical extension $(\mathrm{Cm}\,\mathcal{S}(\Gamma))^\sigma$ (see Definition 14), and hence by Theorem 15 that $\mathrm{Cm}\,\mathcal{S}(\Gamma)$ is representable.

Call a set $X$ of nodes of $\Gamma$ *small* if the induced subgraph of $\Gamma$ on the set of nodes $X$ has finite chromatic number. Call a set *large* if its complement is small. Then the set of all nodes is large, any superset of a large set is large, and the intersection of two large sets is still large (because the union of two small sets is small). By assumption, $\Gamma$ itself is not small, so $\varnothing$ is not large. Now, using Zorn's lemma or the boolean prime ideal theorem, for each colour $\mathsf{c} \in \{\mathsf{r}, \mathsf{g}, \mathsf{b}\}$ the set

$$\{\mathsf{c}_L : L \subseteq \Gamma,\ L \text{ large}\}$$

of $\mathsf{c}$-coloured copies of large sets can be extended to an ultrafilter $\mu_\mathsf{c}$ of $\mathrm{Cm}\,\mathcal{S}(\Gamma)$ — *i.e.*, an atom of the canonical extension $(\mathrm{Cm}\,\mathcal{S}(\Gamma))^\sigma$, if we identify $\mu_\mathsf{c}$ with $\{\mu_\mathsf{c}\}$ again. If $Z \subseteq \Gamma$ and $\mathsf{c}_Z \in \mu_\mathsf{c}$, then $Z$ is not small, and so in particular, not independent.

The three atoms $\mu_\mathsf{r}, \mu_\mathsf{g}, \mu_\mathsf{b}$ are very useful for $\exists$ when playing the game $G_\omega((\mathrm{Cm}\,\mathcal{S}(\Gamma))^\sigma)$. In fact, they allow her to win it. First, a little calculation will establish that any triangle in a network with two edges labelled with the same $\mu_\mathsf{c}$ is consistent:

(∗) *Let* $\gamma$ *be any ultrafilter of* $\mathrm{Cm}\,\mathcal{S}(\Gamma)$*, and let* $\mathsf{c} \in \{\mathsf{r}, \mathsf{g}, \mathsf{b}\}$ *be given. Then* $(\mu_\mathsf{c}, \mu_\mathsf{c}, \gamma)$ *is a consistent triple of atoms of* $(\mathrm{Cm}\,\mathcal{S}(\Gamma))^\sigma$.

This is clear if $\{1'\} \in \gamma$. Assume that $\{1'\} \notin \gamma$. Take $X, Y \in \mu_\mathsf{c}$ and $Z \in \gamma$. Then (by Definition 14) we need to find $x \in X$, $y \in Y$, and $z \in Z$ such that $(x, y, z)$ is a consistent triple of atoms in $\mathcal{S}(\Gamma)$. We can replace these sets by smaller ones in their ultrafilters; so we can suppose that $X = Y \subseteq \mathsf{c}_\Gamma$. Now, $X$ has the form $\mathsf{c}_{X'}$ for some $X' \subseteq \Gamma$.

But $\mathsf{c}_{X'} \in \mu_\mathsf{c}$, so as we saw, $X'$ cannot be independent. Let $p, q \in X'$ be connected by an edge of $\Gamma$. Then $\mathsf{c}_p \in X$, $\mathsf{c}_q \in Y$. We know that $Z$ cannot be $\{1'\}$ or $\varnothing$; take $z \in Z$ with $z \neq 1'$. Then by the definition of $\mathcal{S}(\Gamma)$, $(\mathsf{c}_p, \mathsf{c}_q, z)$ is consistent.

Now let us see how $\exists$ can win the game $G_\omega((\mathrm{Cm}\,\mathcal{S}(\Gamma))^\sigma)$. Suppose that in some round, the current network is $N$, and $\forall$ picks nodes $x, y \in N$ and atoms (ultrafilters) $\alpha, \beta$. If $\exists$ has to extend the network, we will have $\{1'\} \notin \alpha, \beta$. Now since in $\mathrm{Cm}\,\mathcal{S}(\Gamma)$ we have $\{1'\} + \mathsf{r}_\Gamma + \mathsf{g}_\Gamma + \mathsf{b}_\Gamma = 1$, any ultrafilter must contain one of these four sets — in fact, exactly one, since they are pairwise disjoint. So there are $\mathsf{c}, \mathsf{c}' \in \{\mathsf{r}, \mathsf{g}, \mathsf{b}\}$ such that $\mathsf{c}_\Gamma \in \alpha$ and $\mathsf{c}'_\Gamma \in \beta$. Since we have three colours, $\exists$ can pick a colour $\mathsf{c}'' \notin \{\mathsf{c}, \mathsf{c}'\}$ (this is chiefly why we introduced three colours). Then the following holds:

(∗∗) *For any ultrafilter* $\gamma$ *not containing* $\{1'\}$*, the triples* $(\alpha, \mu_{\mathsf{c}''}, \gamma)$ *and* $(\beta, \mu_{\mathsf{c}''}, \gamma)$ *are consistent triples of atoms of* $(\mathrm{Cm}\,\mathcal{S}(\Gamma))^\sigma$.

This is simply because if $X \in \alpha$, $Y \in \mu_{c''}$, and $Z \in \gamma$, we can find $x \in X$ of colour c, $y \in Y$ of colour c″, and $z \in Z$ with $z \neq 1'$. The triple $(x, y, z)$ is not monochromatic, so it is a consistent triple in $\mathcal{S}(\Gamma)$. So by Definition 14, $(\alpha, \mu_{c''}, \gamma)$ is consistent, as claimed. The argument for $(\beta, \mu_{c''}, \gamma)$ is similar.

The player $\exists$ lets the new network be $N'$ with new node $z$. She labels $N'(z,t) = N'(t,z) = \mu_{c''}$ for each node $t$ of $N$ with $t \neq x, y$.

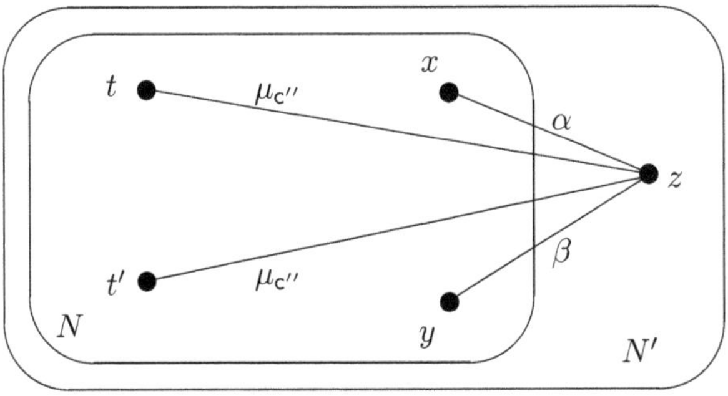


Then $N'$ is a network. We check consistency of triangles in $N'$. The Peircean law in $(\mathrm{Cm}\,\mathcal{S}(\Gamma))^\sigma$ ensures that it is enough to check a triangle in any single orientation. The triangle $xyz$ is consistent, because $\forall$'s move is assumed legal. Pick any $t \in N$ other than $x, y$. As $t \neq x, y$, we have $\{1'\} \notin N(x,t), N(y,t)$ (*cf.* Definition 5). So by $(**)$ above, the triangles $txz$ and $tyz$ are consistent. For any $t, t' \in N$ other than $x, y$, triangle $tt'z$ is consistent by $(*)$ above. All other triangles lie in $N$ and so are (inductively) consistent. Other checks are straightforward. So this gives a winning strategy for $\exists$ in the game. q.e.d.

**Corollary 18.** $\mathcal{S}(\Gamma)$ is strongly representable iff $\Gamma$ has infinite chromatic number (for any infinite graph $\Gamma$).

*Proof.* By Theorems 16 and 17. q.e.d.

## 5.2 Applications

The corollary allows us to translate problems about atom structures into problems about graphs. Graphs seem easier to work with, and far more is known about them.

If we replace $\mathrm{Cm}\,\mathcal{S}(\Gamma)$ by a subalgebra (*e.g.*, $\mathrm{Tm}\,\mathcal{S}(\Gamma)$), the left-to right implication in Theorem 17 can fail. Even if the nodes of $\Gamma$ can be partitioned into independent sets $I_0, \ldots, I_{n-1}$ for some finite $n$, it might be that the element $\{\mathsf{c}_x : x \in I_k\}$ does not belong to the algebra, for some $k < n$ and some colour $\mathsf{c}$. Indeed, taking the graph $Z$ with nodes $\mathbb{Z}$ and edges between consecutive integers only, a not too difficult exercise shows that the term algebra $\mathrm{Tm}\,\mathcal{S}(Z)$ is indeed representable, though the chromatic number of $Z$ is just two. (The first part of the exercise is to calculate exactly which sets of atoms are generated using the relation algebra operations.) Thus, $\mathcal{S}(Z)$ is weakly but (by Corollary 18) not strongly representable, and we conclude:

**Theorem 19.** There exist weakly but not strongly representable atom structures.

A more complicated sequence of graphs is derived from finite graphs $G_n$ $(n < \omega)$ constructed probabilistically by Erdős [Erd59]. Each $G_n$ has chromatic number at least $n$, and no cycles of length $n$ or less. Here, a *cycle* is a sequence $x_1, \ldots, x_l$ of distinct nodes with $(x_1, x_2), \ldots, (x_{l-1}, x_l)$, and $(x_l, x_1)$ all being graph edges. The length of this cycle is $l$.

We can use this wonderful construction in graph theory to answer the last remaining question from those listed above. We set $\Gamma_k$ to be the disjoint union of the graphs $G_n$ for all $n > k$. Each $\Gamma_k$ has infinite chromatic number, but an ultraproduct $\Gamma$ of the $\Gamma_k$ has no cycles, and hence (by well known graph theoretic work: see, *e.g.*, [Die97]) chromatic number just two. It follows from Corollary 18 that every $\mathcal{S}(\Gamma_k)$ is strongly representable, but an ultraproduct $\mathcal{S}(\Gamma)$ of them is not strongly representable. ($\mathcal{S}(-)$ commutes with taking ultraproducts.) By Łoś's theorem (see [Hod93, Theorem 9.5.1] or [ChaKei90, Theorem 4.1.9]), any first-order sentence true in all the $\mathcal{S}(\Gamma_k)$ must also be true in $\mathcal{S}(\Gamma)$. We conclude that:

**Theorem 20.** The class of strongly representable atom structures is not elementary: it cannot be defined by *any* set of first-order axioms.

Probabilistic constructions of graphs have been useful for relation algebras on other occasions. For example, in Theorem 15 we mentioned Monk's result that if $\mathcal{A}$ is a representable relation algebra then so also is its canonical extension $\mathcal{A}^\sigma$. So **RRA** is closed under taking canonical extensions, and we say that it is a *canonical* variety. But does it have an axiomatisation by equations $\varepsilon$ that are *individually* canonical, in the sense that for any relation algebra $\mathcal{A}$, if $\mathcal{A} \models \varepsilon$ then $\mathcal{A}^\sigma \models \varepsilon$? The answer is 'no': [HodVen05] uses a probabilistic graph construction and a 'local' variant of Theorem 17 to show that:

**Theorem 21.** Every first-order axiomatisation of **RRA** has infinitely many non-canonical sentences.

Similar considerations led to a proof [GolHodVen04, GolHodVen03] that not every canonical variety is generated by an elementary class of frames, solving a problem of Fine in modal logic [Fin75]. More details of these and other related results can be found in [HirHod02b, HodVen05] and [HirHod02a, Chapter 14].

# 6 Games in algebraic logic: pros and cons

Games have made a substantial contribution to our understanding of relation algebras. The idea has many precursors, notably in the seminal paper of Lyndon [Lyn50]. Let us end with a rundown of the pros and cons of using games in relation algebras and algebraic logic generally.

## 6.1 Pros

1. Games provide a simple practical test for representability. (They are also very useful for theoretical purposes, as we saw in Theorem 17.)
2. Games can be used to produce axioms as well (with care, they sometimes even yield finite axiomatisations).
3. Sometimes, a winning strategy can be extracted and used for other things, such as decidability, complexity, finite model property.
4. Games on relation algebras generalise to games for other kinds of algebras of relations, such as complex algebras (see, *e.g.*, [HodMikVen01]).
5. Most importantly in our view, games can suggest some fairly sophisticated constructions of relation algebras. These can be used to prove:

    (a) **RRA** is not finitely axiomatisable (first proved in [Mon64], not using games).

    (b) **RRA** is not axiomatisable by equations using finitely many variables altogether ([Jón91], although the result was stated by Tarski in a video made in 1975).

    (c) **RRA** is not closed under Monk completions [Hod97]: the example $\mathrm{Tm}\,\mathcal{S}(Z)$ above shows this, since its completion is isomorphic to $\mathrm{Cm}\,\mathcal{S}(Z)$. Hence, **RRA** is not Sahlqvist-axiomatisable [Ven97b].

    (d) In first-order logic, more 3-variable sentences are provable with $n+1$ variables than with $n$ variables, for all $n \geq 3$.
    See [HirHodMad02a] and for the link to games and relation algebras, see [HirHodMad02b].

    (e) For a finite relation algebra $\mathcal{A}$, it is undecidable whether $\mathcal{A} \in$ **RRA** [HirHod02a].

    (f) **RRA** is canonical (Monk), but any first-order axiomatisation of it has infinitely many non-canonical axioms [HodVen05].

## 6.2 Cons

We use games as a *construction method,* essentially forcing, to build representations of relation algebras. In general, the representations so obtained are *infinite.* These games are not good at building *finite representations.*

For example, suppose that $\mathcal{A}$ is a finite relation algebra with a *'flexible atom'*, $f$, say. This means that $(a, b, f)$ is consistent for all atoms $a, b \neq 1'$. The game $G_\omega(\mathcal{A})$ shows that $\mathcal{A}$ is representable: $\exists$ can win by using $f$ to label network edges wherever needed, and it will always be consistent to do so.

**Problem 22 (Maddux).** Must such an $\mathcal{A}$ have a *finite* representation?

There is a general issue here: *find ways of constructing finite representations.* Can we combine games with, *e.g.*, probabilistic constructions?

Some algebraic logicians avoid games and prefer the traditional 'step by step' approach, enumerating the requirements of a construction and dealing with them one by one. Certainly, games are not needed in simple cases, but when the going gets tougher we believe that they are invaluable, and they bring their own insights. The feeling that games are in some way undignified is addressed by Hodges, who comments:

> 'The notion of a game has to do with people acting together, setting themselves and each other tasks. As a result, game-theoretic versions of mathematical ideas often have a direct intuitive appeal when compared with more formalistic treatments. In the period 1900–1950 logic was fighting to establish itself as a serious branch of mathematics, and if you want your mathematics to be serious you don't start by talking about people setting up competitions or exercise sessions. Today logic has won its battle for recognition, and [we] can afford to make intuitiveness one of [our] chief aims.' [HirHod02a, p. vii]

# References.

[AndMonNém91] Hajnal **Andréka**, J. Donald **Monk**, and István **Németi** (*eds.*), Algebraic Logic, North-Holland 1991 [Colloquia Mathematica Societatis Janos Bolyai 54]

[Bir35] Garrett **Birkhoff**, On the Structure of Abstract Algebras, **Proceedings of the Cambridge Philosophical Society** 31 (1935), p. 433–454

[ChaKei90] Chen C. **Chang** and H. Jerome **Keisler**, Model Theory, 3rd edition, North-Holland 1990

[Die97] Reinhard **Diestel**, Graph Theory, Springer 1997 [Graduate Texts in Mathematics 173]

[Erd59] Paul **Erdős**, Graph Theory and Probability, **Canadian Journal of Mathematics** 11 (1959), p. 34–38

[Fin75] Kit **Fine**, Some Connections Between Elementary and Modal Logic, *in:* [Kan75, p. 15–31]

[GolHodVen03] Robert **Goldblatt**, Ian **Hodkinson**, and Yde **Venema**, On Canonical Modal Logics That Are Not Elementarily Determined, **Logique et Analyse** 181 (2003), p. 77–101

[GolHodVen04] Robert **Goldblatt**, Ian **Hodkinson** and Yde **Venema**, Erdős Graphs Resolve Fine's Canonicity Problem, **Bulletin of Symbolic Logic** 10 (2004), p. 186–208

[HenMonTar71] Leon **Henkin**, J. Donald **Monk**, and Alfred **Tarski**, Cylindric Algebras, Part I, North-Holland 1971

[HirHod97a] Robin **Hirsch** and Ian **Hodkinson**, Axiomatising Various Classes of Relation and Cylindric Algebras, **Logic Journal of the IGPL** 5 (1997), p. 209–229

[HirHod97b] Robin **Hirsch** and Ian **Hodkinson**, Complete Representations in Algebraic Logic, **Journal of Symbolic Logic** 62 (1997), p. 816–847

[HirHod02a] Robin **Hirsch** and Ian **Hodkinson**, Relation Algebras by Games, North-Holland 2002 [Studies in Logic and the Foundations of Mathematics 147]

[HirHod02b] Robin **Hirsch** and Ian **Hodkinson**, Strongly Representable Atom Structures of Relation Algebras, **Proceedings of the American Mathematical Society** 130 (2002), p. 1819–1831

[HirHodMad02a] Robin **Hirsch**, Ian **Hodkinson**, and Roger D. **Maddux**, Provability with Finitely Many Variables, **Bulletin of Symbolic Logic** 8 (2002), p. 348–379

[HirHodMad02b] Robin **Hirsch**, Ian **Hodkinson**, and Roger D. **Maddux**, Relation Algebra Reducts of Cylindric Algebras and an Application to Proof Theory, **Journal of Symbolic Logic** 67 (2002), p. 197–213

[Hod93] Wilfrid **Hodges**, Model Theory, Cambridge University Press 1993 [Encyclopedia of mathematics and its applications 42]

[Hod97] Ian **Hodkinson**, Atom Structures of Cylindric Algebras and Relation Algebras, **Annals of Pure and Applied Logic** 89 (1997), p. 117–148

[HodMikVen01] Ian **Hodkinson**, Szabolcs **Mikulás**, and Yde **Venema**, Axiomatizing Complex Algebras by Games, **Algebra Universalis** 46 (2001), p. 455–478

[HodVen05] Ian **Hodkinson** and Yde **Venema**, Canonical Varieties With No Canonical Axiomatisation, **Transactions of the American Mathematical Society** 357 (2005), p. 4579–4605

[Jón59] Bjarni **Jónsson**, Representation of Modular Lattices and of Relation Algebras, **Transactions of the American Mathematical Society** 92 (1959), p. 449–464

[Jón91] Bjarni **Jónsson**, The Theory of Binary Relations, *in:* [AndMonNém91, p. 245–292]

[JónTar51] Bjarni **Jónsson** and Alfred **Tarski**, Boolean Algebras with Operators I, **American Journal of Mathematics** 73 (1951), p. 891–939

[JónTar52] Bjarni **Jónsson** and Alfred **Tarski**, Boolean Algebras with Operators II, **American Journal of Mathematics** 74 (1952), p. 127–162

[Kan75] Stig **Kanger** (*ed.*), Proceedings of the 3rd Scandinavian Logic Symposium, Uppsala, Sweden, April 9-11, 1973, North Holland 1975

[Kra+97] Marcus **Kracht**, Maarten **de Rijke**, Heinrich **Wansing**, and Michael **Zakharyaschev**, Advances in Modal Logic, volume 1, Proceedings of the 1st International Workshop on Advances in Modal Logic, Berlin, Germany, October 8-10, 1996, CSLI Publications 1997

[Lyn50] Roger C. **Lyndon**, The Representation of Relational Algebras, **Annals of Mathematics** 51 (1950), p. 707–729

[Lyn56] Roger C. **Lyndon**, The Representation of Relation Algebras II, **Annals of Mathematics** 63 (1956), p. 294–307

[Lyn61] Roger C. **Lyndon**, Relation Algebras and Projective Geometries, **Michigan Mathematics Journal** 8 (1961), p. 21–28

[Mad83] Roger D. **Maddux**, A Sequent Calculus for Relation Algebras, **Annals of Pure and Applied Logic** 25 (1983), p. 73–101

[Mad89] Roger D. **Maddux**, Non-Finite Axiomatizability Results for Cylindric and Relation Algebras, **Journal of Symbolic Logic** 54 (1989), p. 951–974

[Mad91] Roger D. **Maddux**, The Origin of Relation Algebras in the Development and Axiomatization of the Calculus of Relations, **Studia Logica** 50 (1991), p. 421–455

[McK66] Ralph N. **McKenzie**, The Representation of Relation Algebras, *PhD thesis*, University of Colorado at Boulder, 1966

[Mon64] J. Donald **Monk**, On Representable Relation Algebras, **Michigan Mathematics Journal** 11 (1964), p. 207–210

[Pei33] Charles S. **Peirce**, Collected Papers, edited by Charles Hartshorne and Paul Weiss, Harvard University Press 1933

[Ram30] Frank P. **Ramsey**, On a Problem of Formal Logic, **Proceedings of London Mathematical Society** 30 (1930), p. 264–286

[Sch95] F. W. K. Ernst **Schröder**, Algebra und Logik der Relative, Part I, Vorlesungen über die Algebra der Logik (exacte Logik), volume 3, B.G. Teubner 1895 *reprint:* Chelsea 1966

[Sto36] Marshall H. **Stone**, The Theory of Representations for Boolean Algebras, **Transactions of the American Mathematical Society** 40 (1936), p. 37–111

[Tar41] Alfred **Tarski**, On the Calculus of Relations, **Journal of Symbolic Logic** 6 (1941), p. 73–89

[Tar55] Alfred **Tarski**, Contributions to the Theory of Models III, **Indagationes Mathematicae** 17 (1955), p. 56–64

[TarGiv87] Alfred **Tarski** and Steven R. **Givant**, A Formalization of Set Theory Without Variables, American Mathematical Society 1987 [AMS Colloquium Publications 41]

[Ven97a] Yde **Venema**, Atom Structures, *in:* [Kra+97, p. 291–305]

[Ven97b] Yde **Venema**, Atom Structures and Sahlqvist Equations, **Algebra Universalis** 38 (1997), p. 185–199

**Received**: March 1st, 2005;
**In revised version**: August 24th, 2005;
**Accepted by the editors**: August 31st, 2005.

FotFS V

Stefan **Bold**, Benedikt **Löwe**,
Thoralf **Räsch**, Johan **van Benthem** (*eds.*)
**Foundations of the Formal Sciences V**
Infinite Games

# Evolutionary Game Theory: Why Equilibrium and Which Equilibrium

WEI-TORNG JUANG AND HAMID SABOURIAN

Institute of Economics
Academia Sinica
Taipei 115, Taiwan
wjuang@econ.sinica.edu.tw

Faculty of Economics
University of Cambridge
Sidgwick Avenue
Cambridge CB3 9DD, United Kingdom
Hamid.Sabourian@econ.cam.ac.uk

ABSTRACT. In this survey paper, we are addressing two central questions about the foundations of game theory (why equilibrium and which equilibrium) within the dynamic evolutionary framework, paying speical attention to the behavioural rules (the type space) in each population and their possible evolution.

## 1 Introduction

Two questions central to the foundations of game theory are [Sam02]:

1. *Why equilibrium*? Should we expect Nash equilibrium play (players choosing best response strategies to the choices of others)?
2. If so, *which equilibrium*? Which of the many Nash equilibria that arise in most games should we expect?

To address the first question the classical game theory approach employs assumptions that agents are rational and they all have common knowledge of such rationality and beliefs. For a variety of reasons there has been wide dissatisfaction with such an approach. One major concern has to do with the plausibility and necessity of rationality and the common knowledge of rationality and beliefs. Are players fully rational, having unbounded computing ability and never making mistakes? Clearly, they are not. Moreover, the assumption of common (knowledge of) beliefs begs another question as to how players come to have common (knowledge of) beliefs. As we shall see in later discussion, under a very broad range of situations, a system may reach or converge to an equilibrium even when players are not fully rational (boundedly rational) and are not well-informed. As a result, it may be, as Samuelson points out, that an equilibrium appears not because agents are rational, "but rather agents appear rational because an equilibrium has been reached. [Sam97]"

The classical approach has also not been able to provide a satisfactory solution to the second question on selection among multiple equilibria that commonly arise. The equilibrium selection problem is particularly acute when the game has many strict Nash equilibria. This is because such equilibria are robust with respect to small perturbations and as a result they survive most refinements of Nash equilibrium proposed in the classical approach (one possible exception is [HarSel88]). Such dissatisfaction with the classical approach has spurred the emergence of the evolutionary approach.

Since Maynard Smith and Price [MayPri73, May82] proposed a new way of thinking about the evolution of animal behavior, sex and population genetics, evolutionary game theory has enjoyed a vibrant development. In the last 20 years, game theorists (longer for mathematical biologists) have addressed these questions in the context of evolutionary/learning frameworks. The hope has been that such an evolutionary approach can provide: first, a justification for Nash equilibrium (by coordinating beliefs so that they are consistent with strategies); second, new insights into which equilibrium will be played.

The biological idea behind the evolutionary approach is that more successful players/strategies will have a better chance to survive or prosper in the population. One typical mechanism consistent with the concept is replicator dynamics in which the population frequency for each strategy changes as a function of the difference between the performance of the strategy and that of the population average. Replicator dynamics are too specific to capture complex human behaviour, however. To better fit human environments, other dynamics such as monotonic dynamics that include replicator dynamics as a special case have been proposed (see below for the definition of the different dynamics). Such selection dynamics in human setting represent the process of learning and imitation (the adjustment rules) of the individuals. We shall give a detailed discussion on these dynamics in the next section.

In this paper, we would like to survey some of the literature on the two questions (why equilibrium and which equilibrium) within the *dynamic* evolutionary framework, paying special attention to the behavioural rules (the type space) in each population and their possible evolution. What follows will not be an exhaustive survey. There are some excellent textbooks and surveys on evolutionary game theory.[1]

In the first part, we shall discuss the conditions under which a deterministic evolutionary dynamics process converges to a Nash equilibrium and potential failure for such convergence. The dynamics in these models are usually determined by some fixed behavioural rule that all agents follow (for example reinforcement learning, imitation, best response, fictitious play).

We focus on Nash equilibrium because firstly it is the central equilibrium concept in non-cooperative games and secondly, in evolutionary models, irrespective of the particular dynamics (adjustment rules agents have), there is strong justification for Nash equilibria in terms of the steady states of these systems and their local dynamics. On the other hand, as Hofbauer and Sigmund point out, "a central result on general 'adjustment dynamics' shows that every reasonable adaptation process will fail, for some games, to lead to a Nash equilibrium [HofSig03]" globally. We shall briefly outline Hart and Mas-Colell's explanation for the lack of global convergence to Nash equilibria in such set-ups [HarMas03]. However, almost all deterministic evolutionary models that have looked at the issue of convergence to Nash have considered dynamics in which all agents adopt one fixed behavioural rule. We shall also consider the question of convergence to Nash when multiple rules coexist, agents can adopt different rules and the selection is at the level of rules. Here, we will argue that convergence to Nash may be easier to establish if the set of rules allowed are sufficiently rich.

In the second part of this paper, we shall address the "which equilibrium" question by introducing small perpetual random shocks to the evolutionary dynamics in finite populations. This approach has been very successful in selecting between different strict Nash equilibria in specific applications. However, with this approach, it turns out that the specific equilibria selected depend very much on the adjustment rules allowed. For example, best response dynamics seems to select the risk dominant equilibrium in a $2 \times 2$ coordination game [KanMaiRob93, You93a] while the imitation one seems to favour the efficient equilibrium [RobVeg96]. Therefore, to answer the "which equilibrium" question, we need to explore the dynamics in which multiple rules coexist and compete with each other. This is important not only for addressing the question of equilibrium selection but also because in a world with heterogeneous individuals it is not plausible that a single adjustment rule can capture all important properties of human behaviour. As Young points out, "we would guess that people adapt their adaptive behav-

[1] *E.g.*, [FudLev98], [HofSig98], [HofSig03], [Sam97], [Veg96], [Veg03], [Wei95], and [You98].

ior according to how they classify a situation (competitive or cooperative, for example) and reason from personal experience as well as from knowledge of others' experiences in analogous situations. In other words, learning can be very complicated indeed. [You98, p. 29]" For the most of this section we compare the issue of equilibrium selection in the context of specific games for the different dynamics/rules. We then conclude the paper by considering some specific models with small perpetual random shocks in which multiple rules coexist, and in particular we discuss some of the justification provided for some specific rules such as the imitation one.

## 2 Why Equilibrium? Deterministic Evolutionary Models

The standard evolutionary game with discrete or continuous time consists of $n$ large (often infinite) populations of myopic and unsophisticated agents playing some underlying one-shot (normal form) $N$-player game $G = (A_i, \pi_i)_{i=1}^n$ infinitely often, where, for each player $i = 1, .., n$, $A_i$ is a finite set of (pure) strategies (henceforth also called actions) and $\pi_i : \prod_i A_i \to \mathbf{R}$ is the payoff function at each date.

Each population $i$ consists of a set of (countable) types. In the most of the literature a type in population $i$ is usually identified with an action $a_i \in A_i$ in the one-shot game $G$. That is, he is programmed to *always* execute $a_i$.

Before describing the dynamics, we introduce some standard definitions. Let $A \equiv \prod_i A_i$ be the set of action profiles for the one-shot game. We adopt the convention that for any profile $y = (y_1, .., y_n)$, $y_{-i}$ refers to $y$ without its $i$th component. Next, for any strategy profile $a \in A$, let $B_i(a) = \{a_i' \in A_i \mid \pi_i(a_i', a_{-i}) \geq \pi_i(a_i'', a_{-i}) \; \forall a_i'' \in A_i\}$ be the set of best responses for $i$ given $a$. Then a profile of strategies $a^* \in A$ is a Nash equilibrium if $a_i^* \in B_i(a^*)$ for all $i$. A Nash equilibrium $a^* \in A$ is strict if $B_i(a^*)$ is unique for all $i$. Similarly, we can define a Nash equilibrium for the game $G$ in the space of mixed strategies.

**Dynamics.** At each date $t$ each member of population $i$ is randomly matched with one member of every other population to play the game $G$ and receives a payoff depending on his action and the actions taken at $t$ by those with whom he is matched.

Time could be discrete and with $t = 0, 1, 2, ...$ or continuous where $t \in [0, \infty)$.[2] At each date $t$ the state of the system is the proportion (probability)

[2]Since any dynamic evolutionary system involves playing a one-shot game infinitely often, such models, by definition, have an infinite aspect about them. In addition to this infinity of the time horizon, in this survey paper we also consider cases in which the size of the population, the set of types and/or strategies are infinite.

of each type/strategy/action in each population; thus it can be described by $n$ probability distributions $P_i^t = \{P_i^t(a_i)\}_{a_i \in A_i}$ where $P_i^t(a_i)$ denotes the proportion (probability) of strategy $a_i$ in population $i$ at date $t$. The state of the system evolves according to some dynamics describing selection (and mutation in models with randomness). In the discrete case, the dynamics (with no mutation) is given by

$$P^{t+1} = \Gamma(P^t) \tag{1}$$

where $P^t = (P_1^t, .., P_n^t)$. With continuous time the law of motion is described by

$$\dot{P^t} = \Gamma(P^t) \tag{2}$$

The selection dynamics for the case of discrete time are such that at each $t+1$ the proportion of each strategy changes as a function of how well the type has done on average at $t$, in terms of payoff, relative to the payoff of the other strategies. Note that the proportions at each date $t+1$ has a stringent Markovian property of depending only on the environment (payoffs) in the previous period $t$ and not on outcomes prior to $t$. The most canonical model of evolutionary dynamics that embodies the idea of Darwinian selection is the replicator dynamics. Here, the growth of each strategy in any population $i$ is assumed to be an increasing function of its average (expected) payoff minus population $i$'s average payoff. Thus, in the case of continuous time the replicator dynamics is described by

$$\dot{P_i^t} = P_i^t(a_i)(E\pi_i^t(a_i) - E\overline{\pi}_i^t) \tag{3}$$

where, with some abuse of notations, $E\pi_i^t(a_i)$ denotes the average (expected) payoff to strategy $a_i$ at time $t$ and $E\overline{\pi}_i^t = \Sigma_{a_i' \in A_i} E\pi_i^t(a_i')P_i^t(a_i')$ is the average payoff in population $i$ at date $t$. Note that since each population is 'large' number and there is no aggregate uncertainty, $E\pi_i^t(a_i)$ is simply $\Sigma_{a_{-i} \in A_{-i}} \pi_i(a_i, a_{-i})P_{-i}^t(a_{-i})$. We can also describe a discrete analogue for replicator dynamics.

There are several properties of replicator dynamics to note. First there is no birth of new types or death of existing types in finite time: for all $t$, $P_i^0(a_i) > 0$ implies $P_i^t(a_i) > 0$. Second, the dynamics is determined by relative payoffs in the sense that types that do better in terms of average payoffs grow faster than those that do less well.

In the biological literature, the selection dynamics represents reproduction based on fitness. In fact, if the evolutionary model is such that agents live for only one period, reproduction is asexual, $E\pi_i^t(a_i)$ represents the number of offspring of type $a_i$, each offspring of $a_i$ type does exactly the same as $a_i$ and there is no mutation, then replicator dynamics, defined by (3), precisely describes the dynamical system for the continuous time. In a non-biological (*e.g.*, of economics, political science or sociology) setting, the selection represents learning and imitation. Ideally, one would like to build up the selection dynamics from a precise and acceptable theory of how individual players switch between different actions/types. The evolutionary approach often avoids such a difficult task and instead it places assumptions directly on the selection dynamics and hopes that these properties are general enough to include processes produced by a variety of learning and imitation theories.[3]

Nachbar, Björnerstedt and Weibull, Schlag, and Borgers and Sarin[4] and others show that one may also be able to provide a justification for replicator dynamics based on imitation and learning. In general, however, these learning/imitation models are too specific and there is often no compelling reason for adopting such a precise imitation/learning specifications. Thus, the replicator dynamics framework simply cannot capture the complexity of learning and imitation in social and economic contexts and one needs a broader type of dynamics than replicator dynamics .

Fortunately, it turns out in the deterministic evolutionary set-ups a large number of dynamics share many essential features. One very large class of dynamics that includes replicator dynamics as a special case is defined by the property that strategies that do better in terms of payoffs grow faster relative to those that do less well. More precisely, a selection evolutionary dynamics is called monotonic if at every date $t$ the growth rates of the different strategies are ranked by their average payoffs:

$$G_i^t(a_i') > G_i^t(a_i'') \Leftrightarrow E\pi_i^t(a_i') > E\pi_i^t(a_i'') \; \forall a_i', a_i'' \in A_i \tag{4}$$

where $G_i^t(a_i)$ is the growth rate of strategy $a_i$ at $t$, and thus satisfies

$$G_i^t(a_i)P_i^t(a_i) = P_i^{t+1}(a_i) - P_i^t(a_i)$$

in the discrete time model and $G_i^t(a_i)P_i^t(a_i) = \dot{P}_i^t(a_i)$ when time is continuous.

Monotonic dynamics is consistent with fairly general learning and imitation models. Consider for example an imitation dynamics in which at any date a player has the chance to sample another player at random and adopt his behaviour; and the rate at which a player of a certain type switches behaviour to another type depends on the current average payoffs of the two types. Monotonic dynamics is consistent with such imitation models as

[3] *Cf.* [FudLev98, You04].
[4] *Cf.* [Nac90, BjöWei96, Sch98, BörSar97], respectively.

long as the rate at which players switch depends positively on the success of the sampled type's payoff relative one's own.

Often in describing monotonic dynamics it is assumed that the dynamics satisfy two further properties. First it is assumed that the function describing the adjustment, $\Gamma$ is continuous (Lipschitz continuous for the continuous time case) and second there is no birth of new types or death of existing types: for all $t$, $P_i^0(a_i) > 0$ implies $P_i^t(a_i) > 0$. The first condition is for technical reasons; however, note that it excludes certain type of discontinuous imitation dynamics in which players switch their behaviour if the sampled behaviour has a strictly better payoff. The no birth assumption is however very much consistent with the imitation story because imitative behaviour has the property that players only sample amongst existing types; as a result only types that existed in the past survive. The assumption that types are not created or destroyed can be relaxed in some cases; however, in order to simplify the discussion we shall also assume, unless stated otherwise, that monotonic dynamics has this property.

The properties of monotonic dynamics, discussed in the next section, broadly holds, for even larger classes of dynamics. For example, one generalisation of monotonic dynamics is *weakly positive dynamics*: for any population, if there exists a type that has a higher payoff than the average of the population, then some type must have a positive growth.

Another type of dynamics is the (myopic) best response dynamics [Mat92]. Here, at any date a fixed proportion of each population chooses a best response strategy (one that maximises the player's one-period payoff) given the average strategy in every other population in the previous period. Another important variation of best response dynamics is the fictitious play dynamics where each player chooses a best response to the historical frequency of past plays. Such dynamics are often described in the context of finite populations. They are clearly inconsistent with monotonic dynamics because they require players to switch to best responses (when the opportunity arises) whereas monotonic dynamics require the ranking of growth rates of *all* strategies according to payoffs. Moreover, these dynamics may require discontinuous shift in behaviour. Nevertheless, it turns out that many of the qualitative features of these dynamics in terms of Nash are similar to those of the monotonic dynamics in the deterministic settings.

Before turning to properties of the dynamics there are three further points to note. First, the literature on evolutionary game theory often considers single population models in which all agents play a *symmetric* one-shot game repeatedly. Here, we described a more general set-up with multiple populations to allow for the possibility that the underlying game $G$ may be asymmetric.[5] For some specific analysis, to simplify the discussions, we shall at times limit ourselves to single population models. Second, much of the literature on deterministic evolutionary dynamics considers continuous time. The results for the discrete and continuous dynamics sometimes differ (the former dynamics is often less "nice"). For ease of exposition, we shall at times limit the discussion to one kind of time dynamics. However, often there are analogous results for the other kind of time dynamics. Finally, the models in this section are mainly deterministic while those in the "which equilibrium" section are stochastic. The reason is that deterministic environments suffice for our discussion on the "why equilibrium" part while introducing mutations here only complicates unnecessarily without adding much insight. In contrast, in the "which equilibrium" part in the next section, to select between equilibria, we introduce mutations or noises to perturb a dynamic system in such a way that the underlying system may drift among each equilibrium; this then allows us to examine at which equilibrium the system will spend most of the time.

### 2.1 Convergence and Stability in Standard Deterministic Models

**Folk Theorem.** Before describing the properties of deterministic evolutionary models, we need to introduce two *local* stability concepts that are common in the literature: Lyapunov stability and asymptotic stability. A stationary state is *Lyapunov stable* if small perturbations do not result in dynamics that takes the system far from it. A stationary state is *asymptotically stable* if it is Lyapunov stable and there exists a neighbourhood of it such that for any initial state belonging to this neighbourhood the path converges to it. When the underlying game $G$ is finite ($A_i$ is finite) and a type refers to a pure action $a_i$ in $G$ we have the following Folk Theorem of the evolutionary game theory.

**Theorem 1 (Hofbauer and Sigmund, [HofSig98]).** For any 'reasonable dynamics':

1. Any (pure or mixed) Nash equilibrium of $G$ is a stationary state of the evolutionary dynamics.[6]
2. Any Lyopunov stable stationary state is a Nash equilibrium (possibly mixed).
3. A strict Nash equilibrium is an asymptotically stable stationary state.

[5] *Cf.* [Wei95] for some discussion of the difficulties with multi-population models that are not present in single population models.

[6] Note that although here agents do not randomise and choose pure strategies, the proportion of each population that take different strategies (actions) corresponds to a mixed strategy.

4. The limit of any interior convergent path is a Nash equilibrium (possibly mixed).

There are a few remarks concerning this theorem. First, the term 'reasonable dynamics' refers to most of the dynamics found in the literature. However, for the purpose of illustrating the ideas here, define it to be dynamics that are continuous and monotonic (with discrete or continuous time).

Second, at a stationary state of such a dynamic all strategies played by various members must give the same payoffs; since at any Nash equilibrium any strategy that is played with a positive probability must be optimal (best response) it follows that any Nash equilibrium is a stationary state (condition 1 in Theorem 1). However a stationary state need not be a Nash equilibrium in models with no birth (as in the case with imitation) because there may be a better strategy that is not played and hence its population may not grow. However, a Lyapunov stable stationary state must be a Nash equilibrium because any superior strategy that is not played in the stationary state will be played once a perturbation occurs; once this occurs the superior strategy will grow leading the system away from the stationary state (condition 2 in Theorem 1).

Third, if a state corresponds to a strict Nash equilibrium then it must be asymptotically stable (condition 3 in Theorem 1) because in the case of a strict Nash equilibrium there exists a neighbourhood of the equilibrium such that the equilibrium strategy for each player is the unique best strategy.

Fourth, even when the dynamics are continuous and monotonic there is no guarantee that such dynamic will converge. If the path does converge, however, then the limiting state of this convergent path is a Nash equilibrium (condition 4 in Theorem 1); otherwise in the limit of this path there is some strategy that does better than others and hence must grow, but this contradicts the path being convergent.

Fifth, note that Nash equilibria in the Folk Theorem can equally refer to mixed ones even though no agent randomises. This is simply a reflection of the fact that in an evolutionary setting the state of the system refers to a probability distribution over the pure strategies.

**Global Convergence.** The above arguments that an outcome is a Nash equilibrium if it is stable, or a Nash equilibrium is the limit of any convergent path, or the path converges to a strict Nash equilibrium if the process starts from states that are sufficiently close to that equilibrium, provide a somewhat qualified answer to "*why equilibrium*". We may want a stronger result of the form that an evolutionary dynamics must produce *global* convergence to a Nash equilibrium. However, starting with Shapely's early example on fictitious play,[7] this stronger result that the evolutionary dynamics converges to Nash globally from any (interior) initial state is missing. Indeed, the dynamic trajectories of evolutionary processes in general do not converge; periodic cycles as well as all the other complexities of arbitrary dynamical systems (limit cycles and chaos) are all possible.

One simple example for which the unique Nash equilibrium may not be even asymptotically stable is the 2-player zero-sum generalised "Rock-Paper-Scissors" game in which each player has three strategies $R, S, P$ and the payoff for the row player is described by the following matrices:

$$\begin{bmatrix} 1 & 2+\varepsilon & 0 \\ 0 & 1 & 2+\varepsilon \\ 2+\varepsilon & 0 & 1 \end{bmatrix}$$

where $\varepsilon$ is some real number. This game has a unique mixed Nash equilibrium $\mu^*$ in which each strategy is chosen with a probability $1/3$. The simplest evolutionary dynamics to consider is a single population replicator dynamics with continuous time. Here it is not difficult to verify that at any $t$ the derivative of the product of the probabilities of the three strategies, $P^t(R)P^t(S)P^t(P)$, is positive (negative, zero) if $\varepsilon$ is positive (negative, zero). Thus, depending on the value of $\varepsilon$ this product term is always increasing, constant or decreasing. But this implies the following.

- For $\varepsilon > 0$ all trajectories from all interior initial state spiral inwards towards the unique Nash equilibrium $\mu^*$; thus the dynamics is globally stable.
- For $\varepsilon = 0$ all paths are cycles; thus the Nash equilibrium is Lyapunov stable but not asymptotically stable.
- for $\varepsilon < 0$ all trajectories from all interior initial state except $\mu^*$ spiral outwards; thus the Nash equilibrium is unstable.[8]

The above example is of course very specific. However, the possibility of cycles, global instability and complex dynamics is even more prevalent when there are multiple populations, time is discrete, each player has a large number of strategies or other standard dynamics are considered.

In [HarMas03], Hart and Mas-Colell try to provide an answer for why it is difficult to formulate sensible dynamics that always guarantee (global) convergence to a Nash equilibrium. They consider an adaptive stationary dynamics (not strictly a standard evolutionary model) with a finite population

[7] *Cf.* [Sha64].

[8] *Cf.* [Wei91] for detailed discussion on the dynamics of more generalised "Rock-Paper-Scissors" games.

in which the adjustment in a player's strategy does not depend on the payoff function of other players (it may depend on the other player's strategies and his own payoff function). They call this dynamics uncoupled and claim that any sensible heuristic dynamics must have this property (most standard dynamics based on agents with limited rationality found in the literature, such as best-reply, fictitious play, better reply, payoff-improving, monotonic are all uncoupled). They show that stationary uncoupled dynamics cannot be guaranteed to converge to a Nash equilibrium in a deterministic setting with continuous time.[9] They therefore conclude that the lack of convergence is due to the informational requirement of uncoupledness which precludes too much coordination of behaviour amongst agents.[10]

To illustrate the nature of the Hart-Mas-Colell result consider the general dynamics for the continuous time described in (2). The precise dynamics clearly depend on the payoff functions and can thus be rewritten, with some abuse of notation, as

$$\dot{P_i^t} = \Gamma_i(P^t, \pi) \text{ for all } i = 1, .., n$$

Hart and Mas-Colell call the above dynamic system uncoupled if $\dot{P_i^t}$ does not depend on $\pi_j$ for all $j \neq i$. Thus $\Gamma_i(P^t, \pi) = \Gamma_i(P^t, \pi')$ for any two profile of payoffs $\pi$ and $\pi'$ such that $\pi_i = \pi_i'$.

To facilitate the analysis they restrict the dynamics $\Gamma = (\Gamma_1, .., \Gamma_n)$ to be $C^1$. Suppose further that the dynamics is hyperbolic (the eigenvalues corresponding to the Jacobian of $\Gamma$ have non-zero real parts), so that it behaves locally like a linear system $\dot{P^t} = D\Gamma(P^t, \pi)P^t$, where for any $P$, $D\Gamma(P, \pi)$ is the Jacobian matrix of $\Gamma(., \pi)$ computed at $P$. Then (asymptotic) stability imposes conditions on the Jacobian matrix $D\Gamma$. Uncoupledness also imposes conditions on the $\Gamma$. Unfortunately, these two sets of conditions may not be consistent for many games.

One example provided by Hart and Mas-Colell to illustrate such inconsistency is a family of three player games in which each player has two strategies, and the payoffs are given by the following two matrices:

$$\begin{bmatrix} 0, 0, 0 & \varepsilon^1, 1, 0 \\ 1, 0, \varepsilon^3 & 0, 1, \varepsilon^3 \end{bmatrix} \text{ and } \begin{bmatrix} 0, \varepsilon^2, 1 & \varepsilon^1, 1, 0 \\ 1, \varepsilon^2, 0 & 0, 0, 0 \end{bmatrix},$$

[9] There exist uncoupled dynamics that are not convergent to Nash equilibria but are close to them most of the time; *cf.* [FosYou02].

[10] Hart and Mas-Colell and others have studied specific learning processes that allow stochastic moves (regret-matching) that always converge to a correlated equilibrium [HarMas00, HarMas01]. They argue that learning from past history induces correlation in behaviour and therefore one cannot expect dynamic processes of this kind to induce to Nash equilibrium. *Cf.* also [You04] for further discussion of these issues.

where all $\varepsilon^i$ are some numbers in a small neigbourhood of 1, and player 1 chooses the row, player 2 the column and player 3 the matrix. (Thus, the family of games considered is defined by the above two matrices where for each $i$, $\varepsilon^i \in (1-\eta, 1+\eta)$ for some small $\eta > 0$.)

Denote the game (the profile of payoff functions) when $\varepsilon^i = 1$ for all $i$ by $\pi^*$. Next note that for any $\varepsilon^i$, each such game in this family has a unique Nash equilibrium; in particular the Nash equilibrium of the game $\pi^*$ is given by each player randomising with probability half.

Next, consider any dynamics $\Gamma$ that is uncoupled. The state of the dynamics of this game is described by $P_1, P_2, P_3 \in [0,1]$ denoting respectively the probability of top row, left column and left matrix. Now, by way of contradiction suppose that the dynamics converges to Nash in this family of games. Then since $(1/2, 1/2, 1/2)$ is the Nash equilibrium when $\varepsilon = 1$ it must hold that

$$\Gamma_1(1/2, 1/2, 1/2, \pi^*) = 0. \tag{5}$$

Now for any $a$ close to $1/2$ consider another game $\pi^a$ where $\varepsilon^1 = \varepsilon^2 = 1$ and $\varepsilon^3 = \frac{a}{1-a}$; thus the payoffs of players 1 and 2 are the same as in $\pi^*$ and the payoff of 3 is different. By simple calculation one can show that the state $(a, 1/2, 1/2)$ is the unique Nash equilibrium of $\pi^a$. Again since the dynamics is convergent it must be that $\Gamma(a, 1/2, 1/2, \pi^a) = 0$. But since the dynamics is uncoupled and player 1 has the same payoff in $\pi^a$ and $\pi^*$, it follows that

$$\Gamma_1(a, 1/2, 1/2, \pi^*) = 0. \tag{6}$$

Since (6) holds for any $a$ close to $1/2$ it follows from (5) that $\partial\Gamma_1/\partial P_1 = 0$ at $(1/2, 1/2, 1/2, \pi^*)$. By the same argument one can show that entire diagonal of the Jacobian of function $D\Gamma(., \pi^*)$ vanishes at $(1/2, 1/2, 1/2)$. Now if the dynamics is hyperbolic this implies that the Jacobian has an eigenvalue with positive real part; but then we have a contradiction to the claim that in the game $\pi^*$ the dynamics converges ($(1/2, 1/2, 1/2)$ is asymptotically stable).

The negative results on convergence are for an arbitrary set of games. There are, however, some noteworthy families of games that do have stable equilibria (at least with continuous time) for some of the standard (uncoupled) dynamics such as replicator dynamics, monotonic dynamics, best response, fictitious play.[11]

One such family of games are those that are strictly dominance solvable (games in which there is unique solution to iterative deletion of strictly dominated strategies). Here, with monotonic dynamics the strictly dominance solution is globally stable for any initial interior state. This is simply because with monotonic dynamics it can be shown that the share of any

[11] *Cf.* [Hof00, San$\infty$].

pure strategy that does not survive iterative deletion of strictly dominated strategies goes to zero asymptotically. The basic idea for this result is that if a pure strategy is strictly dominated by another pure strategy then its growth rate is always less than the other and hence its share must go to zero.[12] By the same argument, the result extends to iterative deletion of strictly dominated strategies: Once the shares of the strictly dominated strategies are close to zero then strategies that are removed in the second round of iterative deletion of strictly dominated strategies must have a lower payoff than some other strategies and so their share starts to shrink to zero and so on.

The above results on strict domination do not extend to weak dominance and iterative deletion of weakly dominated strategies. Samuelson and Zhang give an example of weakly dominated pure strategy that is not eliminated with even continuous time replicator dynamics [SamZha92].

Other families of games with globally stability for some standard dynamics include zero-sum, two-player games, nondegenerate $2 \times 2$ games and (weighted) potential games. For the fictitious play dynamics, the beliefs of players (empirical distribution of the actions of the individual players) converge to a Nash equilibrium in finite, zero-sum, two-player games and in nondegenerate $2 \times 2$ games.[13] Also, for best response and fictitious play dynamics the solution path in (weighted) potential games is globally stable and converges to Nash [MonSha96b]. The class of potential games include many noteworthy games such as common interest games and classical Cournot Oligopoly with constant marginal cost. In such games players effectively strive to maximize a common function — the potential function. Thus, in potential games, as well as in two-person zero-sum games, the payoff of one player can be used to determine the payoff of other players. This may explain why Hart and Mas-Colell's negative result on convergence, based on the informational content of uncoupled dynamics, does not apply to such games.

To obtain global convergence for a more general class of games we may have to consider other plausible evolutionary processes that are not adaptive and/or uncoupled as in Hart and Mas-Colell. Next, we shall do precisely this by allowing richer set of types (in particular types that can choose

[12]This result does not hold for mixed strategies. Even for the case of continuous time dynamics stronger conditions than monotonicity are needed to ensure that strategies that are dominated by mixed strategies are eliminated (*e.g.*, aggregate monotonicity in [SamZha92], convex monotonicity in [HofWei96]). In the discrete case even these stronger conditions do not suffice. Dekel and Scotchmer give an example to show that discrete time replicator dynamics does not eliminate all strategies strictly dominated by a mixed strategies [DekSco92].

[13]*Cf.* [Rob51, Miy61, MonSha96a].

different actions at different dates) and allowing selection at the level of rules. Before turning to this issue, would like to make a few brief remarks on the evolutionary dynamics above when the strategy space $A_i$ of the underlying game is infinite.

**Stage-games with infinite strategies.** Infinite (one-shot) games are of interest for many reasons. First, when mixed strategies are allowed the strategy space is trivially continuous. Second, there are many games in economic contexts that are modelled with an infinite or continuous strategy set (*e.g.*, bargaining games, the War of Attrition and Cournot duopoly, to name a few). Third, one may be interested in the question of whether the infinite case is the limit of successively finer finite approximations (the infinite case may then be regarded as a reference point).

It turns out that whether the strategy set for the one-shot game is infinite or finite does matter in terms of developing the evolutionary dynamics on the set of probability distributions over the set of strategies (*e.g.*, the replicator equation), ensuring that the solutions are well-defined, stability of the stationary states and its relationship with static game-theoretic equilibrium concepts such as Nash. There are some recent papers on this topic (*e.g.*, [Sey00, OecRie01, OecRie02, Cre04]). Most of this recent literature assumes continuous time. The immediate issue here is, not surprisingly, what constitutes an appropriate notion of closeness and/or convergence for probability distributions on the set of strategies. Here there are several different possible topological extensions of the finite strategy space set-up, and conclusions on stability depend critically on what definitions are adopted. Here, for reasons of space we shall not consider this issue any further except to mention that even one of the results in Theorem 1 does not extend to the infinite case. Oechssler and Riedel provide an example showing that strict equilibria need not be asymptotically stable or even Lyapunov stable when the stage game is infinite [OecRie02].

**Global Convergence and Richer Type Space.** In standard evolutionary models types in a population refer to strategy in the underlying one shot-game. In [AndSab96], Anderlini and Sabourian have looked at the question of global convergence with a richer type space. In particular, they allow a type to be a rule of behaviour taking different actions at different dates (or histories) and selection is applied to the set of rules. More formally, a type $x_i : H \to A_i$ for population $i$ is now a rule mapping from the set of past information $H$ into the set of pure actions $A_i$. It assumed that there are a countable number of types/rules. If $H$ is sufficiently informative then all standard adjustment rules such as best-response or imitation can be included in the set of allowable types. The critical assumption in [AndSab96] is that the informational content of $H$ is such that when a type makes a decision the dates are known; thus types may condition their actions at least on time and thus can take different actions at different times.

The evolutionary dynamics is the same as the discrete dynamics described above except that it operates by selecting amongst types (rules) according to how well the type has done at that date relative to the average payoffs of others. Formally, let $Q_i^t(x_i)$ be the proportion type $x_i$ in population $i$ at date $t$. The dynamics prescribes how $Q_i^t$ evolves according to how well $x_i$ does at $t$. Since payoffs depend on actions, without any loss of generality one can write the growth of each type at each date $t$ as function of the action he takes and the distributions of actions in all the populations at $t$; thus the growth rate of type $x_i$ at $t$, $g_i^t(x_i) = \frac{Q_i^{t+1}(x_i) - Q_i^t(x_i)}{Q_i^t(x_i)}$, can be written as

$$g_i^t(x_i) = \gamma_i(P^t, a_i^t(x_i))$$

where, as before, $P^t$ is the distribution over the set of action profiles at $t$, $a_i^t(x_i)$ denotes the strategy taken by $x_i$ at $t$ and $\gamma_i$ is the growth function for population $i$.

Here, as in the standard model, every limit point of the distributions of actions of the players corresponds to a Nash equilibrium (possibly mixed) of the one-shot game if the dynamics, defined by the function $\gamma_i$, is continuous and monotonic. This result is similar to the existing Folk Theorem of evolutionary game theory mentioned above (condition 4 in Theorem 1): if a path converges it must be to Nash.

The main result of [AndSab96] concerns the global convergence of the distributions over types. First, define a type to be *feasible* if it has a positive initial probability. Anderlini and Sabourian show that for every $i$ and $x_i$, $Q_i^t(x_i)$ converges if for each population $i$ there exists a feasible type $\overline{x}_i$ ($Q_i^0(\overline{x}_i) > 0$), referred to as the 'smart' type, which can grow as fast as any other type in population $i$ at *each* date.[14] The existence of such a smart type with a positive initial probability ensures that the system converges. The intuition for this is as follows. Since $\overline{x}_i$ has the maximum growth rate it follows that $Q_i^t(\overline{x}_i)$ is increasing. Since $Q_i^t(\overline{x}_i)$ is also bounded, it then follows that the growth rate of $\overline{x}_i$ must go to zero. But the growth rates of other types in population $i$ do not exceed that of the smart type and the sum of the growth rates of all types in population $i$ at each date is zero. Therefore, it follows that the growth rate of others must also go to zero. (Effectively, since $Q_i^t(\overline{x}_i)$ is monotonically increasing, it is like a Lyapunov function.)

[14]The assumption of positive initial probability for the smart types is more than is needed to establish the result. In fact, Anderlini and Sabourian assume only that for each population a smart type is born at some finite date with a positive probability.

There are four immediate points to note concerning this result. First, since the system is deterministic, it is always possible to construct a smart type as long as types are allowed to condition their actions at least on time. This is because at any date $t$, a smart type needs to take an action $\overline{a}_i^t$ that is best in terms of growth:

$$\gamma_i(P^t, \overline{a}_i^t) \geq \gamma_i(P^t, a_i), \forall a_i. \tag{7}$$

Now, given the parameters of the system $Q_i^0$ and the selection function $\gamma_i$, for any $t$ it is possible, by computing recursively forward (simulating) the system, to compute first $P^t$ and then find an action $\overline{a}_i^t$ that satisfies (7).

Second, note that each smart type may need to take different actions at different times to ensure that it grows faster than others in the same population.

Third, the result does not depend on any particular assumption about the shape of the selection dynamics.

Fourth, note that although probabilities over types converge, probabilities over actions do not necessarily converge. This is because each type can condition on the past and can take different actions at different times. Therefore, even when probabilities over types converge, if the underlying game $G$ has more than one Nash equilibrium, it is possible that in the limit different types could be switching between different Nash equilibria of $G$. An immediate corollary is that if $G$ has a unique Nash equilibrium then probabilities over actions converge as well.

The main assumption in [AndSab96] needed to establish the convergence result is the existence of a *feasible* smart type for each population. This assumption, however, is not primitive.[15] The identity of any smart type depends on the initial profile of distributions over types $Q^0 = (Q_1^0, .., Q_n^0)$; but then there is no guarantee that any smart type $\overline{x}_i$ for population $i$ corresponding to $Q^0$ is such that $Q_i^0(\overline{x}_i) > 0$. An attractive set-up would be a set of assumptions on the of feasible types for the initial distributions $Q_i^0$ for each $i$ that would automatically ensure that each population has a smart type with a positive initial probability. This would require a closure property on the set of feasible types.

Clearly, we cannot ensure that the smart types are feasible if we restrict the set of types in population $i$ to be a finite set (*e.g.*, a finite automata with a bound) because the smart types (corresponding to any arbitrary initial distribution $Q^0$ with this finite support) for population $i$ may not belong to this finite set. Another possible approach is to make the set of feasible types sufficiently rich so that it includes the smart types. However, this cannot go

[15]Note that this assumption has some similarity with the grain of truth assumption in the Bayesian learning literature [KalLeh93].

too far: if there is a continuum of feasible types, then it may not be possible to have an initial distribution for each population which attaches positive weights to a smart type. Thus, some restriction on the cardinality of the set of rules seems necessary. An obvious restriction is to make the space of allowable types to be *countable infinity*.

Anderlini and Sabourian explore the case in which the set of types is restricted to be Turing-computable machines. The set of (Turing-)computable functions have nice closure properties (in particular the existence of a computable Universal program that can 'simulate' other computable functions); moreover, the cardinality of this set is countable infinite. However, restricting types to be Turing machines introduces new difficulties all related to the halting problem in the space of Turing machines. In particular, now one needs to construct a computable smart type that halts; but the existence of such a halting machine may be problematic because simply simulating the system and taking the action that maximizes the growth rate at each date may not be possible: in simulating the system the smart type may end up simulating other types that do not halt.

This problem cannot be trivially avoided by simply assuming that all types always halt. This is because to simulate the system the initial distribution $Q_i^0$ needs to be computable but this may not be consistent with the support of $Q_i^0$ being the same as the set of all halting machines.

## 3 Which Equilibrium: Stochastic Model with a Finite Population

The critique that standard game theory imposes too strong rationality on players is one reason for the interest in evolutionary game theory. Another reason, as we mentioned before, has been that the standard game theory does not have good predictive power; a game may have too many Nash equilibria.

Does the evolutionary approach help to select amongst the multiplicity of Nash equilibria? By only considering asymptotic stability of a Nash equilibrium in a deterministic model the answer may have to be "no" in many interesting classes of games. In particular, according to the Folk Theorem of evolutionary game theory (condition 3 in Theorem 1) every strict Nash equilibrium is asymptotically stable in the standard deterministic evolutionary model. Therefore, such models cannot be used to select between Nash equilibria (and thus cannot predict which equilibrium will occur) when the game has multiple strict Nash equilibria.

However, asymptotic stability is about one-shot mutation, it stays silent for the cases in which (small) mutants invade perpetually. Utilizing techniques from Freidlin, Wentzell, Foster and Young,[16] consider a selection dynamics that is subject to stochastic shocks at every date. These perpetual stochastic shocks provide momentum for the system to escape from one strict equilibrium to another. However, as the probability that a random shock occurs becomes smaller, some equilibria may be visited more often. When such probability approaches zero, the system will be attracted into a much smaller set of equilibria, the stochastically stable set.

Kandori, Mailath, Rob, Young, Robson, Vega-Redondo, and many others have successfully applied the same approach (with small perpetual mutations) to select amongst the set of strict Nash equilibria in evolutionary models with a finite population.[17] In these papers the rules the agents follow are fixed (*e.g.*, best reply, imitation). However, it turns out that the particular equilibrium selected depends precisely on the rules allowed. Specifically, in the $2 \times 2$ coordination game with one equilibrium being risk dominant and the other efficient, the best reply rule tends to select the former and the imitation rule tends to select the latter. As in the deterministic case above, we ask how the predictions of these models are affected if multiple rules are present in the system and rules that evolve are endogenously determined through some evolutionary selection. Juang provides a result that favors imitation when both the best response and imitation co-exist [Jua02].

In this section we will discuss mainly the question of equilibrium selection in stochastic evolutionary models in the specific contexts of the $2 \times 2$ coordination game mentioned above and the classical Cournot oligopoly. We shall explain how the selection results depend crucially on the rules allowed and we will describe some attempts at resolving the issue of which rule by allowing multiple rules and applying evolutionary selection to the set of rules. We shall also briefly discuss other models with multiple rules and in particular some of the justifications provided for some specific rules such as imitation.

## 3.1 Mutation, Long-Run Equilibrium and Stochastic Stability

### 3.1.1 Mathematical background

Consider an evolutionary model in which a population of $M$ ($M$ is even) players are randomly paired to play a stage game repeatedly. After observing the outcome of the plays at the previous period, a fixed share of the players is randomly chosen to revise their actions. Under quite general set-ups this dynamics can be described by a stationary Markov process $P$ on a finite state space $S$.

---

[16] *Cf.* [FreWen84, FosYou90].
[17] *Cf.* [KanMaiRob93, You93a, RobVeg96].

In general if there exist multiple strict Nash equilibria in the stage game then there exist multiple recurrent classes in the underlying Markov process $P$. (Henceforth, if a recurrent class consists of exactly one state, we shall refer to it as a recurrent state.) In this case we say such dynamics are non-ergodic and the limit points of the dynamics are described by the set of recurrent classes of $P$, denoted by $\text{RC} = \{X_1, .., X_r\}$. Specifically, depending on the initial starting state and the random process that follows, the dynamics may be absorbed into one of the recurrent classes and the system will stay there forever. Thus, we cannot say anything about the "which equilibrium" question.

To deal with the non-ergodicity problem, we introduce mutation by letting players take, with small probability, actions that differ from what they intend to take originally. This may represent players making mistakes, doing experimentation or population renewal. More specifically, suppose the dynamic is perturbed by allowing each agent $i$ to mutate independently with a small probability $\varepsilon > 0$ such that the perturbed process $P^{\varepsilon}$ is also a stationary Markov chain with the state space $S$. Here mutation is parameterised by $\varepsilon$ and $P^{\varepsilon}$ converges to $P$ as $\varepsilon \to 0$.

The introduction of mutation allows the system to switch between any two states (and between any two recurrent classes of $P$) with a positive probability. As a result, for any $\varepsilon > 0$ the perturbed process $P^{\varepsilon}$ is ergodic; and therefore it has a *unique* invariant distribution $\mu^{\varepsilon}$ that summarises the "long-run" behaviour of the perturbed process from any initial condition.

We are interested in cases where the amount of mutation is arbitrarily small. It turns out that as the mutation rate $\varepsilon$ approaches zero, the corresponding invariant distribution $\mu^{\varepsilon}$ converges to an invariant distribution $\mu$ of the unperturbed dynamics $P$. This often results in selection among multiple recurrent classes (equilibria) of the unperturbed dynamics $P$. That is, with mutation we allow the system to drift among different recurrent classes/states of $P$, but with the mutation rate going to zero, the system will concentrate on a much smaller set of states.

The states that $\mu$ attaches a positive probability are called the stochastically stable states and the set of such states is denoted by

$$\Omega \equiv \{s \mid \mu(s) > 0\}.$$

Clearly, the set of stochastically stable $\Omega$ is a subset of the set of recurrent states $\text{RC} = \{X_1, .., X_r\}$ of $P$ and are the states that would be observed in the long-run with arbitrarily small amounts of mutation. As we shall see in later discussion, in many cases, the stochastically stable set $\Omega$ is singleton; in such cases this technique of perturbing the dynamics is very powerful in selecting a unique state.

In general, the characterization of $\Omega$ involves locating the recurrent class(es) with the least *stochastic potential*, as proposed by Freidlin, Wentzell, Foster and Young [FreWen84, FosYou90]. When there are only two recurrent classes in the system, this can be done by just counting the minimum number of mutations necessary for the system to transit from one recurrent class to the other. For illustration, suppose there are two recurrent classes $X_1$ and $X_2$. Denote the least number of mutation for the system to switch from $X_2$ to $X_1$ (from $X_1$ to $X_2$) as $\gamma_1$ ($\gamma_2$). Suppose $\gamma_1 < \gamma_2$ ($\gamma_1 > \gamma_2$). This means it is easier for the system to transit from $X_2$ to $X_1$ (from $X_1$ to $X_2$) than the other way round. We can then conclude that the recurrent class $X_1$ ($X_2$) is stochastically stable and we expect to observe $X_1$ ($X_2$) almost all the time in the long run.

For cases with more than two recurrent states, we need to "count" the number of mutations in a way that is a little more complicated. Suppose, for example, that there are four recurrent states ($S_1, S_2, S_3, S_4$) in a system. To compute the stochastic potential of any recurrent state, say, $S_4$, we have to construct a "tree" with root $S_4$. To be specific, for any other recurrent class, find a unique directed path from that state to $S_4$. Here we illustrate in Figure 1 below four different trees with root $S_4$. Sum up the total number of mutations for each feasible $X_4$-tree. The least one among all the $X_4$-trees is the stochastic potential of $S_4$. Similarly we can compute the stochastic potential for all other recurrent classes. The recurrent states/classes that have the least stochastic potential are stochastically stable.

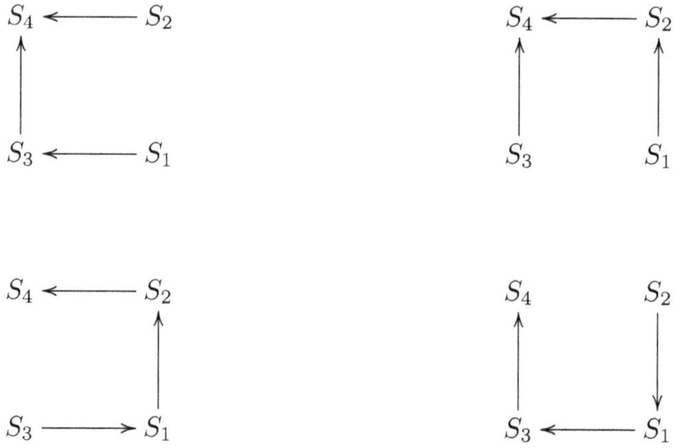


Figure 1. Four different $S_4$-trees.

### 3.1.2 Risk Dominance versus Efficiency

The classical example of equilibrium selection is a $2 \times 2$ coordination game with two strict Nash equilibria; one is risk dominant and the other efficient. The payoff matrix of such a game is given by Figure 2 where

$$a > c, d > b, a > d \text{ and } a + b < c + d. \tag{8}$$

| | $A$ | $B$ |
|---|---|---|
| $A$ | $a, a$ | $b, c$ |
| $B$ | $c, b$ | $d, d$ |

Figure 2. A $2 \times 2$ coordination game.

Note that given (8) the game has two pure strategy strict Nash equilibria $(A, A)$ and $(B, B)$, the former is efficient and the latter is risk dominant.

**Best response rule.** Kandori, Mailath, and Rob construct a model with a single finite population of players $M$.[18] The players are repeatedly matched to play the $2 \times 2$ coordination game as in Figure 2. Before they play the game in each period, players can revise their action over time. In particular, at each date the players are assumed to adopt the best response rules (actions that are best responses to the distribution of actions in the previous period.

The state of the system $s$ is defined to be the number of players that choose action $A$, thus $s \in \{0, 1, ..., M\}$. For any state $s$ realized in the last period, we can define the best response rule followed by all the players by

$$r^{\mathrm{BR}}(s) = \begin{cases} A & \text{if } \frac{s}{M} > p^*; \\ B & \text{if } \frac{s}{M} < p^*; \\ \text{randomize between } A \text{ and } B & \text{if } \frac{s}{M} = p^*, \end{cases}$$

where $p^* = (d - b)/(a + d - b - c)$.

Without mutation, it is straightforward to see that the distribution of actions converges to one of the two strict Nash equilibria (to states 0 or $M$), depending on the initial state. To see this, suppose the system starts at some state $s_0$ such that $s_0 > Mp^*$. Then playing $A$ is a best response and whenever players are revising their actions they will play $A$ and the system will converge to $s = M$. On the other hand, if the system starts at $s_0 < Mp^*$, then the system will converge to the state in which all play $B$ $(s = 0)$.

[18] *Cf.* [KanMaiRob93]. Note that here since the game is symmetric there is a single population of agents.

Next Kandori, Mailath and Rob introduce mutation by allowing players to make small mistakes *independently* in choosing their actions. Thus, for any given mutation rate $\varepsilon$, they obtain a perturbed Markov process that is ergodic and thus has a unique stationary distribution $\mu^{\varepsilon}$. Kandori, Mailath and Rob find that in the long-run (as $\varepsilon \to 0$) the risk-dominant equilibrium $B$ will be selected. More formally, they show that with best response rules the unique stochastically stable state is $s = 0$.

To see this note that there are exactly two recurrent classes/states 0 and $M$. Since the players follow the best response rule, it is easy to check that one needs $[(1 - p^*)M]^+$ action mutations to transit from state $M$ to state 0 and $[p^*M]^+$ action mutations for the other way around, where the notation $[x]^+$ denotes the smallest integer that is not less than $x$. Since $[(1 - p^*)M]^+ < [p^*M]^+$ it follows that it is easier for the system to transit from state $M$ to state 0 than the other way round (0 has a lower stochastic potential). Thus, $s = 0$ is stochastically stable.

Young (1993a) has an analogous set of results for a similar setup except that players randomly choose a best response to a sample from finite histories.

Before considering rules of behaviour other than the best response rule, note that in the above (and in the other models or perturbed dynamics we consider below) the probability that a mutation/error occurs is independent of the date, the state and the identity of the players. In the absence of a good explanation for mutations such a uniformity assumption on the mutation rates seems like a reasonable starting point.

However, it should be noted that the above sharp predictions of Kandori, Mailath and Rob (and the models below) depend crucially on this uniformity assumption. Bergin and Lipman introduce mutations rates that are state-dependent in the Kandori-Mailath-Rob set-up and show that any strict Nash equilibrium can be selected in the long-run as the mutation rates go to zero, depending on the relative probability of the mutation rates in the different states [BerLip96]. The intuition behind this result is easy to see. The stochastic stability of a recurrent state depends on its stochastic potential, which involves "counting" the number of mutations for relevant transitions. Suppose now that the mutation rates are state-dependent; in particular suppose that the mutation rate in each state $s \in \{[p^*M]^+, ..., M\}$ is $\varepsilon$ and the mutation rate in each state $s \in \{0, 1, ..., [p^*M]^+ - 1\}$ is $\varepsilon^{\alpha}$, for some constant $\alpha$. The system with no mutation has, as before, two recurrent states 0 and $M$. But now with mutation, the probability of the transition from state 0 to state $M$ is of the order $\varepsilon^{\alpha[p^*M]^+}$ while the probability of the transition from state $M$ to state 0 is of the order $\varepsilon^{[(1-p^*)M]^+}$. Also, denote the unique invariant distribution by $\mu^{\varepsilon}$. Then it is not difficult to show that there exist $\alpha^*$ such that as $\varepsilon \to 0$ the ratio $\mu^\varepsilon(s=0)/\mu^\varepsilon(s=M)$ goes to infinity if $\alpha > \alpha^*$ and to zero if $\alpha < \alpha^*$. In other words, the assumption of state-dependent mutations allows us to support any of the two recurrent states by adjusting the appropriate mutation rate for each state.

**Imitation rule.** In [RobVeg96], Robson and Vega-Redondo construct a model similar to Kandori-Mailath-Rob except that players adopt the imitation rule to play the $2 \times 2$ coordination game described in Figure 2. Specifically, each player, whenever revising actions, chooses the one that performed the best (in terms of average payoff) in the previous period. The state of the system $s \in \{0, 1, ..., M\}$ is again the number of players that choose action $A$. The imitation rule is given by

$$r^{\mathrm{I}}(s) = \begin{cases} A & \text{if } \pi^A(s) > \pi^B(s); \\ B & \text{if } \pi^A(s) < \pi^B(s); \\ \text{randomize between } A \text{ and } B & \text{if } \pi^A(s) = \pi^B(s), \end{cases}$$

where $\pi^A(s)$ and $\pi^B(s)$ are the average payoffs for using action $A$ and action $B$ respectively in state $s$.

Again without mutation, the distribution of actions converges to one of the two strict Nash equilibria (to states 0 or $M$). With small mutations Robson and Vega-Redondo show that the efficient equilibrium, rather than the risk dominant one, is selected in the long run. More formally, they show that if players adopt the imitation rule then there exist some $\overline{M} > 0$ such that for all $M > \overline{M}$ the unique stochastically stable state is $s = M$.

To see this note that with the imitation rule there are also exactly two recurrent classes 0 and $M$. Moreover, the system can switch from state 0 to state $M$ if two agents mutate to playing $A$ and are paired together to play the game in the same period. Since $c > b$ it also follows that to transit from state $M$ to state 0 requires strictly more than two mutations if $M > 2(a-b)/(a-c)$. Thus state $M$ is stochastically stable.

The striking difference between the results in [KanMaiRob93] and [RobVeg96] lies in the role of the rule adopted by players. In the Kandori-Mailath-Rob model, players best respond to the action frequency in the last period. For an action to be a best response, there must be at least a fixed share of the population playing the same action. In a $2 \times 2$ coordination game illustrated in Figure 2, if all players are playing $A$, then they will continue to do so unless at least $[(1-p^*)M]^+$ players mutate to $B$, in which case playing $B$ is the best response rather than playing $A$. Since $B$ is risk dominant ensures it follows that $(1-p^*) < 1/2$. Thus it is easier to switch from all playing $A$ to all playing $B$ than the other way round. When players imitate each other in the Robson-Vega-Redondo model, players only care about the performance of an action, rather than its frequency. This gives the efficient equilibrium an advantage. Consider the transition from all playing $B$ to all playing $A$. We actually need only two mutations plus some "luck" in random matching: Two players mutate to $A$ and these two are paired to play the stage game. In this case, the average payoff from $A$ is $a$ while the average payoff from $B$ remains $d$. Although such "luck" in pairing occurs with a small probability, it is far more likely than players mutate to the other action as the mutation rate $\varepsilon$ approaches zero. As for the transition the other direction, it is not difficult to see that a number of mutations proportional to the population size ($\frac{(a-c)}{(a-b)}M$) is indispensable. Thus if the population is reasonably large, the former transition is more likely to occur.

**Rule Evolution and Equilibrium Selection.** The Kandori-Mailath-Rob and Robson-Vega-Redondo results illustrate the effect that different rules have on the selection amongst multiple equilibria. What if players are free to choose from different rules? In [Jua02] —as a preliminary step towards rule evolution and equilibrium selection— Juang looks at a model very similar to those in [KanMaiRob93] and [RobVeg96] except that players may be able to choose either best response or imitation rules to play the stage games. Specifically, at each period one randomly chosen player can revise his rule by choosing one of the existing rules according to how each rule has performed in the previous period. Thus, he will make no change if only one rule is remaining.

Juang adopts a fairly general set-up of rule selection. The only assumption he makes on the rule selection criterion is what he calls *experimental*: If at any date $t$ all players play some Nash equilibrium then at period $t+1$ the selection assigns a positive probability to each existing rule.

There are two points to note concerning the this assumption. First, it does not impose any restrictions when agents are not playing a Nash equilibrium. Second, at any Nash equilibrium all agents (and thus rules) are doing as well as each other; hence selection has no force and therefore it is reasonable to assume that every existing rule receives a positive probability as in the experimental assumption.

Juang introduces "mutation" at two levels. The first one refers to the action level. As before, a player makes mistakes with probability $\varepsilon > 0$ when choosing actions (implementing his rules). The second is at the rule level. It is assumed that, when updating rules, each player is also prone to making mistakes with probability $\varepsilon^\eta$ where $0 < \eta < \infty$. Thus, in term of probability, one rule mutation is equivalent to $\eta$ action mutations.

In this set-up, a state is a 3-tuple denoting the number of players playing $A$; the number of players playing $A$ and adopting the imitation rule; and the total number of players adopting the imitation rule.

Note that with no mutation, because of the assumption that rule selection criteria are experimental, any state in which both rules coexist cannot be recurrent. This is because in any recurrent class players must be playing a Nash equilibrium and therefore if both rules coexisted, by the experimental property, both will be chosen with a positive probability. This allows the total number of players adopting the imitation rule to drift between 1 and $M-1$ until it reaches either 0 or $M$.

Since any state in which more than one rule is present is not recurrent and in any recurrent class agents must be playing the same Nash equilibrium, it follows that the system with no mutation has four recurrent states $S_1, S_2, S_3$ and $S_4$. In state $S_1$ all adopt the imitation rule and play $A$; in $S_2$ all adopt the imitation rule and play $B$; in $S_3$ all adopt the best response rule and play $A$; and in $S_4$ all adopt the best response rule and play $B$.

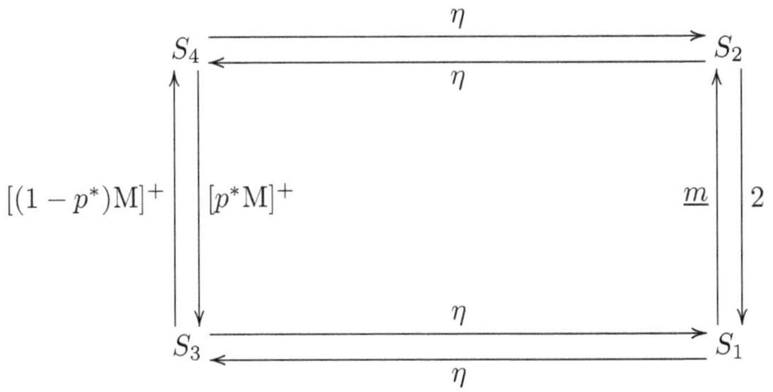


Figure 3. A Preliminary Tree. The number attached to each arrow indicates the resistance for that transition.

Figure 3 demonstrates the relationship between the setup of Juang with and those in [KanMaiRob93] and [RobVeg96]. The transitions between $S_3$ and $S_4$ describe the ones in [KanMaiRob93] and the transitions between $S_1$ and $S_2$ describe the ones in [RobVeg96].

Juang's demonstrate the following result on the long-run (as $\varepsilon \to 0$) behaviour of the system. The states $S_1$ and $S_3$ in which all agents play $A$ (the action corresponding to the efficient equilibrium) are the only stochastically stable states in the above model provided the population size M is sufficiently large.

A sketch of the proof of Juang's result is as follows. First, consider the transition from any state where all agents adopt one rule to any state where all agents adopt the other, while using the same Nash equilibrium action profile. Given the assumption that all rule selection criteria are *experimental*, one rule mutation is sufficient for any such transition. This is because once one player mutates to adopt a rule that is different from the one adopted by all players, then both rules are present in the population. Since all players are playing the same Nash equilibrium, the rule selection criterion will prescribe both rules with a positive probability next period. The above arguments illustrate that one rule mutation suffices for a system to transit from one state in which all players adopt one rule to another state in which all players adopts the other rule, while they are playing the same Nash equilibrium.

Second, with four recurrent states, we need to compute the stochastic potential for each recurrent state in the way stated in subsection 3.1.1. Figure 3 describes the transitions between recurrent states and corresponding number of mutations. It is easy to see that states $S_1$ and $S_3$ are stochastically stable provided the population size M is sufficiently large. The intuition behind this is as follows. For transition from a state where all agents play $A$ to another state where all agents use the same rule and play $B$, more than two action mutations are needed if $M$ is above some parameter. But, as in [RobVeg96], when all players are adopting the imitation rule, two action mutations are sufficient for the transition from all $B$'s to all $A$'s. Therefore any tree with the least stochastic potential must contain the transition from $S_2$ to $S_1$. This property gives the efficient equilibrium an advantage over the risk-dominant one. (It is easy to see that both states $S_1$ and $S_3$ have the stochastic potential $2\eta + 2$ while both states $S_2$ and $S_4$ have the stochastic potential $2\eta + \underline{m}$ where $\underline{m}$ refers to the least number of action mutation needed for the system to switch from all $A$'s to all $B$'s, when all players are adopting the imitation rule).

### 3.1.3 Cournot Oligopoly model: Nash or Walrasian equilibrium

Classical Cournot Oligopoly set-up has been another area in which the technique of stochastic evolutionary dynamics described above has been applied. Here, again the theory has sharp predictions which depend on the nature of the rules of behaviour allowed.

A symmetric discretised version of the Cournot model is as follows. There is a single of population consisting of $n$ identical firms. In the one-shot game they set their output quantities simultaneously. They face an inverse demand function given by $P(.)$; thus if the total output brought to the market is $Q$ the price will be given by $P(Q)$. Each firm has a constant marginal cost $c > 0$. The strategy of firm $i$ is simply the quantity of output $q_i$ it produces. To keep the finiteness of the model we assume that the set of output choices is finite.

In this set-up the payoff of firm $i$ is simply

$$\pi_i(q_i, q_{-i}) = q_i \left\{ P(\sum_{j=1}^{n} q_j) - c \right\}$$

Two reference benchmark outcomes for this one-shot game are the symmetric Cournot-Nash equilibrium and the Walrasian (also called Competitive) equilibrium. The former, denoted by a symmetric n-tuple of outputs $\mathbf{q}^c = (q^c, .., q^c)$ refers to the Nash equilibrium of the one-shot game. The latter refers to the outcome in a competitive market in which firms cannot influence prices and is defined by a symmetric n-tuple of outputs $\mathbf{q}^w = (q^w, .., q^w)$ where $q^w$ satisfies

$$q^w \{P(nq^w) - c\} \geq q_i \{P(nq^w) - c\} \text{ for all output levels } q_i.$$

Assume that these two equilibria exist and are unique. Assume also that $P(.)$ is decreasing so that $q^c < q^w$.

Now suppose that the above Cournot game is played repeatedly and at any date each firm has an opportunity to adjust its behaviour with a positive probability.

**Best response dynamics.** First, let us consider the case in which the firms follow a myopic best response rule. The selection problem in this case is trivial. Since the Cournot game is a potential game, the best response dynamics for this game with no mutation/perturbation is globally stable, and therefore the output decisions of the firms converge to the unique Cournot-Nash equilibrium $\mathbf{q}^c$. When the dynamics is perturbed the set of stochastically stable states belong to the limit points (recurrent classes) of the unperturbed dynamics. Since the latter is unique, the stochastically stable state corresponds to the Cournot-Nash equilibrium.

**Imitation dynamics.** If all firms follow the imitation rule then whenever a firm receives the opportunity to revise its actions it follows the action of one of the firms which obtained a higher payoff in the previous period. Imitation implies that the dynamics with no mutation converges a.s. to a monomorphic state in which all firms produce the same output. The recurrent classes of the dynamics are simply the set of all symmetric n-tuples $\mathbf{q}= (q, q, .., q)$.

Next, consider the same model with small mutation. Vega-Redondo shows that the stochastically stable set of this imitation dynamics is unique and is given by all firms producing the Walrasian output $q^w$ [Veg96]. The basic idea here is as follows. First at any recurrent state $\mathbf{q}$ a single mutation by one firm to $q^w$ is sufficient to induce the unperturbed dynamics towards the Walrasian equilibrium $\mathbf{q}^w$ with a positive probability. The intuition for this is that after the single mutation to $q^w$ the mutant earns a higher profit than the other firms. But then by imitation all firms can end up producing $q^w$. Second, it requires more than one mutation to switch from $\mathbf{q}^w$ to any other recurrent state $\mathbf{q}$. This is because if in state $\mathbf{q}^w$only one firm mutates to another output level $q$, the mutant ends up earning a lower profit that the other firms; therefore the unperturbed imitation dynamics must lead the process back to state $\mathbf{q}^w$.

The Imitation rule in [Veg96] has one period memory. In [Alό04], Alós-Ferrer considers the case of imitators with long but finite memory in the set-up of [Veg96]. He shows that the process converges to a set of monomorphic states in which the firms produce an output between the Cournot and the Walrasian outcome.

**Imitation and best response rules.** What happens if different firms adopt different rules. Schipper considers a model in which a population of imitators and best responders play repeatedly the above symmetric Cournot game [Sch02]. He shows that the long run distribution converges to a recurrent set of states in which imitators are better off than are best responders. His finding is robust even when best responders are more sophisticated. In his model the players can not change their rules. Thijssen relaxes this restriction and studies a similar environment as Schipper but allows players to change their rules [Thi05]. At the rule evolution level, he assumes that firm imitate the rule of the firm with the highest profit; thus the rule selection criterion is imitating in this set-up. Thijssen shows that Walrasian behavior is the unique stochastically stable state in this set-up with competing rules. He establishes his result by using the theory of nearly-complete decomposability to disentangle the action evolution and the rule evolution. Intuitively, this means one uses the limit distribution of the action evolution to obtain the limit distribution of rule evolution.

### 3.1.4 Other applications of stochastic stability

There are also many other applications of the stochastic stability approach in economic models such as bargaining [You93b], social contracts [You98], and social networks [GoyVeg05], to name a few. In most of these applications players choose a single rule — best response rule. (Some exceptions are Saez-Marti and Weibull [SaeWei99], and Matros [Mat03] who extend Young's [You93b] result on selection in 2-player Nash bargaining problem to two rules set-up in which some individuals use the best response rule and others use best response to best response).

In [Hur95] and [You98], Hurkens and Young extend [You93a] to general finite games and characterise the stochastic stable states of the perturbed dynamics in which at any date all individuals have finite memory and choose a (myopic) best response to a sample of past plays of their opponents. The difficulty with dealing with more general games is that the recurrent classes of the unperturbed games with minimal stochastic potential may involve cycles and stochastic stable states need not be Nash equilibria of the underlying games. For generic games Hurkens and Young characterise the stochastic stable states by showing that they belong to pure-strategy profiles in minimal sets closed under best response.

On the other hand, in [JosMat04], Josephson and Matros characterise the equilibrium of the same model as those in [Hur95] and [You98] but with a different behavioural rule from best response. They assume that all individuals follow an imitation rule of choosing the most attractive strategy in their sample. First, not surprisingly, they show, in contrast to the results with best-reply dynamics, that with imitation rules a state is a limit point of the unperturbed dynamics if and only if it is a repetition of a single pure-strategy profile, a monomorphic state. Second, they show that all stochastically stable states of the perturbed dynamics are monomorphic and belong to a minimal set closed under single better replies (a minimal set of strategy profiles such that no single player can obtain a weakly better payoff by deviating unilaterally and playing a strategy outside the set).

Young [You93a, You98], Hurkens [Hur95], and Josephson and Matros [JosMat04] consider general games with single rules. Matros [Mat04] allows introduces of multi-rules in the same evolutionary setting. Here, as in [AndSab96] discussed in the previous section, a rule refers to an action choice for each sample of past (finite) observations. He defines *weakly rational rules* as the rules that are "consistent with an equilibrium." This set includes all belief-based rules (those that are best responses to some belief) as well all imitation rules. He then shows that if players use only weakly rational rules including the best response rule then any limit point of the unperturbed dynamics belong to a minimal set closed under best response. Thus, very informally, without perturbation the dynamics with the set of weakly rational rules select the same outcome as do the dynamics with only the best response rules as in [Hur95] and [You98]. Matros then considers the perturbed dynamics and restricts his analysis to a subset of the weakly rational set, which he calls boundedly rational rules. The definition of this subset is quite complicated and rather ad hoc; nevertheless it still includes all belief-based and imitation rules. Very informally, he shows that when players use boundedly rational rules, including the best response and the imitation rule, the long-run outcomes of the perturbed dynamics as the mutation becomes small are the same as those that will be obtained when agents choose only best response and imitation dynamics.

## 3.2 Why Imitate?

The literature on evolutionary framework with small but perpetual random shocks has been successful in selecting amongst multiplicity of equilibria in some well-known applications but as we have explained above the precise selection in many instances has been dependent on which rules of behaviour are allowed. In particular, as we have discussed, the best response rule and the imitation rule often have very different predictions. The important question is to ask why one rule is chosen rather than another. A less ambitious question may be why players imitate, rather than best respond, as the latter seems to make more sense in terms of rationality and expected utility maximization. The papers [Jua02] and [Thi05], discussed above, address the question of selection between best response and the imitation rules in the context of specific games by introducing a selection dynamics between the two rules and letting them compete. However, the games they consider are very specific. In the remainder of this section we will briefly discuss some justifications of why individuals imitate in contexts other than the standard evolutionary game.

### 3.2.1 Imitate to maximize expected payoff

Schlag shows that through imitation, most individuals will learn to choose an expected payoff maximizing action with probability arbitrarily close to one, provided the population is sufficiently large [Sch98]. In his model finitely many agents repeatedly choose among actions that yield uncertain payoffs (multi-armed bandits). Periodically, new agents replaces some existing ones. Before entering the population each individual must commit to a behavioral rule that determines his next choice. When choosing an action in each period, an individual, guided by his rule, will sample another individual and is informed of the action chosen and the payoff received by that sampled agent in the last period. By comparing the performance of that sample with his own, the player decides to either switch to the action chosen by that sampled player or stick to his own action. Such a rule is "*imitating*" in that the player only selects either the action he chose in the last period or the action adopted by the individual he samples.

Schlag shows that an imitating rule is improving (payoff increasing in each bandit) if and only if, when two agents using different actions happen to sample each other, the difference in the probabilities of switching is proportional to the difference in their realized payoffs[19]. By allowing agents to select among improving rules, the author characterises the unique optimal *Proportional Imitation Rule.* Schlag then shows that, for any initial state

[19]Since *Imitate if Better* cannot distinguish between lucky and certain (or highly probable) payoffs, this rule might generate negative expected improvement and is not improving.

with all actions present, the above optimal rule will guide most individuals to select the expected payoff maximizing action after finite periods.

### 3.2.2 Imitate to survive

The above discussion illustrates that through imitation individuals may "learn" to choose the optimal action. Imitation performs well even when competing with others. Blume and Easley study a model in which investors distribute their wealth among a bundle of assets [BluEas92]. Asset $s$ pays off a positive amount of revenue if and only if state $s$ occurs. It is easy to see that any investor putting zero share of his wealth on any asset $s$ with state $s$ occurring with positive probability will almost surely be bankrupt in finite time. Moreover, if we assume all traders have identical saving rates, then to survive in the market each investor must learn to distribute his wealth on each asset $s$ with a share proportional to the probability that state $s$ occurs. Blume and Easley try to find out what kinds of rules could survive in the market. Obviously the simple rule that invest in asset $s$ with a fixed wealth share equal to the probability that state $s$ occurs will survive. That is, an investor holding the correct belief on the states of the world and invest according to his belief will survive in the market.

They also show that some other rules may survive in the market. For example, a Bayesian learning rule whose prior has finite support containing the true model may survive. This is not surprising as a Bayesian will learn the true model if his prior contains the true model and he could observe all histories and update his belief according to his observation. Another rule that may survive is a search rule. An investor who adopts the search rule selects at date $t$ the portfolio rule that would have maximized his date $t$ wealth share given the actual prices and states up to date $t$. Both the Bayesian learning rule and the search rule involve relatively sophisticated behavior. For example, a Bayesian's prior must contain the true model and, he must update his posterior probability on each model in each date. For the search rule, the investor must remember the history of the state-price process, compute the objective function and maximize it. Consider an imitator as follows. Use at date $t$ the rule adopted at date $t-1$ by the trader with the largest wealth share. It turns out that such a imitation rule will survive: Imitating the trader with the largest wealth share is equivalent to imitating the most successful trader in the long run. The trader with the largest wealth share may not be adopting the most successful rule initially, but then his wealth share will shrink and will eventually be dominated by another trader who performs better. In the long run the investor with the largest wealth share must adopt the most successful rule. Thus an imitator will eventually settle down to the most successful rule and will survive in the market.

### 3.2.3 Imitation to efficiency

Ellison and Fudenberg construct a social learning model in which a continuum of agents repeatedly select between two competing alternatives with one being superior to the other but subject to random shocks [EllFud93]. Therefore, the inferior choice may do better than the superior one some of the time. In this model, if only a share of the agents may revise their choice each time and when revising, they just choose the alternative that performed better in the last period, then the society will settle down to a state in which the share of the agents adopting the superior choice is equivalent to the probability that the better alternative outperforms the other, the so-called *probability-matching* behavior.

Ellison and Fudenberg then assume that an agent, when revising his choice, also takes into account how many agents adopting each alternative. Specifically, he considers not only the performance of both alternatives but also their popularity. Thus the weight an agent assigns to the popularity of an alternative plays an important role in the social learning environment. Intuitively, if the weight is too small, then we shall observe a similar outcome as the probability-matching case, and thus there will always be a non-zero share of the society who adopt the inferior choice. On the other hand, if the weight is too large, then the society will settle down to either the efficient choice or the inferior one, depending on the initial state and the realized random shocks. Only with the weight within an optimal range will a society learn to choose the efficient choice.

Juang considers a heterogeneous society in which agents may have different popularity weights and introduce a replicator-like dynamic to differentiate agents according to their performance [Jua01]. He characterises two conditions, called *precision* and *diversity,* each of which guarantees all agents in the society herd on the superior alternative. Precision requires that at least non-negligible share of agents use popularity weight that is within the optimal range in Ellison and Fudenberg's [EllFud93]. This is straightforward as agents with such a weight will learn the superior choice precisely in the long run and those who fail to do so will be outperformed and will vanish in the long run. The diversity condition requires that at least one "low" popularity weight and one "high" popularity weight co-exist in the society (low and high are relative to the optimal range stated above). Intuitively, agents with lower (higher) weight care more about performance (popularity). If the inferior choice is more popular, then agents with low weight will outperform those with high weight, and help the society escape from inefficient herding and reach the state in which the superior choice is more popular. Once this occurs, agents with high weight will adopt the superior choice with greater proportion and outperform those with low weight and lead the society into the efficient herding in the limit. An interpretation for diversity is that "*rational*" agents locate the optimal choice for the society while it is "imitating" agents who make the profits.

## References.


[Al604] Carlos **Alós-Ferrer**, Cournot versus Walras in Dynamic Oligopolies with Memory, **International Journal of Industrial Organisation** 22 (2004), p. 193–217

[AndSab96] Luca **Anderlini** and Hamid **Sabourian**, The Evolution of Algorithmic Learning Rules: A Global Stability Result, **EconWPA Game Theory and Information** 9510001 (1996)

[Arr+96] Kenneth J. **Arrow**, Enrico **Colombatto**, Mark **Perlman**, and Christian **Schmidt** (*eds.*), The Rational Foundations of Economic Behaviour, Macmillan Press 1996

[BerLip96] James **Bergin** and Barton L. **Lipman**, Evolution with State-Dependent Mutations, **Econometrica** 64 (1996), p. 943–956

[BjöWei96] Jonas **Björnerstedt** and Jörgen W. **Weibull**, Nash Equilibrium and Evolution by Imitation, *in:* [Arr+96, p. 155–171]

[BluEas92] Lawrence **Blume** and David **Easley**, Evolution and Market Behavior, **Journal of Economic Theory** 58 (1992), p. 9–40

[BörSar97] Tilman **Börgers** and Rajiv **Sarin**, Learning through Reinforcement and Replicator Dynamics, **Journal of Economic Theory** 77 (1997), p. 1–14

[Cre04] Ross **Cressman**, Stability of the Replicator Equation with Continuous Strategy Space, Mimeo 2004

[DekSco92] Eddie **Dekel** and Suzanne **Scotchmer**, On the Evolution of Optimizing Behavior, **Journal of Economic Theory** 57 (1992), p. 392–406

[DreShaTuc64] Melvin **Dresher**, Lloyd S. **Shapley**, and Albert W. **Tucker** (*eds.*), Advances in Game Theory, Princeton University Press 1964 [Annals of Mathematics Studies 52]

[EllFud93] Glenn **Ellison** and Drew **Fudenberg**, Rules of Thumb for Social Learning, **Journal of Political Economy** 101 (1993), p. 612–643

[FosYou90] Dean **Foster** and H. Peyton **Young**, Stochastic Evolutionary Game Dynamics, **Theoretical Population Biology** 38 (1990), p. 219–232

[FosYou02] Dean **Foster** and H. Peyton **Young**, Learning, Hypothesis Testing, and Nash Equilibrium, Mimeo 2002

[FreWen84] Mark I. **Freidlin** and Alexander D. **Wentzell**, Random Perturbations of Dynamical Systems, Springer 1984

[FudLev98] Drew **Fudenberg** and David K. **Levine**, The Theory of Learning in Games, MIT Press 1998

[GoyVeg05] Sanjeev **Goyal** and Fernando **Vega-Redondo**, Learning, Network Formation, and Coordination, **Games and Economic Behavior** 50 (2005), p. 178–207

[HarSel88] John C. **Harsanyi** and Reinhard **Selten**, A General Theory of Equilibrium Selection in Games, MIT Press 1988

[HarMas00] Sergiu **Hart** and Andreu **Mas-Colell**, A Simple Adaptive Procedure Leading to Correlated Equilibrium, **Econometrica** 68 (2000), p. 1127–1150

[HarMas01] Sergiu **Hart** and Andreu **Mas-Colell**, A General Class of Adaptive Strategies, **Journal of Economic Theory** 98 (2001), p. 26–54

[HarMas03] Sergiu **Hart** and Andreu **Mas-Colell**, Uncoupled Dynamics do not Lead to a Nash Equilibrium, **American Economic Review** 93 (2003), p. 1830–1836

[Hof00] Josef **Hofbauer**, From Nash and Brown to Maynard Smith: Equilibria, Dynamics, and ESS, **Selection** 1 (2000), p. 81–88

[HofSig98] Josef **Hofbauer** and Karl **Sigmund**, Evolutionary Games and Population Dynamics, Cambridge University Press 1998

[HofSig03] Josef **Hofbauer** and Karl **Sigmund**, Evolutionary Game Dynamics, **Bulletin of the American Mathematical Society** 40 (2003), p. 479–519

[HofWei96] Josef **Hofbauer** and Jörgen W. **Weibull**, Evolutionary Selection against Dominated Strategies, **Journal of Economic Theory** 71 (1996), p. 558–573

[Hur95] Sjaak **Hurkens**, Learning by Forgetful Players, **Games and Economic Behavior** 11 (1995), p. 304–329

[JosMat04] Jens **Josephson** and Alexander **Matros**, Stochastic Imitation in Finite Games, **Games and Economic Behavior** 49 (2004), p. 244–259

[Jua01] Wei-Torng **Juang**, Learning from Popularity, **Econometrica** 69 (2001), p. 735–747

[Jua02] Wei-Torng **Juang**, Rule Evolution and Equilibrium Selection, **Games and Economic Behavior** 39 (2002), p. 71–90

[KalLeh93] Ehud **Kalai** and Ehud **Lehrer**, Rational Learning Leads to Nash Equilibrium, **Econometrica** 61 (1993), p. 1019–1045

[KanMaiRob93] Michihiro **Kandori**, George J. **Mailath**, and Rafael **Rob**, Learning, Mutation and Long Run Equilibria in Games, **Econometrica** 61 (1993), p. 29–56

[Mat03] Alexander **Matros**, Clever Agents in Adaptive Learning, **Journal of Economic Theory** 111 (2003), p. 110–124

[Mat04] Alexander **Matros**, Simple Rules and Evolutionary Selection, Mimeo 2004

[Mat92] Akihiko **Matsui**, Best Response Dynamics and Socially Stable Strategies, **Journal of Economic Theory** 57 (1992), p. 343–362

[May82] John **Maynard Smith**, Evolution and the Theory of Games, Cambridge University Press 1982

[MayPri73] John **Maynard Smith**, and George R. **Price**, The Logic of Animal Conflicts, **Nature** 246 (1973), p. 15–18

[Miy61] Koichi **Miyasawa**, On the Convergence of the Learning Process in a $2 \times 2$ Non-Zero-Sum Two Person Game, Research Memorandum No. 33, Economic Research Program, Princeton University 1961

[MonSha96a] Dov **Monderer** and Lloyd S. **Shapley**, Fictitious Play Property for Games with Identical Interests, **Journal of Economic Theory** 68 (1996), p. 258–265

[MonSha96b] Dov **Monderer** and Lloyd S. **Shapley**, Potential Games, **Games and Economic Behavior** 14 (1996), p. 124–143

[Nac90] John H. **Nachbar**, Evolutionary Selection Dynamics in Games: Convergence and Limit Properties, **International Journal of Game Theory** 19 (1990), p. 59–89

[OecRie01] Jörg **Oechssler** and Frank **Riedel**, Evolutionary Dynamics on Infinite Strategy Spaces, **Economic Theory** 17 (2001), p. 141–162

[OecRie02] Jörg **Oechssler** and Frank **Riedel**, On the Dynamic Foundation of Evolutionary Stability in Continuous Models, **Journal of Economic Theory** 107 (2002), p. 223–252

[Rob51] Julia **Robinson**, An Iterative Method of Solving a Game, **Annals of Mathematics** 54 (1951), p. 296–301

[RobVeg96] Arthur J. **Robson** and Fernando **Vega-Redondo**, Efficient Equilibrium Selection in Evolutionary Games with Random Matching, **Journal of Economic Theory** 70 (1996), p. 65–92

[SaeWei99] Maria **Saez-Marti** and Jörgen W. **Weibull**, Clever Agents in Young's Evolutionary Bargaining Model, **Journal of Economic Theory** 86 (1999), p. 268–279

[Sam97] Larry **Samuelson**, Evolutionary Games and Equilibrium Selection, MIT Press 1997

[Sam02] Larry **Samuelson**, Evolution and Game Theory, **Journal of Economic Perspectives** 16 (2002), p. 47–66

[SamZha92] Larry **Samuelson** and Jianbo **Zhang**, Evolutionary Stability in Asymmetric Games, **Journal of Economic Theory** 57 (1992), p. 363–391

[San∞] William H. **Sandholm**, Population Games and Evolutionary Dynamics, *to be published by* MIT Press

[Sch02] Burkhard C. **Schipper**, Imitators and Optimizers in Cournot Oligopoly, Mimeo 2002

[Sch98] Karl H. **Schlag**, Why Imitate, and If So, How? A Boundedly Rational Approach to Multi-armed Bandits, **Journal of Economic Theory** 78 (1998), p. 130–156

[Sel91] Reinhard **Selten** (*ed.*), Game Equilibrium Models I: Evolution and Game Dynamics, Springer 1991

[Sey00] Rob M. **Seymour**, Dynamics for Infinite Dimensional Games, Mimeo 2000

[Sha64] Lloyd S. **Shapley**, Some Topics in Two-Person Games, *in:* [DreShaTuc64, p. 1–28]

[Thi05] Jacco J.J. **Thijssen**, Nearly-Complete Decomposability and Stochastic Stability with an Application to Cournot Oligopoly, Mimeo 2005

[Veg96] Fernando **Vega-Redondo**, Evolution, Games, and Economic Behavior, Oxford University Press 1996

[Veg03] Fernando **Vega-Redondo**, Economics and the Theory of Games, Cambridge University Press 2003

[Wei95] Jörgen W. **Weibull**, Evolutionary Game Theory, MIT Press 1995

[Wei91] Franz J. **Weissing**, Evolutionary Stability and Dynamic Stability in a Class of Evolutionary Normal Form Games, *in:* [Sel91, p. 29–97]

[You93a] H. Peyton **Young**, The Evolution of Conventions, **Econometrica** 61 (1993), p. 57–84

[You93b] H. Peyton **Young**, An Evolutionary Model of Bargaining, **Journal of Economic Theory** 59 (1993), p. 145–168

[You98] H. Peyton **Young**, Individual Strategy and Social Structure: An Evolutionary Theory of Institutions, Princeton University Press 1998

[You04] H. Peyton **Young**, Strategic Learning and its Limits, Oxford University Press 2004

**Received**: May 1st, 2005;
**In revised version**: January 5th, 2006; November 5th, 2007;
**Accepted by the editors**: November 8th, 2007.

Stefan **Bold**, Benedikt **Löwe**,
Thoralf **Räsch**, Johan **van Benthem** (*eds.*)
**Foundations of the Formal Sciences V**
Infinite Games

# Simplicity, Truth, and the Unending Game of Science

KEVIN T. KELLY*

Department of Philosophy
Carnegie Mellon University
Baker Hall 135
Pittsburgh, PA 15213-3890, USA
kk3n@andrew.cmu.edu

ABSTRACT. This paper presents a new explanation of how preferring the simplest theory compatible with experience assists one in finding the true answer to a scientific question when the answers are theories or models. Inquiry is portrayed as an unending game between science and nature in which the scientist aims to converge to the true theory on the basis of accumulating information. Simplicity is a topological invariant reflecting sequences of theory choices that nature can force an arbitrary, convergent scientist to produce. It is demonstrated that among the methods that converge to the truth in an empirical problem, the ones that do so with a minimum number of reversals of opinion prior to convergence are *exactly* the ones that prefer simple theories. The approach explains not only simplicity tastes in model selection, but aspects of theory testing and the unwillingness of natural science to break symmetries without a reason.

## 1 Introduction

In natural science, one typically faces a situation in which several (or even infinitely many) available theories are compatible with experience. Standard

*I would like to thank Seth Casana, John Taylor, Joseph Ramsey, Richard Scheines, Oliver Schulte, Pieter Adriaans, Teddy Seidenfeld, and Balazs Gyenis for discussions and comments. Special thanks are due to the organizers of the Fifth International Conference on Foundations of the Formal Sciences for such an interesting interdisciplinary conference devoted to infinite game theory. The centrality of determinacy to the main results of this paper reflects the influence of some of the excellent presentations by other authors in this volume.

practice is to choose the *simplest* theory among them and to cite "Ockham's razor" as the excuse (Figure 1). "Simplicity" is understood in a variety

Figure 1. Ockham to the rescue.

of ways in different contexts. For example, simpler theories are supposed to posit fewer entities or causes (Ockham's original formulation), to have fewer adjustable parameters, to be more "unified" and "elegant", to posit more uniformity or symmetry in nature, to provide stronger explanations, or to be more strongly cross-tested by the available data. But in what sense is Ockham's razor truly an excuse? For if you already know that the simplest theory compatible with experience is true, you don't need any help from Ockham (Figure 2). And if you don't, then the true theory

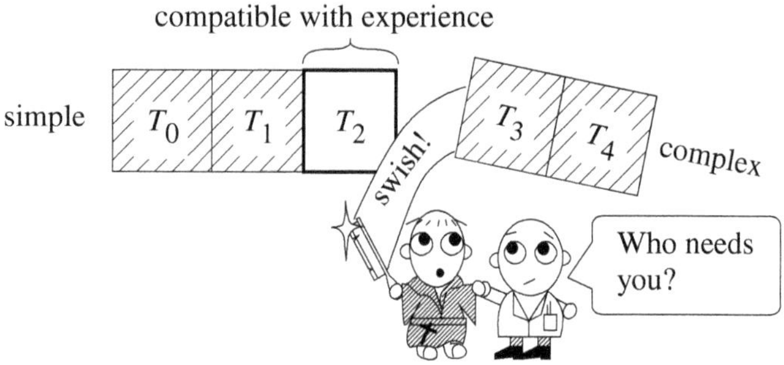


Figure 2. A Platonic Dilemma: Case I.

might be complex, so it is unclear how Ockham helps you find it (Figure 3). Indeed, how could a *fixed* bias toward simplicity indicate the possibly

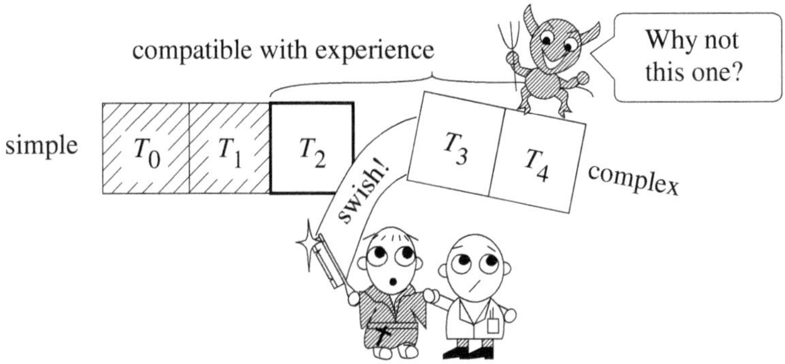


Figure 3. A Platonic Dilemma: Case II.

complex truth any better than a broken thermometer that always reads "zero" can indicate the temperature? You don't have to be a card-carrying skeptic to wonder what the tacit connection between simplicity and truth-finding could possibly be.

This essay explains the connection between simplicity and truth by modelling inquiry as an unending process, in which the scientist's aim is to converge to the truth in a way that minimizes, in a worst-case sense, reversals of opinion prior to convergence to the truth. Scientific methods may then be analyzed formally as strategies in an infinite game of perfect information, which brings to the subject powerful mathematical tools such as Donald Martin's Borel determinacy theorem [Mar75]. The proposed, long-run, strategic perspective on inquiry may appear abstract and remote from the day-to-day nuances of concrete scientific practice. Nonetheless, it is very general and singles out Ockham's razor as the best possible strategy to follow at every stage of inquiry, so its import for short-run practice is both sharp and concrete. Furthermore, the following review of standard attempts to link simplicity with theoretical truth in the short run reveals that they are all either irrelevant or based upon wishful thinking or circular arguments. A relevant, non-circular, long-run explanation may be better than no explanation at all.

## 2 Some Traditional Explanations of Ockham's Razor

Gottfried W. Leibniz explained the elusive connection between simplicity and truth by means of a direct appeal to the grace of God (Figure 4; [Lei14, §§ 55–59]) Since God is omnipotent and infinitely kind (to scientists, at least), it follows that the actual world is the most elegant (*i.e.*, simple)

Figure 4. Gottfried W. Leibniz' Theological Explanation.

universe that could possibly produce such a rich array of effects. Hence, simplicity doesn't track the truth the way a thermometer tracks temperature; truth, by the grace of God, tracks simplicity. This explanation merely underscores the desperate nature of the question.

Immanuel Kant confronted the issue in his *Kritik der Urtheilskraft.*[1] According to Kant, the faculty of judgment must prescribe or presuppose that the diverse laws of nature may be unified under a small set of causes if nature is to be intelligible at all. But theories that involve a few extra causes are also intelligible, so intelligibility falls far short of explaining why one should prefer theories with fewer causes or entities over those that involve more.

Some latter-day philosophers have emphasized that simple theories have various "virtues", most notably, that simpler or more unified theories are more thoroughly tested or confirmed by a given evidence set (*e.g.*, [Pop68, p. 251–281], [Gly80], or [Fri83, p. 236–250]). For if a theory has many free parameters (ways of being true), then new evidence simply "sets" the parameters and there is no risk of the theory, itself, being refuted altogether. But a simple theory does carry the risk of being refuted. It seems only fair to pin a medal of valor on the simple theory for surviving its self-imposed ordeal. The question, however, is truth, not valor, and the true theory might not be simple, in which case it wouldn't be valorous. To assume otherwise amounts to wishful thinking — the epistemic sin of concluding that the truth is as pleasing (intelligible, severely testable, explanatory, unified, uniform, symmetrical) as you would like it to be. Rudolf Carnap sought uniformity of nature in logic itself [Car50, p. 562–567]. This "logic" amounts, however, to nothing more than the imposition of greater prior probabilities on more

[1] *Cf.* the *Akademie-Ausgabe* [NatWin08, p. 185].

uniform worlds, where uniformity is judged with respect to an arbitrarily selected collection of predicates. The argument goes like this (Figure 5). Suppose there are but two predicates, "green" and "blue" and that every-

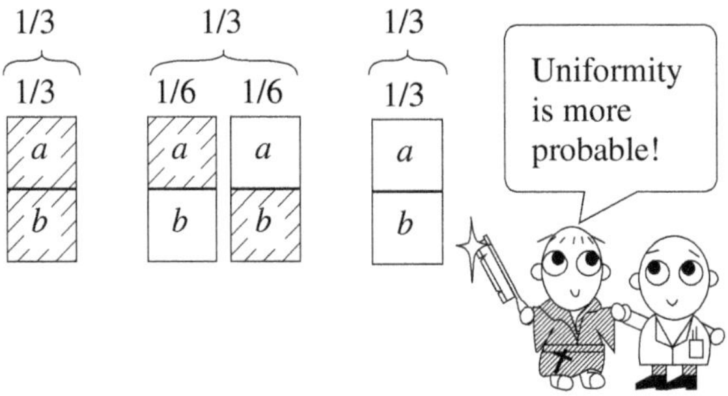


Figure 5. Rudolf Carnap's "Logical" Explanation.

thing is either green or blue. Suppose there are two observable objects, $a$ and $b$. Two worlds are *isomorphic* just in case a one-to-one substitution of names takes you from one world to the other in a way that preserves the basic predicates in your language. Hence the uniform world in which $a$ and $b$ are both green is in its own isomorphism class, as is the uniform world in which $a$ and $b$ are both blue. The two non-uniform worlds in which $a$ and $b$ have different colors can each be reached from the other by a one-to-one, color-preserving substitution of names, so they end up in the same isomorphism class. Now Carnap invokes the principle of indifference to put equal probabilities of one third on each of these three automorphism classes and invokes it again to split the one third probability on the non-uniform class over the two non-uniform worlds. The resulting probability distribution is then biased so that uniform worlds get probability one third and non-uniform worlds get probability one sixth. So uniform worlds are more probable than non-uniform worlds (by a factor of two in this tiny example, but the advantage increases as observable individuals are added).

Nelson Goodman objected that whatever is logical ought to be preserved under translation and that Carnap's uniformity bias based on linguistic syntax isn't [Goo83, p. 59–83]. For uniformly green and uniformly blue experience are uniform. But one can translate green and blue into "grue" and "bleen", where "grue" means "green if $a$ and blue if $b$" and "bleen" means "blue if $a$ and green if $b$" (Figure 6). Then in the grue/bleen language, the worlds that used to be non-uniform are now uniformly grue or uniformly bleen, respectively and the worlds that used to be uniform are non-uniform,

Figure 6. Nelson Goodman's "Grue" Argument.

for "green" means "grue if $a$ and bleen if $b$" and "blue" means "bleen if $a$ and grue if $b$". Since logical inferences are based entirely on syntax and syntactically the situation between green/blue and grue/bleen is entirely symmetrical, uniformity cannot be a feature of logical syntax. The moral is that Carnap's story makes uniformity of nature a mere matter of description. But a guide to truth could not be a mere matter of description, since truth doesn't depend upon how it is described.

## 3 Statistical Explanations

So much for philosophy. Surely, the growing army of contemporary statisticians, machine learning researchers, and industrial "data miners" must have a better explanation based on rigorous, mathematical reasoning. Let's check. A major player in the scientific methodology business today is *Bayesian* methodology. The basic idea is to allow personal biases to enter into statistical inferences, where personal bias is represented as a "prior" probability measure over possibilities. The prior probability of hypothesis $H$ is then combined with experience $E_t$ available at $t$ via *Bayes' theorem* to produce an updated probability of $H$ at $t'$, which represents your updated opinion concerning $H$:

$$P_{t+1}(H) = P_t(H \mid E_t) = \frac{P_t(H) \cdot P_t(E_t \mid H)}{P_t(E_t).}$$

It is clear from the formula that your prior opinion $P_t(H)$ is a factor in your posterior opinion $P_{t+1}(H)$, so that the simplest theory compatible with the new data ends up being most probable in the updated probabilities. Ockham's razor is just a systematic bias toward simpler theories. So to explain

Figure 7. The Circular Bayesian Explanation.

its efficacy by appealing to a prior bias toward simplicity is patently circular (Figure 7).

Bayesians also have a more subtle story. Yes, it begs the question simply to impose a prior bias toward simple theories, so let's be "fair" and impose *equal* prior probabilities on competing theories, be they simple or complex. Now suppose, for concreteness, that we have just two theories, simple theory $S$ and complex theory $C(\theta)$ with free parameter $\theta$ which (again for concreteness) can be set to any value from 1 to $k$ (Figure 8). Suppose, further, that

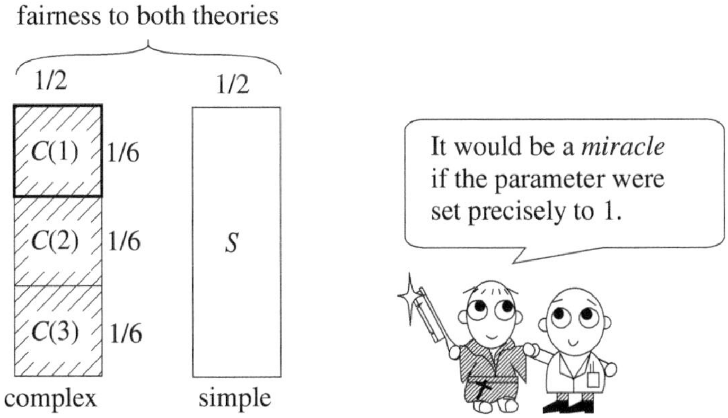


Figure 8. The Miracle Explanation.

$S$ consistently entails $E_t$, as does $C(1)$, but that for all other parameter values $i$, $C(i)$ is refuted by $E_t$. Thus, $P_t(E_t \mid S) = P_t(E_t \mid C(1)) = 1$

but for all $i$ distinct from 1, $P_t(E_t \mid C(i)) = 0$. Suppose, again, that you have no prior idea which parameter value of $C(i)$ would be the case if $C(\theta)$ were true (that's what it means for the parameter to be "free"). So $P_t(\theta \mid C(i))$ is uniform.[2] Turning the crank on Bayes' theorem, one obtains $P_t(S \mid E_t)/P_t(C \mid E_t) = k$. So even though the complex theory could save the data just as well as the simple one, the simple theory that does so without any *ad hoc* fiddling ends up being "confirmed" much more sharply by the same data $E_t$ (*e.g.*, [Ros83, p. 74–75]). Surely that explains how severe testability is a mark of truth, for doesn't the more testable theory end up more probable after a fair contest?

One must exercise caution when Bayesians speak of fairness, however, for probabilistic "fairness" between "blue" and "non-blue" implies a marked bias toward "blue" in a choice among "blue, yellow, red". That is all the more true in the present case: "fairness" between $S$ and $C$ induces a strong bias for $S$ with respect to $C(1), \ldots, C(k)$. One could just as well insist upon "fairness" at the level of parameter settings rather than at the level of theories (Figure 9). In that case, one would have to impose equal probabilities

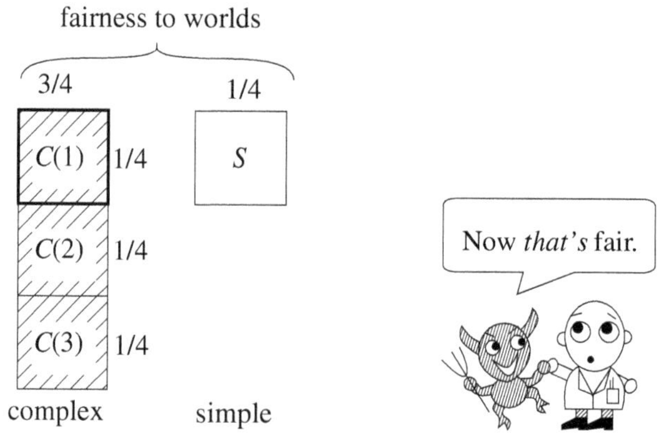


Figure 9. The Miracle Reversed.

of $1/(k+1)$ over the $k+1$ possibilities $\{S, C(1), \ldots, C(k)\}$. Now $C(0)$ and, hence, $C$, will remain forever at least as a probable as $S$ in light of evidence

[2]This is a discrete version of the typical restrictions on prior probability in Bayesian model selection (*cf.* [Was04, p. 220–221]). If the parameters are continuous, each parameter setting receives zero prior probability, but the result is the same because the likelihood of the more complex theory must be integrated over a higher-dimensional space than that of the simpler theory.

agreeing with $S$. Classical statisticians explain Ockham's razor in terms of "overfitting" (*cf.* [Was04, p. 218–225] for a textbook review). "Overfitting" occurs when you want to estimate a sampling distribution by setting the free parameters in some statistical model. In that case, the expected squared predictive error of the estimated model will be higher if the model employed is too complex (*e.g.*, [ForSob94]). This is a kind of objective, short-run connection between simplicity and truth-finding, but it doesn't really address the question at hand, which is how Ockham's razor helps you find the true *theory*, which is quite another matter from which theory or model to use to estimate the underlying sampling distribution. The quickest way to see why is this: suppose that God were to tell you that the true model has fifty free parameters. On a small sample, the overfitting argument would still urge you to use a much smaller model for estimation and prediction purposes (Figure 10). So the argument couldn't be concerned with finding the true theory.

Figure 10. The "Overfitting" Explanation.

More subtly, the sense of approximate error employed in the overfitting literature is wrong for theory selection. Getting "close" to the underlying sampling distribution might not get you "close" to the form of the true model, since distributions arbitrarily close to the true distribution could be generated by models arbitrarily far from the true model.[3] Thus, distance from the theoretical truth is typically discontinuous with distance from the true sampling distribution, so minimizing the latter distance may fail to get you close in the former, as in the case of God informing you that the true model is very complex. Another point about overfitting is that even to the

[3]This is particularly true when the features of the model have counterfactual import beyond prediction of the actual sampling distribution, as in causal inference [GlySchSpi00, p. 47–53].

extent that it does explain the role of simplicity in statistical prediction, it is tied, essentially, to inference problems in which the data are stochastically generated, leaving one to wonder why simplicity should have any role in deterministic inference problems, where it still feels like a good idea.

Finally, there are theorems of the sort that some method equipped with a prior bias toward simplicity is guaranteed to *converge* in some sense to the true model as experience (or sample size) increases (*e.g.*, [LohZhe95]). That would indeed link Ockham's razor with truth-finding *if* it could be shown that *other* possible biases don't converge to the truth. But they do. The logic of convergence results is not that Ockham's advice points at or indicates the truth, but that it is "washed out" or "swamped", eventually, by accumulating experience, even if the advice is so misleading as to throw you off the track for a long time (Figure 11). But alternative biases would

Figure 11. The Convergence Explanation.

also be washed out by experience eventually[4] so that's hardly a ringing endorsement of Ockham's razor. What is required is an argument that Ockham's razor is, in some sense, the *best possible* bias for finding the true theory.

## 4 Post Mortem

To recapitulate, the standard explanations of the mysterious relationship between simplicity and theoretical truth are either circular, wishful, or irrelevant. Still, they provide useful information about how the relationship can't be explained. For indication of the truth is too strong a connection to establish without begging the question at the outset, as Leibniz and the Bayesians do. On the other hand, mere convergence in the limit is too weak

[4] *Cf.* [Hal74, p. 212, Theorem A]. Also, see the critical discussion of Bayesian convergence theorems in [Kel96, p. 302–330].

to single out simplicity as the right bias. The crux of the puzzle, then, is to come up with a notion of "helping to find the truth" that is strong enough to single out simplicity as the right bias to have but that is not so strong as to demand a question-begging appeal to Ockham's razor at the outset in order to establish it. Thus, the appropriate notion of "help" must be stronger than convergence in the limit and weaker than indication in the short run.

The account developed below steers between these two extremes by considering a refined concept of convergence, namely, convergence with a minimum number of reversals of opinion prior to arrival at the goal (Figure 12). This is stronger than mere convergence in the limit, which says nothing

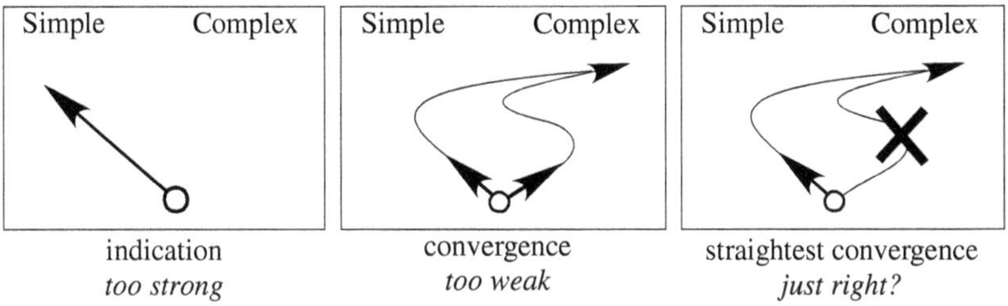


Figure 12. Three Kinds of "Help".

about minimizing reversals of opinion along the path to the truth, and is weaker than indication, which allows for no reversals of opinion whatever. It will be demonstrated that an ongoing bias toward simplicity minimizes kinks in your course to the truth in a certain precise sense. But first, I illustrate the general flavor of the approach by showing that something similar happens almost every time you ask for directions.

## 5 Asking for Directions

Suppose you are headed home on a road trip and get lost in a small town. In frustration, you stop to ask a local resident how to get home (Figure 13). Before you can even say where you are headed, he gives you the usual sort of advice: directions to the nearby freeway entrance ramp, which happens to be a few blocks back toward where you just came from. Now suppose that, in a fit of hubris, you disregard the resident's advice in favor of some intuitive feeling that your home is straight ahead (Figure 14). That ends up being a bad idea (Figure 15). You leave town on a small rural route that winds its wild way over the mountains. At some point, you concede the error of your ways and turn around to follow the resident's directions to the

Figure 13. Asking For Directions.

Figure 14. Hubris!

freeway. The freeway then follows as straight a route home as is practicable through mountainous terrain. As you speed your way homeward, you have ample time to regret: *if you hadn't ignored the local resident's advice, you wouldn't have added that useless, initial U-turn to your otherwise optimal journey home*. Let's take stock of a few striking features of this mundane tale. First, the local resident's advice was indeed *helpful*, since it would have put you on the straightest possible path home. Second, *by disregarding the advice, you incurred an extra U-turn or kink in your route*. What is particularly vexing about the initial U-turn is that it occurs even before you properly begin your journey. It's a sort of navigational "original sin"

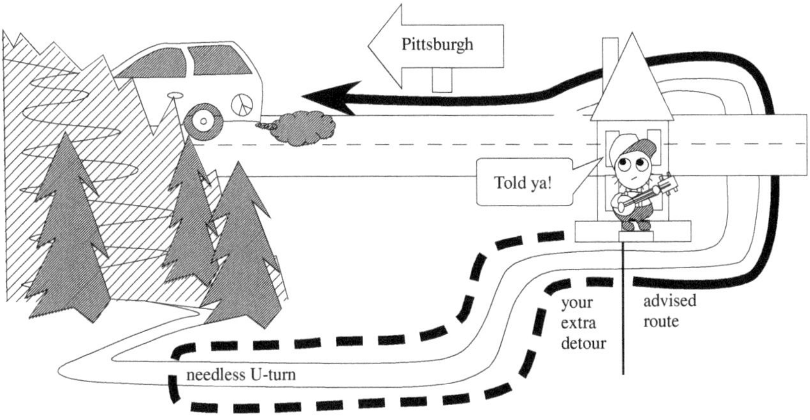


Figure 15. The U-turn Argument.

that you can never be absolved of. Third, the resident *didn't need to know where you were going in advance* in order to give you helpful advice. Any stranger asking for directions in a small, isolated town would do best to get on the freeway. Hence, the resident's ability to provide useful information without knowing where your home is *doesn't require an occult or circular explanation.* Suppose, on the other hand, that the resident could give you a compass course home before knowing where you are headed. That *would* require either a circular or an occult explanation (an Ouija board or divining rod). Fourth, even the freeway is not perfectly straight, so the resident's advice provides *no guarantee against future course reversals*, even if it is the best possible advice. Finally, the resident's advice is the best possible advice *even though it points you away from your goal* at the outset. If help required that you be aimed in the right direction, then the resident would have to give you a compass heading home, which wouldn't be possible unless he already knew where your goal was or had an Ouija board or divining rod.

So the typical situation in which you ask for directions home from a distant, small town has all the fundamental features that an adequate explanation of the truth-finding efficacy of Ockham's razor must have. Perhaps Ockham also provides fixed advice that puts you on the best possible route to the truth without pointing you at the truth and without guarantees against future course reversals along the way. It remains, then, to explain what the freeway to the truth is and how Ockham's advice leads you to it.

## 6 The Freeway to the Truth

Even lexicography suggests an essential connection between freeways and truth-finding, for both changes in course and changes in opinion are called changes in *attitude*. According to this analogy, Ockham's advice should somehow minimize changes of opinion prior to convergence to the right answer.[5] Let's consider how the story goes in the case of a very simple truth-finding problem.

Suppose that there is an emitter of discrete, readily detectable particles at arbitrary intervals and that you know that it can emit at most finitely many particles altogether (Figure 16). The question is how many particles

Figure 16. Counting Particles (Effects).

it will ever emit. What makes the problem interesting is that an arbitrarily long interval without new particles can easily be mistaken for total absence of future particles. This problem has more general significance than might be apparent at first, for think of the particles as detectable *effects* that are arbitrarily hard to detect as parameters in the true theory are tuned toward zero. For example, in curve fitting, the curvature of a quadratic curve may be so slight that it requires a huge number of data to notice that the curve is non-linear.[6] So the theory that the curve is quadratic but

[5]The idea of counting mind-changes already appears in [Put65]. Since then, the idea has been studied extensively by computer scientists interested in computational learning (*cf.* [Jai+99] for a review). The focus, however, is on categorizing the complexities of problems rather than on singling out Ockham's razor as an optimal method. Oliver Schulte and I began looking at retraction minimization as a way to severely constrain one's choice of hypothesis in the short run in 1996 (*cf.* [Sch99a, Sch99b]). Schulte has also applied the idea to the inference of conservation laws in particle physics [Sch01]. The ideas in this essay build upon and substantially simplify and generalize the initial approach taken in [Kel02, GlyKel04, Kel04]). While the present manuscript was in press, the approach was developed further in [Kel07]. Some differences between the two approaches are mentioned in subsequent footnotes.

[6]It is assumed that he data are increasingly precise but inexact; else three points would settle the question [Pop68, p. 131–131]. The same point holds if the data are noisy. In that case, tuning the parameters toward zero makes the effects statistically undetectable at small sample sizes (*cf.* [GlyKel04, Kel04] for an application of the preceding ideas to

not linear predicts the eventual detection of effects that would never appear under the linear theory. Similarly, the curvature of a cubic curve may be so slight that it is arbitrarily hard to distinguish from a quadratic curve. The point generalizes to typical model selection settings regardless of the interpretation of the parameters. So deciding among models or theories with different free parameters is quite similar, after all, to counting particles.

The traditional formulation of "Ockham's razor" is to *not multiply entities without necessity.* It is "necessary" (on pain of outright inconsistency) to assume as many particles as you have seen, but it is not necessary to assume more, so that *if* you conclude anything, you should conclude exactly as many particles as you have seen so far (Figure 17). The most aggressive

Figure 17. Ockham in Action.

Ockham method is the *counting method* that concludes that every particle has been seen at every stage. More realistic Ockham methods refuse to commit themselves to any answer at all until a long time has passed with no novel effects. Ockham's razor, itself, says nothing about how long this "confidence-building" time should last and the following argument for Ockham's razor doesn't imply anything about how long it should be either; it simply requires you to adopt some Ockham method, whether the method waits or not. That is as it should be, since even believers in short-run evidential support (*e.g.*, Rudolf Carnap and the Bayesians) allow for arbitrary individual differences concerning the time required for confidence buildup.

Other intuitive formulations of empirical simplicity conspire with the view that the Ockham answer should be the exact count. First, the Ockham theory that there are no more particles than you have seen is the most *uniform* theory compatible with experience, for it posits a uniformly particle-free future. Second, the Ockham theory is the most *testable* theory compatible with experience, for if it is false, you will see another particle and it will

stochastic problems.

be decisively refuted. Any theory that anticipates more particles than have been seen might be false, because there are fewer particles than anticipated, in which case it will never be refuted decisively, since the anticipated particles might always appear later. Third, the Ockham theory is most explanatory, since the theory that posits extra particles fails to explain the *times* at which those particle appear. The theory that there are no more particles fails to posit extra, unexplained times of appearance. Fourth, the Ockham theory is most *symmetrical*, since the particle-free future is preserved under permutation of times, whereas a future punctuated by new particle appearances would be altered by such permutations. Fifth, the Ockham theory has the *fewest free parameters*, because each time of appearance of a new particle is a free parameter in a theory that posits extra particles. So in spite of its apparent triviality, the problem of counting things that are emitted from a box does illustrate a wide range of intuitive aspects of empirical simplicity. That isn't so surprising in light of the analogy between particles and empirical effects tied to free parameters.

If you follow an Ockham solution to the particle-counting problem, then you change your mind in light of increasing data at most once per particle. If the true count is $k$, then you change your mind at most $k$ times. By way of comparison, suppose that you have a hankering to violate Ockham's razor by producing a different answer (Figure 18). You might reason as follows.

Figure 18. Ockham Violation!

The particle emitter has overturned every successive Ockham answer in the past (*i.e.*, "zero", "one", "two", and "three"), so you expect it will overturn the current Ockham answer "four" as well. So by induction on Ockham's unbroken losing streak in the past, you anticipate failure again and guess "five" (or some greater number of your choosing) rather than the Ockham answer "four". Philosophers of science call this the "negative induction from the history of science" [Lau81]. Why side with Ockham, rather than with the negative induction against him?

Efficiency is future-directed. Slush funds or debts may have been accumulated in the past, but efficiency optimization in the present concerns future costs incurred by future acts over which you have some control. So think of inquiry as starting from scratch at each moment. Accordingly, the *subproblem* entered at a given stage of inquiry consists of the restriction of possibilities to those consistent with current experience and only mind-changes incurred after entering the subproblem are counted in that subproblem.[7] Consider the subproblem entered when you first say "five", upon having seen only four particles. There is no deadline by which the fifth particle you anticipate has to show up, so you may have to wait a long time for it, even if you are right. You wait and wait (Figure 19). Your graduate

Figure 19. The Pressure Builds.

students exhaustively examine the particle emitter for possible malfunctions. Colleagues start talking about the accumulation of "null results" and discuss the "anomalous" failure of the anticipated marble to appear. True, the posited particle could appear (to your everlasting fame) at any time, so your theory isn't strictly refuted. Nonetheless, you feel increasing pressure to switch to the four-particle theory as the anomaly evolves into a full-blown "crisis". This increasing pressure comes not from the "weight of the evidence", as philosophers are wont to say, but from your strategic aim to converge to the truth, regardless of what it happens to be. For if you never change your mind from "five" to "four" and the fifth particle never appears, you will converge for eternity to "five" when the truth is "four". So at some time of your choosing, you must (on pain of converging to the wrong answer) cave in to the pressure from nature's strategic threat and switch back to the (Ockham) theory that the machine will produce just four particles (Figure 20). Won't that make for interesting gossip in the

[7]One might object that a sub-problem should hold past information and past theory choices fixed and sum the total cost from the outset of inquiry. That approach is developed in detail in [Kel07].

Particle Counting Association, where you are feted as the sole defender of the five particle theory?[8]

To summarize, in the subproblem entered when you first say "five", nature can force you to change your mind at least once (from "five" to "four"), in the manner just described, without presenting a fourth particle. The same is not true of Ockham, who enters the same subproblem saying "four" (or nothing at all) and who never changes his mind until the next particle appears. Thereafter, Ockham changes his mind exactly one time per extra particle. But you can be forced by nature to change your mind at *least* once per extra particle (on pain of not converging to the truth) in the same manner already described; for a long period during which there are exactly $i$ particles forces you to say "$i$" on pain of not converging to the truth, after which nature can present the $i+1$st particle, *etc.* (Figure 21).

Figure 20. The Agony of Retreat.

Hence, *if a solution violates Ockham's razor in the particle counting problem, then in the subproblem entered at the time of the violation, whatever sequence of outputs the Ockham solution produces, the violator can be forced by nature to produce a sequence including at least the same mind-changes plus another one (the initial U-turn from "five" to "four").* You should have listened to Ockham!

[8] I am alluding, of course, to Thomas Kuhn's 1962 celebrated historical theory of the structure of scientific revolutions [Kuh62]. Null experiments generate anomalies which evolve after careful consideration into crises that ultimately result in paradigm change. Kuhn concludes, hastily, that the change is an unlawful matter of politics that has little to do with finding the truth. I respond that it is a necessary consequence of the logic of *efficient* convergence to the truth after a violation of Ockham's razor, as will become clear in what follows. Many of the celebrated scientific revolutions in physics have been the results of Ockham violations (*e.g.*, Ptolemy vs. Copernicus, Fresnel vs. Newton, Newton vs. Einstein, and creationism vs. Darwin). In each of these cases, a theory positing extra free parameters (with attendant empirical effects) was chosen first and a simpler theory was thought of later and came to replace the former, often after an accumulation of null experiments.

Figure 21. Ockham Avoids Your Initial U-turn.

The same argument works if you violate Ockham's razor in the other possible way, by saying "three" when four particles have been seen. For nature can refuse to present more particles until you change your mind to "four" on pain of never converging to the right answer if the right answer is "four". But in the same subproblem, Ockham would already have said "four" if he said anything at all and, in either case, you can be forced into an extra mind-change in each answer. So the U-turn argument also explains the need for maintaining consistency with the data.

So there is, after all, a close analogy between particle counting and getting on the freeway. Your initial mind change from "five" to "four" is analogous to your initial U-turn back to the local resident's house *en route* to the highway. Thereafter, no matter what the true answer is, you can be forced to change your mind at least once for each succssive particle, whereas Ockham changes his mind at most once per successive particle. These mind-changes are analogous to the unavoidable curves and bends in the freeway. So no matter what the truth is, you start with a U-turn Ockham avoids and can be forced into every mind-change Ockham performs thereafter. As in the freeway example, you have botched the job before you even properly get started. In both stories, the advice is the best possible. Nonetheless, it does not impose a bound on future course reversals; nor does it point you toward your goal by some occult, unexplained mechanism.

A striking feature of the explanation is that it is entirely game-theoretic. There is no primitive notion of "support" or "confirmation" by data of the sort that characterizes much of the philosophical literature on induction and theory choice (Figure 22).[9] Nor are there prior probabilities that foster the illusion of "support" of general conclusions by a few observations. The

[9] In this respect, my approach is a generalization and justification of the "anti-inductivism" of Karl Popper [Pop68, p. 27–30].

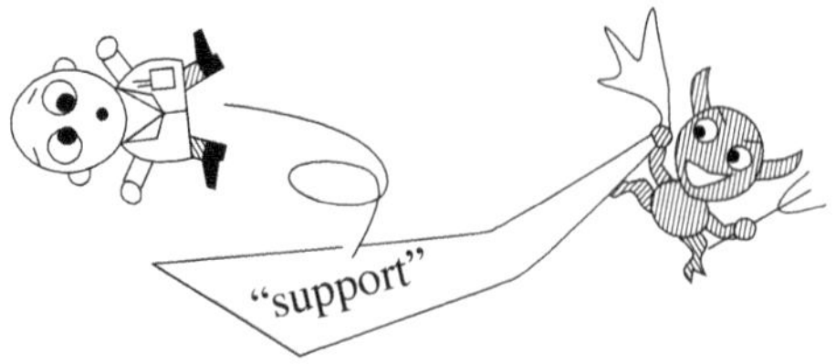


Figure 22. Pulling the Rug Out.

phenomenology of “support” by evidence emerges entirely from the aim of winning this truth-finding game against nature. Furthermore, the game is essentially infinite. For if there were an *a priori* bound on the time by which the next particle would arrive if it arrives at all, then you could simply “outwait” nature and avoid changing your mind altogether. So the argument is situated squarely within the theory of infinite, zero-sum games, which is the topic of this volume.

Here is why the reference to subproblems is essential to the U-turn argument. Suppose that you are asleep when you see the first particle and that when you see the second particle you wake up and guess “three”, expecting that you will also sleep through the third (Figure 23). Thereafter, you

Figure 23. Ockham Still Wins in Subproblem.

always agree with Ockham. If the third particle doesn’t appear right away, you can be forced to change your mind to “two”, but that’s only your second retraction — Ockham wouldn’t have done better. Now that you have “caught up” with Ockham, you match him no matter how many particles you see in the future. But that is only because you “saved” a retraction in the past by sleeping through the first particle. That is like hoarding a “slush fund” to hide future mismanagement from the public. In the subproblem entered when you say "three", the slush fund is emptied and you have to demonstrate your efficiency from scratch. In that subproblem, your first reversal of opinion back to "two" gets added to all your later mind-changes and you never catch up, so Ockham wins. The moral: an arbitrary Ockham solution beats you in the subproblem in which you violate Ockham's razor, but the Ockham solution does as well as you in every subproblem, so the Ockham solution is better. In the best case for the violator, the anticipated fifth particle might appear immediately after the violation, before the method even has a chance to become queasy about over-estimating (Figure 24). In that case, the violator's output sequence in the subprob-

Figure 24. Best Case Fairy to the Rescue.

lem entered at the violation begins with "five", "five", whereas Ockham's output sequence in the same subproblem begins with "four", "five", which is worse. Hence, the Ockham method doesn't weakly dominate the violator's mind-changes in the subproblem in question. But that merely explains why the U-turn argument, which does establish the superiority of an arbitrary Ockham solution over an arbitrary non-Ockham solution, is not a weak dominance argument. The U-turn argument essentially involves a worst-case dimension lacking in weak dominance, for nature can *force* the non-Ockham solution from "five" back to "four" (on pain of convergence to the wrong answer) by withholding particles long enough and can then reveal another particle to make it say "five", "four", "five", which is never produced by any Ockham method and which properly extends the Ockham sequence "four", "five".[10] Nor, for that matter, is the U-turn argument a

[10]Indeed, Ockham methods are weakly dominated by methods that hang on to their original count for an arbitrarily long period of time (Teddy Seidenfeld, personal communication). That isn't so bad, after all, because there are compelling reasons not to under-count (*e.g.*, the undercount couldn't possibly be true). The crux of Ockham's razor is to motivate not over-counting, and over-counters do not dominate the retractions of Ockham methods in this way. More to the point, the alternative theory developed in [Kel07] is not subject to this objection, because tardiness of retractions is also penalized,

standard worst-case or "minimax" argument, for there is no fixed bound on mind-changes for any solution to the counting problem (nature can force an arbitrary solution through any number of mind-changes).

## 7 A General Conception of Scientific Problems

A *scientific problem* specifies a set $\Omega$ of possible worlds the scientist must succeed in together with a question $\mathcal{Q}$ which partitions $\Omega$ into mutually exclusive *potential answers*. The aim is to find the true answer for $w$ no matter which world $w$ in $\Omega$ you happen to live in. If $\mathcal{Q}$ is a binary partition, one thinks of a decision or test problem for one of the two cells vs. the other. If it is a fixed range of alternatives extensionally laid out in advance, one speaks of *theory choice*. If it is an infinite partition latently specified by some criterion determining the kind of theory that would count as success, the situation might be described as *discovering* the truth.

The most characteristic thing about empirical science is that you don't get to see $w$ in its entirety. Instead, you get some incomplete evidence or information about $w$, represented by some subset of $\Omega$ containing $w$. The set of all possible information states you might find yourself in is modelled as the collection of open sets $\mathcal{V}$ in a topological space over $\Omega$. A *scientific problem* is just a triple $(\Omega, \mathcal{V}, \mathcal{Q})$, where $(\Omega, \mathcal{V})$ is a topological space and $\mathcal{Q}$ partitions $\Omega$ (Figure 25). The idea is that, although the scientist never

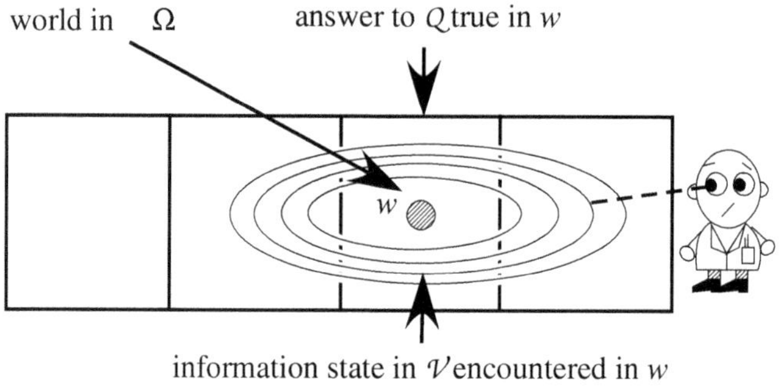


Figure 25. A Scientific Problem.

gets to see the actual world $w$, itself, he does get to see ever smaller open neighborhoods of $w$.

so the delayers do not end up ahead. The present theory has the advantage of greater mathematical elegance, however.

The use of topology to model information states is not a mere stipulation, for information concerns verifiable effects and topology is perhaps best understood as the mathematical theory of verifiability.[11] The point is seen most directly as follows. Identify each proposition with the set of possible worlds or circumstances in which it would be true, so propositions may be modelled as subsets of the set $\Omega$ of possible worlds. Say that a proposition is *verifiable* if and only if there exists a method or procedure that examines experience and that eventually illuminates a light if the proposition is true and that never illuminates the light otherwise. For example, illuminating the light when a particle appears yields a verification procedure for the proposition that at least one particle will appear.

The contradiction is the empty set of worlds (it can't possibly be true) (Figure 26). It is verifiable by the trivial verification procedure that never

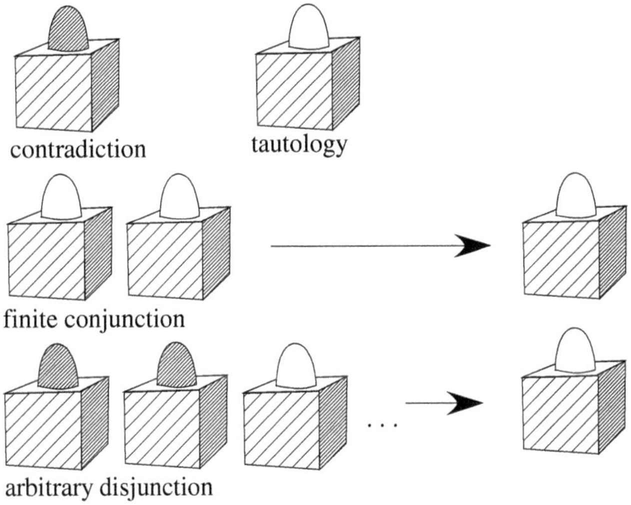


Figure 26. Verifiable Propositions are Open Sets.

illuminates its light. Similarly, the tautologous proposition consists of the whole set $\Omega$ of worlds and is verifiable by the trivial procedure that turns on its light *a priori*. Suppose that two verifiable propositions $A, B$ are given. Their conjunction $A \cap B$ is verifiable by the procedure that turns on its light if and only if the respective verification procedures for $A$ and for $B$ have both turned on their lights. Finally, suppose a collection $\mathcal{D} \subset \Omega$

[11]Topology is also used to model partial information states in denotational semantics [Sco82].

of verifiable propositions is given. Their disjunction $\bigcup \mathcal{D}$ is verifiable by the procedure that turns on its light just in case the procedure for some proposition $A \in \mathcal{D}$ turns on its light (you will see that light eventually as long as each respective procedure is only a finite distance away). Hence, the verifiable propositions $\mathcal{V}$ over $\Omega$ constitute the open sets of a topological space $(\Omega, \mathcal{V})$. So every theorem about open sets in a topological space is also true of ideal empirical verifiability. One of the most characteristic features of topology is that open sets are closed under arbitrary union but only under finite intersection. That is also explainable in terms of verifiability. Suppose you are given an infinite collection $\mathcal{C}$ of verifiable propositions. Is there a verification procedure for $\bigcap \mathcal{C}$? Not always. For the respective verification procedures for the elements of $\mathcal{C}$ may all turn on their lights, but at different times, so that there is no time by which you can be sure that it is safe to to turn on your light for $\bigcap \mathcal{C}$ (Figure 27). That is an instance

Figure 27. The Demon of Arbitrary Conjunction.

of the classical problem of induction: no matter how many lights you have seen go on, the next light might never do so. So not only are the axioms of topology satisfied by empirical verifiability; the characteristic asymmetry in the axioms reflects the problem of induction.

In a given topological space $(\Omega, \mathcal{V})$, the problem of induction arises in a world $w \in \Omega$ with respect to proposition $H$ just in case every information state (open proposition) true in $w$ is compatible both with $H$ and with $\neg H$ (Figure 28).[12] In standard, topological parlance, the problem of induction arises with respect to $H$ in $w$ just in case $w$ is a *boundary point* of $H$. So the demons of induction live in the boundaries of propositions one would like to know the truth values of. In a world that is an *interior point* of $H$, one eventually receives information verifying $H$ (since an interior point of $H$ has a neighborhood contained in $H$). Hence, not every verified proposition is verifiable, since a verified proposition merely has non-empty interior, whereas a verifiable proposition is open. But if a non-verifiable proposi-

[12] Let $\neg H$ denote $\Omega \backslash H$.

Figure 28. Demons Live in Boundaries.

tion is verified, some open proposition entailing it is verified, so information states can still be identified with open sets.

Less abstractly, recall the particle-counting problem. A possible world determines how many particles emerge from the machine for eternity and when each such particle emerges. Thus, one may model possible worlds as $\omega$-sequences of bits, where 1 in position $n$ indicates appearance of a new particle at stage $n$ and 0 indicates that no new particle appears at $n$. Consider the situation in which you have seen the finite bit string $(b_0, \ldots, b_{n-1})$. The corresponding information state is the set of all $\omega$-sequences of bits that extend the finite bit string observed so far. Call this proposition the *fan* with handle $(b_0, \ldots, b_{n-1})$, since all the worlds satisfying the fan agree up to $n$ and then "fan out" in all possible ways from $n$ onward (Figure 29). Any disjunction of verifiable events is verifiable (see above), so any union of fans is also verifiable and, hence, open (just wait for the handle of one of the fans to appear before turning on the light). The resulting space over arbitrary, $\omega$-sequences of bits is very familiar in topology, where it is known as as the *Cantor* space. In the particle-counting problem, it is assumed that at most finitely many particles will appear, so one must restrict Cantor space down to the $\omega$-sequences that converge to 0.

Consider the proposition that exactly two particles will be observed for eternity. This proposition is impossible to verify (no matter what you see, another particle may appear later). Hence, its interior is empty and every element is a boundary point, where the problem of induction arises. In this space, the boundary points are particularly suggestive of the problem of induction (Figure 30). For example, consider the world $(1, 1, 0, \ldots)$ where the dots indicate an infinite tail of zeros. No matter how far you travel down this sequence (*i.e.*, no matter what information you receive in this world), there exist worlds in which more than two particles appear later

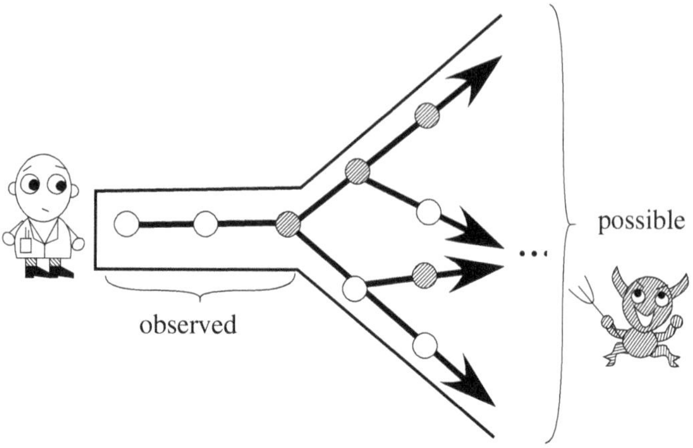


Figure 29. A Fan of Sequential Worlds.

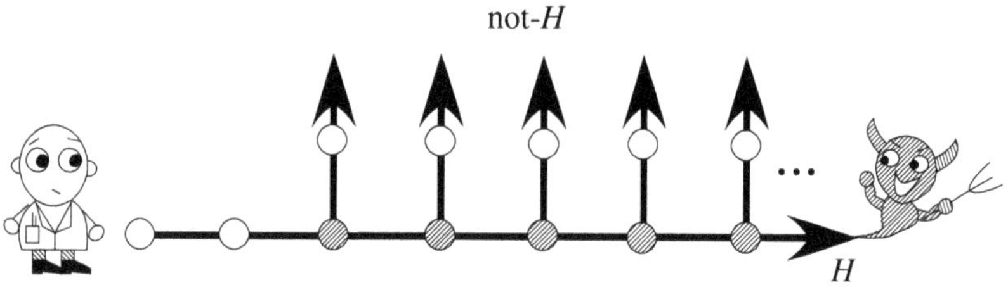


Figure 30. Boundary Points in Cantor Space.

than you have observed so far. So nature is in a position to drag you down the sequence $(1, 1, 0, \ldots)$ until you cave in and say "two" and is still free to show you another particle, as in the U-turn argument. The U-turn argument hinges, therefor, upon the topologically invariant structure of boundary points between answers to a question.

## 8 The Unending Game of Science

Each scientific problem determines an infinite, zero-sum game of perfect information (*cf.* [Kec95, p. 137–148] or [Kel96, p. 121–137]) between the *scientist*, who responds to each information state by selecting an answer (or by refusing to choose), and the impish *inductive demon*, who responds to the scientist's current guess history with a new information state. The demon is not a malicious force in nature; he merely personifies the difficulty of the challenge the scientist poses for himself by addressing a given scientific problem.

In this *truth-finding game*, the demon and the scientist take turns, starting with the scientist (Figure 31). Together, the demon and the scientist

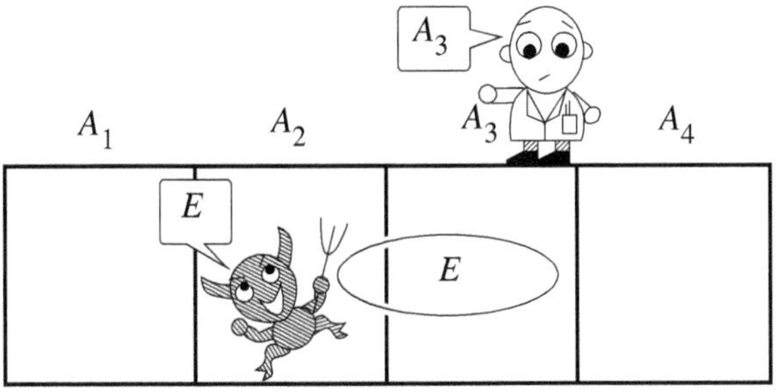


Figure 31. The Players.

produce a pair of $\omega$-sequences, an *information sequence* produced by the demon and an *answer sequence* produced by the scientist. Life would be too easy for the demon if he were allowed to withhold some crucial information for eternity, so the scientist is the victor by default if the demon fails to present complete, true information about some world in $\Omega$ in the limit.[13] In other words, an information sequence $\{E_i : i \in \omega\}$ for the problem should be a downward-nested sequence of open sets whose intersection is non-empty and contained in some answer $A$. Then say that the information sequence is *for* $A$.

The scientist wins the *convergence game* by default if the demon fails to present an information sequence for some world in the problem and by merit if his outputs stabilize eventually, to the answer true in some world $w$ the demon presents true information for. In other words, if the demon presents a legitimate information sequence for $w$, there must exist a stage in the play sequence after which the scientist's answer is correct of $w$. A winning strategy for the scientist in the convergence game is called a *solution* to the underlying empirical problem. For example, the obvious counting strategy solves the particle-counting problem. A problem is *solvable* just in case it has a solution.[14]

[13]One might reply that if it is impossible for the demon to fulfil his duty, the scientist loses since even the total experience received in the limit of inquiry doesn't settle the question. The game could be set up to reflect either viewpoint.

[14]It is interesting to inquire into the topological nature of solvability, since solvability

# 9 Comparing Mind-Changes

Consider two possible sequences of answers, $\sigma$ and $\tau$. Say that $\sigma$ *maps into* $\tau$ (written $\sigma \leq \tau$) just in case there is an answer and order preserving mapping (not necessarily one-to-one) from positions in $\sigma$ to positions in $\tau$, where suspension of judgement is a wild-card in the former sequence that matches any answer in the latter (Figure 32).[15] Since the mapping

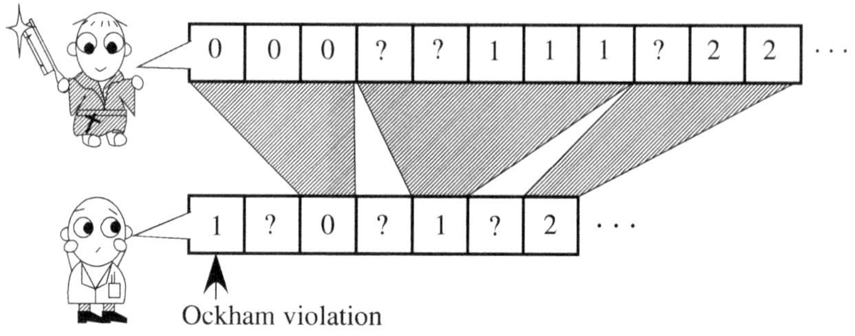


Figure 32. Top Output Sequence Better Than Bottom.

preserves answers and order, it also preserves mind-changes (not counting mere suspension as a mind-change). So when $\sigma$ maps into $\tau$, one may say that $\sigma$ is *as good as* $\tau$ so far as mind-changes are concerned. Say that $\sigma$ maps *properly* into $\tau$ (written $\sigma < \tau$) if, in addition, the latter fails to map into the former, as in Figure 32. Then $\sigma$ is *better than* $\tau$.

One can also say of two *sets* of output sequences that the former is *as good as* the latter just in case each element of the former is as good as some element of the latter (Figure 33) and is *better than* the latter if, in addition, the latter is not as good as the former.[16] The former set is *strongly better* than the latter just in case each of the former's elements is better than

is a topological invariant and must, therefore, be grounded in a problem's topological structure. For example, if the space is separable and the question is a countable partition, then solvability is equivalent to each cell being $\mathbf{\Delta}^0_2$ Borel (*cf.* [Kel96, p. 228, Corollary 9.10]). Such facts are not strictly necessary for understanding Ockham's razor, and are therefore omitted from this essay.

[15]In fact, an Ockham method's output sequences map *injectively* into output sequences the demon can force out of an arbitrary method. In [Kel07], methods are compared in terms of Pareto-dominanace with respect to number and timing of retractions, where a retraction occurs whenever an informative answer is dropped. The streamlined account of costs just presented does not penalize gratuitous question marks or tardiness of retractions, since question marks are wild cards and the mappings employed are many-one.

[16]This is not the same as weak dominance, since the existential quantifier allows for a worst-case pairing of output sequences by the mapping.

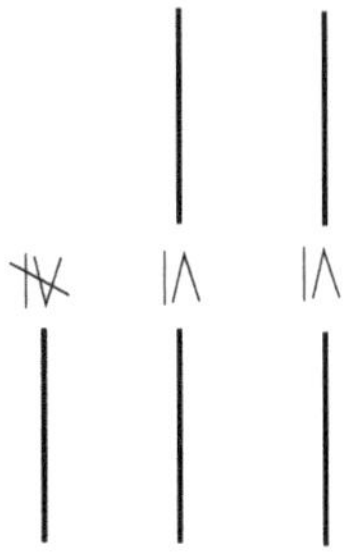

Figure 33. Top Set Better Than Bottom.

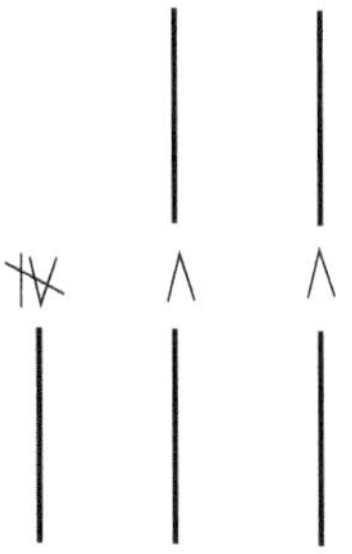

Figure 34. Top Set Strongly Better Than Bottom.

some element of the latter that is not as good as any element of the former (Figure 34).[17] Extend the symbols $\leq$ and $<$ to sets of output sequences accordingly.

The set of output sequences of a solution to a problem is the set of all output sequences $\sigma$ such there exists some information sequence for some answer along which the method produces $\sigma$. Then one can say of two methods that the former is *as good*, *better*, or *strongly better* than the latter just in case their respective sets of output sequences bear the corresponding relation. Finally, say that a solution is *efficient* in a problem just in case it as good as any other solution in each subproblem. Again, the idea is

[17]The requirement that the sequence mapped to is not as good as any of the former method's output sequences precludes cases in which a method is strongly better than itself. For example, if there are only two answers in the particle problem, "even" and "odd", then each output sequence of the obvious method that answers according to whether the number of observed particles is even or odd is better than some other output sequence of the same method (*e.g.*, $(E, O, E, O, \ldots) < (O, E, O, E, O, \ldots)$).

that inefficiency is forward-looking and should not be offset by foibles or stockpiles of credit (slush funds) earned in the past.

By way of illustration, the counting solution is efficient in the particle-counting problem, as is any Ockham solution to this problem (remember that Ockham solutions can suspend belief for artibrary periods of time). That is because the demon can force any solution through any ascending sequence of answers and Ockham methods produce only ascending sequences of answers. Furthermore, any non-Ockham solution is worse than any Ockham solution in the subproblem entered when the violation occurs. Indeed, it was shown that the violator is strongly worse than any Ockham solution in that subproblem, because the demon can force the violator into any ascending sequence after the U-turn back to the Ockham answer. Hence, the counting problem has the remarkable property that *its efficient solutions are exactly its Ockham solutions.* That is surely a result worth pressing as far as possible! But first, Ockham's razor must be defined with corresponding generality.

## 10 What Simplicity Isn't

The concept of simplicity appears, at first, to be a hodge-podge of considerations, including uniformity of nature, theoretical unity, symmetry, testability, explanatory power, and minimization of entities, causes, and free parameters. But in spite of these manifold aspects, it remains possible that simplicity is a deep, unified, concept with multiple manifestations, depending on the particular structure of the problem addressed. It is suggestive in this regard that the trivial particle-counting problem already illustrates all of the intuitive aspects of simplicity just mentioned and that they seem to cluster around the nested problems of induction posed by the repeated possibility that a new particle might appear.

It is easy, at least, to say what simplicity *couldn't* be. It couldn't be anything fixed that does not depend on the structure of the problem. For it is a commonplace in the analysis of formal procedures that different algorithmic approaches are efficient at solving different problems. So if simplicity did not depend, somehow, on the structure of the particular problem addressed, Ockham's razor couldn't possibly be necessary for efficient convergence to the truth in a wide range of distinct problems possessing different structures.

That is the trouble with concepts of simplicity like notational brevity [LiVit97, p. 317–337], uniformity of worlds [Car50, p. 562–567], prior probabilistic biases, and historical "entrenchment" [Goo83, p. 90–100]. Left to themselves, none of these ideas conforms to the essential structural interplay between a problem's question and its underlying informational topology, so none of them could contribute objectively to truth-finding efficiency over a range of different problems. All of them could be made to do so by selecting notation that reflects the relevant structure of the problem addressed [Mit97, p. 174]. But then the essence of simplicity is captured by the rules for selecting appropriate notation, rather than by brevity, uniformity, or the like.

## 11 Simplicity and Ockham's Razor Defined

The task of defining simplicity is facilitated by knowing in advance how Ockham's razor is justified. We can, therefore, "solve backwards" for simplicity, by generalizing the features of particle counting that give rise to the the U-turn argument. The key to the U-turn argument is the demon's ability to *force* a given sequence of mind-changes from an arbitrary solution. In the particle-counting problem, the demon can present information from the zero-particle world until the scientist caves in and concludes that there will be zero particles (on pain of not converging to the true answer) (Figure 35). Then the demon can present a particle followed by no further particles

Figure 35. Demon Forcing a Sequence of Answers.

until the scientist concludes "one particle", again on pain of not converging to the true answer, and so forth. This can't go on forever, though, because the demon must present data from some world in $\Omega$, and all such worlds present at most finitely many particles. Hence, for each finite ascending sequence $\sigma$ of answers, the demon can force an arbitrary solution to the particle-counting problem into an output sequence that $\sigma$ maps into. But the demon has no strategy for dragging an arbitrary solution through any *non-ascending* sequence, say, $(1, 0)$. For the obvious counting method will wait to see the first particle before concluding "one" and, thereafter, the demon can no longer trick it into thinking that there are no particles, since the particle has already been presented. That is a fundamental asymmetry in the problem.

More generally, if $\sigma$ is a finite, non-repetitive sequence of answers, then the *$\sigma$-avoidance game* for a problem is won by the scientist just in case the demon fails to present an appropriate information sequence or the scientist wins the truth-finding game *and* fails to produce a sequence of conjectures as bad as $\sigma$. The demon wins if he presents appropriate information that makes the scientist lose the truth-finding game *or* that somehow lures the scientist into producing an output sequence as bad as $\sigma$. When the demon has a winning strategy in the $\sigma$-avoidance game, one may say that the demon can *force* $\sigma$ from an arbitrary solution to the problem. For example, it was shown that the demon has a winning strategy in the $(0, 1, 2, \ldots, n)$-avoidance game in the particle-counting problem, since every method can be induced to produce that output sequence (or a sequence that is at least as bad). Then say that $\sigma$ is *demonic* in a problem just in case the demon can force it in the problem.

The demonic sequences in a problem reflect a deep relationship between the question $\mathcal{Q}$ and the underlying topology $\mathcal{V}$. The ability of the demon to force demonic sequence $(0, 1, 2, \ldots, n)$ implies that there is a zero particle world that is a limit point of one particle worlds each of which is a limit point of two particle worlds and so forth. So demonic sequences represent iterated problems of induction within the overall problem (Figure 36). According to

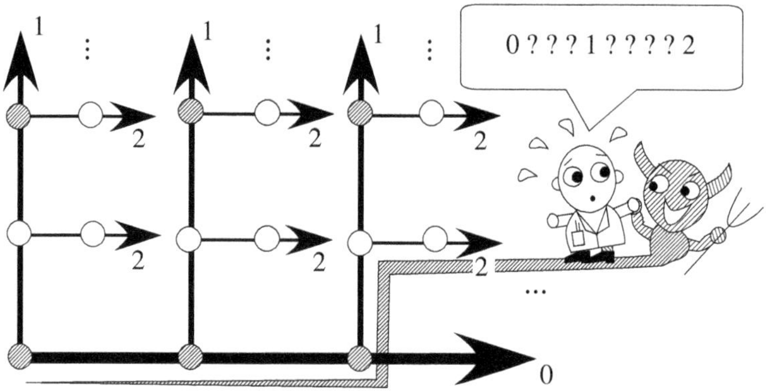


Figure 36. Demonic Sequence in the Particle-Counting Problem.

intuition, simpler answers are associated with the most deeply embedded problems of induction, for starting from 0, the demon can drag a solution through every ascending sequence, but after presenting some particles, he can never drag the counting solution back to 0. That suggests a natural definition of empirical simplicity. If $A$ is a potential answer, then say that the *A-sequences* are the demonic sequences starting with $A$. Say that answer $A$ is *as simple as* $B$ just in case the set of $B$-sequences is as good as the set of $A$-sequences and that $A$ is *simpler than* $B$ just in case the set of $B$-sequences is better than the set of $A$-sequences. This definition agrees with intuition in the counting problem and, hence, in parameter-freeing problems of the usual sort, such as curve-fitting.

The proposed explication of simplicity has a striking, intuitive advantage over the familiar idea that simplicity has something to do with dimensionality of an answer in a continuous parameter space. For if you were to learn that the true parameter values are rational, then the topological dimension of each answer would drop to zero, flattening all simplicity distinctions. But since the rational-valued subspace of the parameter space preserves demonic sequences, it also preserves simplicity in the proposed sense.

Now say, quite naturally, that an answer $A$ is *Ockham* in a problem just in case $A$ has an information sequence and is as simple in the problem as any other answer. This is a way of saying that $A$ is the simplest answer compatible with experience, where compatibility with experience includes the assumption that complete evidence can be presented at all; else there is no use debating whether simplicity helps one find the truth. Ockham's razor is then: *never say an answer unless it is Ockham for the current subproblem.* Finally, a solution is *Ockham* if it solves the problem and always heeds Ockham's razor. "The" Ockham answer is typically unique, but not always. If there is no Ockham answer, an Ockham method must suspend judgment. If there is more than one, an Ockham method may produce any one of them. It may sound odd to allow for an arbitrary choice among Ockham answers,[18] but keep in mind that two hypotheses could be maximally simple (no other answer's demonic sequences are *worse*) without being Ockham (every other answer's demonic sequences map in). The truly objectionable choices turn out to be among maximally simple answers that are not Ockham, as will be explained below.

Here is a handy re-formulation of the Ockham answer concept, where $*$ denotes concatenation. The proofs of all propositions are presented in the Appendix.

**Proposition 1 (Ockham characterization).** If the problem is solvable, $A$ is Ockham if and only if for every demonic sequence $\sigma$ for the problem, $A * \sigma$ is demonic for the problem.

*Proof.* Suppose that $A$ is Ockham. Let $\sigma$ be an arbitrary, demonic sequence. If $\sigma$ is empty, then trivially $A * \sigma$ is demonic, since there is an information sequence for some world in $A$ and a solution must converge to $A$ on this

[18] The theory presented in [Kel07] does not have this property.

sequence. So suppose that $\sigma = B * \gamma$. Since $A$ is Ockham, there exists demonic $A * \tau$ such that $B * \gamma \leq A * \tau$, in virtue of some mapping $\varphi$. If $\varphi(0) = 0$, then $B = A$, so:

$$A * \tau \geq A * \gamma \geq A * A * \gamma = A * (B * \gamma) = A * \sigma.$$

So since $A * \tau$ is demonic, so is $A * \sigma$. If $\varphi(0) > 0$, then since $B * \gamma \leq A * \tau$ and $B \neq A$, it follows that $\sigma = B * \gamma \leq \tau$. Hence: $A * \tau \geq A * \sigma$. So since $A * \tau$ is demonic, so is $A * \sigma$.

Conversely, suppose that for each demonic sequence $\sigma$, $A * \sigma$ is demonic. The empty sequence () is trivially demonic, so $A * () = (A)$ is demonic, so some $\delta \geq (A)$ is forcible, so *there is some information sequence whose intersection is $A$.* For suppose otherwise. Since the problem is solvable, let a solution be given. Since there is no information sequence for $A$, the solution remains a solution if its output is changed to "?" each time it produces $A$. Hence, no sequence as bad as $(A)$ is forcible. Contradiction. Finally, let $B$ be an answer and let $B * \sigma$ be demonic. Then by assumption, $A * B * \sigma$ is demonic. So the $A$ sequences are as bad as the $B$ sequences. Thus, in light of the italicized claim, $A$ is Ockham. q.e.d.

## 12 Efficiency, Games and Determinacy

Lifting the U-turn argument to the general version of Ockham's razor just defined requires a short digression into the nature of efficient solutions. A method is *as good* as a set of sequences just in case the method's set of output sequences is as good as the given set, and similarly for the other ordering relations defined above. Then it is immediate that the demonic sequences are as good as an arbitrary, efficient solution to the subproblem, since each solution can be forced to produce each demonic sequence. It is far less trivial whether an efficient solution must be as good as the set of demonic sequences. This is where Ockham's razor interfaces with recent developments in descriptive set theory (*cf.* [Kec95, p. 137–146] for a succinct development of the following material).

Say that a game is *determined* just in case one player or the other has a winning strategy and that a scientific problem is *forcing-determinate* just in case for each finite answer sequence $\sigma$, the $\sigma$-avoidance game is determined. Forcing-determinacy turns out to be a surprisingly mild restriction. Say that a problem is *typical* just in case the set $\mathcal{Q}$ of possible answers is countable and the set of possible information streams for worlds in $\Omega$ is Borel. Then the following proposition is an easy consequence of Donald Martin's Borel determinacy theorem.[19]

[19] *Cf.* [Mar75].

**Proposition 2.** Typical, solvable problems are forcing-determinate.

*Proof.* Let $(\Omega, \mathcal{V}, \mathcal{Q})$ be a solvable problem, let $\varepsilon$ be an $\omega$-sequence of open sets, let $\gamma$ be an $\omega$-sequence of answers and let $\sigma$ be a finite sequence of answers. The pair $(\varepsilon, \gamma)$ wins for the demon in the $\sigma$-avoidance game in $(\Omega, \mathcal{V}, \mathcal{Q})$ if and only if (i) $\varepsilon$ is an information sequence for some answer in $\mathcal{Q}$ and either (ii) there exists answer $A$ in $\mathcal{Q}$ such that $\varepsilon$ is for $A$ and $\gamma$ does not converge to $A$ or (iii) $\gamma$ is as bad as $\sigma$. Condition (i) is Borel by assumption. Condition (iii) is open and, hence, Borel. In light of condition (i) and the fact that the problem has a solution $M$, condition (ii) reduces to: there exists an $A$ in $\mathcal{Q}$ such that $M$ converges to $A$ along $\varepsilon$ and $\gamma$ does not converge to $A$. Convergence is Borel and $\mathcal{Q}$ is countable, so the overall winning condition for the demon is Borel. Apply Martin's (1975) Borel determinacy theorem, which states that all such games with Borel winning conditions are determined. q.e.d.

The following results all concern solutions and, hence, are vacuously true if the problem in question is unsolvable. Therefore:

**Proposition 3.** Each of the following propositions remains true if "forcing-determinate" is replaced with "typical".

*Proof.* Immediate. q.e.d.

Now it is easy to show that:

**Proposition 4 (Efficiency Characterization).** Let the problem be forcing-determinate. An arbitrary solution is efficient if and only if it is as good as the set of demonic sequences in each subproblem.

*Proof.* Let an efficient solution $M$ to a forcing-determinate problem be given. Then in each subproblem, $M$ is as good as an arbitrary solution. Let $\sigma$ be a finite output sequence of $M$ in a given subproblem. So every solution to the subproblem produces an output sequence as bad as $\sigma$, so the scientist has no winning strategy in the $\sigma$-avoidance game. So by forcing-determinacy, the demon has a winning strategy, so $\sigma$ is demonic. So an efficient method is as good as the demonic sequences in each subproblem. Conversely, suppose that $M$ is as good as the set of demonic sequences in a given subproblem. By definition, the set of demonic sequences is as good as an arbitrary solution in the subproblem. So $M$ is as good as an arbitrary solution in the subproblem and, hence, is efficient in the subproblem. q.e.d.

So not only is an efficient solution as good as any solution, it is as good *because* it is as good as the demonic sequences, which are as good as any solution.[20]

## 13 Efficient Solutions = Ockham Solutions

Here is the main result. Ockham is indeed necessary *and* sufficient for efficiency in an extremely broad range of problems. The hypothesis of forcing-determinacy makes the proof surprisingly easy.

**Proposition 5 (Ockham Equivalence Theorem).** Let the problem be forcing-determinate. Then the efficient solutions are exactly the Ockham solutions.

*Proof.* Let a forcing-determinate problem be given. For the necessity argument, suppose that solution $M$ violates Ockham's razor upon entering some subproblem by producing non-Ockham answer $A$. Let $D$ be the set of demonic sequences for the subproblem. Since $A$ is not Ockham and $M$ is a solution, there exists (by Proposition 1) a demonic sequence $\sigma$ in the subproblem such that $A * \sigma$ does not map into any demonic sequence. Hence, $M \not\leq D$. So by Proposition 4, $M$ is not efficient.

For sufficiency, it suffices to argue that *every finite sequence of Ockham answers encountered in subproblems successively reached as experience increases maps into some demonic sequence in the first subproblem.* For then an Ockham solution, which produces only sequences of Ockham answers interspersed with question marks, is as good as the demonic sequences in an arbitrary subproblem and, hence, is efficient, since the demonic sequences are, by definition, as good as an arbitrary solution. In the base case, each Ockham answer $A$ in a subproblem has an information sequence for it, so the singleton sequence $(A)$ can be forced by the demon in the subproblem and, hence, is demonic in the subproblem. Now consider a finite, downward-nested sequence of non-empty open sets $(E_0, \ldots, E_{n+1})$ determining respective sub-problems with respective Ockham answers $(A_0, \ldots, A_{n+1})$. By the induction hypothesis, $(A_1, \ldots, A_{n+1})$ is demonic in $P_1$. Furthermore, since $E_1$ is a non-empty subset of $E_0$, whatever the demon can force in $P_1$ he can force in $P_0$, so $(A_1, \ldots, A_{n+1})$ is demonic in $P_0$. So since $A_0$ is Ockham in $P_0$, $(A_0, A_1, \ldots, A_{n+1})$ is demonic in $P_0$, which proves the italicized claim and, hence, the theorem. q.e.d.

More can be shown for the particle-counting problem and for others of its attractive kind. For such problems have the special feature that in each

[20] Such a result is called a *universal factorization* [Mac71, p. 1–2].

subproblem, if $A$ is an Ockham violation upon entering the subproblem, then there exists an Ockham answer $U$ upon entering the subproblem such that the binary sequence $A * U$ is not as good as any demonic sequence for the subproblem. Say that such problems are *stacked.*[21] Examples of non-stacked problems illustrate intuitive ideas about empirical symmetry and will be considered in the next section. The result is:

**Proposition 6 (Strong Ockham Necessity for Stacked Problems).** In a stacked, forcing-determinate problem, if a solution violates Ockham's razor upon entering a sub-problem, the solution is stongly worse than each efficient solution in the same sub-problem.

*Proof.* Consider an efficient solution to a stacked, forcing-determinate problem and suppose that $M$ solves a given subproblem but violates Ockham's razor upon entering it by producing $A$. Let $U$ be the Ockham answer promised by the stacking property. Then since $M$ already says $A$ and $U$ is compatible with the current subproblem, the demon can force $M$ to produce $U$ after producing $A$. (That is the initial U-turn resulting from the Ockham violation). Consider an arbitrary, finite output sequence $\tau$ of the efficient solution. Then for some demonic $\delta$, $\tau \leq \delta$ (Proposition 4) and, hence, $\tau \leq A * U * \delta$. Since $U$ is Ockham and $\delta$ is demonic, $U * \delta$ is demonic (by Proposition 1). So since $M$ already says $A$ and $U * \delta$ is demonic, $A * U * \delta$ is as good as one of the output sequences of $M$ in the current subproblem. Of course, $\tau \leq \delta \leq A * U * \delta$, so $M$ is as bad as the efficient solution. Furthermore, $A * U$ maps into no demonic sequence, so neither does $A * U * \delta$. Since all the optimal method's output sequences map into demonic sequences, it follows that $A * U * \delta$ is as good as none of the optimal method's output sequences. Hence, $M$ is *strongly* worse than the efficient solution in the current subproblem. q.e.d.

This fits closely with the spirit of the freeway example and with what is going on in particle counting and curve fitting. The property of being "stacked" can be viewed as the topological essence underlying the very strong Ockham intuitions attending such problems.

[21]To see that the particle-counting problem is stacked, suppose that $A$ is not Ockham upon seeing, say, four particles. Let $U$ be the Ockham answer "four". Then the binary sequence $A * U$ maps into no demonic sequence in the subproblem. For if $A$ posits fewer than four particles, $A$ maps into no demonic sequence since the demon can't force an arbitrary solution into a refuted answer. If $A$ posits more particles, then $(A, U)$ maps into no demonic sequence since all such sequences are ascending.

## 14 Testing as an Instance of Ockham's Razor

Suppose that you want to decide whether some theory is true or not. That poses a binary question: the theory vs. the theory's denial. Typically, the theory is refutable and, hence, closed. Everyone chooses the refutable (*e.g.*, point) hypothesis as the null hypothesis and its denial as the alternative. On the proposed account of simplicity, the decision to accept the refutable hypothesis until it is refuted is an instance of Ockham's razor and is underwritten by the U-turn argument, so that the proposed account of efficient theory choice subsumes this aspect of testing practice as a special case.

First, observe that the demon can force you to conclude the refutable hypothesis $H$ (by showing you a boundary point in the hypothesis, since closed sets contain all of their boundary points). Then he can show you data refuting the theory. So only $(H, \neg H)$ and its subsequences are demonic. Hence, only $H$ is Ockham (Proposition 1), so (by Proposition 5) every efficient solution says $H$ (or suspends) until $H$ is refuted, which reflects practice. Finally, that practice is efficient (since its output sequences are all demonic), so Ockham's razor bites and you should heed his advice.

The trouble with standard conceptions of hypothesis testing is that they ignore the possibility of extra mind-changes. Yes, it is refutable to say that the bivariate mean of a normal distribution is precisely $(0, 0)$, since $\{(0, 0)\}$ is closed (and hence refutable) in the underlying parameter space. But what if you want to test the non-refutable and non-verifiable hypothesis that exactly one component of the mean is zero? Solving this binary question requires multiple mind-changes, as in particle-counting and other model selection problems. For the demon can make it appear that both components are zero until you probably say "no" (as sample size increases) and can then reveal deviation of one component from zero until you probably say "yes" and then can reveal deviation of the other component from zero until you probably say "no" again, for a total of two mind-changes (in probability). Essentially, you are just counting deviations of mean components from zero as you were counting particles before. So the demonic sequences are all the sequences that map into "yes, no, yes", so the obvious method of counting nonzero mean components is efficient and the unique Ockham hypothesis at each stage is the one that agrees with the current nonzero mean count. So you should heed Ockham's advice, (as you naturally would in this example).

Since testing theory usually isn't applied until all the parameters in a model are fixed by point estimates, it appears as though a testing theory for refutable (closed) hypotheses is good enough. Hence, the essential involvement of Ockham's razor in testing theory is missed and so the strong analogy between model selection and testing with multiple mind-changes is missed as well. The proposed account of Ockham's razor, therefore, suggests a new, more unified foundation for classical statistics, whose development lies beyond the scope of this explorative essay.

## 15 Ockham and Respect for Symmetry

When there are two maximally simple answers compatible with the data, Ockham can't help you decide among them and the strong intuition is to wait for nature to "break the symmetry" prior to choosing. For example, modify the particle-counting problem so that particles come in two colors, green and blue and you have to specify the total number of each that will ever be emitted. Assume also that you can hear particles rattle down the faucet before they emerge from the machine (Figure 37). Having seen no

Figure 37. Breaking Symmetry.

particles, you hear the first one coming. What do you conclude? It is hard to say, for both colors of marbles will stop appearing, eventually, so there is no general "pattern" to detect in the data, and there is no obvious primacy of one color over the other so far as the underlying problem is concerned. This is not mere skepticism, since after the next marble is observed, you will eventually have to leap to the bold Ockham hypothesis that no more particles are coming. Instead, it is respect for symmetry, one of the strongest intuitions in science since Greek times. That leads to an intriguing idea. Perhaps the U-turn argument also explains our strong hesitance to break symmetries in experience. Then respect for symmetry would simply be Ockham's razor conforming itself to the structure of symmetrical problems. That is correct.

Consider how Ockham's razor applies to the case at hand. When you hear the rattling that announces the first particle, you have entered a new subproblem. There are two equally simple answers at that point, "one green, zero blue" or symmetrically "zero green, one blue". But *neither* of these answers is Ockham. For each answer constitutes a unit demonic sequence, but neither binary sequence consisting of the two symmetrical competitors is demonic in the subproblem, since the demon can't take back the first particle after its color is observed. So Ockham demands silence, and we already know from Proposition 5 that every efficient solution to the problem must heed this advice. Is there an efficient solution? Indeed, just heed Ockham's advice by counting the total number of particles whose colors are seen and by suspending judgment when the next rattle is heard. Respect for symmetry follows from Ockham's razor.

The symmetrical problem under discussion is a nice example of a non-stacked problem. For consider the answer "zero green, one blue". There is no Ockham answer one can concatenate to this answer in the subproblem entered with the first rattle because there is no Ockham answer at all. And the violator is not strongly worse than the Ockham method just described in that subproblem, because the demon can force even an optimal method to say the same answer the violator chose in advance and the violator produces no output sequence worse than that.

The same argument works even after a run of a thousand exclusively green particles, in which case it might be objected that past experience does break symmetry between blue and green. But the subproblem so entered is topologically equivalent to the original problem prior to seeing any marbles. Hence, no non-circular, efficiency-based account of Ockham's razor could possibly explain why it is more efficient to say "green" rather than "blue" upon entering the subproblem.

In the preceding problem, the counting question slices the problem into topologically invariant simplicity degrees corresponding to particle counts in spite of occasional symmetries (*e.g.*, when the particle rattles and has not yet been seen). In other problems, symmetry is so pervasive that Ockham's razor doesn't bite at all (Figure 38). For example, suppose you have

Figure 38. Overly Symmetrical Problems.

to report not only how many particles will appear, but when each one will appear (forgetting about color). It might seem, at first, that the simplest answer is to say that you have seen all the particles already and that they appear exactly when they were observed to, since if you were asked only how many particles there are, you would only be permitted to say the number seen so far. That is so, if you choose to conceive of the sequence identification problem as a refinement of the particle counting problem. The trouble is that the sequence identification problem also refines alternative problems that lead to incompatible simplicity judgments. For example, *oneicles* are non-particles up to stage one and particles thereafter. There are finitely many particles if and only if there are finitely many oneicles, so the underlying space $\Omega$ is unaltered by the translation. But the answers to the two counting problems are different and the U-turn argument leads to correspondingly different recommendations (*i.e.*, to count particles or to count onecles, respectively). Since the sequence identification problem refines the problems of counting particles, oneticles, twoticles, threeticles, *etc.*, it can't consistently favor one kind of counting over another without making a global, symmetry-breaking choice in favor of one of its possible coarsenings. The only sensible resolution of this Babel of alternative coarsenings is for Ockham to hold his tongue.

And that's just what the proposed theory says. First of all, no answer is Ockham in this problem, since every demonic sequence is of unit length. For consider a single answer. The answer is true in just one world, which the demon can present until you take the bait. So each unit sequence of answers can be forced. For each alternative answer (satisfied by an alternative world), there is a least stage by which the two cease agreeing and diverge. But some solution refuses to be convinced of the first answer (on pain of converging to the wrong answer) until the divergence point is already passed (Figure 39). So the demon can force no binary sequence of answers

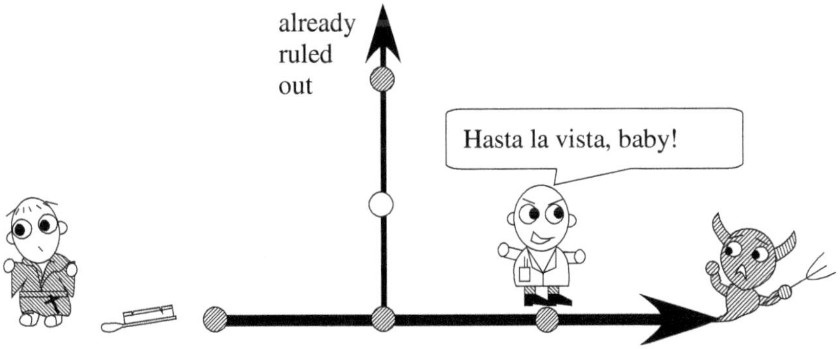


Figure 39. The Trouble With Singleton Answers.

from an arbitrary solution. Hence (Proposition 4), there can be no efficient solution, since no solution to this problem succeeds without mind-changes.

So there are lots of solutions to this problem, but no efficient ones. Hence, even if there were an Ockham answer, there would be no efficient method to put normative teeth into the U-turn argument! Ockham is both mute and toothless in this problem.

Again, that is the correct answer. The sequence-identification problem is completely symmetrical in the sense that any homeomorphism of the space into itself results in the very same problem (since each permuted world still ends up in a singleton answer over the same topological space). So there is no objective, structural sense in which one answer is simpler than another any more than there is any objective physical sense about where zero degrees longitude is. Coordinate systems are not physically real because they aren't preserved under physical symmetries; philosophical notions of simplicity (*e.g.*, brevity, sequential uniformity, entrenchment) are not real because they aren't preserved under problem symmetries. To seek objective, truth-finding efficiency in distinctions that really aren't in the problem is like trying to generate electricity by formally spinning coordinate axes. The situation is different in the counting problem. There exist homeomorphisms of the underlying topological space that materially alter the original problem (*e.g.*, the unique Ockham hypothesis "no particles" would become "no oneicles", which means "one particle at stage 1"). It is precisely this lack of symmetry in the particle-counting problem that allows Ockham to slice it into objective simplicity degrees.

The usual attempts to use coding, "entrenchment", or prior probability to force a foliation of the sequence identification problem into simplicity degrees must involve the imposition of extraneous considerations lacking in the problem's intrinsic structure as presented. Therefore, such considerations couldn't possibly have anything objective to do with solving the problem (as stated) efficiently. So the proposed account yields precisely the right judgment in this example when its true nature is properly understood.

One can also arrive at overly-symmetrical problems by coarsening the particle-counting problem. For example, consider the question whether there is an even or an odd number of particles. Since this coarsens the particle-counting problem, one again expects "even" to be the Ockham answer when an even number of particles have been observed and "odd" to be the right answer otherwise (Figure 40). But the proposed theory of Ockham's razor doesn't agree. Ockham is once again silenced, but this time the difficulty is exactly reversed: every solution is efficient and every answer is Ockham in every subproblem so every method satisfies Ockham's razor and the U-turn argument can't even begin (Figure 41).[22]

[22]In the theory presented in [Kel07], there is no Ockham solution to this problem. Either way, Ockham refuses to choose among potential solutions to the problem.

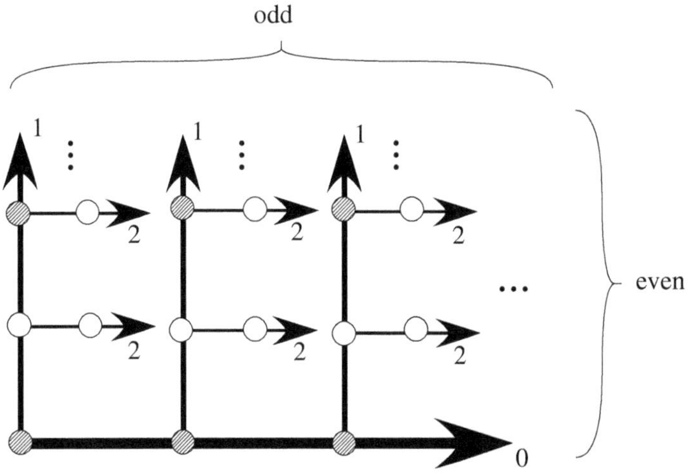


Figure 40. Even/Odd as Particle Counting.

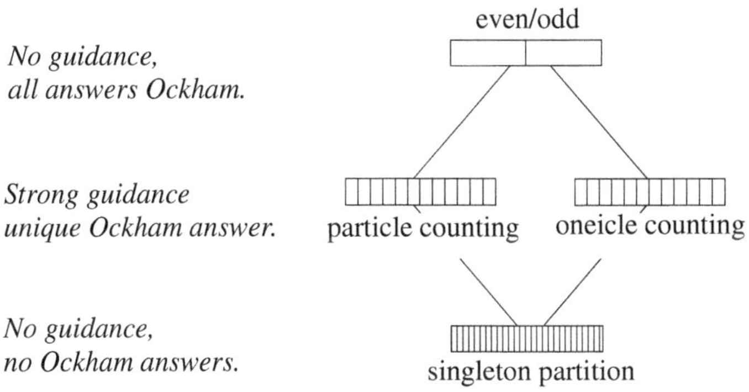


Figure 41. Ockham Under Refinement.

The theory is right. Yes, if one thinks of the problem as a coarsening of particle counting, "even" must come first. But one could also think of it as a coarsening of counting oneicles instead of particles. Then the zero oneicle world is an "odd" world. The one oneicle worlds include the zero particle world as well as all the two particle worlds in which the first appears right away. These are all "even particle" worlds. Continuing in this way one obtains a oneicle-counting simplicity foliation (Figure 42) in

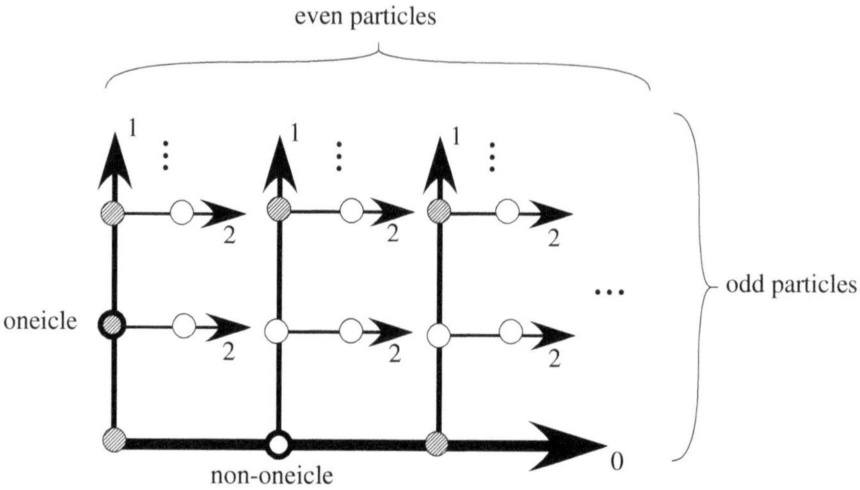


Figure 42. Even/Odd as Oneicle Counting.

which the obvious "first" conjecture is "odd". But the oneicle translation is a homeomorphism of the space that reflects each answer onto the other, so a prior preference for "even" couldn't have anything to do with the objective efficiency of solutions to the even/odd problem as stated.

The urge to extend Ockham's advice to symmetrical problems is understandable — guidance is most precious when there is none. And even in light of the proposed account, nothing prevents us from saying that we are really interested in counting marbles rather than merely saying whether they are even or odd, in which case the problem is no longer symmetrical. But it is quite another matter to smuggle extra structure into a symmetrical problem without acknowledging that one has done so, for such practice is not warranted by truth-finding efficiency in the problem addressed (Figure 43).

## 16 Conclusion: Ockham's Family Secret

Ockham is beloved as an inexhaustible source of free information that somehow parlays the scientist's limited viewpoint into sweeping generalizations about unseen realities (Figure 44). But his very appeal is his undoing, for it is impossible to explain how his fixed advice could be true without assuming exactly what we rely upon him to tell us.

Figure 43. Theft Over Honest Toil.

Figure 44. Ockham's Day Job.

This paper presents an alternative view, according to which Ockham helps us find the truth, but in an unexpected way. He doesn't provide any guarantee that the theory he selects is true or probably true. He doesn't point at the truth. He can't even bound the number of future surprises or U-turns you will have to make in the future on your way to the truth. All he does is save you the trouble of *needless* surprises beyond those arbitrarily many surprises nature is objectively in a position to exact from you. But in that respect, his advice is still uniquely the best.

The proposed explanation is unnerving because it singles out simplicity as the right bias to have, but falls so far short of our craving for certainty, verification, and guarantees against future surprises. That is far harder to dismiss than the usual, academic sort of skepticism, which finds no connection between simplicity and truth and urges rejection of simplicity-based conclusions altogether.

It is also ironic that Ockham is viewed as a comforting figure when, in fact, he is built out of the inductive demon's opportunities to successively force science to reverse course. Indeed, Ockham and the demon work together as a coordinated team, since Ockham changes his recommendations each time the demon uses up one of his opportunities to fool the scientist (Figure 45).

Figure 45. But by Night...

The key to understanding Ockham's razor is to set aside our instinctive appetite for certainty and to focus squarely on the objective complexity properties of empirical problems that underly unavoidable reversals of scientific opinion through time. A similar focus on problem complexity has long been the norm in the mathematical theories of computability, computational complexity, and descriptive set theory. In these established, scientific subjects, nobody would dream of "victory" over complexity. It is late in the day for the philosophy of science and induction to be dreaming still.

## References.

[Car50] Rudolf **Carnap**, Logical Foundations of Probability, University of Chicago Press 1950

[Ear83] John **Earman** (*ed.*), Testing Scientific Theories, Universitiy of Minnesota Press 1983

[ForSob94] Malcolm R. **Forster** and and Elliott **Sober**, How to Tell When Simpler, More Unified, or Less Ad Hoc Theories will Provide More Accurate Predictions, **The British Journal for the Philosophy of Science** 45 (1994), p. 1–35

[Fri83] Michael **Friedman**, Foundations of Space-Time Theories, Princeton University Press 1983

[FriGoeHar07] Michèle **Friend**, Norma B. **Goethe**, and Valentina S. **Harizanov** (*eds.*), Introduction to the Philosophy and Mathematics of Algorithmic Learning Theory, Springer 2007 [Handbook of Philosophical Logic 9]

[Ger75] Carl J. **Gerhardt** Die Philosophischen Schriften von G. W. Leibniz, Volume IV, Weidmann 1875

[Gly80] Clark N. **Glymour**, Theory and Evidence, Princeton University Press 1980

[GlyKel04] Clark N. **Glymour** and Kevin T. **Kelly**, Why Probability Does Not Capture the Logic of Scientific Justification, *in:* [Hit04, p. 94–114]

[GlySchSpi00] Clark N. **Glymour**, Richard **Scheines**, and Peter **Spirtes**, Causation, Prediction, and Search, 2nd edition, MIT Press 2000

[Goo83] Nelson **Goodman**, Fact, Fiction, and Forecast, 4th edition, Harvard University Press 1983

[Hal74] Paul R. **Halmos**, Measure Theory, Springer 1974

[Hit04] Christopher **Hitchcock** (*ed.*), Contemporary Debates in the Philosophy of Science, Blackwell 2004

[Jai+99] Sanjay **Jain**, Daniel **Osherson**, James S. **Royer**, and Arun **Sharma**, Systems That Learn: An Introduction to Learning Theory, MIT Press 1999

[Kec95] Alexander S. **Kechris**, Classical Descriptive Set Theory, Springer 1995 [Graduate Texts in Mathematics 156]

[Kel96] Kevin T. **Kelly**, The Logic of Reliable Inquiry, Oxford 1996

[Kel02] Kevin T. **Kelly**, Efficient Convergence Implies Ockham's Razor, Article for the 2002 International Workshop on Computational Models of Scientific Reasoning and Applications, Las Vegas, USA, June 24-27, 2002

[Kel04] Kevin T. **Kelly**, Justification as Truth-finding Efficiency: How Ockham's Razor Works, **Minds and Machines** 14 (2004), p. 485–505

[Kel07] Kevin T. **Kelly**, How Ockham's Razor Helps You Find the Truth — Without Pointing at It, *in:* [FriGoeHar07, p. 111–143]

[Kuh62] Thomas **Kuhn**, The Structure of Scientific Revolutions, University of Chicago Press 1962

[Lau81] Larry **Laudan**, A Confutation of Convergent Realism, **Philosophy of Science** 48 (1981), p. 19–48

[Lei14] Gottfried W. **Leibniz**, Monadologie, 1714, *in:* [Ger75, p. 607–23]

[LiVit97] Ming **Li** and Paul **Vitanyi**, An Introduction to Kolmogorov Complexity and Its Applications, Springer 1997

[LiVit00] Ming **Li** and Paul **Vitanyi**, Minimum Description Length Induction, Bayeisanism, and Kolmogorov Complexity, **IEEE Transactions on Information Theory** 46 (2000), p. 446–464

[LohZhe95] Wei-Yin **Loh** and Xiaodong **Zheng**, Consistent Variable Selection in Linear Models, **Journal of the American Statistical Association** 90 (1995), p. 151–156

[Mac71] Saunders **MacLane**, Categories for the Working Mathematician, Springer 1971

[Mar75] Donald A. **Martin**, Borel determinacy, **Annals of Mathematics** 102 (1975), p. 363–371

[Mit97] Tom **Mitchell**, Machine Learning, McGraw-Hill 1997

[NatWin08] Paul **Natorp**, Wilhelm **Windelband** (*eds.*), Kant's gesammelte Schriften, Die Akademie-Ausgabe, Abteilung I: Werke, Band V: Kritik der praktischen Vernunft, 1788, Kritik der Urtheilskraft, 1793, Berlin-Brandenburgische Akademie der Wissenschaften 1908

[NieSch82] Mogens **Nielsen** and Erik M. **Schmidt** (*eds.*), Proceedings of the 9th Colloquium on Automata, Languages and Programming, Aarhus, Denmark, July 12-16, 1982, Springer 1982 [Lecture Notes in Computer Science 140]

[Pop68] Karl **Popper**, The Logic of Scientific Discovery, Harper 1968

[Put65] Harry **Putnam**, Trial and Error Predicates and a Solution to a Problem of Mostowski, **Journal of Symbolic Logic** 30 (1965), p. 49–57

[Qui69] Willard Van Orman **Quine**, Natural Kinds, *in:* [Res69, p. 5–23]

[Res69] Nicholas **Rescher** (*ed.*), Essays in Honor of Carl D. Hempel, A Tribute on the Occasion of His Sixty-Fifth Birthday, Reidel 1969

[Ros83] Roger **Rosenkrantz**, Why Glymour is a Bayesian, *in:* [Ear83, p. 69–98]

[Sch99a] Oliver **Schulte**, The Logic of Reliable and Efficient Inquiry, **The Journal of Philosophical Logic** 28 (1999), p. 399–438

[Sch99b] Oliver **Schulte**, Means-Ends Epistemology, **The British Journal for the Philosophy of Science** 50 (1999), p. 1–31

[Sch01] Oliver **Schulte**, Inferring Conservation Laws in Particle Physics: A Case Study in the Problem of Induction, **The British Journal for the Philosophy of Science** 51 (2001), p. 771–806

[Sco82] Dana S. **Scott**, Domains for Denotational Semantics *in:* [NieSch82, p. 577–613]

[Was04] Larry **Wasserman**, All of Statistics: A Concise Course in Statistical Inference, Springer 2004

**Received**: April 12th, 2005;
**In revised version**: December 2nd, 2005;
**Accepted by the editors**: June 6th, 2007.

FotFS V

Stefan **Bold**, Benedikt **Löwe**,
Thoralf **Räsch**, Johan **van Benthem** (*eds.*)
**Foundations of the Formal Sciences V**
Infinite Games

# Random Strategies with Historical Memory for the Robin Hood Game

BOAZ TSABAN

Department of Mathematics
Bar-Ilan University
Ramat-Gan 52900, Israel

Department of Mathematics
Weizmann Institute of Science
Rehovot 76100, Israel

tsaban@math.biu.ac.il

ABSTRACT. The *Robin Hood* game is played as follows: On day $i$, the Sheriff puts $s(i)$ bags of gold in the cave. On night $i$, Robin removes $r(i)$ bags from the cave. The game is played for each $i \in \mathbb{N}$. Robin wins if each bag which was put in the cave is eventually removed from it; otherwise the Sheriff wins.

Gasarch, Golub, and Srinivasan studied the Robin Hood game in the case of random strategies where Robin has no historical memory. We extend their main result to the case of bounded historical memory, and obtain a hierarchy of provably distinct games.

## 1 The Robin Hood game

The *Robin Hood* game $\mathrm{RH}(r, s, A)$ is defined for functions $r, s : \mathbb{N} \to \mathbb{N}$ such that $1 \leq r(i) < s(i)$ for each $i$, and for a set $A$, as follows:

1. On day $i$, the *Sheriff* (of Nottingham) puts $s(i)$ bags of gold in the cave, each labelled by an element of $A$. No label is used twice (over the course of the entire game).
2. On night $i$, *Robin* (Hood) removes $r(i)$ bags from the cave.

The game is played for each $i \in \mathbb{N}$. Robin wins if each bag which was put in the cave is eventually removed from it; otherwise the Sheriff wins.

It is easy to see that if Robin has an unlimited historical memory (knowing at each night $i$ which of the bags in the cave appeared first), then he has a winning strategy: On night $i$ pick $r(i)$ bags out of those which arrived first.

Deterministic strategies for this game were studied, from the set-theoretic point of view, in [Sch94, SchWei97]. Gasarch, Golub and Srinivasan [GasGolSri03] consider the case where Robin has *no historical memory*, that is, he cannot distinguish between the days where the bags were put in the cave. They suggest the following *probabilistic* strategy for Robin: On night $i$, remove random $r(i)$ bags out of the cave (with uniform distribution). They say that Robin wins *almost surely* if for each bag put in the cave, its probability of being eventually removed is 1. The probability is taken over Robin's coin tosses. More precisely, the probability that a bag is not removed is the product of all probabilities $p_i$ that the bag is not removed at night $i$, and the Sheriff's winning (or Robin's losing) probability is the supremum of all probabilities $p_x$ that a bag $x$ put in the cave is never removed.[1] Let $L(i) = \sum_{j=1}^{i} s(j) - r(j)$ denote the number of bags in the cave after night $i$. The main result in [GasGolSri03] is that Robin wins almost surely if, and only if, the series

$$\sum_{i=1}^{\infty} \frac{r(i)}{L(i) + r(i)}$$

diverges; otherwise Robin loses almost surely.

## 2 Strategies with bounded historical memory

We generalize the above mentioned result to the case of bounded historical memory. The typical case is that Robin can, on each night, identify the bags put in the cave on the last $k$ days, where $k$ is constant. It will turn out that already the natural strategy for historical memory $k = 1$ is strictly stronger than the natural strategy in the memoryless case. Moreover $k = 2$ yields a strictly stronger strategy than $k = 1$, *etc.* (Theorem 5). In fact, the game can be analyzed in a much broader family of cases, as will be shown in the sequel.

The most general case is that Robin can, on night $i$, identify the bags put in the cave on the last $b(i)$ days, where $b : \mathbb{N} \to \mathbb{N} \cup \{0\}$ is a function with $b(i) \leq i$ for all $i$.[2] (So that $b(i) \equiv 0$ is the memoryless case studied in [GasGolSri03]). It is natural to denote this game by $\mathrm{RH}(r, s, b, A)$, but our analysis below is independent of the set $A$, so we will simply write

[1] Our definition of the Sheriff's winning probability is simpler than the one given in [GasGolSri03], but both our and the proofs of [GasGolSri03] work for both definitions and actually imply that the definitions are equivalent.

[2] The notation $b(i)$ stands for the *bound* on Robin's historical memory on night $i$.

RH$(r, s, b)$. A key observation is that the following natural restriction leads to a substantial simplification of the analysis of the generalized games.

**Restriction 1.** We pose the restriction that Robin cannot remember anything that he forgot earlier, that is, $b(i+1) \leq b(i)+1$ for each $i$; equivalently, the function $i - b(i)$ is nondecreasing.

We suggest the following deterministic and random strategies for Robin, motivated by the strategy given in [GasGolSri03] for the memoryless case: Call a bag *very old* if Robin cannot tell the day it was put in the cave. An important observation is that Robin can identify the very old bags since he can identify the bags which are *not* very old.

$\mathsf{Oldest}_{\mathrm{DET}}$: On night $i$ Robin chooses any $r(i)$ many bags out of the very old bags. If there are less than $r(i)$ many very old bags, then Robin also chooses some of the oldest bags among the ones he remembers, so as to choose $r(i)$ bags in total.[3]

$\mathsf{Oldest}_{\mathrm{RND}}$: Same as $\mathsf{Oldest}_{\mathrm{DET}}$, but the $r(i)$ bags are chosen at random, with uniform probability, out of the older bags.[4]

Observe that if $\mathsf{Oldest}_{\mathrm{DET}}$ is a winning strategy for Robin, then so is $\mathsf{Oldest}_{\mathrm{RND}}$. Write

$$\tilde{L}(i) = \max\left\{0, \sum_{j=1}^{i-b(i)} s(j) - \sum_{j=1}^{i-1} r(j)\right\}$$

Then $\tilde{L}(i)$ is the number of bags put in the cave on days $1, 2, \ldots, i - b(i)$ and not removed until day $i$.

**Proposition 2.**

1. If $i - b(i)$ is bounded, then $\mathsf{Oldest}_{\mathrm{DET}}$ is a winning strategy for Robin in RH$(r, s, b)$ (for each $r$, and $s$).
2. If there exist infinitely many $i$ such that $\tilde{L}(i) \leq r(i)$, then $\mathsf{Oldest}_{\mathrm{DET}}$ is a winning strategy in RH$(r, s, b)$.

---

[3] To put this more precisely, on night $i$ Robin has a partition of the bags in the cave into disjoint (possibly empty) sets $S_0, S_1, \ldots, S_{r(i)}$ such that $S_0$ is the set of very old bags, and for $k = 1, \ldots, r(i)$, $S_k$ is the set of bags put in the cave $r(i) - k$ days ago. If $|S_0| \geq r(i)$, then Robin chooses any $r(i)$ many bags out of the bags in $S_0$. Otherwise, let $m$ be the minimal such that $r(i) < \sum_{k=0}^{m} |S_k|$. Then Robin takes all bags in the sets $S_0, \ldots, S_{m-1}$, as well as $r(i) - \sum_{k=0}^{m-1} |S_k|$ many bags from $S_m$.

[4] Using the notation of the previous footnote, If $|S_0| \geq r(i)$, then Robin chooses at random (with uniform probability) $r(i)$ many bags out of the bags in $S_0$. Otherwise Robin takes all bags in the sets $S_0, \ldots, S_{m-1}$, as well as $r(i) - \sum_{k=0}^{m-1} |S_k|$ many random bags from $S_m$.

*Proof.* (1) is easy. To prove (2), assume that a bag was put in the cave on day $d$. By (1) we may assume that $i - b(i)$ is unbounded. Let $i$ be such that $d < i - b(i)$. Restriction 1 ensures that this will also hold for all larger $i$'s, so we may assume further that $\tilde{L}(i) \leq r(i)$. This means that on *night* $i$, all bags put on days $\leq i - b(i)$ (in particular, those put on day $d$) were removed from the cave. q.e.d.

We may therefore make the following additional restriction.

**Restriction 3.** $\tilde{L}(i) > r(i)$ for all but finitely many $i$.

**Theorem 4.** Assume that $r$, $s$, and $b$ satisfy Restrictions 1 and 3, and Robin uses the strategy $\mathsf{Oldest}_{\mathrm{RND}}$.

1. If $\sum r(i)/\tilde{L}(i) = \infty$, then Robin wins almost surely.
2. If $\sum r(i)/\tilde{L}(i) < \infty$, then the Sheriff wins almost surely.

*Proof.* If there is $i$ such that $\tilde{L}(i) \leq r(i)$, let $i^*$ be the maximal such $i$. Then on night $i^*$, all bags put in the first $i^* - b(i^*)$ days are removed, and $\tilde{L}(i) > r(i)$ for all $i > i^*$. Since the convergence of the series in question does not depend on the first few elements, this shows that we may assume that $\tilde{L}(i) > r(i)$ for all $i$.

Observe that if $i - b(i)$ is bounded, then $\tilde{L}(i)$ is eventually equal to 0, contradicting our assumption, thus $i - b(i)$ is unbounded. Assume that a bag was put in the cave on day $d$, then for each large enough $i$, $d < i - b(i)$ so that on night $i$, the probability that the bag in question is removed is $r(i)/\tilde{L}(i)$.

Now, the probability that a bag put in the cave on day $d$ is *not* eventually removed is

$$\prod_{i=d}^{\infty} \left(1 - \frac{r(i)}{\tilde{L}(i)}\right). \tag{1}$$

The product (1) converges to 0 (*i.e.*, Robin wins almost surely) if

$$\sum r(i)/\tilde{L}(i) = \infty.$$

If $\sum r(i)/\tilde{L}(i) < \infty$, then the product (1) is positive for $d = 1$. Thus, its limit when $d \to \infty$ is 1. Consequently, the Sheriff wins almost surely. q.e.d.

Note that in Theorem 4, (1) implies that the other direction of (2) also holds, and (2) implies that the other direction of (1) also holds. Thus, the theorem gives an exact characterization of when Robin wins almost surely and when the Sheriff wins almost surely, and shows that it is always the case that one of them wins almost surely. In the case $b(i) \equiv 0$, Theorem 4 reduces to the main result of [GasGolSri03], described at the end of Section 1.

## 3 Historical memory helps

To make sure that the generalization made in 4 is not trivial, we must find instances where additional historical memory changes the strategy's status from a losing strategy to a winning one.

Assume that $b, c : \mathbb{N} \to \mathbb{N}$. We say that $c$ *eventually dominates* $b$ if there exists $m$ such that for all $i > m$, $b(i) < c(i)$.

**Theorem 5.** Assume that $c$ eventually dominates $b$, $i - c(i)$ is unbounded, and $b$ and $c$ satisfy Restrictions 1 and 3. Then there exist functions $r, s : \mathbb{N} \to \mathbb{N}$ such that for Robin, $\mathsf{Oldest}_{\mathrm{RND}}$ is a (surely) winning strategy in $\mathrm{RH}(r, s, c)$ and an almost-surely losing strategy in $\mathrm{RH}(r, s, b)$.

*Proof.* Since convergence of series does not depend on the first few terms, we may assume that for each $i$, $b(i) < c(i)$. Since $\tilde{L}(i)$ also depends on the bounding function $b$, let us denote it here by $\tilde{L}_b(i)$. It follows that for each $r$ and $s$, $\tilde{L}_c(i) < \tilde{L}_b(i)$ for all $i$. It thus suffices to consider the case where $c(i) = b(i) + 1$ for each $i$, and therefore $\tilde{L}_c(i) + s(i - b(i)) = \tilde{L}_b(i)$.

At step $i$ of the construction we have the definition of $s$ at $1, \ldots, i - c(i)$ and $r$ at $1, \ldots, i - 1$, and therefore $\tilde{L}_c(i)$ is defined. Define $r(i) = \max\{i, \tilde{L}_c(i)\}$, and define $s$ on $i - c(i) + 1, \ldots, i + 1 - c(i+1)$ to be $r(i)^3$ (so that $r(i)/\tilde{L}_b(i) = r(i)/(\tilde{L}_c(i) + s(i - b(i))) = r(i)/(r(i) + r(i)^3) < 1/r(i)^2 \leq 1/i^2$).

Since $r(i) \geq \tilde{L}_c(i)$ for each $i$, we have by Proposition 2 that $\mathsf{Oldest}_{\mathrm{RND}}$ is a winning strategy in $\mathrm{RH}(r, s, c)$. Now, $\sum r(i)/\tilde{L}_b(i) \leq \sum 1/i^2 < \infty$, so by Theorem 4, $\mathsf{Oldest}_{\mathrm{RND}}$ is an almost-surely losing strategy in $\mathrm{RH}(r, s, b)$.

q.e.d.

In particular, we have the following.

**Corollary 6.** For each $n = 0, 1, 2, \ldots$, there exist functions $r, s : \mathbb{N} \to \mathbb{N}$ such that for Robin, $\mathsf{Oldest}_{\mathrm{RND}}$ is a (surely) winning strategy in $\mathrm{RH}(r, s, n+1)$ and an almost-surely losing strategy in $\mathrm{RH}(r, s, n)$.

## 4 Random Sheriff

The authors of [GasGolSri03] pose the question of the behavior of the Robin Hood game when the Sheriff's strategy is random as well. Our analysis in Section 2, being independent of the Sheriff's moves, shows that the results apply to this case as well. In addition to the first strategy of [GasGolSri03] which was described in the introduction to the present paper (Section 1), a second random strategy for Robin is sketched in [GasGolSri03], and is conjectured to be an almost-surely winning strategy in the game RH$(1, s, [0,1])$ where $s$ is constant.

While we are unable to analyze the second strategy for lack of some details, we can see that the first strategy already works. Here $b(i) = 0$ (no historical memory) and $L(i) = \sum_{j=1}^{i} s(j) - r(j) = \sum_{j=1}^{i} s - 1 = i(s-1)$, therefore

$$\sum_{i=1}^{\infty} \frac{r(i)}{\tilde{L}(i)} = \sum_{i=1}^{\infty} \frac{r(i)}{L(i)+r(i)} = \sum_{i=1}^{\infty} \frac{1}{L(i)+1} = \sum_{i=1}^{\infty} \frac{1}{i(s-1)} = \infty,$$

thus by Theorem 4, $\mathsf{Oldest}_{\mathrm{RND}}$ is an almost-surely winning strategy in this game. Since $\mathsf{Oldest}_{\mathrm{RND}}$ coincides with the first strategy of [GasGolSri03], we have that the first strategy of [GasGolSri03] is also an almost-surely winning strategy against a *random* Sheriff in RH$(1, s, [0,1])$ (as well as any game RH$(r, s, A)$ with $\sum r(i)/(L(i) + r(i))$ diverging). Theorem 4 is an extension of this phenomenon to the case of nonzero historical memory.

## 5 Open problems

Among the problems which naturally arise, the following two seem to be the most interesting.

**Conjecture 7.** $\mathsf{Oldest}_{\mathrm{RND}}$ is the best random strategy among the strategies which are independent of the index set $A$.

Our analysis repeatedly uses Restriction 1 posed on the bounding function $b$.

**Problem 8.** Analyze the the case where $b$ does not satisfy Restriction 1.

Finally, our strategy $\mathsf{Oldest}_{\mathrm{RND}}$ uses unboundedly many "random bits". The referee suggests the problem whether good strategies exist, in which the number of random bits used at each specific step is bounded by some constant.

## References.


[GasGolSri03] William **Gasarch**, Evan **Golub**, and Aravind **Srinivasan**, When Does a Random Robin Hood Win?, **Theoretical Computer Science** 304 (2003), p. 477–484

[Sch94] Marion **Scheepers**, Variations of a Game of Gale (II): Markov strategies, **Theoretical Computer Science** 129 (1994), p. 385–396

[SchWei97] Marion **Scheepers** and William A. R. **Weiss**, Variations of a Game of Gale (III): Remainder Strategies, **Journal of Symbolic Logic** 62 (1997), p. 1253–1264

**Received**: December 6th, 2004;
**In revised version**: May 23rd, 2005;
**Accepted by the editors**: October 16th, 2005.

Stefan **Bold**, Benedikt **Löwe**,
Thoralf **Räsch**, Johan **van Benthem** (*eds.*)
**Foundations of the Formal Sciences V**
Infinite Games

# On Infinite Ehrenfeucht-Fraïssé Games

Jouko Väänänen*

Department of Mathematics and Statistics
University of Helsinki
Gustaf Hällströmin katu 2b PL 68
00014 University of Helsinki, Finland
jouko.vaananen@helsinki.fi

ABSTRACT. This paper gives a survey of the infinite Ehrenfeucht-Fraïssé game, extending the classical approach to transfinite Ehrenfeucht-Fraïssé games and analysing their basic properties.

## 1 Introduction

The Ehrenfeucht-Fraïssé game is well-known as a versatile tool for studying elementary equivalence of structures. This paper gives a survey of the basic properties of this game with special emphasis on the infinite Ehrenfeucht-Fraïssé game. In particular, we extend the classical approach to the case of a transfinite game. By a transfinite game we mean a game with more than $\omega$ moves. This brings in new phenomena such as non-determinacy.

We commence in Section 2 with basic concepts of games of length $\alpha$. Section 3 and 4 review classical material on the Ehrenfeucht-Fraïssé game of length $\omega$. Section 5 introduces the concept of dynamic Ehrenfeucht-Fraïssé game, by which we mean a finite Ehrenfeucht-Fraïssé game in which the first player has the option of delaying the decision about the actual length of the game. Section 6 is devoted to a survey of the important relationship between elementary equivalence and the Ehrenfeucht-Fraïssé game. In Section 7 the game where players move big chunks of elements at a time is discussed, with emphasis on the main problems and shortcomings of this concept. Section 8 discusses the concept of a strong back-and-forth set, an attempt to remedy some of the weaknesses of the game of Section 7. Section 9 concentrates on

*I am indebted to Ville Nurmi for reading the manuscript carefully and making many suggestions for improvements. Research partially supported by grant 40734 of the Academy of Finland.

the concept of transfinite Ehrenfeucht-Fraïssé game. Investigation of this game has gotten underway only during the last 15 years, while the idea of the Ehrenfeucht-Fraïssé game is already 50 years old. In Section 10 basic results of partially ordered sets, necessary for the study of the dynamic transfinite Ehrenfeucht-Fraïssé game, are introduced. A systematic approach to the transfinite Ehrenfeucht-Fraïssé game is presented in Sections 11 and 12.

We give some elementary proofs in the first half of the paper in order to make the paper more accessible. Many proofs of classical non-trivial results are omitted, but there are excellent reviews of such results in the literature to which we give appropriate references. In the case of transfinite Ehrenfeucht-Fraïssé games we again give more detailed proofs as the material is more recent.

## 2 Basic Concepts of Infinite Games

If $A$ is any non-empty set, we use $A^\alpha$ to denote the set of all $\alpha$-sequences $\langle x_\xi : \xi < \alpha \rangle$ of elements of $A$. Suppose $A_0$ and $A_1$ are arbitrary sets. Let $W \subseteq (A_0 \times A_1)^\alpha$. We define the game $\mathcal{G}_\alpha(A_0, A_1, W)$ as follows. There are two players, Player I and Player II, the first is referred to as "he" and the second as "she". An $\alpha$-sequence

$$\langle x_\xi, y_\xi : \xi < \alpha \rangle = \langle \langle x_0, y_0 \rangle, \ldots, \langle x_\beta, y_\beta \rangle, \ldots \rangle_{\beta < \alpha} \tag{1}$$

of elements of $A_0 \times A_1$ is called a *play* of $\mathcal{G}_\alpha(A_0, A_1, W)$. The play (1) is a *win for* Player II if $\langle x_\xi, y_\xi : \xi < \alpha \rangle \in W$ and otherwise a *win for* Player I.

A *strategy of* Player I in the game $\mathcal{G}_\alpha(A_0, A_1, W)$ is an $\alpha$-sequence $\langle \sigma_\xi : \xi < \alpha \rangle$ of functions $\sigma_\xi : A_1^\xi \to A_0$. We say that Player I has *used the strategy* $\sigma$ *in the play* (1) if for all $\xi < \alpha$ we have $x_\xi = \sigma_\xi(\langle y_\gamma : \gamma < \xi \rangle)$. The strategy $\sigma$ of Player I is a *winning strategy*, if every play where Player I has used $\sigma$ is a win for Player I.

A *strategy of* Player II in the game $\mathcal{G}_\alpha(A_0, A_1, W)$ is an $\alpha$-sequence $\langle \tau_\xi : \xi < \alpha \rangle$ of functions $\tau_\xi : A_0^{\xi+1} \to A_1$. We say that Player II has *used the strategy* $\tau$ *in the play* (1) if for all $\xi < \alpha$ we have $y_\xi = \tau_\xi(\langle x_\gamma : \gamma \leq \xi \rangle)$. The strategy $\tau$ of Player II is a *winning strategy*, if every play where Player II has used $\tau$ is a win for Player II.

A *position* of the game $\mathcal{G}_\alpha(A_0, A_1, W)$ is any initial segment

$$p = ((x_0, y_0), \ldots, (x_\gamma, y_\gamma), \ldots)_{\gamma < \beta} \tag{2}$$

of a play (1). We say that Player I has *used the strategy* $\sigma = \langle \sigma_\xi : \xi < \alpha - \beta \rangle$ *after position* (2) *in the play* (1), if (1) extends (2) as a sequence, and for all $\gamma$ with $\beta \leq \gamma$ we have $x_\gamma = \sigma_{\gamma - \beta}(\langle y_\delta : \beta \leq \delta < \gamma \rangle)$. The strategy $\sigma$ of Player I is a *winning strategy in position* $p$, if every play extending $p$ where Player

I has used $\sigma$ after position $p$ is a win for Player I. We say that Player II has *used strategy* $\tau = \langle \tau_\xi : \xi < \alpha - \beta \rangle$ *after position* (2) *in the play* (1) if (1) extends (2) and for all $\gamma$ with $\beta \leq \gamma$ we have $y_\gamma = \tau_{\gamma-\beta}(\langle x_\delta : \beta \leq \delta \leq \gamma \rangle)$. The strategy $\tau$ of Player II is a *winning strategy in position* $p$, if every play extending $p$ where Player II has used $\tau$ after $p$ is a win for Player II.

Let $A^{<\alpha} = \bigcup_{\beta<\alpha} A^\beta$. If $W \subseteq (A_0 \cup A_1)^{<\alpha}$, the game $\mathcal{G}_{<\alpha}(A_0, A_1, W)$ is defined as $\mathcal{G}_\alpha(A_0, A_1, W)$ above.

An important example of a class of infinite games is the class of closed games. A subset $W$ of $(A_0 \times A_1)^\alpha$ is *closed*, if a play is in $W$ whenever every initial segment of the play extends to a play in $W$. A game is *closed*, if $W$ is.

**Lemma 1.** Suppose $A_0, A_1$ are sets, $W \subseteq (A_0 \times A_1)^\alpha$ and (2) is a position in the game $\mathcal{G}_\alpha(A_0, A_1, W)$. Suppose furthermore that Player I does not have a winning strategy in position (2). Then for every $x_\beta \in A_0$ there is $y_\beta \in A_1$ such that Player I does not have a winning strategy in position

$$p = \langle \langle x_0, y_0 \rangle, \ldots, \langle x_\beta, y_\beta \rangle \rangle. \tag{3}$$

*Proof.* The proof is by contradiction. Suppose there were an $x_\beta \in A_0$ such that for all $y_\beta \in A_1$ Player I has a winning strategy $\sigma^{y_\beta}$ in position (3). We define a strategy $\sigma = \langle \sigma_\xi : \xi < \alpha - \beta \rangle$ of Player I in position (2) as follows: $\sigma_0(\emptyset) = x_\beta$ and for $\gamma > \beta$, $\sigma_{\gamma-\beta}(y_\xi \ : \ \beta \leq \xi < \gamma) = \sigma^{y_\beta}_{\gamma-\beta}(\langle y_\xi : \beta \leq \xi < \gamma \rangle)$. This is a winning strategy of Player I in position (2), contrary to assumption.

q.e.d.

**Theorem 2 (Gale-Stewart, [GalSte53]).** If $A_0, A_1$ are any sets and $W \subseteq (A_0 \times A_1)^\omega$ is closed, then the game $\mathcal{G}_\omega(A_0, A_1, W)$ is determined.

*Proof.* Suppose Player I has no winning strategy. We define a strategy $\tau = (\tau_0, \tau_1, \ldots)$ of Player II in the game $\mathcal{G}_\omega(A_0, A_1, W)$ as follows: Let $a$ be some arbitrary element of $A_1$. For each position $p = ((x_0, y_0), \ldots, (x_{i-1}, y_{i-1}))$ in the game $\mathcal{G}_\omega(A_0, A_1, W)$ such that Player I does not have a winning strategy in position $p$, and for each $x_i \in A_0$, we have by Lemma 1 some $y_i \in A_1$ such that Player I does not have a winning strategy in position $p' = ((x_0, y_0), \ldots, (x_i, y_i))$. Let us denote this $y_i$ by $y_i = f(p, x_i)$. If $p = ((x_0, y_0), \ldots, (x_{i-1}, y_{i-1}))$ is a position in which Player I *does* have a winning strategy, we let $f(p, x_i) = a$. We have defined a function $f$ defined on positions $p$ and elements $x_i \in A_0$. Let $\tau_0(x_0) = f(\emptyset, x_0)$. Assuming $\tau_0, \ldots, \tau_{i-1}$ have been defined already, let $\tau_i(x_0, \ldots, x_i) = f(p, x_i)$, where $p = ((x_0, y_0), \ldots, (x_{i-1}, y_{i-1}))$ and $y_0 = \tau_0(x_0), y_{i-1} = \tau_{i-1}(x_0, \ldots, x_{i-1})$. It is easy to see that in every play in which Player II uses this strategy, every position $p$ is such that Player I does not have a winning strategy in position $p$, and it is also easy to see that, because $W$ is closed, this is a winning strategy of Player II. q.e.d.

Theorem 2 can be vastly generalized. For basic results in this direction, *cf.*, *e.g.*, [Jec97]. However, the result does not generalize to $\mathcal{G}_\alpha(A_0, A_1, W)$ where $\alpha > \omega$. For a stronger assumption which gives a winning strategy for Player II, consider the following: Suppose $W \subseteq (A_0 \times A_1)^\alpha$ is closed and $X$ is a set of positions of the game $\mathcal{G}_\alpha(A_0, A_1, W)$ such that

**(1)** For all $\beta < \alpha$ and all $\langle x_\xi, y_\xi : \xi < \beta\rangle \in X$ and all $x_\beta \in A_0$ there is $y_\beta \in A_1$ such that $\langle x_\xi, y_\xi : \xi \leq \beta\rangle \in X$.

**(2)** If $\langle x_\xi, y_\xi : \xi < \beta\rangle \in X$, then $\langle x_\xi, y_\xi : \xi < \alpha\rangle \in W$ for some sequence $\langle x_\xi, y_\xi : \xi < \alpha\rangle$ extending $\langle x_\xi, y_\xi : \xi < \beta\rangle$.

**(3)** If $\beta < \alpha$ is a limit ordinal, $\langle x_\xi, y_\xi : \xi < \beta\rangle \in (A_0 \times A_1)^\beta$ and $\langle x_\xi, y_\xi : \xi < \gamma\rangle \in X$ for all $\gamma < \beta$, then $\langle x_\xi, y_\xi : \xi < \beta\rangle \in X$.

Then Player II has a winning strategy in the game $\mathcal{G}_\alpha(A_0, A_1, W)$. On the other hand, if Player II has a winning strategy in the game $\mathcal{G}_\alpha(A_0, A_1, W)$, there is no reason to conclude that there is a set $X$ of positions of the game $\mathcal{G}_\alpha(A_0, A_1, W)$ satisfying the above conditions, unless $\alpha = \omega$ (see [HuuHytRau99]).

## 3 Back-and-forth sets

Suppose $L$ is a relational vocabulary. We denote $L$-structures by $\mathcal{M}$, $\mathcal{N}$, $\mathcal{M}'$, $\mathcal{N}'$, *etc.* The universe of $\mathcal{M}$ is $M$, that of $\mathcal{N}$ is $N$, *etc.* We use $\mathrm{Str}(L)$ to denote the class of all $L$-structures. Since $L$ is relational, any subset $N$ of the universe of an $L$-structure $\mathcal{M}$ defines a substructure $\mathcal{N}$ of $\mathcal{M}$, which we denote by $\mathcal{M}\restriction N$. A partial mapping $f$ between $M$ and $N$ is a *partial isomorphism* between $L$-structures $\mathcal{M}$ and $\mathcal{N}$ if it an isomorphism between $\mathcal{M}\restriction \mathrm{dom}(f)$ and $\mathcal{N}\restriction \mathrm{rng}(f)$. We denote the set of all partial isomorphisms between $\mathcal{M}$ and $\mathcal{M}'$ by $\mathrm{Part}(\mathcal{M}, \mathcal{M}')$. The set of all *finite* partial isomorphisms between $\mathcal{M}$ and $\mathcal{M}'$ by $\mathrm{Part}_\omega(\mathcal{M}, \mathcal{M}')$.

Back-and-forth sets are very useful weaker versions of isomorphisms. To get a picture of this, suppose $f : \mathcal{M} \cong \mathcal{M}'$. Then $f \in \mathrm{Part}(\mathcal{M}, \mathcal{M}')$ and we can go back and forth between $\mathcal{M}$ and $\mathcal{M}'$ with $f$ in the following sense:

$$\forall a \in M \exists b \in M'(f(a) = b) \tag{4}$$
$$\forall b \in M' \exists a \in M(f(a) = b) \tag{5}$$

We now generalize this to a situation where we do not quite have an isomorphism but only a set $P$ which reflects the back and forth conditions (4) and (5) of an isomorphism.

**Definition 3.** Suppose $\mathcal{M}$ and $\mathcal{M}'$ are $L$-structures. A *back-and-forth set* for $\mathcal{M}$ and $\mathcal{M}'$ is any non-empty set $P \subseteq \text{Part}(\mathcal{M}, \mathcal{M}')$ such that

$$\forall f \in P \forall a \in M \exists g \in P(f \subseteq g \text{ and } a \in \text{dom}(g)) \tag{6}$$

$$\forall f \in P \forall b \in M' \exists g \in P(f \subseteq g \text{ and } b \in \text{rng}(g)) \tag{7}$$

The structures $\mathcal{M}$ and $\mathcal{M}'$ are said to be *partially isomorphic*, in symbols $\mathcal{M} \simeq_p \mathcal{M}'$, if there is a back-and-forth set for them.

This important concept was introduced by Fraïssé in 1953 [Fra55]. Good presentations of back-and-forth sets and their applications in logic are in [Kue75a, Mak77, Dic75, Bar75].

**Lemma 4.** The relation $\simeq_1$ is an equivalence relation on $\text{Str}(L)$.

*Proof.* The relation $\simeq_1$ is reflexive, because $\{id_M\}$ is a back-and-forth set for $\mathcal{M}$ and $\mathcal{M}'$, where $id_M(x) = x$ for $x \in M$. If $P$ is a back-and-forth set for $\mathcal{M}$ and $\mathcal{M}'$, then $\{f^{-1} : f \in P\}$ is a back-and-forth set for $\mathcal{M}'$ and $\mathcal{M}$. Finally, if $P_1$ is a back-and-forth set for $\mathcal{M}$ and $\mathcal{M}'$ and $P_2$ is a back-and-forth set for $\mathcal{M}'$ and $\mathcal{M}''$, then $\{f_2 \circ f_1 : f_1 \in P_1, f_2 \in P_2\}$ is a back-and-forth set for $\mathcal{M}$ and $\mathcal{M}''$, where we stipulate $\text{dom}(f_2 \circ f_1) = f_1^{-1}(\text{dom}(f_2))$. q.e.d.

**Proposition 5.** If $\mathcal{M} \simeq_1 \mathcal{M}'$, where $\mathcal{M}$ and $\mathcal{M}'$ are countable, then $\mathcal{M} \cong \mathcal{M}'$.

*Proof.* Let us enumerate $M$ as $\langle a_n : n < \omega \rangle$ and $M'$ as $\langle b_n : n < \omega \rangle$. Let $P$ be a back-and-forth set for $\mathcal{M}$ and $\mathcal{M}'$. Since $P \neq \varnothing$, there is some $f_0 \in P$. We define a sequence $(f_n : n < \omega)$ of elements of $P$ as follows: Suppose $f_n \in P$ is defined. If $n$ is even, say $n = 2m$, let $f_{n+1} \in P$ and $y \in M'$ so that $f_n \cup \{\langle a_m, y \rangle\} \subseteq f_{n+1}$. If $n$ is odd, say $n = 2m + 1$, let $f_{n+1} \in P$ and $x \in M$ so that $f_n \cup \{\langle x, b_m \rangle\} \subseteq f_{n+1}$. Finally, let $f = \bigcup_{n=0}^{\infty} f_n$. Clearly, $f : \mathcal{M} \cong \mathcal{M}'$. q.e.d.

This proposition is not true for uncountable structures. Indeed, let $L = \varnothing$ and let $\mathcal{M}$ and $\mathcal{M}'$ be any infinite $L$-structures. Then $\text{Part}_\omega(\mathcal{M}, \mathcal{M}')$ is a back-and-forth set for $\mathcal{M}$ and $\mathcal{M}'$. Thus $\mathcal{M} \simeq_p \mathcal{M}'$. But $\mathcal{M} \ncong \mathcal{M}'$ if, for example, $M = \mathbb{Q}$ and $M' = \mathbb{R}$. The failure of Proposition 5 to generalize, is a major topic later in this paper.

**Proposition 6.** Suppose $\mathcal{M}$ and $\mathcal{M}'$ are dense linear orders without endpoints. Then $\mathcal{M} \simeq_1 \mathcal{M}'$.

*Proof.* Let $P = \text{Part}_\omega(\mathcal{M}, \mathcal{M}')$. It turns out that this straightforward choice works. Clearly, $P \neq \varnothing$. Suppose then $f \in P$ and $a \in M$. Let us enumerate $f$ as $\{\langle a_1, b_1\rangle, \ldots, \langle a_n, b_n\rangle\}$ where $a_1 < \ldots < a_n$. Since $f$ is a partial isomorphism, also $b_1 < \ldots < b_n$. Now we consider different cases. If $a < a_1$, we choose $b < b_1$ and then $f \cup \{\langle a, b\rangle\} \in P$. If $a_i < a < a_{i+1}$, we choose $b \in M'$ so that $b_i < b < b_{i+1}$ and then $f \cup \{\langle a, b\rangle\} \in P$. If $a_n < a$, we choose $b > b_n$ and again $f \cup \{\langle a, b\rangle\} \in P$. Finally, if $a = a_i$, we let $b = b_i$ and then $f \cup \{\langle a, b\rangle\} = f \in P$. We have proved (6). Condition (7) is proved similarly. q.e.d.

Putting Proposition 5 and Proposition 6 together yields the result familiar to students of elementary logic: countable dense linear orders without endpoints are isomorphic.

## 4 The Ehrenfeucht-Fraïssé Game

The back-and-forth feature of a back-and-forth set can be formulated neatly in terms of a game. Suppose $\mathcal{M}$ and $\mathcal{M}'$ are $L$-structures for some relational $L$. We imagine a situation in which two mathematicians argue about whether $\mathcal{M}$ and $\mathcal{M}'$ are isomorphic or not. The mathematician whom we call Player II claims that they are isomorphic, while the other mathematician whom we call Player I claims the models have an intrinsic structural difference and they cannot possibly be isomorphic.

The matter would be quickly resolved if Player II was required to show the claimed isomorphism. But the rules of the game are different. The rules are such that Player II is required to show only small pieces of the claimed isomorphism.

More exactly, Player I asks what is the image of an element $a_1$ of $M$ that he chooses at will. Then Player II is required to respond with some element $b_1$ of $M'$ so that

$$\{\langle a_1, b_1\rangle\} \in \text{Part}(\mathcal{M}, \mathcal{M}'). \tag{8}$$

Alternatively, Player I might have chosen an element $b_1$ of $M'$ and then Player II would have been required to produce an element $a_1$ of $M$ such that (8) holds. The one-element mapping $\{\langle a_1, b_1\rangle\}$ is called the *position* in the game after the first move.

Now the game goes on. Again Player I asks what is the image of an element $a_2$ of $M$ (or alternatively he can ask what is the pre-image of an element $b_2$ of $M'$). Then Player II produces an element $b_2$ of $M'$ (or in the alternative case an element $a_2$ of $M$). In either case the choice of Player II has to satisfy

$$\{\langle a_1, b_1\rangle, \langle a_2, b_2\rangle\} \in \text{Part}(\mathcal{M}, \mathcal{M}'). \tag{9}$$

Again, $\{\langle a_1, b_1\rangle, \langle a_2, b_2\rangle\}$ is called the position after the second move.

We go on until the position $\{\langle a_1, b_1\rangle, \ldots, \langle a_n, b_n\rangle\} \in \text{Part}(\mathcal{M}, \mathcal{M}')$ after the n'th move has been produced. If Player II has been able to play all the moves according to the rules she is declared the winner. Let us call this game $\text{EF}_n(\mathcal{M}, \mathcal{M}')$. If Player II can win repeatedly whatever moves Player I plays, we say that Player II has a *winning strategy*.

**Example 7.** Suppose $\mathcal{M}$ and $\mathcal{M}'$ are two $L$-structures and $L = \varnothing$. Thus the structures $\mathcal{M}$ and $\mathcal{M}'$ consist merely of a universe with no structure on it. In this singular case any one-to-one mapping is a partial isomorphism. The only thing Player II has to worry about, say in (9), is that $a_1 = a_2$ if and only if $b_1 = b_2$. Thus Player II has a winning strategy in $\text{EF}_n(\mathcal{M}, \mathcal{M}')$ if $M$ and $M'$ both have the same number or at least $n$ elements. So Player II can have a winning strategy even if $M$ and $M'$ have different cardinality and there could be no isomorphism between them for the trivial reason that there is no bijection. The intuition here is that by playing a small number of elements, or even $\aleph_0$ many, it is not possible to get hold of the cardinality of the universe if it is large.

**Example 8.** Let $\mathcal{M}$ be a linear order of length 3 and $\mathcal{M}'$ a linear order of length 4. How many moves does Player I need to beat Player II? Suppose $M = \{a_1, a_2, a_3\}$ in increasing order and $M' = \{b_1, b_2, b_3, b_4\}$ in increasing order. Clearly, if Player I plays at any moment the smallest element, also Player II has to play the smallest element or face defeat on the next move. Also, if Player I plays at any moment the smallest but one element, also Player II has to play the smallest but one element or face defeat in two moves. Now in $\mathcal{M}$ the smallest but one element is the same as the largest but one element, while in $\mathcal{M}'$ they are different. So if Player I starts with $a_2$, Player II has to play $b_2$ or $b_3$, or else she loses in one move. Suppose she plays $b_2$. Now Player I plays $b_3$ and Player II has no good moves left. To obey the rules, she must play $a_3$. That is how long she can play, for when Player I now plays $b_4$, Player II cannot make a legal move any more. In fact Player II has a winning strategy in $\text{EF}_2(\mathcal{M}, \mathcal{M}')$ but Player I has a winning strategy in $\text{EF}_3(\mathcal{M}, \mathcal{M}')$.

We now proceed to a more exact definition of the game.

**Definition 9.** Suppose $L$ is a vocabulary and $\mathcal{M}, \mathcal{M}'$ are $L$-structures such that $M \cap M' = \varnothing$. The *Ehrenfeucht-Fraïssé game* $\text{EF}_\alpha(\mathcal{M}, \mathcal{M}')$ is the game $G_\alpha(M \cup M', M \cup M', W_\alpha(\mathcal{M}, \mathcal{M}'))$, where $W_\alpha(\mathcal{M}, \mathcal{M}') \subseteq ((M \cup M')^2)^\alpha$ is the set of $p = \langle x_\xi, y_\xi : \xi < \alpha\rangle$ such that:

**(G1)** For all $\beta < \alpha$: $x_\beta \in M \iff y_\beta \in M'$.

**(G2)** If we denote $v_\beta = \begin{cases} x_\beta & \text{if } x_\beta \in M \\ y_\beta & \text{if } y_\beta \in M \end{cases}$ $v'_\beta = \begin{cases} x_\beta & \text{if } x_\beta \in M' \\ y_\beta & \text{if } y_\beta \in M', \end{cases}$

then $f_p = \{(v_\xi, v'_\xi) : \xi < \alpha\}$ is a partial isomorphism $\mathcal{M} \to \mathcal{M}'$.

We call $v_\beta$ and $v'_\beta$ above *corresponding* elements.

This game was introduced (for $\alpha \leq \omega$) by Ehrenfeucht in [Ehr60]. Note that the game $\mathrm{EF}_\alpha$ is a closed game, since our relation symbols are finitary.

One of the main themes of this paper is the question: Given two structures $\mathcal{M}$ and $\mathcal{N}$, how to measure how close they are to being isomorphic? They may be non-isomorphic for a totally obvious reason, like two graphs one of which has a triangle while the other does not. They may also be non-isomorphic for an extremely subtle reason which involves the use of axiom of choice (see later). One of the basic tools in trying to answer this question is the concept of Ehrenfeucht-Fraïssé-game.

**Proposition 10.** Suppose $L$ is a vocabulary and $\mathcal{M}$ and $\mathcal{M}'$ are two $L$-structures. The following conditions are equivalent:

1. $\mathcal{M} \simeq_1 \mathcal{M}'$
2. Player II has a winning strategy in $\mathrm{EF}_\omega(\mathcal{M}, \mathcal{M}')$.

*Proof.* Assume $M \cap M' = \varnothing$. Let $P$ be first a back-and-forth set for $\mathcal{M}$ and $\mathcal{M}'$. We define a winning strategy $\tau = \langle \tau_i : i < \omega \rangle$ for Player II. Since $P \neq \varnothing$ we can fix an element $f$ of $P$. Condition (6) tells us that if $a_1 \in M$, then there is $b_1 \in M'$ such that

$$f \cup \{\langle a_1, b_1 \rangle\} \in P \tag{10}$$

and we can let $\tau_0(a_1)$ be one such $b_1$. Likewise, if $b_1 \in M'$, then there are $a_1 \in M$ such that (10) holds and we can let $\tau_0(b_1)$ be some such $a_1$. We have defined $\tau_0(c_1)$ whatever $c_1$ is. To define $\tau_1(c_1, c_2)$, let us assume Player I played $c_1 = a_1 \in M$. Thus (10) holds with $b_1 = \tau_0(a_1)$. If $c_2 = a_2 \in M$ we can use (6) again to find $b_2 = \tau_1(a_1, a_2) \in M'$ such that $f \cup \{\langle a_1, b_1 \rangle, \langle a_2, b_2 \rangle\} \in P$. The pattern should be clear now. The back-and-forth set $P$ guides Player II to always find a valid move. Let us then write the proof in more details: Suppose we have defined $\tau_i$ for $i < j$ and we want to define $\tau_j$. Suppose Player I has played $x_0, \ldots, x_{j-1}$ and Player II has followed $\tau_i$ during each round $i < j$. During the inductive construction of $\tau_i$ we took care to define also a partial isomorphism $f_i \in P$ such that $\{v_0, \ldots, v_{i-1}\} \subseteq \mathrm{dom}(f_{i-1})$. Now Player I plays $x_j$. By assumption there is $f_j \in P$ extending $f_{j-1}$ such that if $x_j \in M$, then $x_j \in \mathrm{dom}(f_j)$ and if $x_j \in M'$, then $x_j \in \mathrm{rng}(f_j)$. We let $\tau_j(x_0, \ldots, x_j) = f_j(x_j)$ if $x_j \in M$ and

$\tau_j(x_0, \ldots, x_j) = f_j^{-1}(x_j)$ otherwise. This ends the construction of $\tau_j$. This is a winning strategy because every $f_p$ is a partial isomorphism $\mathcal{M} \to \mathcal{M}'$.

For the converse, suppose $\tau = \langle \tau_n : n < \omega \rangle$ is a winning strategy of Player II. Let $Q$ consist of all plays of $\mathrm{EF}_\omega(\mathcal{M}, \mathcal{M}')$ in which Player II has used $\tau$. Let $P$ consist of all possible $f_p$ where $p$ is a position in the game $\mathrm{EF}_\omega(\mathcal{M}, \mathcal{M}')$ with an extension in $Q$. It is clear that $P$ is non-void and has the properties (6) and (7). q.e.d.

## 5 The Dynamic Ehrenfeucht-Fraïssé Game

If Player I has a winning strategy in $\mathrm{EF}_\omega(\mathcal{M}, \mathcal{M}')$, then the game is essentially decided every time in finitely many moves. This means that the tree of the plays of the game, where Player I uses his winning strategy, is well-founded and has a rank. This observation leads to the following refinement of the game:

**Definition 11.** Let L be a relational vocabulary and $\mathcal{M}, \mathcal{M}'$ L-structures such that $M \cap M' = \varnothing$. Let $\alpha$ be an ordinal. The Dynamic Ehrenfeucht-Fraïssé game $\mathrm{EFD}^\alpha(\mathcal{M}, \mathcal{M}')$ is the game

$$\mathrm{G}_{<\omega}((\mathrm{M} \cup \mathrm{M}') \times \alpha, (\mathrm{M} \cup \mathrm{M}'), \mathrm{W}_{<\omega,\alpha}(\mathcal{M}, \mathcal{M}')),$$

where $\mathrm{W}_{<\omega,\alpha}(\mathcal{M}, \mathcal{M}')$ is the set of

$$p = \langle\langle\langle x_0, \alpha_0\rangle, y_0\rangle, \ldots, \langle\langle x_{n-1}, \alpha_{n-1}\rangle, y_{n-1}\rangle\rangle$$

such that if

(D1) $\alpha > \alpha_0 > \ldots > \alpha_{n-1}$.

then

(D2) For all $i < n : x_i \in \mathrm{M} \leftrightarrow y_i \in \mathrm{M}'$.

(D3) If we denote $v_i = \begin{cases} x_i \text{ if } x_i \in \mathrm{M} \\ y_i \text{ if } y_i \in \mathrm{M} \end{cases}$ $v'_i = \begin{cases} x_i \text{ if } x_i \in \mathrm{M}' \\ y_i \text{ if } y_i \in \mathrm{M}' \end{cases}$

then $f_p = \{(v_0, v'_0), \cdots, (v_{n-1}, v'_{n-1})\}$ is a partial isomorphism $\mathcal{M} \to \mathcal{M}'$.

The only new feature is condition (D1). Thus $\mathrm{EFD}^\alpha(\mathcal{M}, \mathcal{M}')$ is more difficult for Player I to play than the old $\mathrm{EF}_\omega(\mathcal{M}, \mathcal{M}')$, but easier than any $\mathrm{EF}_n(\mathcal{M}, \mathcal{M}')$ (if $\alpha \geq \omega$).

The analog of the back-and-forth set in the context of the dynamic game is the concept of a back-and-forth sequence. Back-and-forth sequences constitute a systematic way of representing a winning a strategy of Player II in the game $\mathrm{EFD}^\alpha$.

**Definition 12 (Karp, [Kar65]).** A *back-and-forth sequence* $(P_\beta : \beta \leq \alpha)$ of length $\alpha$ is characterized by the conditions

$$\varnothing \neq P_\alpha \subseteq \ldots \subseteq P_0 \subseteq \mathrm{Part}(\mathcal{M}, \mathcal{M}') \tag{11}$$
$$\forall f \in P_{\beta+1} \forall a \in M \exists b \in M'(f \cup \{(a,b)\} \in P_\beta) \text{ for } \beta < \alpha. \tag{12}$$
$$\forall f \in P_{\beta+1} \forall b \in M' \exists a \in M(f \cup \{(a,b)\} \in P_\beta) \text{ for } \beta < \alpha. \tag{13}$$

We write $\mathcal{M} \simeq_p^\alpha \mathcal{M}'$ if there is a back-and-forth sequence of length $\alpha$ for $\mathcal{M}$ and $\mathcal{M}'$.

The following proposition demonstrates that back-and-forth sequences indeed capture the winning strategies of Player II in $\mathrm{EFD}^\alpha(\mathcal{M}, \mathcal{M}')$:

**Proposition 13.** Suppose $L$ is a vocabulary and $\mathcal{M}$ and $\mathcal{M}'$ are two $L$-structures. The following conditions are equivalent:

1. $\mathcal{M} \simeq_p^\alpha \mathcal{M}'$
2. Player II has a winning strategy in $\mathrm{EFD}^\alpha(\mathcal{M}, \mathcal{M}')$.

*Proof.* Let us assume $M \cap M' = \varnothing$. Let $(P_i : i \leq \alpha)$ be a back-and-forth sequence for $\mathcal{M}$ and $\mathcal{M}'$. We define a winning strategy $\tau = \langle \tau_i : i < \omega \rangle$ for Player II. Suppose we have defined $\tau_i$ for $i < j$ and we want to define $\tau_j$. Suppose Player I has played $x_0, \alpha_0, \ldots, x_{j-1}, \alpha_{j-1}$ and Player II has followed $\tau_i$ during round $i < j$. During the inductive construction of $\tau_i$ we took care to define also a partial isomorphism $f_i \in P_{\alpha_i}$ such that $\{v_0, \ldots, v_{i-1}\} \subseteq \mathrm{dom}(f_i)$. Now Player I plays $x_j$ and $\alpha_j < \alpha_{j-1}$. By assumption there is $f_j \in P_{\alpha_j §}$ extending $f_{j-1}$ such that if $x_j \in M$, then $x_j \in \mathrm{dom}(f_j)$ and if $x_j \in M'$, then $x_j \in \mathrm{rng}(f_j)$. We let $\tau_j(x_0, \ldots, x_j) = f_j(x_j)$ if $x_j \in M$ and $\tau_j(x_0, \ldots, x_j) = f_j^{-1}(x_j)$ otherwise. This ends the construction of $\tau_j$. This is a winning strategy.

For the converse, suppose $\tau = \langle \tau_n : n < \omega \rangle$ is a winning strategy of Player II. Let $Q$ consist of all plays of $\mathrm{EFD}^\alpha(\mathcal{M}, \mathcal{M}')$ in which Player II has used $\tau$. Let $P_\beta$ consist of all possible $f_p$ where

$$p = (((x_0, \alpha_0), y_0), \ldots, ((x_{i-1}, \alpha_{i-1}), y_{i-1}))$$

is a position in the game $\mathrm{EFD}^\alpha(\mathcal{M}, \mathcal{M}')$ with an extension in $Q$ and $\alpha_{i-1} \geq \beta$. It is clear that $(P_\beta : \beta \leq \alpha)$ has the properties (11), (12) and (13). q.e.d.

**Lemma 14.**

(1) If Player II has a winning strategy in $\mathrm{EFD}^\alpha(\mathcal{M}, \mathcal{M}')$ and $\beta \leq \alpha$, then Player II has a winning strategy in $\mathrm{EFD}^\beta(\mathcal{M}, \mathcal{M}')$.

(2) If Player I has a winning strategy in $\mathrm{EFD}^{\alpha}(\mathcal{M},\mathcal{M}')$ and $\alpha \leq \beta$, then Player I has a winning strategy in $\mathrm{EFD}^{\beta}(\mathcal{M},\mathcal{M}')$.
(3) If $\alpha$ is a limit ordinal $\neq 0$ and Player II has a winning strategy in $\mathrm{EFD}^{\beta}(\mathcal{M},\mathcal{M}')$ for each $\beta < \alpha$, then Player II has a winning strategy in $\mathrm{EFD}^{\alpha}(\mathcal{M},\mathcal{M}')$.

*Proof.* (1) Any move of Player I in $\mathrm{EFD}^{\beta}(\mathcal{M},\mathcal{M}')$ is a legal move of Player I in $\mathrm{EFD}^{\alpha}(\mathcal{M},\mathcal{M}')$. Thus if Player II can beat Player I in $\mathrm{EFD}^{\alpha}$ she can beat him in $\mathrm{EFD}^{\beta}$.
(2) If Player I knows how to beat Player II in $\mathrm{EFD}^{\alpha}$, he can use the very same moves to beat Player II in $\mathrm{EFD}^{\beta}$.
(3) In his opening move Player I plays $\alpha_0 < \alpha$. Now Player II can pretend we are actually playing the game $\mathrm{EFD}^{\alpha_0+1}(\mathcal{M},\mathcal{M}')$, and she has a winning strategy for that game. q.e.d.

**Definition 15.** An ordinal $\alpha$ such that Player II has a winning strategy in $\mathrm{EFD}^{\alpha}(\mathcal{M},\mathcal{M}')$ and Player I has a winning strategy in $\mathrm{EFD}^{\alpha+1}(\mathcal{M},\mathcal{M}')$ is called the *Scott watershed* of $\mathcal{M}$ and $\mathcal{M}'$.

By Lemma 14 the Scott watershed is uniquely determined, if it exists. In two extreme cases the Scott watershed does not exist. Firstly, maybe Player I has a winning strategy even in $\mathrm{EF}_0(\mathcal{M},\mathcal{M}')$. This may happen if we have 0-place relation symbols, one of which may be true in $\mathcal{M}$ but false in $\mathcal{M}'$. If we allow constant symbols in the vocabulary, then two constants may be identical in one model but non-identical in the other. In these cases $\mathrm{Part}(\mathcal{M},\mathcal{M}') = \varnothing$. Secondly, Player II may have a winning strategy even in $\mathrm{EF}_{\omega}(\mathcal{M},\mathcal{M}')$, so Player I has no chance in any $\mathrm{EFD}^{\alpha}(\mathcal{M},\mathcal{M}')$, and there is no Scott watershed. In any other case the Scott watershed exists. The bigger it is, the closer $\mathcal{M}$ and $\mathcal{M}'$ are to being isomorphic. Respectively, the smaller it is, the farther $\mathcal{M}$ and $\mathcal{M}'$ are from being isomorphic. If the watershed is as small as finite, the structures $\mathcal{M}$ and $\mathcal{M}'$ are not even elementarily equivalent (if the vocabulary is finite).

**General problem**: Given $\mathcal{M}$ and $\mathcal{M}'$, find the Scott watershed!

How far afield do we have to go to find the Scott watershed? It is very natural to try first some small ordinals. But if we try big ordinals, it would be nice to know how high we have to go. There is a simple answer given by the next proposition: If the models have infinite cardinality $\kappa$, and the Scott watershed exists, then it is $< \kappa^{+}$. Thus for countable models we only need to check countable ordinals. For finite models this is not very interesting:

if the models have at most $n$ elements, and there is a watershed, then it is at most $n$.

**Proposition 16.** If Player II has a winning strategy in $\mathrm{EFD}^{\alpha}(\mathcal{M},\mathcal{M}')$ for all $\alpha < (|\mathrm{M}|+|\mathrm{M}'|)^+$ then Player II has a winning strategy in $\mathrm{EF}_{\omega}(\mathcal{M},\mathcal{M}')$.

The strategy of Player II is to stay throughout the game in a position in which she has a winning strategy in $\mathrm{EFD}^{\alpha}(\mathcal{M},\mathcal{M}')$ for arbitrary large $\alpha < (|\mathrm{M}| + |\mathrm{M}'|)^+$. She can maintain this strategy for mere cardinality reasons as follows. Suppose Player I picks $a$ in, say, $M$. For each of the arbitrary large $\alpha < (|\mathrm{M}| + |\mathrm{M}'|)^+$ the winning strategy of Player II gives a response $b^{\alpha} \in M'$. By the Pigeon Hole Principle, there is one fixed $b \in M'$ such that for arbitrary large $\alpha$ we have $b^{\alpha} = b$. Now Player II plays $b$. In the new position Player II has a winning strategy in $\mathrm{EFD}^{\alpha}(\mathcal{M},\mathcal{M}')$ for arbitrary large $\alpha < (|\mathrm{M}| + |\mathrm{M}'|)^+$.

The above theorem is particularly important for countable models since countable partially isomorphic structures are isomorphic. Thus the countable ordinals provide a complete hierarchy of thresholds all the way from not being even elementarily equivalent to being actually isomorphic. For uncountable models the hierarchy of thresholds reaches only to partial isomorphism which may be far from actual isomorphism.

An important early result on high Scott watersheds is the following result of Karp [Kar65]:

**Theorem 17.** Suppose $\delta$ satisfies the condition $\alpha < \delta \Longrightarrow \omega^{\alpha} < \delta$ and $\mathcal{M}$ is any linear order with a first element, then $\delta \simeq_p^{\alpha} \delta \times \mathcal{M}$.

The above result gives for any $\alpha$ of cardinality $\kappa \geq \aleph_0$ structures $\mathcal{M}$ and $\mathcal{M}'$ of cardinality $\kappa$ such that the watershed of them exists and is at least $\alpha$.

## 6 Relation Between Games and Logic

Let us write $\mathcal{M} \equiv_n \mathcal{M}'$, if $\mathcal{M}$ and $\mathcal{M}'$ satisfy the same sentences of $L_{\omega\omega}$ of quantifier-rank $\leq n$. Suppose $L$ is an arbitrary (relational) vocabulary, $\mathcal{M}$ and $\mathcal{M}'$ are $L$-structures, and $n < \omega$. The following conditions are equivalent:

**(i)** $\mathcal{M} \equiv_n \mathcal{M}'$.

**(ii)** $\mathcal{M}\restriction_{L'} \simeq_p^n \mathcal{M}'\restriction_{L'}$ for all finite $L' \subseteq L$.

Extending this to the whole logic $L_{\omega\omega}$, the following are equivalent for all $\mathcal{M}$ and $\mathcal{M}'$:

**(iii)** $\mathcal{M} \equiv_{\omega\omega} \mathcal{M}'$

**(iv)** $\mathcal{M}\restriction_{L'} \simeq_p^{\omega} \mathcal{M}'\restriction_{L'}$ for all finite $L' \subseteq L$.

The proofs are straightforward once we observe that the number of non-equivalent first order formulas of quantifier-rank $\leq n$ of a fixed finite vocabulary and a fixed finite set of free variables is finite. This leads to the following important characterization of first order definability, due to Fraïssé: For all $n < \omega$ the equivalence relation $\mathcal{M} \equiv_n \mathcal{M}'$ divides $\mathrm{Str}(L)$ into finitely many equivalence classes $C_i^n$, $i = 1, \dots, m_n$, such that for each $C_i^n$ there is a sentence $\phi_i^n$ of $L_{\omega\omega}$ of quantifier-rank $n$ with the properties:

1. For all $L$-structures $\mathcal{M}$: $\mathcal{M} \in C_i^n \iff \mathcal{M} \models \phi_i^n$.
2. If $\phi$ is an $L$-sentence of quantifier rank $\leq n$, then there are $i_1, \dots, i_k$ such that $\models \phi \leftrightarrow (\phi_{i_1}^n \vee \ldots \vee \phi_{i_k}^n)$

From this we can get the following characterization of first order logic, due to Fraïssé [Fra55]:

**Theorem 18.** Suppose $K$ is a class of $L$-structures for a finite relational vocabulary $L$. Then $K$ is $L_{\omega\omega}$-definable if and only if there is $n \in \omega$ such that $K$ is closed under $\simeq_p^n$.

In infinitary logic, let us write $\mathcal{M} \equiv_{\infty\omega}^{\alpha} \mathcal{M}'$, if $\mathcal{M}$ and $\mathcal{M}'$ satisfy the same sentences of $L_{\infty\omega}$ of quantifier-rank $\leq \alpha$, and $\mathcal{M} \equiv_{\infty\omega} \mathcal{M}'$, if $\mathcal{M}$ and $\mathcal{M}'$ satisfy the same sentences of the whole $L_{\infty\omega}$. We can relax the restriction on finite vocabularies and get for all $\mathcal{M}$ and $\mathcal{M}'$ the equivalence

$$\mathcal{M} \equiv_{\infty\omega}^{\alpha} \mathcal{M}' \iff \mathcal{M} \simeq_1^{\alpha} \mathcal{M}' \tag{14}$$

and

$$\mathcal{M} \equiv_{\infty\omega} \mathcal{M}' \iff \mathcal{M} \simeq_1 \mathcal{M}'. \tag{15}$$

These equivalence reveal the close relationship between infinitary logic and the Ehrenfeucht-Fraïssé-game. The relationship goes in fact much deeper. To see this we go on to define so called *Scott sentences* $\sigma_{\mathcal{M}}^{\alpha}$ from [Sco65]: Let $L$ be a vocabulary, $\mathcal{M}$ an $L$-structure, and $a_0, \dots, a_{n-1} \in M$. We call a formula *basic* if it is atomic or negated atomic. Then we define

$$\begin{aligned}
\sigma^0_{\mathcal{M},a_0,\dots,a_{n-1}} &= \bigwedge\{\varphi(x_0,\dots,x_{n-1}) : \varphi(x_0,\dots,x_{n-1}) \\
&\qquad \text{is a basic } L\text{-formula and } \mathcal{M} \models \varphi(a_0,\dots,a_{n-1})\} \\
\sigma^{\alpha+1}_{\mathcal{M},a_0,\dots,a_{n-1}} &= (\forall x_n \bigvee_{a_n \in M} \sigma^{\alpha}_{\mathcal{M},a_0,\dots,a_n}) \wedge (\bigwedge_{a_n \in M} \exists x_n \sigma^{\alpha}_{\mathcal{M},a_0,\dots,a_n}) \\
\sigma^{\nu}_{\mathcal{M},a_0,\dots,a_{n-1}} &= \bigwedge_{\alpha<\nu} \sigma^{\alpha}_{\mathcal{M},a_0,\dots,a_{n-1}}, \text{ for limit } \nu \\
\sigma^{\alpha}_{\mathcal{M}} &= \sigma^{\alpha}_{\mathcal{M},\varnothing}.
\end{aligned}$$

The following conditions are equivalent, almost by definition:

(1) $\mathcal{M}' \models \sigma^{\alpha}_{\mathcal{M},a_0,\ldots,a_{n-1}}(b_0,\ldots,b_{n-1})$
(2) $(\mathcal{M},a_0,\ldots,a_{n-1}) \simeq^{\alpha}_{1} (\mathcal{M}',b_0,\ldots,b_{n-1})$

The *Scott height* $\mathrm{SH}(\mathcal{M})$ of a model $\mathcal{M}$ is the supremum of all ordinals $\alpha+1$, where $\alpha$ is the Scott watershed of a pair $(\mathcal{M},a_1,\ldots,a_n) \not\simeq_1 (\mathcal{M},b_1,\ldots,b_n)$ and $a_1,\ldots,a_n,b_1,\ldots,b_n \in \mathrm{M}$. The *Scott-sentence* of a structure $\mathcal{M}$ is the $L_{\infty\omega}$-sentence

$$\sigma_{\mathcal{M}} = \sigma^{\mathrm{SH}(\mathcal{M})}_{\mathcal{M},\varnothing} \wedge \bigwedge_{\substack{n\in\omega \\ a_0,\ldots,a_{n-1}\in M}} \forall x_0 \ldots \forall x_{n-1}(\sigma^{\mathrm{SH}(\mathcal{M})}_{\mathcal{M},a_0,\ldots,a_{n-1}} \rightarrow \sigma^{\mathrm{SH}(\mathcal{M})+1}_{\mathcal{M},a_0,\ldots,a_{n-1}}).$$

Now the equivalence

$$\mathcal{M}' \models \sigma_{\mathcal{M}} \Longleftrightarrow \mathcal{M}' \simeq_1 \mathcal{M}$$

gives a more nuanced version of (15). Note, that if $\mathcal{M}$ is a well-ordered set, then it is, up to isomorphism, the only model of $\sigma_{\mathcal{M}}$. The situation is the same with countable models:

**Theorem 19 (Scott's Isomorphism Theorem, [Sco65]).** Suppose $\mathcal{M}$ is a countable model. Then for all countable $\mathcal{M}'$

$$\mathcal{M}' \models \sigma_{\mathcal{M}} \quad \Longleftrightarrow \quad \mathcal{M}' \cong \mathcal{M}.$$

This is a remarkable result. It puts countable models on levels of a well-ordered hierarchy according to the Scott height. On each level there is an invariant, the Scott sentence of the model, that characterizes the model up to isomorphism. These invariants need, of course, not be simple in any way, but they have a uniform tree-structure, the differences occurring only at the leaves of the tree. The invariants provide a way to systematize and classify countable models according to syntactic properties of the Scott sentence.

It is relatively easy to prove by induction on $\alpha$ that for every $\alpha$ there is only a set of non-equivalent formulas of $L_{\infty\omega}$ of quantifier-rank $\leq \alpha$. Thus, for example, for any $\alpha$ there are only a set of non-equivalent $\sigma^{\alpha}_{\mathcal{M}}$, while there is a proper class of non-equivalent $\sigma_{\mathcal{M}}$. From this it follows that for all ordinals $\alpha$ the equivalence relation $\mathcal{M} \equiv^{\alpha}_{\infty\omega} \mathcal{M}'$ divides the class $\mathrm{Str}(L)$ of all $L$-structures into a set of equivalence classes $C^{\alpha}_{i}, i \in \mathrm{I}$, such that if we choose any representatives $\mathcal{M}_i \in C^{\alpha}_{i}$, then:

1. For all $L$-structures $\mathcal{M}$: $\mathcal{M} \in C^{\alpha}_{i} \Longleftrightarrow \mathcal{M} \models \sigma^{\alpha}_{\mathcal{M}_i}$
2. If $\varphi$ is an $L$-sentence of $L_{\infty\omega}$ of quantifier-rank $\leq \alpha$, then there is a set $I_0 \subseteq I$ such that $\models \varphi \leftrightarrow \bigvee_{i\in I_0} \sigma^{\alpha}_{\mathcal{M}_i}$.

Note again that if we tried to prove the above for the finer relation $\simeq_1$, we would run into the trouble that there are a proper class of equivalence classes. Analogously to Theorem 18 we get a characterization of $L_{\infty\omega}$:

**Theorem 20 (Karp, [Kar65]).** Suppose $L$ is an arbitrary vocabulary and $K$ is a class of $L$-structures. $K$ is definable in $L_{\infty\omega}$ if and only if $K$ is closed under $\simeq_1^\alpha$ for some $\alpha$.

The above theorem gives a kind of normal form for sentences of $L_{\infty\omega}$: everything is a disjunction of sentences $\sigma^\alpha_{\mathcal{M}}$, which in turn have a very canonical form. For finite $\alpha$ and finite relational vocabulary the formulas $\sigma^\alpha_{\mathcal{M}}$ are first-order.

## 7 Moving many elements at a time

An essential feature of first order logic and its infinitary extensions $\mathrm{L}_{\kappa\omega}$ and $L_{\infty\omega}$ is that quantification is over individuals only. This makes these logics set-theoretically "absolute", *i.e.*, the truth or falsity of $\mathcal{M} \models_s \varphi$ depends only on elements of $M$, not on subsets of $M$ or sets of subsets of $M$. We now introduce infinitary logics which allow quantification over infinite sequences of elements of the domain of discourse. We will be able to say new things, but we pay a price: model theory becomes more complex and many questions depend on set-theoretical assumptions.

We first define the appropriate version of the Ehrenfeucht-Fraïssé game. In this game the players play sequences of a given length. Each round consists of a choice of a sequence by Player I followed by a choice of a sequence by Player II. The goal of Player II is to make sure the played sequences form, element by element, a partial isomorphism. Thus if Player I plays a sequence $x_0 = (x_0(0), \ldots, x_0(n), \ldots)$ which is a descending sequence relative to a linear order $<$ in one of the models, Player II tries to play likewise a sequence $y_0 = (y_0(0), \ldots, y_0(n), \ldots)$ which constitutes a descending sequence relative to $<$ in the other model. If that other model is well-ordered by $<$, she loses right away.

For another example, suppose one of the models is countable while the other is uncountable. If Player I is allowed to play infinite sequences he can immediately let $x_0$ enumerate the countable model. Whatever Player II plays, Player I wins during the next round.

To define the new game exactly, we fix some notation. A function $s : \alpha \to M$ is called a *sequence* of *length* $\mathrm{len}(s) = \alpha$. The set of all sequences of length $\alpha$ of elements of $M$ is denoted by $M^\alpha$. Finally $M^{<\alpha} = \bigcup_{\beta<\alpha} M^\beta$.

**Definition 21.** Suppose $\kappa$ is a regular cardinal. The Ehrenfeucht-Fraïssé game with moves of size $< \kappa$ on $\mathcal{M}$ and $\mathcal{M}'$, $\mathrm{EF}^\kappa_\omega(\mathcal{M}, \mathcal{M}')$, is the game in which Player I plays $x_n \in M^{<\kappa} \cup (M')^{<\kappa}$ and Player II responds with

$x_n \in M^{<\kappa} \cup (M')^{<\kappa}$. Player II wins if for all $n$,

(1) $\mathrm{len}(x_n) = \mathrm{len}(y_n)$
(2) $x_n \in M^{<\kappa} \leftrightarrow y_n \in (M')^{<\kappa}$

and if we denote

$$v_n = \begin{cases} x_n & \text{if } x_n \in M^{<\kappa} \\ y_n & \text{if } y_n \in M^{<\kappa} \end{cases} \quad v'_n = \begin{cases} x_n & \text{if } x_n \in (M')^{<\kappa} \\ y_n & \text{if } y_n \in (M')^{<\kappa}, \end{cases}$$

(3) $\{(v_n(\xi), v'_n(\xi)) : \xi < \mathrm{len}(x_n), n \in \omega\} \in \mathrm{Part}(\mathcal{M}, \mathcal{M}')$.

The *dynamic version of* $\mathrm{EF}^{\kappa}_{\omega}$, denoted by $\mathrm{EFD}^{\alpha}_{\kappa}$, is defined by requiring that the moves of Player I are pairs $(x_n, \alpha_n)$ where $x_n \in M^{<\kappa} \cup (M')^{<\kappa}, \alpha_n < \alpha$ To win, Player I has to play $\alpha > \alpha_0 > \alpha_1 > \ldots$ just as in $\mathrm{EFD}^{\alpha}$.

**Definition 22.** Let $\eta$ denote the ordertype of the rationals. For $A \subseteq \omega_1$ let

$$r^A_\alpha = \begin{cases} 1+\eta & \text{if} \quad \alpha \in A \setminus \{0\} \\ \eta & \text{if} \quad \alpha \notin A \setminus \{0\} \end{cases}$$

$$\phi(A) = \sum_{\alpha<\omega_1} (r^A_\alpha \times \{\alpha\}).$$

The elements of $r^A_\alpha \times \{\alpha\}$ in $\phi(A)$ are said to have rank $\alpha$.

Clearly, $\phi(A)$ is an $\aleph_1$-like dense linear order.

**Lemma 23.** $\phi(\varnothing) \ncong \phi(\omega_1)$.

*Proof.* Suppose $f : \phi(\varnothing) \cong \phi(\omega_1)$. For each $a \in \phi(A)$ let $\varrho(a)$ be the unique $\alpha$ such that $a = (q, \alpha)$ for some $q$. Let us define a sequence $\{a_n : n \in \omega\}$ as follows:

$$\begin{array}{lcl} a_0 & \in & \phi(\varnothing) \text{ arbitrary} \\ a_1 & \in & \phi(\omega_1) \text{ such that } \varrho(a_1) > f(a_0) \\ a_2 & \in & \phi(\varnothing) \text{ such that } \varrho(a_2) > f^{-1}(a_1) \\ a_3 & \in & \phi(\omega_1) \text{ such that } \varrho(a_3) > f(a_2) \\ \mathit{etc.} & & \end{array}$$

Let $\delta = \sup_n \varrho(a_n)$. Let $a$ be the smallest element in $r^{\omega_1}_\delta$. Now $a$ is the supremum of the set $\{a_{2n+1} : n \in \omega\}$ in $\phi(\omega_1)$, but $f^{-1}(a)$ cannot be the supremum of $\{a_{2n} : n \in \omega\}$ in $\phi(\varnothing)$ as $r^{\varnothing}_\delta$ has no first element. This shows that $f$ cannot be an isomorphism. q.e.d.

The above proof actually shows that $\phi(A) \not\cong \phi(B)$ if $A \triangle B$ is stationary.

**Lemma 24.** Player II has a winning strategy in $\mathrm{EF}_\omega^{\aleph_1}(\mathcal{M}, \mathcal{M}')$ whenever $\mathcal{M}$ and $\mathcal{M}'$ are $\aleph_1$-like dense linear orders without first element.

*Proof.* During the game Player II maintains an element $u_n \in M$, an element $v_n \in M'$ and an isomorphism $\pi_n$ between the initial segment of $\mathcal{M}$ determined by $u_n$ and the initial segment of $\mathcal{M}'$ determined by $v_n$. Player II takes care that the partial isomorphism of played elements is a subfunction of $\pi_n$. Suppose then Player I plays a countable sequence $s_n$ in one of the models. Now Player II moves her $u_n$ and $v_n$ to new positions $u_{n+1}, v_{n+1}$ above the elements in $s_n$. The intervals $(u_n, u_{n+1})$ of $\mathcal{M}$ and $(v_n, v_{n+1})$ of $\mathcal{M}'$ are countable dense linear orderings without endpoints, hence isomorphic. So Player II can maintain her strategy. q.e.d.

**Corollary 25 (Morley).** There are non-isomorphic models $\mathcal{M}$ and $\mathcal{M}'$ of cardinality $\aleph_1$ such that Player II has a winning strategy in $EF_\omega^{\aleph_1}(\mathcal{M}, \mathcal{M}')$.

We use $[A]^{<\lambda}$ to denote the set $\{a \subseteq A : |a| < \lambda\}$.

**Definition 26.** Suppose $\mathcal{M}$ and $\mathcal{M}'$ are $L$-structures. A $\lambda$-*back-and-forth set* for $\mathcal{M}$ and $\mathcal{M}'$ is any non-empty set $P \subseteq \mathrm{Part}(\mathcal{M}, \mathcal{M}')$ such that

(1) $\forall f \in P \forall a \in [M]^{<\lambda} \exists g \in P(f \subseteq g \text{ and } a \subseteq \mathrm{dom}(g))$
(2) $\forall f \in P \forall b \in [M']^{<\lambda} \exists g \in P(f \subseteq g \text{ and } b \subseteq \mathrm{rng}(g))$.

The structures $\mathcal{M}$ and $\mathcal{M}'$ are said to be $\lambda$-*partially isomorphic*, in symbols $\mathcal{M} \simeq_\lambda \mathcal{M}'$, if there is a $\lambda$-back-and-forth set for them.

It is easy to see that $\simeq_\lambda$ is an equivalence relation on $\mathrm{Str}(L)$ for any $L$. The big drawback of $\simeq_\lambda$ in comparison to $\simeq_1$ is that for $\lambda > \omega$ there is no guarantee that $\mathcal{M} \simeq_\lambda \mathcal{M}', |M| \leq \lambda, |M'| \leq \lambda$ implies $\mathcal{M} \cong \mathcal{M}'$.

**Theorem 27.** The following conditions are equivalent:

(1) $\mathcal{M} \equiv_{\infty\lambda} \mathcal{N}$ (*i.e.*$\mathcal{M}$ and $\mathcal{N}$ satisfy the same sentences of $\mathrm{L}_{\infty\lambda}$).
(2) Player II has a winning strategy in $\mathrm{EF}_\omega^\lambda(\mathcal{M}, \mathcal{N})$.
(3) $\mathcal{M} \simeq_\lambda \mathcal{N}$.

The following is a dynamic version of Theorem 27.

**Theorem 28.** Suppose $\kappa = |M^{<\lambda}| + |M'^{<\lambda}|$ and Player II has a winning strategy in $\mathrm{EFD}_\lambda^\alpha(\mathcal{M}, \mathcal{N})$ for each $\alpha < \kappa^+$. Then Player II has a winning strategy in $\mathrm{EF}_\omega^\lambda(\mathcal{M}, \mathcal{N})$.

For working with winning strategies of Player II in $\mathrm{EFD}_\lambda^\alpha(\mathcal{M},\mathcal{N})$ we have the appropriate version of back-and-forth sequence: A *$\lambda$-back-and-forth sequence* $(P_\beta : \beta \leq \alpha)$ is characterized by the conditions

$$\begin{array}{l}\emptyset \neq P_\alpha \subseteq \cdots \subseteq P_0 \subseteq \mathrm{Part}(\mathcal{M},\mathcal{M}')\\ \forall f \in P_{\beta+1} \forall a \in [M]^{<\lambda} \exists g \in P_\beta (a \subseteq \mathrm{dom}(g) \text{ and } f \subseteq g)\\ \forall f \in P_{\beta+1} \forall b \in [M']^{<\lambda} \exists g \in P_\beta (b \subseteq \mathrm{ran}(g) \text{ and } f \subseteq g).\end{array}$$

We write $\mathcal{M} \simeq_\lambda^\alpha \mathcal{M}'$ if there is a $\lambda$-back-and-forth-sequence of length $\alpha$ for $\mathcal{M}$ and $\mathcal{M}'$.

**Theorem 29.** The following conditions are equivalent:

(1) $\mathcal{M} \equiv_{\infty\lambda}^\alpha \mathcal{N}$ (*i.e.*$\mathcal{M}$ and $\mathcal{N}$ satisfy the same sentences of $\mathrm{L}_{\infty\lambda}$ of quantifier rank $\leq \alpha$).
(2) Player II has a winning strategy in $\mathrm{EFD}_\lambda^\alpha(\mathcal{M},\mathcal{N})$.
(3) $\mathcal{M} \simeq_\lambda^\alpha \mathcal{N}$.

We can define the *$\lambda$-Scott watershed* of $\mathcal{M}$ and $\mathcal{M}'$ as the least ordinal $\alpha$ such that Player II has a winning strategy in $\mathrm{EFD}_\lambda^\alpha(\mathcal{M},\mathcal{M}')$ and Player I has a winning strategy in $\mathrm{EFD}_\lambda^{\alpha+1}(\mathcal{M},\mathcal{M}')$. In the case $\lambda = 2$ the Scott watershed of countable models was countable and every pair of countable non-isomorphic models had a Scott watershed. For $\lambda > \omega$ the situation is less satisfactory. For non-isomorphic models of size $\lambda$ the $\lambda$-Scott watershed need not exist, as it is possible that $\mathcal{M} \simeq_\lambda \mathcal{N}$ and $\mathcal{M} \not\cong \mathcal{N}$. Also, if ,*e.g.*, $\lambda = \aleph_1$, even if the $\lambda$-Scott watershed exists, we only know it is $< (2^\omega)^+$, which may be very big.

The *$\lambda$-Scott height* $\lambda$-SH$(\mathcal{M})$ of a model $\mathcal{M}$ is the least $\alpha$ such that if $a_1, \ldots, a_n, b_1, \ldots, b_n \in [M]^{<\lambda}$ and $\mathrm{len}(a_i) = \mathrm{len}(b_i)$ for $1 \leq i \leq n$

$$(\mathcal{M}, a_1, \ldots, a_n) \simeq_\lambda^\alpha (\mathcal{M}, b_1, \ldots, b_n)$$

then $(\mathcal{M}, a_1, \ldots, a_n) \simeq_\lambda^{\alpha+1} (\mathcal{M}, b_1, \ldots, b_n)$. (Here $(\mathcal{M}, a_1, \ldots, a_n)$ means $(\mathcal{M}, (a_1(\xi))_{\xi<\mathrm{len}(a_1)}, \ldots, (a_n(\xi))_{\xi<\mathrm{len}(a_n)})$.) If none exist, then $\lambda$-SH$(\mathcal{M}) = 0$. Note that $\lambda$-SH$(\mathcal{M}) < (|M^{<\lambda}|)^+$.

**Theorem 30.** If $\mathcal{M} \simeq_\lambda^{\lambda\text{-SH}(\mathcal{M})+\omega} \mathcal{M}'$, then $\mathcal{M} \simeq_\lambda \mathcal{M}'$.

The $\lambda$-Scott spectrum of a first order theory $T$ is $\lambda$-ss$(T) = \{\lambda\text{-SH}(\mathcal{M}) : \mathcal{M} \models T\}$ The $\lambda$-Scott spectrum is a rather little studied subject, due to the lack of a "$\lambda$-version" of Theorem 19. For a recent study, see [LasShe01].

## 8 Strong Back-And-Forth Sets

To overcome the failure of $\simeq_\lambda$ to characterize isomorphism in models of size $\lambda$, a stronger form of $\simeq_\lambda$ was proposed by Dickmann [Dic75] and Kueker [Kue75a]:

**Definition 31.** Suppose $\mathcal{M}$ and $\mathcal{M}'$ are $L$-structures. A *strong $\lambda$-back-and-forth set* for $\mathcal{M}$ and $\mathcal{M}'$ is a $\lambda$-back-and-forth set $P$ which satisfies the following additional condition:

(3) If $\{f_\alpha : \alpha < \beta\}, \beta < \lambda$, is an $\subseteq$-increasing sequence in $P$, then there is $f \in P$ such that $f_\alpha \subseteq f$ for all $\alpha < \beta$.

If there is a strong $\lambda$-back-and-forth set for $\mathcal{M}$ and $\mathcal{M}'$, we write $\mathcal{M} \simeq_{\lambda,s} \mathcal{M}'$.

**Theorem 32.** If $|M| \leq \lambda$ and $|M'| \leq \lambda$, then the following conditions are equivalent:

(1) $\mathcal{M} \cong \mathcal{M}'$.
(2) $\mathcal{M} \simeq_{\lambda,s} \mathcal{M}'$.

*Proof.* Suppose $P$ is a strong $\lambda$-back-and-forth set for $\mathcal{M}$ and $\mathcal{M}'$. Let

$$\begin{aligned} M &= \{a_\alpha : \alpha < \lambda\} \\ M' &= \{b_\alpha : \alpha < \lambda\}. \end{aligned}$$

Let $f_0 \in P$. If $f_\alpha$ has been defined for $\alpha < \gamma$ and $(\alpha < \beta < \gamma \implies f_\alpha \subseteq f_\beta)$ then let $g \in P$ such that $f_\alpha \subseteq g$ for all $\alpha < \gamma$. Then, let $f_\gamma \in P$ such that $g \subseteq f_\gamma, a_\gamma \in \mathrm{dom}(f_\gamma)$ and $b_\gamma \in \mathrm{ran}(f_\gamma)$. Finally, $\bigcup_{\alpha<\lambda} f_\alpha$ is an isomorphism $\mathcal{M} \to \mathcal{M}'$. q.e.d.

**Example 33.** A linear order $(M, <)$ is an $\eta_\alpha$-set if for all sets $M \subseteq M$ and $M' \subseteq M$ such that $|M| < \aleph_\alpha, |M'| < \aleph_\alpha$ and $a < b$ for all $a \in M$ and $b \in M'$, there is $c \in M$ such that $a < c < b$ for all $a \in M$ and $b \in M'$. If $\mathcal{M}$ and $\mathcal{M}'$ are $\eta_\alpha$-sets and $\aleph_\alpha$ is regular, then $\mathcal{M} \simeq_{\aleph_\alpha,s} \mathcal{M}'$.

A model $\mathcal{M}$ is *$\lambda$-homogeneous* if for all $\{a_\alpha : \alpha < \beta\}$ and $\{b_\alpha : \alpha < \beta\}$ such that $\beta < \lambda$ and $(\mathcal{M}, a_\alpha)_{\alpha<\beta} \equiv (\mathcal{M}, b_\alpha)_{\alpha<\beta}$ and every $a_\beta \in M$ there is $b_\beta \in M$ such that $(\mathcal{M}, a_\alpha)_{\alpha\leq\beta} \equiv (\mathcal{M}, b_\alpha)_{\alpha\leq\beta}$. The ordered sets $(\mathbb{Q}, <)$ and $(\mathbb{R}, <)$ are $\aleph_0$-homogeneous. Every countable consistent first order theory has for every $\lambda$ a $\lambda$-homogeneous model.

**Proposition 34.** If $\mathcal{M}$ and $\mathcal{M}'$ are $\lambda$-homogeneous, $\lambda$ is regular and $\mathcal{M} \simeq_\lambda \mathcal{M}'$, then $\mathcal{M} \simeq_{\lambda,s} \mathcal{M}'$.

*Proof.* We let $P$ consist of all $f = \{(a_\alpha, b_\alpha) : \alpha < \beta\}$ such that $\beta < \lambda$ and

$$(*) \quad (\mathcal{M}, a_\alpha)_{\alpha<\beta} \equiv (\mathcal{M}', b_\alpha)_{\alpha<\beta}.$$

$P$ is non-empty, as $\varnothing \in P$. Let us prove that $P$ is a $\lambda$-back-and-forth set. Suppose (*) holds and $a_\beta \in M$ (we treat this case first as the case

$a_\beta \in [M]^{<\lambda}$ then follows). By $\mathcal{M} \simeq_\lambda \mathcal{M}'$ there is $b'_\alpha, \alpha \leq \beta$ such that $(\mathcal{M}, a_\alpha)_{\alpha\leq\beta} \equiv (\mathcal{M}', b'_\alpha)_{\alpha\leq\beta}$. Thus $(\mathcal{M}', b_\alpha)_{\alpha<\beta} \equiv (\mathcal{M}', b'_\alpha)_{\alpha<\beta}$. By $\lambda$-homogeneity there is $b_\beta$ such that $(\mathcal{M}', b_\alpha)_{\alpha\leq\beta} \equiv (\mathcal{M}', b'_\alpha)_{\alpha\leq\beta}$. Thus

$$(\mathcal{M}, a_\alpha)_{\alpha\leq\beta} \equiv (\mathcal{M}', b_\alpha)_{\alpha\leq\beta}.$$

The property (*) is clearly preserved under unions of chains, *i.e.*, if $f_\alpha \in P$ for $\alpha < \gamma$, where $\gamma < \lambda$, and $\alpha < \beta < \gamma \Longrightarrow f_\alpha \subseteq f_\beta$ then $\bigcup_{\alpha<\gamma} f_\alpha \in P$ (here we use regularity of $\lambda$). Thus the above back-and-forth argument extends to $a_\beta \in [M]^{<\lambda}$. q.e.d.

A model $\mathcal{M}$ is $\lambda$-*saturated* if for all $a_\alpha \in M, \alpha < \beta$, where $\beta < \lambda$, every type of the structure $(\mathcal{M}, a_\alpha)_{\alpha<\beta}$ is realized in $(\mathcal{M}, a_\alpha)_{\alpha<\beta}$. $\lambda$-saturated models are always $\lambda$-homogeneous. Every countable consistent first order theory has for every $\lambda$ a $\lambda$-saturated model.

**Proposition 35.** Suppose $\mathcal{M}$ and $\mathcal{M}'$ are $\lambda$-saturated structures and $\mathcal{M} \equiv \mathcal{M}'$. Then $\mathcal{M} \simeq_\lambda \mathcal{M}'$. If moreover $\lambda$ is regular, then $\mathcal{M} \simeq_{\lambda,s} \mathcal{M}'$.

*Proof.* Let $P$ consist of all $f = \{(a_\alpha, b_\alpha) : \alpha < \beta\}$ such that $\beta < \lambda$ and

$$(*) \quad (\mathcal{M}, a_\alpha)_{\alpha<\beta} \equiv (\mathcal{M}', b_\alpha)_{\alpha<\beta}.$$

By assumption, $\varnothing \in P$. Suppose then (*) holds and $a_\beta \in M$ (The case $a_\beta \in [M]^{<\lambda}$ follows from this). Let $\Sigma$ be the set of first order sentences, $\varphi(x_{\alpha_1}, \ldots, x_{\alpha_n}, x_\beta)$, $\alpha_1, \ldots, \alpha_n < \beta$, satisfying $\mathcal{M} \models \varphi(a_{\alpha_1}, \ldots, a_{\alpha_n}, a_\beta)$. Condition (*) guarantees that for any finite $\Sigma_0 \subseteq \Sigma$ there is $b_\beta \in M'$ such that

$$(**) \quad \mathcal{M}' \models \varphi(b_{\alpha_1}, \ldots, b_{\alpha_n}, b_\beta)$$

for all $\varphi \in \Sigma_0$. By $\lambda$-saturation, there is $b_\beta \in M'$ such that (**) holds for all $\varphi \in \Sigma$. We have proved one half of the $\lambda$-back-and-forth criterion for $P$. The other half is similar. q.e.d.

A theory is $\kappa$-categorical if all its models of cardinality $\kappa$ are isomorphic. The first order theory of $(\mathbb{Q}, <)$ is $\aleph_0$-categorical. The first order theory of $(\mathbb{C}, +, \cdot, 0, 1)$ is $\aleph_1$-categorical. A countable first order theory which is $\kappa$-categorical for some $\kappa$ is necessarily complete. By a deep theorem of Morley, if a countable first order theory is $\kappa$-categorical for some uncountable $\kappa$, it is $\kappa$-categorical for all uncountable $\kappa$.

**Corollary 36.** Let $\lambda$ be a regular cardinal $> \omega$. If $T$ is a countable complete first order theory which is $\kappa$-categorical for some (hence all) uncountable $\kappa$, then for any $\mathcal{M} \models T$ and $\mathcal{N} \models T$ of cardinality $\geq \lambda$ we have $\mathcal{M} \simeq_{\lambda,s} \mathcal{N}$.

*Proof.* For such $T$ and $\lambda$ we know that $\mathcal{M}$ and $\mathcal{N}$ have to be $\lambda$-saturated.

q.e.d.

The above results attest to the applicability the the concept $\mathcal{M} \simeq_{\lambda,s} \mathcal{N}$ in comparison to the weaker $\mathcal{M} \simeq_\lambda \mathcal{N}$. But there is something fundamentally puzzling about $\mathcal{M} \simeq_{\lambda,s} \mathcal{N}$. We do not know whether it divides structures to equivalence classes, since we do not know whether it is transitive:

**Open Problem:** Is the relation $\mathcal{M} \simeq_{\lambda,s} \mathcal{N}$ transitive?

There are many partial results about the transitivity of $\mathcal{M} \simeq_{\lambda,s} \mathcal{N}$ but the general question is open. Thus it is not known whether there is some version $L^*$ of infinitary logic such that $\mathcal{M} \simeq_{\lambda,s} \mathcal{N}$ is equivalent to $\mathcal{M} \models \phi \iff \mathcal{N} \models \phi$ for all $\phi \in L^*$. What is known, is, for example, that $\mathcal{M} \simeq_{\aleph_1,s} \mathcal{N}$ is transitive among trees of height $\omega_1$, among models of (at most) continuum size, and among models of different cardinality [VääVel04].

## 9 The Transfinite Ehrenfeucht-Fraïssé game

We shall now take up seriously the possibility that the game $\mathrm{EF}_\alpha(\mathcal{M}, \mathcal{M}')$ may last for more than $\omega$ moves. So what a game of, say, length $\omega + \omega$ would look like: it would be like playing two games of length $\omega$ one after the other. For example, it is by now well-known to the reader that the second player has a winning strategy in the Ehrenfeucht-Fraïssé game of length $\omega$ on $(\mathbb{R}, <)$ and $(\mathbb{R} \setminus \{0\}, <)$. But if Player I is allowed one more move after the first $\omega$ moves, he wins.

For a more enlightening example, suppose $\mathcal{M}$ and $\mathcal{N}$ are equivalence relations such that $\mathcal{M}$ has $\aleph_1$ countable classes and $\aleph_0$ uncountable classes while $\mathcal{N}$ has $\aleph_1$ countable classes and $\aleph_1$ uncountable classes. Does Player II have a winning strategy in $\mathrm{EF}_\omega$. Yes! She just keeps matching different equivalence classes with different equivalence classes. But she can actually win the game of length $\omega + \omega$, too! During the first $\omega$ moves she matches countable equivalence classes with countable ones and uncountable equivalence classes with uncountable ones. After the first $\omega$ moves she may have to match a countable equivalence class with an uncountable class, but Player I will not be able to reveal this bluff. It is only when Player I has $\omega + \omega + 1$ moves that he has a winning strategy: During the first $\omega$ moves play one element from each uncountable class of $\mathcal{M}$. Then play one element $b$ from an unused uncountable equivalence class of $\mathcal{N}$. Player II will match this element with an element $c$ from a countable equivalence class of $\mathcal{M}$. During the next $\omega$ rounds Player I enumerates the countable equivalence class of $c$.

Finally he plays an unplayed element equivalent to $b$. Player II loses as all elements equivalent to $c$ have been played already.

There are some basic properties of transfinite Ehrenfeucht-Fraïssé games that are easy to prove but still worth noting: If Player II has a winning strategy in $\mathrm{EF}_\alpha(\mathcal{M}_0, \mathcal{M}_1)$ and $\beta < \alpha$, then Player II has the same winning strategy in $\mathrm{EF}_\beta(\mathcal{M}_0, \mathcal{M}_1)$. If Player I has a winning strategy in $\mathrm{EF}_\alpha(\mathcal{M}_0, \mathcal{M}_1)$ and $\alpha < \beta$, then Player I has the same winning strategy also in $\mathrm{EF}_\beta(\mathcal{M}_0, \mathcal{M}_1)$. There is clearly no $\alpha$ such that both Player II and Player I have a winning strategy in $\mathrm{EF}_\alpha(\mathcal{M}_0, \mathcal{M}_1)$, as we can let the strategies play against each other. If $\mathcal{M}_0 \cong \mathcal{M}_1$, then Player II has a winning strategy in $\mathrm{EF}_\alpha(\mathcal{M}_0, \mathcal{M}_1)$ for all $\alpha$ by playing the isomorphism. Thus, if Player I has a winning strategy in $\mathrm{EF}_\alpha(\mathcal{M}_0, \mathcal{M}_1)$, then $\mathcal{M}_0 \not\cong \mathcal{M}_1$. If $\mathcal{M}_0 \not\cong \mathcal{M}_1$, then Player I has a winning strategy in $\mathrm{EF}_\alpha(\mathcal{M}_0, \mathcal{M}_1)$ for $\alpha \geq |M_0| + |M_1|$ by simply enumerating all elements of both models. Thus, if Player II has a winning strategy in $\mathrm{EF}_\alpha(\mathcal{M}_0, \mathcal{M}_1)$, where $\alpha \geq |M_0|+|M_1|$, then $\mathcal{M}_0 \cong \mathcal{M}_1$.

There is always at least one $\alpha$ for which Player II has a winning strategy in $\mathrm{EF}_\alpha(\mathcal{M}_0, \mathcal{M}_1)$, namely $\alpha = 0$. (Even this may fail if we have 0-place relation symbols or constant symbols, as observed after Definition 15.) If $\mathcal{M}_0 \cong \mathcal{M}_1$, then there cannot be any $\alpha$ for which Player I has a winning strategy in $\mathrm{EF}_\alpha(\mathcal{M}_0, \mathcal{M}_1)$. But if $\mathcal{M}_0 \not\cong \mathcal{M}_1$, then Player I has a winning strategy in $\mathrm{EF}_\alpha(\mathcal{M}_0, \mathcal{M}_1)$ from some $\alpha$ onwards.

There may be ordinals $\alpha$ for which neither player has a winning strategy. To get an example, note that $S \subseteq \omega_1$ contains a cub[1] if and only if Player I has a winning strategy in $\mathrm{EF}_{\omega+2}(\phi(S), \phi(\varnothing))$, and $S \subseteq \omega_1$ is disjoint from a cub if and only if Player II has a winning strategy in $\mathrm{EF}_{\omega+2}(\phi(S), \phi(\varnothing))$. Thus for bistationary[2] $S \subseteq \omega_1$ the game $\mathrm{EF}_{\omega+2}(\phi(S), \phi(\varnothing))$ is non-determined [Hyt87].

There may also be a limit ordinal $\alpha$ such that Player II has a winning strategy in $\mathrm{EF}_\beta(\mathcal{M}_0, \mathcal{M}_1)$ for each $\beta < \alpha$ but not in $\mathrm{EF}_\alpha(\mathcal{M}_0, \mathcal{M}_1)$. The simplest example is $\mathcal{M}_0 = (\mathbb{Z}, <)$, $\mathcal{M}_1 = (\mathbb{Z} + \mathbb{Z}, <)$, and $\alpha = \omega$.

**Lemma 37.** Let $L$ be a vocabulary and $\alpha$ an ordinal. The relation

$$\mathcal{M}_0 \sim_\alpha \mathcal{M}_1 \Leftrightarrow \text{Player II has a winning strategy in } \mathrm{EF}_\alpha(\mathcal{M}_0, \mathcal{M}_1)$$

is an equivalence relation on $\mathrm{Str}(L)$.

*Proof.* Reflexivity of $\sim_\alpha$ is clear, but since we have assumed models in the game have disjoint domains, we have to prove reflexivity for isomorphic structures, *i.e.*, if $f : \mathcal{M}_0 \cong \mathcal{M}_1$, then Player II has a winning strategy in $\mathrm{EF}_\alpha(\mathcal{M}_0, \mathcal{M}_1)$. In fact, Player II wins then $\mathrm{EF}_\alpha(\mathcal{M}_0, \mathcal{M}_0)$ with the trivial

[1] "Cub" abbreviates " closed and unbounded".

[2] *I.e.*, stationary and co-stationary.

strategy $\tau_\xi(\langle x_\eta : \eta \leq \xi \rangle) = f(x_\xi)$. Symmetricity is also trivial: Suppose Player II wins $\mathrm{EF}_\alpha(\mathcal{M}_0, \mathcal{M}_1)$ with $\boldsymbol{\tau} = \langle \tau_\xi : \xi < \alpha \rangle$. In fact, the very same strategy is winning for Player II in $\mathrm{EF}_\alpha(\mathcal{M}_1, \mathcal{M}_0)$. To prove transitivity of $\sim_\alpha$, suppose $\boldsymbol{\tau} = \langle \tau_\alpha : \xi < \alpha \rangle$ is a winning strategy of Player II in $\mathrm{EF}_\alpha(\mathcal{M}_0, \mathcal{M}_1)$ and $\boldsymbol{\tau}' = \langle \tau'_\xi : \xi < \alpha \rangle$ is a winning strategy of Player II in $\mathrm{EF}_\alpha(\mathcal{M}_1, \mathcal{M}_2)$. We describe a winning strategy $\boldsymbol{\tau}'' = \langle \tau''_\xi : \xi < \alpha \rangle$ of Player II in $\mathrm{EF}_\alpha(\mathcal{M}_0, \mathcal{M}_2)$. The idea is that Player II plays $\mathrm{EF}_\alpha(\mathcal{M}_0, \mathcal{M}_1)$ and $\mathrm{EF}_\alpha(\mathcal{M}_1, \mathcal{M}_2)$ simultaneously. Suppose $\mathbf{z}'' = \langle x''_\eta : \eta \leq \xi \rangle \in (M_0 \cup M_2)^{\xi+1}$. We define by induction over $\eta \leq \xi$ the sequences $\mathbf{z} = \langle x_\eta : \eta \leq \xi \rangle$, $\mathbf{z}' = \langle x'_\eta : \eta \leq \xi \rangle$, and $\boldsymbol{\tau}'' = \langle \tau''_\xi : \xi < \alpha \rangle$ as follows:

| If | $x''_\eta$ | $\in$ | $M_0$ | $M_2$ |
|---|---|---|---|---|
| Then | $x_\eta$ | $=$ | $x''_\eta$ | $\tau'_\eta(\mathbf{z}' \restriction_{\eta+1})$ |
| | $x'_\eta$ | $=$ | $\tau_\eta(\mathbf{z} \restriction_{\eta+1})$ | $x''_\eta$ |
| | $\tau''_\eta(\mathbf{z}'' \restriction_{\eta+1})$ | $=$ | $\tau'_\eta(\mathbf{z}' \restriction_{\eta+1})$ | $\tau_{\eta+1}(\mathbf{z} \restriction_{\eta+1})$ |

Now $\langle \tau''_\xi : \xi < \alpha \rangle$ is a winning strategy of Player II in $\mathrm{EF}_\alpha(\mathcal{M}_0, \mathcal{M}_2)$. q.e.d.

The relations $\sim_\alpha$ form a sequence of finer and finer partitions of $\mathrm{Str}(L)$, starting from the one-class partition $\sim_0$ and eventually approaching the ultimate refinement $\cong$ of every $\sim_\alpha$. Already $\sim_\omega$ separates all ordinals from each other, so the number of equivalence classes becomes a proper class. This is a serious setback.

How close can two structures be to being isomorphic without actually being isomorphic? The Ehrenfeucht-Fraïssé-game provides a method to answering this question. For countable models the Scott watershed gives a perfect scale in terms of countable ordinals as to how close the models are to being isomorphic. For models of cardinality $\lambda > \omega$ the analogous $\lambda$-Scott watershed does not yield as satisfactory a scale. In this case we may try the relations $\sim_\alpha$, $\alpha < \lambda$, as an alternative. After all, in this case $\sim_\lambda$ implies isomorphism. However, we immediately run into an open problem:

**Open problem:** Are there (in ZFC) two non-isomorphic structures $\mathcal{M}$ and $\mathcal{N}$ of cardinality $\aleph_1$ such that II has a winning strategy in $\mathrm{EF}_\alpha(\mathcal{M}, \mathcal{N})$ for all $\alpha < \omega_1$.

A positive answer is known only for the case $\alpha < \omega^2$, but even the case $\alpha = \omega^2$ is open. There is no problem if we assume CH (see Proposition 61). Assuming $V = L$ we can even get non-isomorphic ($\omega_1$-separable) Abelian groups $G$ and $H$ of size $\aleph_1$ such that II has a winning strategy in $\mathrm{EF}_\alpha(\mathcal{M}, \mathcal{N})$ for all $\alpha < \omega_1$. On the other hand, the Proper Forcing Axiom together with the assumption that II has a winning strategy in $\mathrm{EF}^{\omega_1}_{\omega^2+\omega}(G,H)$ implies $G \cong H$ for all $\omega_1$-separable abelian groups of cardinality $\aleph_1$ [EklForShe95]. For more on constructing models very close to being isomorphic, see [HytTuu91, Hyt92, HytSheTuu93, HytShe94, HytShe95, HytShe99].

Of special interest are games where the length $\alpha$ satisfies strong closure properties, as in the case that $\alpha$ is a cardinal, in particular in the case $\alpha = \omega_1$. It turns out that the statement that any two structures $\mathcal{M}, \mathcal{N}$ of cardinality $\aleph_2$ are isomorphic if and only if II has a winning strategy in $\mathrm{EF}_{\omega_1}(\mathcal{M}, \mathcal{N})$, is equiconsistent with the existence of a weakly compact cardinal [MekSheVää93, HytSheVää02].

## 10 On Partially Ordered Sets

In order to define the dynamic version of the transfinite Ehrenfeucht-Fraïssé game, and thereby make a further step towards obtaining a scale of how close two uncountable structures can be to being isomorphic, we need some preliminaries on po-sets.

A *po-set* (or a partial order) is a binary structure $\mathcal{P} = \langle P, \leq_{\mathcal{P}} \rangle$ which is transitive and anti-reflexive. A po-set $\mathcal{P}$ is a *tree* if it has a unique least element (the *root*) and the set of predecessors of any element is well-ordered by $\leq_{\mathcal{P}}$. Suppose $\mathcal{P}$ and $\mathcal{P}'$ are po-sets. We define $\mathcal{P} \leq \mathcal{P}'$ if there is a mapping $f : P \to P'$ such that for all $x, y \in P$: $x <_{\mathcal{P}} y \to f(x) <_{\mathcal{P}'} f(y)$. We write $\mathcal{P} < \mathcal{P}'$, if $\mathcal{P} \leq \mathcal{P}'$ and $\mathcal{P}' \not\leq \mathcal{P}$, and we write $\mathcal{P} \equiv \mathcal{P}'$, if $\mathcal{P} \leq \mathcal{P}'$ and $\mathcal{P}' \leq \mathcal{P}$. Note that $\leq$ is a transitive relation among po-sets.

For any ordinal $\alpha$ let $M'_\alpha$ be the tree of descending sequences $\beta_0 > \ldots > \beta_n$ of elements of $\alpha$ ordered by end-extension. Then $\alpha \leq \beta$ (as ordinals) if and only if $M'_\alpha \leq M'_\beta$ (as po-sets). Suppose $T$ is a tree. $T$ has no infinite branches if and only if there is an ordinal $\alpha$ so that $T \equiv \langle \alpha, > \rangle$.

Suppose $\mathcal{P}$ is a po-set. The tree $\sigma\mathcal{P}$ is defined as follows. Its domain is the set of functions $s$ with $\mathrm{dom}(s) \in \mathrm{Ord}$ such that for all $\alpha, \beta \in \mathrm{dom}(s)$ we have $\alpha < \beta \to s(\alpha) <_{\mathcal{P}} s(\beta)$. The order is $s \leq s' \leftrightarrow s = s' \restriction_{\mathrm{dom}(s)}$. $\sigma'\mathcal{P}$ is the suborder of $\sigma\mathcal{P}$ consisting of sequences $s \in \sigma\mathcal{P}$ of successor length. The $\sigma$-operation was introduced by Kurepa [Kur56], see also [TodVää99].

**Lemma 38.**

**(i)** $\sigma'\mathcal{P} \leq \mathcal{P}$.

**(ii)** $\sigma\mathcal{P} \not\leq \mathcal{P}$.

**(iii)** $\sigma'\mathcal{P} < \sigma\mathcal{P}$.

**(iv)** If $T$ is a tree, then $T \equiv \sigma' T$.

*Proof.* (i) If $s \in \sigma'\mathcal{P}$, let $f(s) = s(\mathrm{dom}(s) - 1)$. Then $f : \sigma'\mathcal{P} \to \mathcal{P}$ is order-preserving. (ii) Suppose $f : \sigma\mathcal{P} \to \mathcal{P}$ were order-preserving. Define inductively $s : \mathrm{Ord} \to \mathcal{P}$ by $s(\alpha) = f(s\restriction_\alpha)$. Since $\alpha < \beta$ implies $s(\alpha) <_{\mathcal{P}} s(\beta)$, we get the result that $\mathcal{P}$ is a proper class, a contradiction.

(iii) $\sigma'\mathcal{P} \leq \sigma\mathcal{P}$ trivially. If $\sigma\mathcal{P} \leq \sigma'\mathcal{P}$, then $\sigma\mathcal{P} \leq \mathcal{P}$ contrary to (ii),

(iv) We know already $\sigma' T \leq T$. Suppose $t \in T$ and $\langle t_\alpha : \alpha \leq \beta \rangle$ is the set of $t' \in T$ with $t' \leq_T t$ in ascending order. Let $\mathrm{dom}(s) = \beta + 1$ and $s_t(\alpha) = t_\alpha$. then $s_t \in \sigma' T$ and $t \mapsto s_t$ is order-preserving. q.e.d.

We have that $Q \not\leq \sigma Q$ since $\sigma Q$ is well-founded while $Q$ is not. In particular $Q \not\leq \sigma' Q$. Hence $\sigma' Q < Q$. Note that $\sigma' Q$ is a special tree while $\sigma Q$ is non-special. (Kurepa, [Kur56]).

**Lemma 39.** There is no sequence $\mathcal{P}_0, \mathcal{P}_1, \ldots$ so that $\sigma\mathcal{P}_{n+1} \leq \mathcal{P}_n$ for all $n < \omega$.

*Proof.* Suppose $f_n : \sigma\mathcal{P}_{n+1} \to \mathcal{P}_n$ is order-preserving. For each fixed $\alpha$, let $s^n_\alpha \in \mathcal{P}_n$ so that $f_n(\langle s^{n+1}_\beta : \beta < \alpha \rangle) = s^n_\alpha$. We get the result that each $\mathcal{P}_n$ is a proper class, a contradiction. q.e.d.

We now define a useful comparison game for posets: Suppose $\mathcal{P}$ and $\mathcal{P}'$ are po-sets. The game $G(\mathcal{P}, \mathcal{P}')$ is defined as follows. Player I plays $p_0 \in \mathcal{P}$, then Player II plays $p'_0 \in \mathcal{P}'$. After this Player I plays $p_1 \in \mathcal{P}$ with $p_0 <_{\mathcal{P}} p_1$, and then Player II plays $p'_1 \in \mathcal{P}'$ with $p'_0 <_{\mathcal{P}'} p'_1$, and so on. At limits Player I moves first $p_\nu \in \mathcal{P}$ with $p_\alpha <_{\mathcal{P}} p_\nu$ for all $\alpha < \nu$. Then Player II moves $p'_\nu \in \mathcal{P}'$ with $p'_\alpha <_{\mathcal{P}'} p'_\nu$ for all $\alpha < \nu$. If a player cannot move, he loses and the other player wins. Since $\mathcal{P}$ and $\mathcal{P}'$ are sets, one of the players eventually wins. Now we have the following two equivalences:

**(i)** $\sigma'\mathcal{P} \leq \mathcal{P}'$ if and only if Player II has a winning strategy in $G(\mathcal{P}, \mathcal{P}')$.

**(ii)** If $\mathcal{P}$ is a tree, then $\mathcal{P} \leq \mathcal{P}'$ if and only if Player II has a winning strategy in $G(\mathcal{P}, \mathcal{P}')$.

*Proof.* (i) Suppose $f : \sigma'\mathcal{P} \to \mathcal{P}'$ is order-preserving. If Player I has played $p_0 < \ldots < p_\alpha$ in $G(\mathcal{P}, \mathcal{P}')$, Player II plays $p'_\alpha = f(\langle p_0, \ldots, p_\alpha \rangle)$. In this way she ends up the winner. Conversely, suppose Player II has a winning strategy in $G(\mathcal{P}, \mathcal{P}')$ and $s \in \sigma'\mathcal{P}$ with $\mathrm{dom}(s) = \alpha + 1$. Let us play $G(\mathcal{P}, \mathcal{P}')$ so that Player I plays $p_\beta = s(\beta)$ for $\beta \leq \alpha$ and Player II uses his winning strategy. After Player I plays $p_\alpha$, Player II plays $p'_\alpha$. If we define $f(s) = p'_\alpha$, we get an order-preserving mapping $\sigma'\mathcal{P} \to \mathcal{P}'$. (i) is proved. (ii) follows from (i). q.e.d.

**Lemma 40.** $\sigma\mathcal{P}' \leq \mathcal{P}$ if and only if Player I has a winning strategy in $G(\mathcal{P},\mathcal{P}')$.

*Proof.* Suppose $f : \sigma\mathcal{P}' \to \mathcal{P}$ is order-preserving. If Player II has played

$$p'_0 < \ldots < p'_\beta < \ldots \ (\beta < \alpha) \tag{16}$$

in $G(\mathcal{P},\mathcal{P}')$, Player I plays $p_\alpha = f(\langle p'_0, \ldots, p'_\beta, \ldots\rangle)$ in $\mathcal{P}'$. In this way Player I has a winning strategy in $G(\mathcal{P},\mathcal{P}')$. On the other hand, if Player I has a winning strategy in $G(\mathcal{P},\mathcal{P}')$ and (16) is an ascending chain in $\mathcal{P}'$, we can let Player I play against the moves $p'_0, \ldots, p'_\beta, \ldots$ of Player II in $G(\mathcal{P},\mathcal{P}')$. Finally Player I plays $p_\alpha$ according to his winning strategy. We let $f(\langle p'_0, \ldots, p'_\beta, \ldots\rangle) = p_\alpha$. Now $f : \sigma\mathcal{P}' \to \mathcal{P}$ is order-preserving. q.e.d.

**Example 41.** Suppose $S \subseteq \omega_1$. Let $T(S)$ be the tree of closed ascending sequences of elements of $S$. Choose disjoint bistationary sets $S_1$ and $S_2$ (it suffices that $S_1 \cap S_2$ is non-stationary). Then $T(S_1) \not\leq T(S_2)$ and $T(S_2) \not\leq T(S_1)$. Thus the game $G(T(S_1), T(S_2))$ is non-determined. $T(S) \leq \mathbb{Q}$ if and only if $S$ is non-stationary. If $S \subseteq \omega_1$ is bistationary and $T$ is Aronszajn, then $T(S) \not\leq T$ and $T \not\leq T(S)$. [TodVää99].

If $T_i$, $i \in I$ is a family of trees, let $\bigoplus_{i\in I} T_i$ be the tree which consists of a union of disjoint copies of $T_i$, $i \in I$, identified at the root. Then $\bigoplus_{i\in I} T_i$ is the supremum of $\{T_i : i \in I\}$ in the sense that $T_i \leq \bigoplus_{i\in I} T_i$ for all $i \in I$ and if $T_i \leq T$ for all $i \in I$, then $\bigoplus_{i\in I} T_i \leq T$. On the other hand, let $\prod_{i\in I} T_i$ be the product tree

$$\prod_{i\in I} T_i = \{s : \mathrm{dom}(s) = I, \forall i \in I(s(i) \in T_i)\}.$$

$$s \leq s' \iff \forall i \in I(s(i) \leq_{T_i} s'(i)).$$

Let $\bigotimes_{i\in I} T_i$ be the subtree

$$\bigotimes_{i\in I} T_i = \{s \in \prod_{i\in I} T_i : \forall i \in I \forall j \in I(\mathrm{ht}_{T_i}(s(i)) = \mathrm{ht}_{T_j}(s(j)))\}.$$

Then $\bigotimes_{i\in I} T_i$ is the infimum of $\{T'_i : i \in I\}$, that is, $\bigotimes_{i\in I} T_i \leq T_i$ for each $i \in I$, and if $T \leq T_i$ for all $i \in I$, then $T \leq \bigotimes_{i\in I} T_i$.

The order $\leq$ between trees with no uncountable branches is quite interesting and not at all completely known. Unlike the order $\leq$ among well-founded trees, which has the property $T \leq T'$ or $T' \leq T$ for all $T$, the order $\leq$ between trees with no uncountable branches has big blobs of incomparable trees. This order has been studied in [TodVää99, Tod07].

## 11 The Transfinite Dynamic Ehrenfeucht-Fraïssé game

In this section we introduce a more general form of the Ehrenfeucht-Fraïssé-game. In this game Player I moves up a po-set, move by move. The game goes on as long as Player I can move. If $\mathcal{P}$ is a po-set, let $\mathbf{r}(\mathcal{P})$ denote the least ordinal $\delta$ so that $\mathcal{P}$ does not have an ascending chain of length $\delta$. This game generalizes at the same time the games $\mathrm{EF}_\alpha(\mathcal{M}_0, \mathcal{M}_1)$ and $\mathrm{EFD}^\delta(\mathcal{M}_0, \mathcal{M}_1)$.

**Definition 42.** Suppose $\mathcal{M}_0$ and $\mathcal{M}_1$ are $L$-structures and $\mathcal{P}$ is a po-set. The game $\mathrm{EF}_\mathcal{P}(\mathcal{M}_0, \mathcal{M}_1)$ is like $\mathrm{EF}_\delta(\mathcal{M}_0, \mathcal{M}_1)$ except that on each round Player I chooses an element $x_\alpha \in M_0 \cup M_1$ and an element $p_\alpha \in \mathcal{P}$. It is required that $p_0 <_\mathcal{P} \ldots <_\mathcal{P} p_\alpha <_\mathcal{P} \ldots$. Finally Player I cannot play a new $p_\alpha$ anymore. Suppose Player I has played $\mathbf{z} = \langle \langle x_\beta, p_\beta \rangle : \beta < \alpha \rangle$ and Player II has played $\mathbf{y} = \langle y_\beta : \beta < \alpha \rangle$. If the arising mapping is a partial isomorphism between $\mathcal{M}_0$ and $\mathcal{M}_1$, Player II has won the game, otherwise Player I has won.

Thus a winning strategy of Player I in $\mathrm{EF}_\mathcal{P}(\mathcal{M}_0, \mathcal{M}_1)$ is a sequence $\boldsymbol{\varrho} = \langle \varrho_\alpha : \alpha < \mathbf{r}(\mathcal{P}) \rangle$ and a strategy of Player II is a sequence $\boldsymbol{\tau} = \langle \tau_\alpha : \alpha < \mathbf{r}(\mathcal{P}) \rangle$. Note that the old $\mathrm{EF}_\alpha(\mathcal{M}_0, \mathcal{M}_1)$ is the same game as $\mathrm{EF}_{\langle \alpha, < \rangle}(\mathcal{M}_0, \mathcal{M}_1)$, and the old $\mathrm{EFD}^\alpha(\mathcal{M}_0, \mathcal{M}_1)$ is the same game as $\mathrm{EF}_{\langle \alpha, > \rangle}(\mathcal{M}_0, \mathcal{M}_1)$. Naturally, the games $\mathrm{EF}_{\langle \alpha, < \rangle}(\mathcal{M}_0, \mathcal{M}_1)$ and $\mathrm{EF}_{\langle \alpha, > \rangle}(\mathcal{M}_0, \mathcal{M}_1)$ are one and the same game, if $\alpha$ is finite. But if $\alpha$ happens to be infinite, there is a big difference. Then the first is a transfinite game while the second can only go on for a finite number of moves.

It turns out that po-sets $\mathcal{P}$ with $\mathbf{r}(\mathcal{P}) < \lambda$ give a perfect scaling for $\mathrm{EF}_\lambda$: Player II has a winning strategy in $\mathrm{EF}_\lambda(\mathcal{M}, \mathcal{N})$ if and only if she has a winning strategy in $\mathrm{EF}_\mathcal{P}(\mathcal{M}, \mathcal{N})$ for every $\mathcal{P}$ with $\mathbf{r}(\mathcal{P}) < \lambda$ (Proposition 48). In particular, for models $\mathcal{M}$ and $\mathcal{N}$ of cardinality $\lambda$, the condition "II has a winning strategy in $\mathrm{EF}_\mathcal{P}(\mathcal{M}, \mathcal{N})$", when $\mathcal{P}$ ranges over all $\mathcal{P}$ with $\mathbf{r}(\mathcal{P}) < \lambda$, provide a perfect scale of non-isomorphism. First we need some elementary properties, making sure that the ordering $\mathcal{P} \leq \mathcal{P}'$ of po-sets is neatly connected to the game $\mathrm{EF}_\mathcal{P}(\mathcal{M}_0, \mathcal{M}_1)$:

**Lemma 43.** If Player II has a winning strategy in $\mathrm{EF}_{\mathcal{P}'}(\mathcal{M}_0, \mathcal{M}_1)$ and $\mathcal{P} \leq \mathcal{P}'$, then Player II has a winning strategy in $\mathrm{EF}_\mathcal{P}(\mathcal{M}_0, \mathcal{M}_1)$. If Player I has a winning strategy in $\mathrm{EF}_\mathcal{P}(\mathcal{M}_0, \mathcal{M}_1)$ and $\mathcal{P} \leq \mathcal{P}'$, then Player I has a winning strategy in $\mathrm{EF}_{\mathcal{P}'}(\mathcal{M}_0, \mathcal{M}_1)$.

**Proposition 44.** Suppose Player II has a winning strategy in $\mathrm{EF}_\mathcal{P}(\mathcal{M}_0, \mathcal{M}_1)$ and Player I has a winning strategy in $\mathrm{EF}_{\mathcal{P}'}(\mathcal{M}_0, \mathcal{M}_1)$. Then $\sigma\mathcal{P} \leq \mathcal{P}'$.

*Proof.* Suppose Player II has a winning strategy $\boldsymbol{\tau}$ in $\mathrm{EF}_\mathcal{P}(\mathcal{M}_0, \mathcal{M}_1)$ and

Player I has a winning strategy $\varrho$ in $\mathrm{EF}_{\mathcal{P}'}(\mathcal{M}_0, \mathcal{M}_1)$. We describe a winning strategy of Player I in $G(\mathcal{P}', \mathcal{P})$. Suppose $\varrho_0(\emptyset) = \langle x_0, p'_0 \rangle$. This $p'_0$ is the first move of Player I in $G(\mathcal{P}', \mathcal{P})$. Suppose Player II plays $p_0 \in \mathcal{P}$. Let

$$\begin{aligned} y_0 &= \tau_0(\langle x_0, p_0 \rangle), \\ \langle x_1, p'_1 \rangle &= \tau_1(\langle y_0 \rangle). \end{aligned}$$

This $p'_1$ is the second move of Player I in $G(\mathcal{P}', \mathcal{P})$. More generally the equations

$$\begin{aligned} y_\beta &= \tau_\beta(\langle \langle x_\gamma, p_\gamma \rangle : \gamma \leq \beta \rangle) \\ \langle x_\alpha, p'_\alpha \rangle &= \tau_\alpha(\langle y_\beta : \beta < \alpha \rangle) \end{aligned}$$

define the move $p'_\alpha$ of Player I in $G(\mathcal{P}', \mathcal{P})$ after Player II has played $\langle p_\beta : \beta < \alpha \rangle$. This game can only end if Player II cannot move $p_\alpha$ at some point, so Player I wins. q.e.d.

If $T$ is a tree, $T + 1$ is the tree which is obtained from $T$ by adding a new element at the end of every maximal branch of $T$.

**Lemma 45.** Suppose $S \subseteq \omega_1$ is bistationary, $\mathcal{M}_0 = \phi(S)$ and $\mathcal{M}_1 = \phi(\emptyset)$. Then Player I has a winning strategy in $\mathrm{EF}_{\sigma\mathcal{P}}(\mathcal{M}_0, \mathcal{M}_1)$, if $\mathcal{P} = T(\omega_1 \backslash S)+1$.

*Proof.* Suppose Player I has to decide how to play $\langle x_\alpha, p_\alpha \rangle$ in $\mathrm{EF}_{\mathcal{P}}(\mathcal{M}_0, \mathcal{M}_1)$. Suppose he has played already $\langle x_\beta, p_\beta \rangle$ and Player II has played $y_\beta$ for $\beta < \alpha$. We assume that Player I has played so that

1. $p_\beta = \langle \langle \delta_\delta : \delta \leq \gamma \rangle : \gamma < \beta \rangle$ $(\in \sigma(T(\omega_1 \setminus S) + 1))$.
2. $y_{\nu+2n} < x_{\nu+2n+1}$ in $\mathcal{M}_0$.
3. $y_{\nu+2n+1} < x_{\nu+2n+2}$ in $\mathcal{M}_1$.
4. $\mathrm{rank}(y_{\nu+2n+1}) < \delta_{\nu+2n+1} < \mathrm{rank}(x_{\nu+2n+2})$.
5. $\mathrm{rank}(y_{\nu+2n}) < \delta_{\nu+2n} < \mathrm{rank}(x_{\nu+2n+1})$.
6. $\delta_\beta < \delta_\gamma$ in $S$ for $\beta < \gamma$.
7. $\delta_\nu = \sup_{\beta<\nu} \delta_\beta$ for limit $\nu$.

It is clear that Player I can continue playing so that the above conditions hold until $\delta_\nu \in S$ for some limit $\nu$. At this point Player I plays the supremum $x_\nu$ of the previous moves in $M_0$. Suppose Player II moves $y_\nu \in M_1$. Now $y_\nu$ cannot be the supremum of the previous moves in $M_1$ (as $\delta_\nu \in S$), so Player I wins with his next move. The "+1" part of $\mathcal{P}$ guarantees that Player I does indeed have one more move. q.e.d.

The proof of the following lemma is similar:

**Lemma 46.** Suppose $S \subseteq \omega_1 \setminus \{0\}$ is bistationary, $\mathcal{M}_0 = \phi(S)$ and $\mathcal{M}_1 = \phi(\varnothing)$. Then Player I does not have a winning strategy in $\mathrm{EF}_{\mathcal{P}}(\mathcal{M}_0, \mathcal{M}_1)$, if $\mathcal{P} = T(\omega_1 \setminus S) + 1$.

## 12 Karp and Scott Trees

Suppose $\mathcal{M}_0 \not\cong \mathcal{M}_1$. Then there is a

$$\delta \leq \mathrm{Card}(M_0) + \mathrm{Card}(M_1)$$

such that Player II does not have a winning strategy in $\mathrm{EF}_\delta(\mathcal{M}_0, \mathcal{M}_1)$. We examine now the case that the least such $\delta$ is a limit ordinal. Thus for all $\alpha < \delta$ there is a winning strategy for Player II in $\mathrm{EF}_{\alpha+1}(\mathcal{M}_0, \mathcal{M}_1)$. In [Hyt87] a tree $K(= K(\mathcal{M}_0, \mathcal{M}_1))$, consisting of all winning strategies of Player II in $\mathrm{EF}_{\alpha+1}(\mathcal{M}_0, \mathcal{M}_1)$ for $\alpha < \delta$, is introduced. We can make $K$ a tree by letting

$$\langle \tau_\xi : \xi \leq \alpha \rangle \leq \langle \tau'_\xi : \xi \leq \alpha' \rangle$$

if and only if $\alpha \leq \alpha'$ and $\forall \xi \leq \alpha(\tau_\xi = \tau'_\xi)$. We call $K$ the *canonical Karp tree* of the pair $(\mathcal{M}_0, \mathcal{M}_1)$. Note that $K$ does not have a branch of length $\delta$, for otherwise Player II would win $\mathrm{EF}_\delta(\mathcal{M}_0, \mathcal{M}_1)$.

**Lemma 47.** Suppose $\mathcal{P}$ is a po-set. Then

$$\text{Player II has a winning strategy in } \mathrm{EF}_{\mathcal{P}}(\mathcal{M}_0, \mathcal{M}_1) \iff \sigma'\mathcal{P} \leq K.$$

*Proof.* $\Rightarrow$ Suppose Player II has a winning strategy $\boldsymbol{\tau}$ in $\mathrm{EF}_{\mathcal{P}}(\mathcal{M}_0, \mathcal{M}_1)$. If $s = \langle s_\xi : \xi \leq \alpha \rangle \in \sigma'\mathcal{P}$, we can define a strategy $\boldsymbol{\tau}'$ of Player II in $\mathrm{EF}_{\alpha+1}(\mathcal{M}_0, \mathcal{M}_1)$ as follows

$$\tau'_\xi(\langle x_\eta : \eta \leq \xi \rangle) = \tau_\xi(\langle \langle x_\eta, s_\eta \rangle : \eta \leq \xi \rangle).$$

Since $K$ does not have a branch of length $\delta$, $\alpha < \delta$, and hence $\boldsymbol{\tau}' \in K$. The mapping $s \mapsto \boldsymbol{\tau}'$ is an order-preserving mapping $\sigma'\mathcal{P} \to K$.

$\Leftarrow$ Suppose $f : \sigma'\mathcal{P} \to K$ is order-preserving. We can define a winning strategy of Player II in $\mathrm{EF}_{\mathcal{P}}(\mathcal{M}_0, \mathcal{M}_1)$ by the equation

$$\tau_\alpha(\langle \langle x_\xi, s_\xi \rangle : \eta \leq \xi \rangle) = f(\langle s_\xi : \xi \leq \alpha \rangle)(\langle \langle x_\xi, s_\eta \rangle : \xi \leq \alpha \rangle).$$

q.e.d.

**Proposition 48 (Hyttinen, [Hyt87]).** Suppose $\delta$ is a limit ordinal such that Player II has a winning strategy in $\mathrm{EF}_\alpha(\mathcal{M}_0, \mathcal{M}_1)$ for all $\alpha < \delta$. The following conditions are equivalent:

**(i)** Player II has a winning strategy in $\mathrm{EF}_\delta(\mathcal{M}_0, \mathcal{M}_1)$.

**(ii)** Player II has a winning strategy in $\mathrm{EF}_\mathcal{P}(\mathcal{M}_0, \mathcal{M}_1)$ for every po-set $\mathcal{P}$ with no branches of length $\delta$.

*Proof.* (i)$\rightarrow$(ii) is trivial. To prove (ii)$\rightarrow$(i), suppose Player II does not have a winning strategy in $\mathrm{EF}_\delta(\mathcal{M}_0, \mathcal{M}_1)$. Let $\mathcal{P} = K(\mathcal{M}_0, \mathcal{M}_1)$. Then $\sigma\mathcal{P}$ does not have branches of length $\delta$, hence by (ii) Player II has a winning strategy in $\mathrm{EF}_{\sigma\mathcal{P}}(\mathcal{M}_0, \mathcal{M}_1)$ and we get $\sigma\mathcal{P} \leq \mathcal{P}$ from Lemma 47, a contradiction.

q.e.d.

**Note:** Player II has a winning strategy in $\mathrm{EF}_\mathcal{P}(\mathcal{M}_0, \mathcal{M}_1)$ if and only if Player II has a winning strategy in $\mathrm{EF}_{\sigma'\mathcal{P}}(\mathcal{M}_0, \mathcal{M}_1)$. So from the point of view of the winning of Player II we could always assume that $\mathcal{P}$ is a tree.

**Corollary 49.** Player II has never a winning strategy in $\mathrm{EF}_{\sigma K}(\mathcal{M}_0, \mathcal{M}_1)$

**Definition 50.** A po-set $\mathcal{P}$ is a *Karp-po-set* of the pair $(\mathcal{M}_0, \mathcal{M}_1)$ if Player II has a winning strategy in $\mathrm{EF}_\mathcal{P}(\mathcal{M}_0, \mathcal{M}_1)$ but not in $\mathrm{EF}_{\sigma\mathcal{P}}(\mathcal{M}_0, \mathcal{M}_1)$. If a Karp-po-set is a tree, we call it a *Karp-tree* [HytVää90].

By the above, there are always Karp-trees for pairs of non-isomorphic structures. Suppose Player I has a winning strategy $\boldsymbol{\varrho}$ in $\mathrm{EF}_\mathcal{P}(\mathcal{M}_0, \mathcal{M}_1)$. Let $S_{\boldsymbol{\varrho}}$ be the set of sequences $\mathbf{y} = \langle y_\xi : \xi \leq \alpha\rangle \in \mathrm{dom}(\boldsymbol{\varrho})$ such that

$$p_{\boldsymbol{\varrho}\restriction\alpha+1,y} \in \mathrm{Part}(\mathcal{M}_0, \mathcal{M}_1).$$

Thus $S_{\boldsymbol{\varrho}}$ is the set of sequences of moves of Player II before she loses $\mathrm{EF}_\mathcal{P}(\mathcal{M}_0, \mathcal{M}_1)$, when Player I plays $\boldsymbol{\varrho}$. We can make $S_{\boldsymbol{\varrho}}$ a tree by ordering it as follows

$$\langle y_\xi : \xi \leq \alpha\rangle \leq \langle y'_\xi : \xi \leq \alpha'\rangle$$

if and only if $\alpha \leq \alpha'$ and $\forall\xi \leq \alpha(y_\xi = y'_\xi)$.

**Lemma 51.** Player I has a winning strategy in $\mathrm{EF}_{\sigma S_{\boldsymbol{\varrho}}}(\mathcal{M}_0, \mathcal{M}_1)$.

*Proof.* The following equation defines a winning strategy $\boldsymbol{\varrho}'$ of Player I in $\mathrm{EF}_{\sigma S_{\boldsymbol{\varrho}}}(\mathcal{M}_0, \mathcal{M}_1)$:

$$\varrho'_\alpha(\langle y_\xi : \xi < \alpha\rangle) = \langle x_\alpha, \langle\langle y_\xi : \xi \leq \beta\rangle : \beta < \alpha\rangle\rangle,$$

where

$$\varrho_\alpha(\langle y_\xi : \xi < \alpha\rangle) = \langle x_\alpha, p_\alpha\rangle.$$

q.e.d.

**Lemma 52.** $\sigma S_{\varrho} \leq \mathcal{P}$.

*Proof.* Suppose $s = \langle\langle y_\xi : \xi \leq \beta_\eta\rangle : \eta < \alpha\rangle \in \sigma S_{\varrho}$, where

$$\beta_0 < \beta_1 < \ldots < \beta_\eta < \ldots (\eta < \alpha).$$

Let $\delta = \sup_{\eta<\alpha} \beta_\eta$ and

$$\varrho_\delta(\langle y_\xi : \xi < \delta\rangle) = \langle x_\delta, p_\delta\rangle.$$

We define $f(s) = p_\delta$. Then $f : \sigma S_{\varrho} \to \mathcal{P}$ is order-preserving. q.e.d.

Note that Lemma 52 implies $\mathcal{P} \not\leq S_{\varrho}$. In particular, if Player I has a winning strategy $\varrho$ in $\mathrm{EF}_\delta(\mathcal{M}_0, \mathcal{M}_1)$, then $S_{\varrho}$ is a tree with no branches of length $\delta$.

Suppose $\mathcal{P}_0$ is such that $\sigma\mathcal{P}_0 \leq \mathcal{P}$ and Player I has a winning strategy in $\mathrm{EF}_{\sigma\mathcal{P}_0}$. So $\mathcal{P}_0$ could be $S_{\varrho}$. Suppose furthermore that there is no $\mathcal{P}_1$ such that $\sigma\mathcal{P}_1 \leq \mathcal{P}_0$ and Player I has a winning strategy in $\mathrm{EF}_{\sigma\mathcal{P}_1}$. Lemma 39 implies that this assumption can always be satisfied.

**Lemma 53.** Player I does not have a winning strategy in $\mathrm{EF}_{\mathcal{P}_0}(\mathcal{M}_0, \mathcal{M}_1)$.

*Proof.* Suppose Player I has a winning strategy $\varrho'$ in $\mathrm{EF}_{\mathcal{P}_0}(\mathcal{M}_0, \mathcal{M}_1)$. Then Player I has a winning strategy in $\mathrm{EF}_{\sigma S_{\varrho'}}(\mathcal{M}_0, \mathcal{M}_1)$ and $\sigma S_{\varrho'} \leq \mathcal{P}_0$, contrary to the choice of $\mathcal{P}_0$. q.e.d.

**Definition 54.** A po-set $\mathcal{P}$ is a *Scott-po-set* of $(\mathcal{M}_0, \mathcal{M}_1)$ if Player I has a winning strategy in $\mathrm{EF}_{\sigma\mathcal{P}}(\mathcal{M}_0, \mathcal{M}_1)$ but not in $\mathrm{EF}_{\mathcal{P}}(\mathcal{M}_0, \mathcal{M}_1)$. If a Scott-po-set is a tree, we call is a *Scott-tree* [HytVää90].

By the above, $S_{\varrho}$ is always a Scott-tree of $(\mathcal{M}_0, \mathcal{M}_1)$, so Scott-trees always exist. Note that

$$\mathrm{Card}(S_{\varrho}) \leq \sup_{\alpha<b(\mathcal{P})} (\mathrm{Card}(M_0) + \mathrm{Card}(M_1))^\alpha.$$

Note that if Player I has a winning strategy in $\mathrm{EF}_{T_i}(\mathcal{M}_0, \mathcal{M}_1)$ for all $i \in I$, then Player II does not have a winning strategy in $\mathrm{EF}_{\bigotimes_{i\in I} T_i}(\mathcal{M}_0, \mathcal{M}_1)$. The families of Scott and Karp trees of $(\mathcal{M}_0, \mathcal{M}_1)$ are closed under suprema. If $\mathcal{P}$ is a Scott-po-set of $(\mathcal{M}_0, \mathcal{M}_1)$, where $\mathrm{Card}(M_0), \mathrm{Card}(M_1) \leq 2^{\aleph_0}$, then there is a Scott-tree of $(\mathcal{M}_0, \mathcal{M}_1)$ such that $T \leq \mathcal{P}$ and $\mathrm{Card}(T) \leq 2^{\aleph_0}$.

**Lemma 55.** Suppose Player I has a winning strategy $\varrho$ in $\mathrm{EF}_{\mathcal{P}}(\mathcal{M}_0, \mathcal{M}_1)$ and $K = K(\mathcal{M}_0, \mathcal{M}_1)$. Then $K \leq S_{\varrho}$.

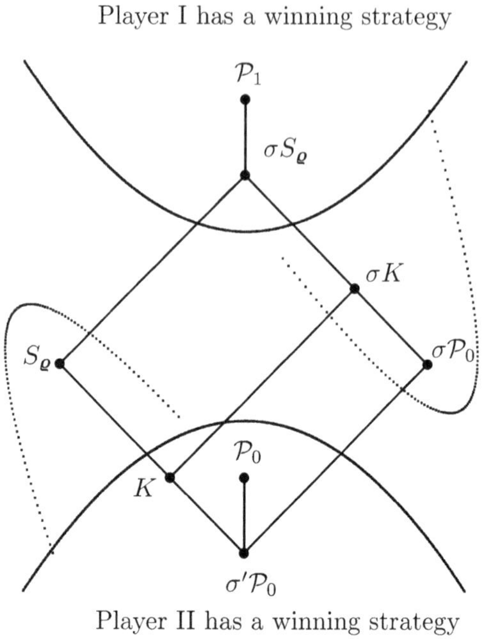


Figure 1. The boundary between Player II having a winning strategy and Player I having a winning strategy.

*Proof.* Suppose $\boldsymbol{\tau} \in K$. Let Player II play $\boldsymbol{\tau}$ against $\boldsymbol{\varrho}$ in $\mathrm{EF}_{\mathcal{P}}(\mathcal{M}_0, \mathcal{M}_1)$. The resulting sequence $\mathbf{y}$ of moves of Player II is an element of $S_{\boldsymbol{\varrho}}$. The mapping $\boldsymbol{\tau} \mapsto \mathbf{y}$ is order-preserving. q.e.d.

Suppose Player II has a winning strategy in $\mathrm{EF}_{\mathcal{P}_0}(\mathcal{M}_0, \mathcal{M}_1)$ and Player I has a winning strategy $\boldsymbol{\varrho}$ in $\mathrm{EF}_{\mathcal{P}_1}(\mathcal{M}_0, \mathcal{M}_1)$. Figure 1 shows the resulting picture.

In summary, we have proved:

**Theorem 56.** Suppose Player II has a winning strategy in $\mathrm{EF}_{\mathcal{P}_0}(\mathcal{M}_0, \mathcal{M}_1)$ and Player I has a winning strategy in $\mathrm{EF}_{\mathcal{P}_1}(\mathcal{M}_0, \mathcal{M}_1)$. Then there are trees $T_0$ and $T_1$ such that

(i) $\sigma'\mathcal{P}_0 \leq T_0 \leq T_1 \leq \mathcal{P}_1$.
(ii) Player II has a winning strategy in $\mathrm{EF}_{T_0}(\mathcal{M}_0, \mathcal{M}_1)$ but not in $\mathrm{EF}_{\sigma T_0}(\mathcal{M}_0, \mathcal{M}_1)$.

(iii) Player I has a winning strategy in $\mathrm{EF}_{\sigma T_1}(\mathcal{M}_0, \mathcal{M}_1)$ but not in $\mathrm{EF}_{T_1}(\mathcal{M}_0, \mathcal{M}_1)$.

**Example 57.** Suppose Player I has a winning strategy in $\mathrm{EF}_\omega(\mathcal{M}_0, \mathcal{M}_1)$. There is a unique $\delta = \delta(\mathcal{M}_0, \mathcal{M}_1)$ such that Player II has a winning strategy in $\mathrm{EF}_{\omega,\delta}(\mathcal{M}_0, \mathcal{M}_1)$ and Player I has a winning strategy in $\mathrm{EF}_{\omega,\delta+1}(\mathcal{M}_0, \mathcal{M}_1)$. Then $\langle \delta, > \rangle$ is both a Karp- and a Scott-po-set for $\mathcal{M}_0$ and $\mathcal{M}_1$.

**Example 58.** Suppose Player II has a winning strategy in $\mathrm{EF}_\alpha(\mathcal{M}_0, \mathcal{M}_1)$ but not in $\mathrm{EF}_{\alpha+1}(\mathcal{M}_0, \mathcal{M}_1)$. Then $\langle \alpha, < \rangle$ is a Karp-tree (in fact a Karp-well-order) of $\mathcal{M}_0$ and $\mathcal{M}_1$. This follows from the fact that $\sigma\langle \alpha, < \rangle \equiv \langle \alpha+1, < \rangle$.

**Example 59.** Suppose Player I has a winning strategy in $\mathrm{EF}_{\alpha+1}(\mathcal{M}_0, \mathcal{M}_1)$ but not in $\mathrm{EF}_\alpha(\mathcal{M}_0, \mathcal{M}_1)$. Then $\langle \alpha, < \rangle$ is a Scott-tree (in fact a Scott-well-order) of $\mathcal{M}_0$ and $\mathcal{M}_1$.

**Example 60.** Let $S \subseteq \omega_1 \setminus \{0\}$ be a bistationary set. Then $\langle \omega + 1, < \rangle$ is a Karp-tree of $\phi(S)$ and $\phi(\varnothing)$ (see remark before Lemma 37), and $T(\omega_1 \setminus S)+1$ is a Scott-tree of $\phi(S)$ and $\phi(\varnothing)$ (Example 45). In this example all Karp po-sets $\mathcal{P}_0$ must satisfy $\mathbf{r}(\mathcal{P}_0) \leq \omega + 1$ and all Scott po-sets $\mathcal{P}_1$ must satisfy $\mathbf{r}(\mathcal{P}_1) \geq \omega_1$, so the Karp po-sets and Scott po-sets are far apart.

We end with two examples of pairs of structures where the Karp- and Scott-po-set are large but close to each other. In fact in both cases they coincide with each other so we can locate exactly where the boundary between Player I winning and Player II winning is.

**Theorem 61.** There are trees $\mathcal{M}_0$ and $\mathcal{M}_1$ of cardinality $2^\omega$ such that $\langle \mathbb{Q}, < \rangle$ is both a Karp- and Scott-po-set of $\mathcal{M}_0$ and $\mathcal{M}_1$.

*Proof.* Let $\mathcal{M}_0$ be the tree of sequences $s = \langle \alpha_0, q_0, \ldots, \alpha_\gamma, q_\gamma \rangle$, where $q_0 < \cdots < q_\gamma$ in $\mathbb{Q}$ and $\alpha_\xi \in \omega_1$, ordered by end-extension. Let $\mathcal{M}_1$ be the tree of sequences $s = \langle \alpha_0, q_0, \ldots, \alpha_\gamma \rangle$ where $q_0 < q_1 < \cdots$ in $\mathbb{Q}$, $\sup q_\xi < \infty$ and $\alpha_\xi \in \omega_1$, ordered by end-extension. If $r \in \mathbb{Q}$, let $\lfloor r \rfloor$ be the integer part of $r$. If $s \in M_0 \cup M_1$, $\lfloor s \rfloor$ is the maximum value of $q_\xi$ in s. Similarly, $\sup(s)$ and $\max(s)$ are defined as $\sup(q_\xi)$ and $\max(q_\xi)$.

We need a new po-set for the proof. This po-set, which we denote by $\mathbb{P}_q$ consists of triples $\langle q, \alpha, \beta \rangle$, where $q \in \mathbb{Q}$ and $\beta < \alpha < \omega_1$. The order is as follows: $\langle q, \alpha, \beta \rangle \leq \langle q', \alpha', \beta' \rangle$ iff $q < q'$ or $q = q', \alpha = \alpha'$ and $\beta \leq \beta'$. Thus $\mathbb{P}_q$ can be obtained from $\langle \mathbb{Q}, < \rangle$ by replacing rationals by well-ordered sets $< \omega_1$, in all possible ways. It is not hard to show that $\mathbb{P}_q \leq \langle \mathbb{Q}, < \rangle$. Let $\mathrm{EF}'_{\mathbb{P}}(\mathcal{M}_0, \mathcal{M}_1)$ be the variant of $\mathrm{EF}_{\mathbb{P}}(\mathcal{M}_0, \mathcal{M}_1)$ in which Player I can only play $\langle x_\alpha, p_\alpha \rangle$ in $M_c$, $c \in \{0, 1\}$, if for all predecessors $z$ of $x_\alpha$ in $M_c$ there is $\beta < \alpha$ such that $x_\beta = z$ or $y_\beta = z$. The point of the new game is that if Player II has a winning strategy in $\mathrm{EF}'_{\mathbb{P}_q}(\mathcal{M}_0, \mathcal{M}_1)$, she has a winning strategy in $\mathrm{EF}_{\mathbb{Q}}(\mathcal{M}_0, \mathcal{M}_1)$. So we only need to show she has a winning strategy in $\mathrm{EF}'_{\mathbb{P}_q}(\mathcal{M}_0, \mathcal{M}_1)$. But since $\mathbb{P}_q \leq \langle \mathbb{Q}, < \rangle$, we only need to show Player II has a winning strategy in $\mathrm{EF}'_{\mathbb{Q}}(\mathcal{M}_0, \mathcal{M}_1)$. Suppose the moves $\langle x_\beta, r_\beta \rangle, \beta \leq \alpha$, have already been made by Player I and the moves $y_\beta, \beta < \alpha$, by Player II. Now Player II should decide what her next move $y_\alpha$ is. During the game Player II has maintained a certain strategy. To describe it, let $g_n : \mathbb{Q} \to \mathbb{Q} \cap [n, n+1)$ be order-preserving. The conditions are

(1) $\lfloor y_\beta \rfloor = \lfloor x_\beta \rfloor + 1$
(2) $\sup y_\beta \leq g_{\lfloor y_\beta \rfloor}(r_\beta)$

Assume first the rank of $x_\alpha$ is a limit ordinal. Thus there is an ascending sequence $a_{\alpha_\beta}, \beta < \nu$, of predecessors of $x_\alpha$ in the relevant model $\mathcal{M}_{c_\alpha}$ and they have been played already. Here $\nu = \cup\nu$. Thus $a_{\alpha_\beta} = x_{\alpha_\beta}$ or $a_{\alpha_\beta} = y_{\alpha_\beta}$ for each $\beta < \nu$. Because of (1) there is $\beta_0 < \nu$ such that $a_{\alpha_\beta} = x_{\alpha_\beta}$ for $\beta_0 \leq \beta < \nu$ or $a_{\alpha_\beta} = y_{\alpha_\beta}$ for $\beta_0 \leq \beta < \nu$. We may also assume $\lfloor a_{\alpha_\beta} \rfloor = N$ for some constant $N \in \omega$ for $\beta_0 \leq \beta < \nu$.

**Case 1.** $a_{\alpha_\beta} = x_{\alpha_\beta}$ for $\beta_0 \leq \beta < \nu$. Let $b_{\alpha_\beta} = y_{\alpha_\beta}$. By (1), $\lfloor b_{\alpha_\beta} \rfloor = N + 1$. By (2),

$$\sup_{\beta<\nu} \sup b_{\alpha_\beta} \leq \sup_{\beta<\nu} g_{N+1}(r_\beta) \leq g_{N+1}(r_\alpha).$$

Thus we can find $y_\alpha$ such that

(3) $\forall \beta < \nu (y_{\alpha_\beta} < y_\alpha)$
(4) $\sup y_\alpha \leq g_{N+1}(r_\alpha)$
(5) $\lfloor y_\alpha \rfloor = N + 1$
(6) $\forall \beta < \alpha (y_\alpha = y_\beta \leftrightarrow x_\alpha = x_\beta)$
(7) $\forall \beta < \alpha (y_\alpha = x_\beta \leftrightarrow x_\alpha = y_\beta)$.

**Case 2.** $a_{\alpha_\beta} = y_{\alpha_\beta}$ for $\beta_0 \leq \beta < \nu$. Let $b_{\alpha_\beta} = x_{\alpha_\beta}$. By (1), $\lfloor b_{\alpha_\beta} \rfloor = N - 1$. Let $y_\alpha$ be such that $\forall \beta < \nu (x_{\alpha_\beta} < y_\alpha)$ and (3)-(7) above hold.

The other case is that $x_\alpha$ has an immediate predecessor $x_\alpha^-$ in $\mathcal{M}_{c_\alpha}$. The predecessor $x_\alpha^-$ is $x_\beta$ or $y_\beta$ for some $\beta < \alpha$. Let us first assume $x_\alpha^- = x_\beta$. Let $N = \lfloor x_\alpha^- \rfloor$. Player chooses an immediate successor $y_\alpha$ of $y_\beta$ in such a way that (5)-(7) hold. Note that

$$\sup y_\beta \leq g_{N+1}(r_\beta) < g_{N+1}(r_\alpha)$$

so (4) can be satisfied as well. The case $x_\alpha^- = y_\beta$ is similar to 2.1.

Now we prove that Player I has a winning strategy in $\mathrm{EF}_{\sigma\mathbb{Q}}(\mathcal{M}_0, \mathcal{M}_1)$. Let

$$f : \langle \mathbb{Q}, < \rangle \to \langle \mathbb{Q} \cap ]0, 1[, < \rangle$$

be order-preserving. The strategy of Player I is to play an increasing sequence in $\mathcal{M}_1$, imitating the moves of Player II. Player I uses $f$ to translate the moves of Player II into his own moves in $\mathbb{Q}\cap]0,1[$. Then eventually Player II cannot move anymore because $\sup_{\beta<\alpha} \max y_\beta = \infty$. At this point Player I jumps to $[1,\infty[\cap\mathbb{Q}$ and wins.

More exactly, suppose Player I has played $\langle x_\beta, s_\beta\rangle$, $\beta < \alpha$, so far and Player II has responded with $y_\beta$, $\beta < \alpha$. The idea of Player I is to maintain the conditions:

(8) He keeps playing in $\mathcal{M}_1$

(9) $x_\beta = \langle 0, \max(y_0), 0, \max(y_1), \ldots, 0, \max(y_\gamma), \ldots, 0\rangle \quad (\gamma < \beta)$

(10) $s_\beta = \langle f(\max y_\xi) : \xi < \beta\rangle \quad (\in \sigma\mathbb{Q})$

It is evident that Player I can always choose $\langle x_\alpha, s_\alpha\rangle$ so that (8)-(10) continue to hold. Since the game cannot go on for $\omega_1$ moves, a point is reached where Player II cannot choose $y_\alpha$ anymore, because $\sup_{\beta<\alpha} \max y_\beta = \infty$.

q.e.d.

**Theorem 62 (Tuuri, [Tuu90]).** There are structures $(\mathcal{M}_0, \mathcal{M}_1)$ of cardinality $2^\omega$ so that $\mathbb{R}$ is both a Karp and a Scott po-set of $(\mathcal{M}, \mathcal{M}')$.

*Proof.* Let $P_0$ be the set of functions $f : \mathbb{R} \to 3$. If $f \in P_0$, let $\mathrm{Supp}(f) = \{r \in \mathbb{R} : f(r) \neq 0\}$. Let

$$P = \{f \in P_0 : \mathrm{Supp}(f) \text{ is a well-ordered subset of } \mathbb{R}\}.$$

If $d \in 3$, let $\#(d)$ be defined by $\#(d) = 0$ if $d = 0$ and $\#(d) = 1$ otherwise. If $f \in P$ let $\#(f)$ be the function $\#(f)(r) = \#(f(r))$. Let $F$ be the set of functions $f$ with $\mathrm{dom}(f)$ an open initial segment of $\mathbb{R}$ and $\mathrm{rng}(f) \subseteq \{0,1\}$. For $f \in F$ we define $(1-f)(r) = 1 - f(r)$, when $r \in \mathrm{dom}(f)$. The construction that follows depends on certain decisions at limit stages and these decisions are not canonically determined by earlier stages of the construction. For this reason we introduce a decision-making function $m$. Let $m : F \to 2$ be a function so that $m(f)$ is arbitrarily chosen, subject to the conditions that if $f \in F$ is eventually constant $d$, then $m(f) = d$, and for all $f$, $m(1-f) = 1 - m(f)$, and if $f$ and $g$ eventually agree, then $m(f) = m(g)$.

Our models will have $P \cup \mathbb{R}$ as the universe, an auxiliary predicate $E$, and a unary predicate $U$ the interpretation of which makes the models different. The predicate $U$ is interpreted by defining two "smashed" versions $f^{\mathcal{M}}$ and $f^{\mathcal{M}'}$ of every $f \in P$. These are defined by induction along $\mathrm{Supp}(f)$. At successor stages the idea is to use the smash function $\#$ to let $f(r)$ generate a value $\#(f(r))$ or $1-\#(f(r))$, according to what decisions have been made before, for $f^{\mathcal{M}}(r')$ and $f^{\mathcal{M}'}(r'), r' < r$. At limit stages we use the function $m$.

Suppose $f \in P$ and $\mathrm{Supp}(f) = \langle x_\alpha : \alpha < \beta \rangle$ in increasing order. Suppose $\mathcal{M}'' \in \{\mathcal{M}, \mathcal{M}'\}$. We define $f^{\mathcal{M}''}(x)$ by cases. If $g$ is a function with $\mathrm{dom}(g) \subseteq \mathbb{R}$, denote the restriction of $g$ to $(-\infty, x)$ by $g\restriction x$ and the restriction to $(-\infty, x]$ by $g\restriction\leq x$.

If $x \in \mathbb{R}$ and $\mathrm{Supp}(f) = \emptyset$ or $x < x_0$, we let $f^{\mathcal{M}''}(x) = 0$ if $\mathcal{M}'' = \mathcal{M}$ and $f^{\mathcal{M}''}(x) = 1$ otherwise. For other $x$ we let

$$f^{\mathcal{M}''}(x) = \begin{cases} \#(f(x)), & \text{if } x = x_\alpha, \text{ and } m(f^{\mathcal{M}''}\restriction x) = 0 \\ 1-\#(f(x)), & \text{if } x = x_\alpha, \text{ and } m(f^{\mathcal{M}''}\restriction x) = 1 \\ f^{\mathcal{M}''}(x_\alpha), & \text{if } x_\alpha < x < x_{\alpha+1} \\ m(f^{\mathcal{M}''}\restriction x), & \text{if } x = \sup_{\alpha<\nu} x_\alpha < x_\nu, \ \nu = \cup\nu \\ f^{\mathcal{M}''}(x_\alpha), & \text{if } x > \sup_{\alpha<\beta} x_\alpha \text{ and } \beta = \alpha + 1 \\ m(f^{\mathcal{M}''}\restriction x^*), & \text{if } x \geq x^* = \sup_{\alpha<\beta} x_\alpha \text{ and } \beta = \cup\beta. \end{cases}$$

Now we are ready to define the models needed in the theorem:

$$\begin{aligned} \mathcal{M}_0 &= \langle P \cup \mathbb{R}, E, \{f \in P : f^{\mathcal{M}} \text{ is eventually } 0\}\rangle \\ \mathcal{M}_1 &= \langle P \cup \mathbb{R}, E, \{f \in P : f^{\mathcal{M}'} \text{ is eventually } 0\}\rangle, \end{aligned}$$

where $E = \{\langle f, g, r\rangle : f, g \in P, r \in \mathbb{R} \text{ and } f\restriction r = g\restriction r\}$.

**Claim:** Player I has a winning strategy in $EF_{\sigma\mathbb{R}}(\mathcal{M}, \mathcal{M}')$.

The strategy of Player I is the following. Player II will play elements $h^0_\alpha$ and $h^1_\alpha$ together with $s^0_\alpha$ and $s^1_\alpha = s^0_\alpha{}^\frown\langle r_\alpha\rangle$ in $\sigma\mathbb{R}$. We construe round $\alpha$ of the game as consisting of actually two rounds. First Player I plays $h^0_\alpha$ and $s^0_\alpha$, and after this $h^1_\alpha$ and $s^1_\alpha$. The responses of Player II are respectively $k^0_\alpha$ and $k^1_\alpha$. As part of his strategy Player I then chooses one of $h^0_\alpha$ and $h^1_\alpha$, say $h^{d_\alpha}_\alpha$, to be denoted by $a_\alpha$, if the move $h^{d_\alpha}_\alpha$ was made in $\mathcal{M}$ and the corresponding $k^{d_\alpha}_\alpha$ in $\mathcal{M}'$ will then be denoted by $b_\alpha$. If the move $h^{d_\alpha}_\alpha$ was made in $\mathcal{M}'$, then $h^{d_\alpha}_\alpha$ will be denoted by $b_\alpha$ and the corresponding $k^{d_\alpha}_\alpha$ by $a_\alpha$. Eventually a limit ordinal $\delta$ will emerge such that $\sup_{\alpha<\delta} r_\alpha = \infty$ and then the unions $f = \cup_\alpha a_\alpha$ and $f' = \cup_\alpha b_\alpha$ will be elements of $P$. The point is that Player I takes care that $f^{\mathcal{M}}$ and $f'^{\mathcal{M}'}$ will not be both eventually 0, so when he finally plays $f$ forcing Player II (because of the predicate $E$) to play $f'$, he has won the game.

Suppose $a_\alpha$, $r_\alpha$ and $b_\alpha$ for $\alpha < \beta$ have been played as above and $\sup_{\alpha<\beta}(r_\alpha) < \infty$. Part of the strategy of Player I is to maintain the condition $a^{\mathcal{M}}_\alpha\restriction\leq r_\alpha = 1 - b^{\mathcal{M}'}_\alpha\restriction\leq r_\alpha$.

**Case 1.** $\beta$ is a successor $\alpha+1$. Since $a_\alpha^{\mathcal{M}} \lceil \leq r_\alpha = 1 - b_\alpha^{\mathcal{M}'} \lceil \leq r_\alpha$ there is a smallest $r > r_\alpha$ such that $a_\alpha(r) \neq 0$ or $b_\alpha(r) \neq 0$, for otherwise Player II has lost the game already. Now Player I plays $h_\beta^0$ and $h_\beta^1$, choosing carefully $h_\beta^0(r)$ and $h_\beta^1(r)$, together with $s_\beta^0$ and $s_\beta^1 = s_\beta^0 {}^\frown \langle r_\beta \rangle$, $r_\beta = r$, in such a way that, to avoid immediate loss, Player II has to play $k_\beta^0$ and $k_\beta^1$ so that $a_\beta^{\mathcal{M}} \lceil \leq r_\beta = 1 - b_\beta^{\mathcal{M}'} \lceil \leq r_\beta$.

**Case 2.** $\beta$ is limit. Let $r_\beta = \sup_{\alpha<\beta} r_\alpha$. Now Player I plays $h_\beta^0$ and $h_\beta^1$, choosing carefully $h_\beta^0(r_\beta)$ and $h_\beta^1(r_\beta)$, together with $s_\beta^0 = \bigcup_{\alpha<\beta} s_\alpha^1$ and $s_\beta^1 = s_\beta^0 {}^\frown \langle r_\beta \rangle$ in such a way that, to avoid immediate loss, Player II has to play $k_\beta^0$ and $k_\beta^1$ so that $a_\beta^{\mathcal{M}} \lceil \leq r_\beta = 1 - b_\beta^{\mathcal{M}'} \lceil \leq r_\beta$

**Claim:** Player II has a winning strategy in $EF_{\mathbb{R}}(\mathcal{M}, \mathcal{M}')$.

Suppose $h_\alpha$ and $r_\alpha \in \mathbb{R}$ have been played by Player I and $k_\alpha$ by Player II for $\alpha < \beta$, and $r_\beta = \sup_{\alpha<\beta} \sup(s_\alpha) < \infty$. Let $a_\alpha = h_\alpha$ if $h_a$ was played in $\mathcal{M}$ and $a_\alpha = k_\alpha$ otherwise. Similarly, let $b_\alpha = h_\alpha$ if $h_a$ was played in $\mathcal{M}'$ and $b_\alpha = k_\alpha$ otherwise. The strategy of Player II is to keep $a_\alpha^{\mathcal{M}}(r_\alpha) = b_\alpha^{\mathcal{M}'}(r_\alpha)$ and $a_\alpha(x) = b_\alpha(x)$ for $x > r_\alpha$. If Player II can keep playing like this he wins since then $a_\alpha^{\mathcal{M}}$ is eventually 0 if and only if $b_\alpha^{\mathcal{M}'}$ is. q.e.d.

Scott- and Karp-po-sets are analogues of the Scott watershed of a pair of models. Due to non-determinacy, the concept of Scott watershed splits into two different concepts. The concept of Scott rank likewise splits into two concepts that are called universal Scott-tree and universal Karp-tree. For more on these, see [HytSheTuu93, Vää95].

We end our survey of the infinite Ehrenfeucht-Fraïssé-game here. Developing methods for the dynamic transfinite Ehrenfeucht-Fraïssé-game has led us to the set-theoretic problem of structure of trees and po-sets. On the other hand, the po-sets and trees provide a perfect measure of non-isomorphism of uncountable models in the same sense as Scott-watersheds measure non-isomorphism of countable models. The difference is that while Scott-watersheds can build on the relatively well-established theory of ordinals, the properties of Karp- and Scott-trees are interwoven with set-theoretical properties of trees and po-sets. More structure theory for trees has to be developed before the analysis of uncountable structures in terms of the infinite Ehrenfeucht-Fraïssé-game can make more progress. On the other hand, interesting model-theoretic problems, related to stability theoretic properties of the models involved, arise already at the level of Ehrenfeucht-Fraïssé-game of length $\omega^2$ [HytTuu91].

# References.


[AddHenTar65] John W. **Addison**, Leon **Henkin**, and Alfred **Tarski**, Theory of Models, Proceedings of the 1963 International Symposium at Berkeley, North-Holland 1965

[Bar75] Jon **Barwise**, Admissible Sets and Structures: An Approach to Definability Theory, Springer 1975 [Perspectives in Mathematical Logic]

[Bar77] Jon **Barwise**, Handbook of Mathematical Logic, Part A, North-Holland 1977 [Studies in Logic and the Foundations of Mathematics 90]

[Dic75] Max A. **Dickmann**. Large Infinitary Languages: Model theory, North-Holland 1975 [Studies in Logic and the Foundations of Mathematics 83]

[Ehr60] Andrzej **Ehrenfeucht**, An Application of Games to the Completeness Problem for Formalized Theories, **Fundamenta Mathematicæ** 49 (1960), p. 129–141

[EklForShe95] Paul C. **Eklof**, Matthew **Foreman**, and Saharon **Shelah**, On Onvariants for $\omega_1$-Separable Groups, **Transactions of the American Math. Society** 347 (1995), p. 4385–4402

[Fra55] Roland **Fraïssé**, Sur quelques classifications des systèmes de relations, **Publications Scientifiques de l'Université d'Alger** 1 (1955), p. 35–182

[GalSte53] David **Gale** and Frank M. **Stewart**, Infinite Games with Perfect Information, *in:* [KuhTuc53, p. 245–266]

[HuuHytRau99] Taneli **Huuskonen**, Tapani **Hyttinen**, and Mika **Rautila**, On the $\kappa$-Cub Game on $\lambda$ and $I[\lambda]$, **Archive for Mathematical Logic** 38 (1999), p. 549–557

[Hyt87] Tapani **Hyttinen**, Games and Infinitary Languages, **Annales Academiae Scientiarum Fennicae**64 (1987), p. 1–32

[Hyt92] Tapani **Hyttinen**, On Nondetermined Ehrenfeucht-Fraïssé Games and Unstable Theories, **Zeitschrift für Mathematische Logik und Grundlagen der Mathematik** 38 (1992), p. 399–408

[HytShe94] Tapani **Hyttinen** and Saharon **Shelah**, Constructing Strongly Equivalent Nonisomorphic Models for Unsuperstable Theories, Part A, **Journal of Symbolic Logic** 59 (1994), p. 984–996

[HytShe95] Tapani **Hyttinen** and Saharon **Shelah**, Constructing Strongly Equivalent Nonisomorphic Models for Unsuperstable Theories, Part B, **Journal of Symbolic Logic** 60 (1995), p. 1260–1272

[HytShe99] Tapani **Hyttinen** and Saharon **Shelah**, Constructing Strongly Equivalent Nonisomorphic Models for Unsuperstable Theories, Part C, **Journal of Symbolic Logic** 64 (1999), p. 634–642

[HytSheTuu93] Tapani **Hyttinen**, Saharon **Shelah**, and Heikki **Tuuri**, Remarks on Strong Nonstructure Theorems, **Notre Dame Journal of Formal Logic** 34 (1993), p. 157–168

[HytSheVää02] Tapani **Hyttinen**, Saharon **Shelah**, and Jouko **Väänänen**, More on the Ehrenfeucht-Fraïssé Game of Length $\omega_1$, **Fundamenta Mathematicæ** 175 (2002), p. 79–96

[HytTuu91] Tapani **Hyttinen** and Heikki **Tuuri**, Constructing Strongly Equivalent Nonisomorphic Models for Unstable Theories, **Annals of Pure and Applied Logic** 52 (1991), p. 203–248

[HytVää90] Tapani **Hyttinen** and Jouko **Väänänen**, On Scott and Karp Trees of Uncountable Models, **Journal of Symbolic Logic** 55 (1990), p. 897–908

[Jec97] Thomas **Jech**, Set Theory, 2nd edition, Springer 1997

[Kar65] Carol R. **Karp**, Finite-Quantifier Equivalence, *in:* [AddHenTar65, p. 407–412]

[Kry95] Michal **Krynicki** Quantifiers: Logics, Models and Computation, volume 1, Kluwer 1995

[Kue75a] David W. **Kueker**, Back-and-Forth Arguments and Infinitary Logics, *in:* [Kue75b, p. 17–71]

[Kue75b] David W. **Kueker** (*ed.*), Infinitary Logic: In Memoriam Carol Karp, Springer 1975 [Lecture Notes in Mathematics 492]

[KuhTuc53] Harold W. **Kuhn** and Albert W. **Tucker** (*eds.*), Contributions to the Theory of Games, volume 2, Princeton University Press 1953 [Annals of Mathematics Studies 28]

[Kur56] Georges **Kurepa**, Ensembles ordonnés et leurs sous-ensembles bien ordonnés, **Comptes Rendus de l'Académie des Sciences de Paris** 242 (1956), p. 2202–2203

[LasShe01] Michael C. **Laskowski** and Saharon **Shelah**, The Karp Complexity of Unstable Classes, **Archive for Mathematical Logic** 40 (2001), p. 69–88

[Mak77] Michael **Makkai**, Infinitary Logic, *in:* [Bar77, p. 3–313]

[MekSheVää93] Alan **Mekler**, Saharon **Shelah**, and Jouko **Väänänen**, The Ehrenfeucht-Fraïssé-Game of Length $\omega_1$, **Transactions of the American Mathematical Society** 339 (1993), p. 567–580

[Sco65] Dana S. **Scott**, Logic with Denumerably Long Formulas and Finite Strings of Quantifiers, *in:* [AddHenTar65, p. 329–341]

[Tod07] Stevo **Todorčević**, Lipschitz Maps on Trees, **Journal of the Institute of Mathematics of Jussieu** 6 (2007), p. 527–556

[TodVää99] Stevo **Todorčević** and Jouko **Väänänen**, Trees and Ehrenfeucht-Fraïssé Games, **Annals of Pure and Applied Logic** 100 (1999), p. 69–97

[Tuu90] Heikki **Tuuri**, Infinitary Languages and EF-Games, *PhD thesis*, University of Helsinki 1990

[Vää95] Jouko **Väänänen**, Games and Trees in Infinitary Logic: A Survey, *in:* [Kry95, p. 105–138]

[VääVel04] Jouko **Väänänen** and Boban **Veličković**, Games Played on Partial Isomorphisms, **Archive for Mathematical Logic** 43 (2004), p. 19–30


**Received**: April 22nd, 2005;
**In revised version**: September 10th, 2005;
**Accepted by the editors**: October 16th, 2005.

Stefan **Bold**, Benedikt **Löwe**,
Thoralf **Räsch**, Johan **van Benthem** (*eds.*)
**Foundations of the Formal Sciences V**
Infinite Games

# On Concept Lattices of Coalitional Game Forms

STEFANO VANNUCCI

Dipartimento di Economia Politica
Universitá di Siena
Piazza S. Francesco 7
53110 Siena, Italia
vannucci@unisi.it

ABSTRACT. Some basic properties of the concept lattice of a coalitional game form and of the desirability relations it induces on the latter through a natural rank function are studied. The use of concept-latticial lengths and widths as basic complexity measures of the underlying game correspondences is also presented by means of several examples.

## 1 Introduction

Game forms are games without outcome evaluation structures: they only describe the relevant interaction structures or 'rules of the game'. Coalitional game forms are meant to represent such interaction structures by focusing on coalitional power, while ignoring the actions available to players and coalitions: coalitions of players (including of course single players) are characterized by the complete list of events they are able to force within the relevant outcome space. Hence a coalitional game form may be regarded as a classification of both coalitions and families of coalitions in terms of their decision power[1]. This view of coalitional game formats as coalitional-power-classifying data structures in turn suggests to focus on *the concept lattice of a coalitional game form.*

The concept lattice of a coalitional game form —a complete lattice— is made up of 'concepts', namely of pairs consisting of 'closed' families of

[1]Notice that coalitional game forms can be attached in a most natural way to any strategic game form and, for that matter, to any extensive game form. Therefore, any extensive or strategic game form can be represented by a suitable coalitional game form.

coalitions and outcome-subsets, ordered by set-inclusion (the interpretation being that the coalitions of a 'closed' family attached to a certain 'closed' family of outcome-subsets or events are precisely those coalitions that are able to force the final outcome within any event of the latter family). In a concept lattice an ordered pair of concepts denotes a hierarchical relation between two distinct types of coalitions (*e.g.*, a coalition belonging to the smaller family but not to the larger is more powerful of a coalition belonging to both). Conversely, an unordered pair of concepts denotes two types of coalitions that are endowed with specialized decision power. Thus *lengths* and *widths* of concept lattices of game forms provide two natural structural complexity measures of the latter, and suggest some new interesting classifications of well-known game forms and games according to the degree of hierarchy and specialization of the allocations of decision power they embody. In particular concept lattices of game forms suggest the possibility to introduce (a few versions of) a new notion of '*concept-latticial infinity*' for game forms and games as based upon suitable concept-latticial parameters (*e.g.*, size, length or width).

Of course the classification of game forms and games provided by concept lattices is a purely structural one, since it is in principle unrelated to behavioural aspects. That classification is at variance with the bulk of the extant game-theoretic literature which is almost invariably concerned with the behaviour of games or game forms with respect to certain solution concepts. There are exceptions, however, the most prominent one being the theory of simple games and their desirability relations (recall that 'power indices' for simple games are mostly monotonic with respect to individual desirability relations). Therefore the present paper also includes a tentative exploration of ways to extend the analysis of desirability relations to general coalitional game forms by means of their concept lattices. It is shown that a few desirability relations for coalitional game forms may be defined in a natural way through the notion of concept-latticial coalitional *height*. In particular it is also proved that the symmetric components of the coalitional desirability relations induced by the concept-latticial rank function are in fact *normal* in Taylor-Zwicker's sense (*cf.* [TayZwi99]). A few significant examples of easily computable concept lattices of game forms, including demand game forms, $2 \times 2$ strategic game forms, and several voting game forms, are also presented and discussed in order to illustrate the structural notions previously introduced.

It should also be remarked here that introducing concept-latticial classifications of coalitional game form raises in turn the issue of the relationships between such 'structural' and 'behavioural' classifications. This is a very important topic which is not addressed in the current paper. A few observations on this theme, however, are provided in the concluding remarks.

The structure of the paper is as follows: Section 2 is devoted to the introduction of some basic results. Section 3 describes the concept lattices of several widely used game forms, and discusses the ensuing classifications of the latter in terms of concept-latticial parameters. Section 4 offers some short concluding remarks.

## 2 Model and Results

A *coalitional game form* is a triple $\mathbf{G} = (N, X, E)$ where $N$ and $X$ are non-empty sets denoting the sets of players and outcomes, respectively, and $E : \mathcal{P}(N) \to \mathcal{P}(\mathcal{P}(X))$ is the coalitional power function: the 'power-value' $E(S)$ of coalition $S \subseteq N$ is the collection of all events $A \subseteq X$ coalition $S \subseteq N$ is able to 'force' (under some suitable interpretation of the latter notion). We also assume $\#N \geq 2$ and $\#X \geq 2$ in order to avoid trivialities. A coalitional game form $(N, X, E)$ is a (standard) *effectivity function* if $E$ satisfies the following boundary conditions:

**EF1)** (Sovereignty) $E(N) \supseteq \mathcal{P}(X) \setminus \{\varnothing\}$.

**EF2)** (Null Set Normalization) $E(\varnothing) = \varnothing$.

**EF3)** (Exhaustiveness) $X \in E(S)$ for any $S$, $\varnothing \neq S \subseteq N$.

**EF4)** (Null Event Unenforceability) $\varnothing \notin E(S)$ for any $S$, $\varnothing \subset S \subseteq N$.

A coalitional game form is *monotonic* if for any $S, T \subseteq N$ and any $A, B \subseteq X$

$$[A \in E(S) \text{ and } S \subseteq T \text{ entail } A \in E(T)] \text{ and}$$

$$[A \in E(S) \text{ and } A \subseteq B \text{ entail } B \in E(S)] .$$

In what follows we shall confine ourselves to *monotonic* coalitional game forms.

A monotonic coalitional game form $\mathbf{G} = (N, X, E)$ is *regular* if $\varnothing \neq A \in E(S)$ entails $X \setminus A \notin E(N \setminus S)$ for any $S \subseteq N$ and $B \subseteq X$, *maximal* if $A \notin E(S)$ entails $(X \setminus A) \in E(N \setminus S)$ for any $\varnothing \neq S \subseteq N$ and $\varnothing \neq A \subseteq X$, *essential* if for any $i \in N$: $E(\{i\}) \neq \{A \subseteq X : A \neq \varnothing\}$, *consensual* if for any $S \subset N$: $E(S) \neq \{A \subseteq X : A \neq \varnothing\}$, and *linear* if $E(S) \supseteq E(T)$ or $E(T) \supseteq E(S)$ for any $S, T \subseteq N$. Moreover, it is *superadditive* if for any $S, T \subseteq N$ and $A, B \subseteq X$, $A \in E(S)$, $B \in E(T)$ and $S \cap T = \varnothing$ entail $A \cap B \in E(S \cup T)$, *convex* if for any $S, T \subseteq N$ and $A, B \subseteq X$, $A \cap B \in E(S \cup T)$ or $A \cup B \in E(S \cap T)$ whenever $A \in E(S)$ and $B \in E(T)$, and *additive* if there exist positive probability measures $p, q$ on $\mathcal{P}(N), \mathcal{P}(X)$, respectively, such that $A \in E(S)$ if and only if $p(S) + q(A) > 1$ (an additive effectivity function is also convex).

A *(monotonic) simple game* on $N$ is an order filter of $(\mathcal{P}(N), \supseteq)$, *i.e.*, a set $W$, $\mathcal{P}(N) \supseteq W \neq \varnothing$, such that $S \in W$ and $T \supseteq S$ entail $T \in W$. The coalitions belonging to $W$ are meant to represent the *winning* or all-powerful ones. Finally, a coalitional game form $(N, X, E)$ is *simple* if there exists an order filter $W$ of $(\mathcal{P}(N), \supseteq)$ such that for any $S \subseteq N$, $A \subseteq X$, $A \in E(S)$ if and only if either $A = X$ and $S \neq \varnothing$ or $A \neq \varnothing$ and $S \in W$ (notice that a simple effectivity function is —by definition— both and monotonic). Indeed, simple effectivity functions amount to simple games as endowed with a fixed outcome set.

In many relevant contexts, one may be focussed on effectivity functions that treat "symmetrically" players and/or outcomes. Such requirements are embodied in the following properties. A coalitional game form $(N, X, E)$ is *anonymous* if for any $A \subseteq X$, and $S, T \subseteq N$ such that $\#S = \#T$: $A \in E(S)$ if and only if $A \in E(T)$, and *neutral* if for any $S \subseteq N$ and any $A, B \subseteq X$ such that $\#A = \#B$: $A \in E(S)$ if and only if $B \in E(S)$.

We are mainly interested in those coalitional game forms —and effectivity functions— that can represent the decision power of coalitions under a certain decision mechanism, or *strategic game correspondence*. A *strategic game correspondence* on $(N, X)$ is a correspondence $G : D \to \mathcal{P}(X)$ where $D \subseteq \prod_{i \in N} S_i$, and $S_i$ is the set of "interactive behaviours" available to player $i \in N$. A *strategic game form* is a single-valued strategic game correspondence.

Now, the notion of decision power admits at least two distinct interpretations, namely "guaranteeing power" and "counteracting power" that in turn correspond to the ability to force maximin and minimax outcomes, respectively. Thus, the allocation of "guaranteeing power" under game correspondence $G$ with domain $D$ is represented by the $\alpha$-effectivity function of $G$ —denoted by $E_\alpha(G)$— as defined by the following rule: for any non-empty $S \subseteq N$, we let

$$(E_\alpha(G))(S) = \{A \subseteq X \quad : \quad \text{a } t^S \in \prod_{i \in S} S_i \text{ exists such that } (t^S, s^{N \setminus S}) \in D \text{ and } G(t^S, s^{N \setminus S}) \subseteq A \text{ for any } s^{N \setminus S} \in \prod_{i \in N \setminus S} S_i\}.$$

Conversely, the allocation of "counteracting power" under game correspondence $G$ with domain $D$ is represented by the $\beta$-effectivity function of $G$ —denoted by $E_\beta(G)$— and defined as follows: for any non-empty

$S \subseteq N$, we let

$$(E_\beta(G))(S) = \{A \subseteq X \ : \ \text{for any } s^{N \setminus S} \in \prod_{i \in N \setminus S} S_i \ \text{some } t^S \in \prod_{i \in S} S_i \text{ exists such that } (t^S, s^{N \setminus S}) \in D \text{ and } G(t^S, s^{N \setminus S}) \subseteq A\}.$$

It is easily checked that $(N, X, E_\alpha(G))$ is regular, $(N, X, E_\beta(G))$ is maximal, and both of them are *monotonic* and —provided that $G$ is non-empty valued— satisfy *null-event-unenforceability*. Also it is well-known that superadditivity and monotonicity of an effectivity function $(N, X, E)$ imply that a game correspondence $G$ exists such that $E = E_\alpha(G)$.[2] Indeed, monotonicity of $\alpha$-effectivity functions and $\beta$-effectivity functions of game correspondences is our main reason for confining the ensuing analysis to monotonic effectivity functions (as mentioned previously). Furthermore, the foregoing distinction between $\alpha$-effectivity functions and $\beta$-effectivity functions brings us to the general notion of a *polarity operator* for effectivity functions, implicitly defined as follows (*cf.*, *e.g.*, [AbdKei91]): the *polar* $\mathbf{G}^* = (N, X, E^*)$ of a monotonic effectivity function $E$ on $\mathbf{G} = (N, X, E)$ is the unique effectivity function such that for any *non-empty* $S \subset N$, $A \subset X$, $A \in E^*(S)$ if and only if $X \setminus A \notin E(N \setminus S)$. (It should be noticed here that $E = E^*$ if and only if $E$ is both regular and maximal.)

The *concept lattice* of a coalitional game form $\mathbf{G}$ can be defined through the following steps. First, define the functions $\curlyvee_E : \mathcal{P}(\mathcal{P}(N)) \to \mathcal{P}(\mathcal{P}(X))$ and $\curlywedge_E : \mathcal{P}(\mathcal{P}(X)) \to \mathcal{P}(\mathcal{P}(N))$ for any $\mathbf{S} \subseteq \mathcal{P}(N)$, $\mathbf{A} \subseteq \mathcal{P}(X)$:

$$\curlyvee_E(\mathbf{S}) = \{A \subseteq X : A \in E(S) \text{ for all } S \in \mathbf{S}\} \ \text{ and}$$

$$\curlywedge_E(\mathbf{A}) = \{S \subseteq N : A \in E(S) \text{ for all } A \in \mathbf{A}\}\,.$$

It is easily seen that $(\curlyvee_E, \curlywedge_E)$ is a *Galois connection* between $(\mathcal{P}(\mathcal{P}(N)), \subseteq)$ and $(\mathcal{P}(\mathcal{P}(X)), \subseteq)$, *i.e.*, for any $\mathbf{S}, \mathbf{T} \subseteq \mathcal{P}(N)$ and $\mathbf{A}, \mathbf{B} \subseteq \mathcal{P}(X)$,

i) if $\mathbf{S} \subseteq \mathbf{T}$ then $\curlyvee_E(\mathbf{S}) \subseteq \curlyvee_E(\mathbf{T})$, and if $\mathbf{A} \subseteq \mathbf{B}$ then $\curlywedge_E(\mathbf{B}) \subseteq \curlywedge_E(\mathbf{A})$, and

ii) $(\curlywedge_E \circ \curlyvee_E)(\mathbf{S}) \supseteq \mathbf{S}$,$(\curlyvee_E \circ \curlywedge_E)(\mathbf{A}) \supseteq \mathbf{A}$.

Now, consider

$$\mathbb{C}(\mathbf{G}) = \{(\mathbf{S}, \mathbf{A}) \in \mathcal{P}(\mathcal{P}(N)) \times \mathcal{P}(\mathcal{P}(X)) : \mathbf{S} = \curlywedge_E(\mathbf{A}), \ \text{and } \mathbf{A} = \curlyvee_E(\mathbf{S})\}\,.$$

[2] *Cf.* [Mou83, Bor+95].

In the language of formal concept analysis (*cf.*, *e.g.*,[GanWil99]) an element $(\mathbf{S}, \mathbf{A})$ of $\mathbb{C}(\mathbf{G})$ is said to be a *concept* of the context $\mathbf{G}$, with *extent* $\mathbf{S}$ and *intent* $\mathbf{A}$ (the latter notions are amenable to straightforward dualizations).

Thus, the (dual)[3] *concept lattice* of $\mathbf{G}$ (sometimes also referred to as its *Galois lattice*) is $\mathbf{L}(\mathbf{G}) = (\mathbb{C}(\mathbf{G}), \geqslant)$ where for any $(\mathbf{S}_1, \mathbf{A}_1), (\mathbf{S}_2, \mathbf{A}_2) \in \mathbb{C}(\mathbf{G})$ we have

$$(\mathbf{S}_1, \mathbf{A}_1) \geqslant (\mathbf{S}_2, \mathbf{A}_2) \text{ iff } \mathbf{A}_1 \supseteq \mathbf{A}_2,$$

which is provably equivalent to $\mathbf{S}_2 \supseteq \mathbf{S}_1$,

$$(\mathbf{S}_1, \mathbf{A}_1) \wedge (\mathbf{S}_2, \mathbf{A}_2) = (\rightthreetimes_E(\leftthreetimes_E(\mathbf{S}_1 \cup \mathbf{S}_2)), \mathbf{A}_1 \cap \mathbf{A}_2),$$

and

$$(\mathbf{S}_1, \mathbf{A}_1) \vee (\mathbf{S}_2, \mathbf{A}_2) = (\mathbf{S}_1 \cap \mathbf{S}_2, \leftthreetimes_E(\rightthreetimes_E(\mathbf{A}_1 \cup \mathbf{A}_2)).$$

It is also well-known and easily shown that both $(\rightthreetimes_E \circ \leftthreetimes_E) : \mathcal{P}(\mathcal{P}(N)) \to \mathcal{P}(\mathcal{P}(N))$ and $(\leftthreetimes_E \circ \rightthreetimes_E) : \mathcal{P}(\mathcal{P}(X)) \to \mathcal{P}(\mathcal{P}(X))$ are closure operators with respect to set-inclusion (recall that a *closure operator* $K$ on a preordered set $(Y, \geqslant)$ is a function $K : Y \to Y$ such that for any $y, x \in Y : K(y) \geqslant y, K(y) \geqslant K(x)$ whenever $y \geqslant x$, and $K(y) \geqslant K(K(y))$ ), and *extents* and *intents* of concepts are precisely the *closed* elements —or fixed points— of $(\rightthreetimes_E \circ \leftthreetimes_E)$and $(\leftthreetimes_E \circ \rightthreetimes_E)$, respectively, (*i.e.*, $(\mathbf{S}, \mathbf{A}) \in \mathbb{C}(\mathbf{G})$ if and only if $\mathbf{S} = \rightthreetimes_E(\leftthreetimes_E(\mathbf{S}))$ and $\mathbf{A} = \leftthreetimes_E(\rightthreetimes_E(\mathbf{A})))$. We shall also denote $(\rightthreetimes_E \circ \leftthreetimes_E)$ and $(\leftthreetimes_E \circ \rightthreetimes_E)$ by $K_{\mathbf{G}}$ and $K^*_{\mathbf{G}}$, respectively. The sets of all fixed points of $K_{\mathbf{G}}$ and $K^*_{\mathbf{G}}$ are also called the *(Galois) closure systems* of coalitional game form $\mathbf{G}$, and denoted by $\mathcal{C}$ and $\mathcal{C}^*$, respectively.

Clearly enough the concept lattice $\mathbf{L}(\mathbf{G})$ —that is also sometimes called the *Galois lattice* of $\mathbf{G}$ (*cf.*, *e.g.*, [BarMon70, Vol.2, Chapter V])— is lattice-isomorphic to the lattices of inclusion-ordered closure systems $(\mathcal{C}, \subseteq)$ and $(\mathcal{C}^*, \supseteq)$, respectively, (*cf.* [DavPri90, p. 227]). Hence $\mathbf{L}(\mathbf{G})$ is complete, has a unique atom if $\mathbf{G}$ is null-set-normalized and a unique co-atom if $\mathbf{G}$ satisfies null-event-unenforceability. Moreover, if $\mathbf{G}$ is linear then $\mathbf{L}(\mathbf{G})$ is also linearly ordered.

---

[3]As pointed out to me by an anonymous referee, the ordering of the concept lattice as defined below is indeed endowed with the reverse ordering of the concept lattice as usually defined in the literature. Therefore, what is referred to as a 'concept lattice' in the text is in fact the *dual* of a concept lattice as usually defined. The reason I insist on dual concept lattices is my intention to focus on rankings of coalitions in terms of decision power, relying on the ability to 'force' events as the relevant criteria/attributes. By contrast, the concept lattice of a coalitional game form in the standard sense is best regarded as a classification of the 'resilience' of (families of) events with respect to coalitional capabilities to act.

Those basic facts concerning $\mathbf{L}(\mathbf{G})$ —which rely on the classic Birkhoff's theorem on concrete, *i.e.*, polarity-induced, Galois connections (*cf.* [Bir67])— can be summarized by the following:[4]

**Proposition 1.** Let $\mathbf{G} = (N, X, E)$ be a coalitional game form. Then a complete lattice $\mathbf{L}(\mathbf{G})$ —the concept lattice of $\mathbf{G}$, uniquely defined up to isomorphisms— can be canonically attached to $\mathbf{G}$. Moreover,

(i) if $\mathbf{G}$ is null-set-normalized, then $\mathbf{L}(\mathbf{G})$ is dense, *i.e.*, has a minimum that is meet-irreducible;
(ii) if $\mathbf{G}$ satisfies null-set-unenforceability, then $\mathbf{L}(\mathbf{G})$ is co-dense, *i.e.*, has a maximum that is join-irreducible;
(iii) $\mathbf{L}(\mathbf{G})$ is finite whenever either $N$ or $X$ is finite;
(iv) if $\mathbf{G}$ is linear then $\mathbf{L}(\mathbf{G})$ is a chain.

**Remark 2.** Proposition 1.ii) implies that $\mathbf{L}(\mathbf{G}) = \mathbf{1} \oplus B(\mathbf{L}(\mathbf{G})) \oplus \mathbf{1}$ for some lattice $B(\mathbf{L}(\mathbf{G}))$ if $\mathbf{G}$ satisfies null-set-unenforceability and null-set-normalization and $\mathbf{L}(\mathbf{G}) = \mathbf{1} \oplus B(\mathbf{L}(\mathbf{G}))$ if it is just null-set-normalized (where $\mathbf{1}$ denotes the degenerate 1-element lattice, and $\oplus$ denotes the linear or ordinal sum operation: *cf.*, *e.g.*, [Bir67] or [DavPri90]). In any case we shall refer to the lattice $B(\mathbf{L}(\mathbf{G}))$ as the *bulk* of $\mathbf{L}(\mathbf{G})$.

Thus the situation can be summarized as follows: as mentioned above, concept lattices of coalitional game forms —and their bulks— provide us with a representation of game correspondences that allows some significant new 'structural' lattice-theoretic classifications. (Of course, a game correspondence $G$ is entitled to —at least— two concept lattices, the $\alpha$-concept lattice and the $\beta$-concept lattice, that correspond to $E_\alpha(G)$ and $E_\beta(G)$, respectively. We shall refer to *the* concept lattice of game correspondence $G$ when $E_\alpha(G) = E_\beta(G)$).

Let us then turn to a more detailed consideration of a few basic concept-latticial parameters of a coalitional game form. To begin with, the concept lattice $\mathbf{L}(\mathbf{G})$ provides —at least— two natural complexity measures for the underlying coalitional game form $\mathbf{G} = (N, X, E)$, namely its *length* and *width* which provide ordinal scales for the degree of hierarchy and dispersion of coalitional decision power embodied in $\mathbf{G}$[5]. We recall here the relevant definitions (*cf.* [Bir67, DavPri90]).

[4] A similar result for the slightly more specialized case of effectivity functions is presented and discussed in [Van99].

[5] Of course this is not meant to deny that sensible game-theoretic or social choice-theoretic interpretations might be possibly suggested for other latticial parameters (*i.e.*, number of join irreducibles) or properties (*e.g.*, modularity or distributivity) of the concept lattice of a coalitional game form.

**Definition 3.** The *length* $\ell(L)$ of a lattice $L$ is the least upper bound of the set of lengths of chains included in $L$ (a chain is a totally ordered set; the length of a chain of $k+1$ elements is $k$).

**Definition 4.** The *width* $\mathrm{w}(L)$ of a lattice $L$ is the size or cardinality of its largest antichain (an antichain is a set of pairwise incomparable elements).

A most useful evaluation of the comparative power of coalitions (and persistence of events) can be introduced relying on the *lengths* of chains in the concept lattice of an coalitional game form $\mathbf{G} = (N, X, E)$ through the following.

**Definition 5.** Let $\mathbf{G} = (N, X, E)$ be a coalitional game form. The *height* $\mathrm{h}_E(x)$ of $x = (\mathbf{C}, \mathbf{C}') \in \mathbf{L}(\mathbf{G})$ is the least upper bound of the set of sizes of chains in $\mathbf{L}(\mathbf{G})$ having $x$ as their maximum. The *height* $\mathrm{r}_E(S)$ of a coalition $S \subseteq N$ is the height $\mathrm{h}_E(x)$ of the highest $x = (\mathbf{C}, \mathbf{C}') \in \mathbf{L}(\mathbf{G})$ such that $S \in \mathbf{C}$ (a dual definition applies to an issue $A \subseteq X$).

A few fundamental properties of a coalitional game form $(N, X, E)$ (hence of a game form $G$ when $E = E_\alpha(G) = E_\beta(G)$) can also be readily expressed using the foregoing notion of *height*:

(i) $(N, X, E)$ is *fully distributed* if $\mathrm{r}_E(\{i\}) > 1$ for any $i \in N$ (in words, $(N, X, E)$ is *fully distributed* if each single player is endowed with some non-communal decision power);

(ii) $(N, X, E)$ is *unspecialized* if $\mathrm{r}_E(S) = \mathrm{r}_E(T)$ entails $E(S) = E(T)$ for any $S, T \subseteq N$, and *specialized* otherwise (in words, $E$ is *unspecialized* if having the same height entails having exactly the same decision power).

(iii) $(N, X, E)$ is *strictly hierarchical* if $\mathrm{r}_E(S) = \mathrm{r}_E(T)$ entails $S = T$ for any $S, T \subseteq N$ (in words, the coalitions are linearly ordered with respect to their height in $(N, X, E)$).

Moreover, both *simple* and *consensual* coalitional game forms as defined above can also be characterized in terms of heights. Namely, a coalitional game form $(N, X, E)$ is *simple* if $\mathrm{r}_E(.)$ is two-valued and monotonic with respect to $(P(N), \supseteq)$, and *consensual* if $\mathrm{r}_E(N) > \mathrm{r}_E(S)$ for any coalition $S \neq N$ (in words, $(N, X, E)$ is *consensual* if the grand coalition is uniquely endowed with maximum height).

Clearly enough the foregoing properties are —generally speaking— not independent. In particular, *strictly hierarchical* entails *unspecialized* and *not simple*; *simple* entails *unspecialized*; *consensual* and *fully distributed* jointly entail *not simple.*

Concerning simple effectivity functions, the following basic result can be easily established (see also the first example in Section 4 below):

**Proposition 6.** Let $\mathbf{G} = (N, X, E)$ be a monotonic effectivity function. Then,

(i) $\mathbf{G}$ is simple and consensual if and only if $B(\mathbf{L}(\mathbf{G}^*)) = \mathbf{1}$;
(ii) $\mathbf{G}$ is simple and not consensual if and only if $B(\mathbf{L}(\mathbf{G}^*)) = \mathbf{2}$ (the two-element Boolean lattice);
(iii) if $\mathbf{G}$ is simple, regular, and not consensual then $B(\mathbf{L}(\mathbf{G})) = B(\mathbf{L}(\mathbf{G}^*)) = \mathbf{2}$.

*Proof.* (i): Let $B(\mathbf{L}(\mathbf{G}^*)) = \mathbf{1}$. Then for any $S \in \mathcal{P}(N)\backslash \{\varnothing\}$, $E^*(S) = \mathcal{P}(X)\backslash \{\varnothing\}$. Now —by definition— $B \in E^*(S)$ if and only if ($S = N$ and $B \neq \varnothing$) or ($N \neq S \neq \varnothing$, $B \neq \varnothing$ and $X\backslash B \notin E(N\backslash S)$). It follows that $E(N\backslash S) = \{X\}$ for any non-empty $S \neq N$ (while obviously $E(N) = \mathcal{P}(X)\backslash \{\varnothing\}$ since $E$ is an effectivity function). Therefore $E$ is indeed both simple and consensual.

Conversely, let $E$ be simple and consensual, *i.e.*, —for any $S \subseteq N, B \subseteq X$— $B \in E(S)$ if and only if $S = N$ and $B \neq \varnothing$ or $S \neq \varnothing$ and $B = X$. Then for any $S \subseteq N, B \subseteq X$ such that $\varnothing \neq S, \varnothing \neq B$, $B \in E^*(S)$ since $C \in E(N\backslash S)$ only if $C = X$. Thus $B(\mathbf{L}(\mathbf{G}^*)) = \mathbf{1}$.

(ii): Notice that —by definition— for any simple $E$ either $B(\mathbf{L}(\mathbf{G})) = \mathbf{2}$ or $B(\mathbf{L}(\mathbf{G})) = \mathbf{1}$. Then ii) follows immediately from (i) above and from the fact that —as it is easily checked— $E$ is simple if and only if $E^*$ is.

(iii): In view of points (i) and (ii) above, it only remains to be shown that if $E$ is simple, not consensual and regular then $B(\mathbf{L}(\mathbf{G})) = \mathbf{2}$. Indeed, suppose not. Then $B(\mathbf{L}(\mathbf{G})) = \mathbf{1}$ or equivalently $E(S) = \mathcal{P}(X)\backslash \{\varnothing\}$ for any $S \subseteq N, S \neq \varnothing$. Hence $\#N \geq 2$ and $\#X \geq 2$ entail $A \in E(S)$, $X\backslash A \in E(T)$ for some $A \neq \varnothing$, $A \neq X$, $S \neq \varnothing$, $T \neq \varnothing$ such that $S \cap T = \varnothing$, which contradicts regularity. q.e.d.

Thus the *unanimity*[6] *effectivity function* turns out to be —somehow— uniquely connected to the degenerate lattice $\mathbf{1}$, while simple and not consensual effectivity functions are —somehow— uniquely connected to the simple[7] Boolean lattice $\mathbf{2}$.

---

[6]The unanimity effectivity function is a simple effectivity function such that the grand coalition $N$ is the only winning or all-powerful coalition, while any other non-empty coalition can only force the universal event $X$. Clearly the unanimity effectivity function is uniquely simple and consensual among effectivity functions.

[7]We recall here that a lattice (or, for that matter, *any* algebra) is *simple* if and only if its congruences (or operation-consistent equivalences) reduce to the trivial ones, *i.e.*, the identity congruence and the universal congruence.

Furthermore, the notion of (concept-latticial) height of a coalition enables a few new notions of *individual and coalitional desirability relations* of a monotonic coalitional game form $(N, X, E)$ as defined below (*cf.* also [Ein85, TayZwi99]). Indeed, let $\mathbf{G}(N, X, E)$ be a monotonic coalitional game form $(N, X, E)$. For any $S \subseteq N$ the *S-reduced* coalitional game form of $(N, X, E)$ —written $(N, X, E_S)$— is defined by the following prescription: for any $T \subseteq N, T \neq \varnothing$, $E_S(T) = E(T \cup S)$, and $E_S(\varnothing) = \varnothing$.

The *individual desirability relation* of $(N, X, E)$ is the binary relation $(N, \succcurlyeq_E)$ defined as follows: for any $i, j \in N$,

$$i \succcurlyeq_E j \text{ iff } E_i(S) \supseteq E_j(S) \text{ for all } S \subseteq N \text{ such that } S \cap \{i, j\} = \varnothing.$$

(Notice that we write $E_i$ for $E_{\{i\}}$, with a slight abuse of notation).

The *(Lapidot) coalitional desirability relation* of $(N, X, E)$ is the binary relation $(\mathcal{P}(N), \succcurlyeq_E^*)$ defined as follows: for any $S, T \subseteq N$,

$$S \succcurlyeq_E^* T \text{ iff } E_S(U) \supseteq E_T(U) \text{ for all } U \subseteq N \backslash (S \cup T).$$

Similar definitions may be obtained in natural ways by using the *height function* of the concept lattice of $\mathbf{G} = (N, X, E)$.

Let us consider as an example[8] the following set of definitions:
The *(normalized) individual* $\boldsymbol{L}$*-desirability relation* of $(N, X, E)$ is the binary relation $(N, \succcurlyeq_{\boldsymbol{L}(E)})$ $[(N, \succcurlyeq^{\circ}_{\boldsymbol{L}(E)})$, respectively$]$ defined by the following rules: for any $i, j \in N$,

$$\begin{aligned} i \succcurlyeq_{\boldsymbol{L}(E)} j \quad &\text{iff} \quad \mathrm{r}_{E_i}(S) \geq \mathrm{r}_{E_j}(S) \\ &\qquad \text{for all } S \subseteq N \text{ such that } S \cap \{i, j\} = \varnothing, \text{ and} \\ i \succcurlyeq^{\circ}_{\boldsymbol{L}(E)} j \quad &\text{iff} \quad \mathrm{r}_{E_i}(S)/\mathrm{r}_{E_i}(N)) \geq (\mathrm{r}_{E_j}(S)/\mathrm{r}_{E_j}(N) \\ &\qquad \text{for all } S \subseteq N \text{ such that } S \cap \{i, j\} = \varnothing. \end{aligned}$$

While the *(normalized) strong —or Lapidot— coalitional* $\boldsymbol{L}$*-desirability relation* of $(N, X, E)$ is the binary relation $(\mathcal{P}(N), \succcurlyeq^*_{\boldsymbol{L}(E)})$ $[(\mathcal{P}(N), \succcurlyeq^{*\circ}_{\boldsymbol{L}(E)})$,

[8] It should be noticed that the desirability relations presented below are by no means the only 'natural' versions of concept-latticial desirabilities based upon heights. As an alternative, one might decree coalition $S \subseteq N$ to be more desirable than coalition $T \subseteq N$ in terms of the concept-latticial heights attached to coalitional game form $(N, X, E)$ whenever $\mathrm{r}_E(S \cup U) \geqslant \mathrm{r}_E(T \cup U)$ for any $U \subseteq N$ such that $U \cap (S \cup T) = \varnothing$.

Incidentally, such a version of concept-latticial coalitional desirability would induce a proper *extension* of $(\mathcal{P}(N), \succcurlyeq_E^*)$, which is not the case for $(\mathcal{P}(N), \succcurlyeq^*_{L(E)})$ as defined below (see Proposition 7.i)). A comparative assessment of several concept-latticial desirabilities is however best left as a topic for further research. My point here is rather to show by some examples the wide range of interesting opportunities offered by introducing height-based concept-latticial desirabilities.

respectively] defined as follows: for any $S, T \subseteq N$,

$$S \succcurlyeq^{*}_{\boldsymbol{L}(E)} T \quad \text{iff} \quad \mathrm{r}_{E_S}(U) \geq \mathrm{r}_{E_T}(U) \text{ for all } U \subseteq N \backslash (S \cup T), \text{ and}$$
$$S \succcurlyeq^{*\circ}_{\boldsymbol{L}(E)} T \quad \text{iff} \quad \mathrm{r}_{E_S}(U)/\mathrm{r}_{E_S}(N)) \geq (\mathrm{r}_{E_T}(U)/\mathrm{r}_{E_T}(N) \text{ for all } U \subseteq N \backslash (S \cup T).$$

Clearly enough —and indulging in a slight abuse of notation— one may regard $\succcurlyeq^{*}_{E}$, $\succcurlyeq^{*}_{\boldsymbol{L}(E)}$ and $\succcurlyeq^{*\circ}_{\boldsymbol{L}(E)}$ as concept-latticial counterparts to $\succcurlyeq_{E}$,$\succcurlyeq_{\boldsymbol{L}(E)}$ and $\succcurlyeq^{\circ}_{\boldsymbol{L}(E)}$, respectively.

It should be emphasized here that standard desirability functions embody a certain notion of dominance with respect to set-inclusion, whereas **L**-desirability relations are also meant to rank coalitions which do not dominate each other in the standard desirability sense, by comparing their relative position with respect to other coalitions (in terms of decision power as summarized by the concept lattice). *Thus, **L**-desirability relations capture the idea that comparative evaluations of 'positional' decision power status across coalitions are a significant factor for the assessment of coalitional decision power under a given coalitional game form.*

The following proposition makes precise our previous claim to the effect that concept-latticial desirability relations are indeed a new, *distinct* class of desirability relations, which can also be regarded as an extension of the standard desirability relations for essential simple games to general monotonic effectivity functions:

**Proposition 7.** Let $(N, X)$ be a pair of non-empty sets, with $N$ finite. Then,

(i) there exist monotonic effectivity functions $(N, X, E), (N, X, E')$ such that:
$\succcurlyeq_{\boldsymbol{L}(E)} \nsubseteq \succcurlyeq_{E}$, $\succcurlyeq^{\circ}_{\boldsymbol{L}(E)} \nsubseteq \succcurlyeq_{E}$, $\succcurlyeq^{*}_{\boldsymbol{L}(E)} \nsubseteq \succcurlyeq^{*}_{E}$, $\succcurlyeq^{*\circ}_{\boldsymbol{L}(E)} \nsubseteq \succcurlyeq^{*}_{E}$, $\succcurlyeq_{E'} \nsubseteq \succcurlyeq_{\boldsymbol{L}(E')}$, $\succcurlyeq_{E'} \nsubseteq \succcurlyeq^{\circ}_{\boldsymbol{L}(E')}$, $\succcurlyeq^{*}_{E'} \nsubseteq \succcurlyeq^{*}_{\boldsymbol{L}(E')}$, and $\succcurlyeq^{*}_{E'} \nsubseteq \succcurlyeq^{*\circ}_{\boldsymbol{L}(E')}$;

(ii) if $(N, X, E)$ is a *simple* effectivity function, then $\succcurlyeq_{E} = \succcurlyeq_{\boldsymbol{L}(E)} = \succcurlyeq^{\circ}_{\boldsymbol{L}(E)}$ if $E$ is *essential*, and $\succcurlyeq^{*}_{E} = \succcurlyeq^{*}_{\boldsymbol{L}(E)} = \succcurlyeq^{*\circ}_{\boldsymbol{L}(E)}$ if $E$ is *consensual.*

*Proof.* (i): Take $X = \{x, y, z, w, v\}$, $\#X = 5$, $N = \{1, 2, 3, 4\}$, and consider the effectivity function $\mathbf{G} = (N, X, E)$ defined as follows:

$$\begin{aligned} A \in E(S) \quad \text{iff} \quad & [A \neq \varnothing \text{ and } S = N] \text{ or} \\ & [A \supseteq X \backslash \{x, y, z\} \text{ and } S \supseteq \{1, 2, 3\}] \text{ or} \\ & [A \supseteq X \backslash \{x, y\} \text{ and } S \supseteq \{1, 2\}] \text{ or} \\ & [A \supseteq X \backslash \{x, z\} \text{ and } S \supseteq \{1, 3\}] \text{ or} \end{aligned}$$

$$[A \supseteq X \setminus \{y,z\} \text{ and } S \supseteq \{2,3\}] \text{ or}$$
$$[A = X \text{ and } S \neq \varnothing].$$

It is easily checked that the $\{1\}$-reduced and $\{2\}$-reduced effectivity functions of $E$, that is $\mathbf{G}_1 = (N, X, E_1)$ and $\mathbf{G}_2 = (N, X, E_2)$, are as follows:

$$\begin{aligned} A \in E_1(S) \quad \text{iff} \quad & [A \neq \varnothing \text{ and } S \supseteq \{2,3,4\}] \text{ or} \\ & [A \supseteq X \setminus \{x,y,z\} \text{ and } S \supseteq \{2,3\}] \text{ or} \\ & [A \supseteq X \setminus \{x,y\} \text{ and } 2 \in S] \text{ or} \\ & [A \supseteq X \setminus \{x,z\} \text{ and } 3 \in S] \text{ or} \\ & [A \supseteq X \setminus \{x\} \text{ and } S \neq \varnothing]\ , \end{aligned}$$

$$\begin{aligned} A \in E_2(S) \quad \text{iff} \quad & [A \neq \varnothing \text{ and } S \supseteq \{1,3,4\}] \text{ or} \\ & [A \supseteq X \setminus \{x,y,z\} \text{ and } S \supseteq \{1,3\}] \text{ or} \\ & [A \supseteq X \setminus \{x,y\} \text{ and } 1 \in S] \text{ or} \\ & [A \supseteq X \setminus \{y,z\} \text{ and } 3 \in S] \text{ or} \\ & [A \supseteq X \setminus \{y\} \text{ and } S \neq \varnothing]. \end{aligned}$$

Hence

$$\mathrm{r}_{E_1}(\{3\}) = \mathrm{r}_{E_2}(\{3\}) = 1, (\mathrm{r}_{E_1}(\{3\})/\mathrm{r}_{E_1}(N)) = (\mathrm{r}_{E_2}(\{3\})/\mathrm{r}_{E_2}(N)) = 1/4,$$

while $E_1(\{3\}) \nsubseteq E_2(\{3\})$ and $E_2(\{3\}) \nsubseteq E_1(\{3\})$: it follows that $1 \sim_{\boldsymbol{L}(E)} 2$, $1 \sim^{\circ}_{\boldsymbol{L}(E)} 2$ (whence $1 \sim^{*}_{\boldsymbol{L}(E)} 2$ and $1 \sim^{*\circ}_{\boldsymbol{L}(E)} 2$ as well) but $1 \not\succcurlyeq^{*}_{E} 2$ and $2 \not\succcurlyeq^{*}_{E} 1$.

Next, consider the (not regular) effectivity function $E'$ on ($N = \{1,2,3,4\}$, $X = \{x,y,z,w\}$) with $A \in E'(S)$ if and only if [$A \neq \varnothing$ and $1 \in S$, or $A \supseteq X \setminus \{x,y\}$ and $S \cap \{1,2\} \neq \varnothing$, or $A \supseteq X \setminus \{x\}$ and $S \cap \{1,2,3\} \neq \varnothing$, or else $A = X$ and $S \neq \varnothing$]. Then $A \in E'_1(S)$ if and only if [$A \neq \varnothing$ and $S \neq \varnothing$], while $A \in E'_3(S)$ if and only if [ $A \neq \varnothing$ and $1 \in S$, or $A \supseteq X \setminus \{x,y\}$ and $S \cap \{1,2\} \neq \varnothing$, or $A = X$ and $S \neq \varnothing$]. Therefore $E'_1(S) \supseteq E'_3(S)$ for any $S \subseteq N$, by definition, while it is immediately checked that $\mathbf{L}(\mathbf{G}_1) = \mathbf{2}$ (that is the two-element chain), $\mathbf{L}(\mathbf{G}_2) = \mathbf{4}$ (*i.e.*, the four-element chain), and $\mathrm{r}_{E_1}(\{2\}) = 0$, $\mathrm{r}_{E_3}(\{2\}) = 1$, whence *not* $1 \succcurlyeq_{\boldsymbol{L}(E)} 3$ and *not* $1 \succcurlyeq^{\circ}_{\boldsymbol{L}(E)} 3$.

(ii): By contradiction. Let $E$ be a simple and essential effectivity function on $(N, X)$ and $i, j \in N$ such that for any $S \subseteq N, S \cap \{i,j\} = \varnothing : \mathrm{r}_{E_i}(S) \geq \mathrm{r}_{Ej}(S)$, and suppose that there exists $T \subseteq N$ such that $T \cap \{i,j\} = \varnothing$ and $E_i(T) \nsupseteq E_j(T)$. Then $E_i(T) = E(T \cup \{i\}) \neq \{A \subseteq X : A \neq \varnothing\}$, hence —by simplicity of $E$— $E_i(T) = \{X\}$ and $E_j(T) = E(T \cup \{j\}) = \{A \subseteq X : A \neq \varnothing\}$. Now, $E_j(\{j\}) = E(\{j\}) \neq \{A \subseteq X : A \neq \varnothing\}$ since $E$

is essential. It follows that $\mathbf{L}(\mathbf{G}_j) = \mathbf{4}$ and $\mathrm{r}_{E_j}(T) = 1$, while clearly $\mathrm{r}_{E_i}(T) = 0$: contradiction.

Now, let us suppose that $E$ is simple and consensual, and that there exist $U, V \subseteq N$ and $T \subseteq N$ such that $T \cap (U \cup V) = \varnothing, E_U(S) \supseteq E_V(S)$ for any $S \subseteq N, S \cap (U \cup V) = \varnothing$, and $\mathrm{r}_{E_V}(T) > \mathrm{r}_{E_U}(T)$ (hence $T \neq \varnothing$, which entails $U \neq N$). Then notice that by definition $E_S[\mathcal{P}(N)] \subseteq E[\mathcal{P}(N)]$ for any $S \subseteq N$. Therefore it follows from simplicity of $E$, Proposition 1 above, and definition of $\mathrm{r}_{E_S}(.)$ that $1 \geq \mathrm{r}_{E_V}(T) > \mathrm{r}_{E_U}(T) \geq 0$, *i.e.*, $[\mathrm{r}_{E_V}(T) = 1$ and $\mathrm{r}_{E_U}(T) = 0]$, which in turn entails —by simplicity of $E$ and definition of $\mathrm{r}_{E_S}(.)$ again— $E_V(T) = \{A \subseteq X : A \neq \varnothing\}$. Hence in view of our hypothesis, $E_U(T) = \{A \subseteq X : A \neq \varnothing\}$ as well: thus $E_U(T) \supseteq E_U(S)$ for any $S \subseteq N$. But then $\mathrm{r}_{E_U}(T) = 0$ entails $E_U(S) = E_U(T) = \{A \subseteq X : A \neq \varnothing\}$ for any $S \subseteq N, S \neq \varnothing$. In particular, $E(U) = E_U(U) = \{A \subseteq X : A \neq \varnothing\}$, *i.e.*, $E$ is *not* consensual (since $U \neq N$): contradiction. A similar argument applies to simple essential effectivity functions, by taking $U = \{i\}$, $V = \{j\}$ for some $i, j \in N, i \neq j$.

We have therefore established that $\succcurlyeq_E = \succcurlyeq_{\boldsymbol{L}(E)}$ whenever $E$ is simple and essential, and $\succcurlyeq^*_E = \succcurlyeq^*_{\boldsymbol{L}(E)}$ if $E$ is simple and consensual.

Now, observe that essentiality of a simple $E$ entails $E_i(\{i\}) = E(\{i\}) \subset \{A \subseteq X : A \neq \varnothing\} = E(N) = E_i(N)$ for any $i \in N$, whence by definition $0 \leq \mathrm{r}_{E_i}(\{i\}) < \mathrm{r}_{E_i}(N) \leq 1$, that is $\mathrm{r}_{E_i}(N) = 1$. It follows that $\succcurlyeq_{\boldsymbol{L}(E)} = \succcurlyeq^{\circ}_{\boldsymbol{L}(E)}$. Similarly, consensuality of a simple $E$ entails $E_T(T) = E(T) \subset \{A \subseteq X : A \neq \varnothing\} = E(N) = E_T(N)$ for any $T \subset N$, whence $\mathrm{r}_{E_T}(N) = 1$, and $\succcurlyeq^*_{\mathbf{L}(E)} = \succcurlyeq^{*\circ}_{\mathbf{L}(E)}$. q.e.d.

**Remark 8.** It should be noticed that Proposition 7.ii) above is tight. To check that, consider the simple effectivity functions $(N, X, E^a), (N, X, E^b)$ defined as follows: for any $A \subseteq X = \{x, y\}$, $S \subseteq N = \{1, 2, 3, 4\}$ $A \in E^a(S)$ if and only if $[(A \neq \varnothing$ and $1 \in S)$ or $(A = X$ and $S \neq \varnothing)]$ and $A \in E^b(S)$ if and only if $[(A \neq \varnothing$ and $S \supseteq \{1, 2\})$ or $(A = X$ and $S \neq \varnothing)]$. Then it is immediately checked that there exist $S \subseteq N$ such that $S \cap \{1, 3\} = \varnothing$, $E^a_3(S) \not\supseteq E^a_1(S)$, while $\mathrm{r}_{E^a_1}(S) = \mathrm{r}_{E^a_3}(S) = 0$, and $S' \subseteq N$ such that $S' \cap (\{1, 2\} \cup \{3\}) = \varnothing$, $E^b_3(S') \not\supseteq E^b_{\{1,2\}}(S')$ while $\mathrm{r}_{E^b_{\{1,2\}}}(S') = \mathrm{r}_{E^b_3}(S') = 0$.

It is worth mentioning here that a few basic properties of $\boldsymbol{L}$-desirability relations suggest a further analogy with desirabilities for simple games and their most standard extensions. Indeed, in their recent monograph on simple games Taylor and Zwicker introduce a notion of *normality* for (classes of) desirability *indifference*[9] relations of simple games (*cf.* [TayZwi99]). Taylor-

[9] A desirability indifference relation $\sim$ is the symmetric component of a desirability relation $\succcurlyeq$, *i.e.*, for any pair $S, T$ of coalitions or players $S \sim T$ if and only if both $S \succcurlyeq T$

Zwicker's definition can be easily extended to desirability indifference relations for effectivity functions as follows.

**Definition 9 (Taylor-Zwicker-normal classes of coalitional desirability indifference relations for effectivity functions).** Let $\mathbf{E}$ be a class of effectivity functions, and $\equiv = \{\equiv_E : E \in \mathbf{E}\}$ a class of coalitional desirability indifference relations (*i.e.*, for any effectivity function $E$ on $(N, X)$ such that $E \in \mathbf{E}$, $\equiv_E \subseteq \mathcal{P}(N) \times \mathcal{P}(N)$ is reflexive and symmetric). Then, class $\equiv$ is *Taylor-Zwicker-normal* if and only if the following criteria are satisfied:

(i) (*Anonymity-Consistency*) If $(N, X, E)$ is an anonymous effectivity function and $S, T \subseteq N$ are such that $\#S = \#T$ then $S \equiv_E T$;

(ii) (*Minimal Discriminatory Power*) Let $(N = \{1,2,3,4\}, X = \{x, y\}, E)$ be an additive effectivity function induced by probability measures $p, q$ with $p(1) = 2/5$, $p(2) = p(3) = p(4) = 1/5$, $q(x) = q(y) = 1/2$. Then *not* $\{2,3\} \equiv_E \{2\}$;

(iii) (*Merging Invariance*) Let $X$ be a non-empty set, $N, N'$ two non-empty finite sets, $f : N \to N'$ a *surjective* function, $(N, X, E)$ and $(N', X, E')$ effectivity functions such that for any $A \subseteq X$, $S' \subseteq N'$, $A \in E'(S')$ if and only if $A \in E(f^{-1}(S'))$. Then for any $S', T' \subseteq N'$ such that $S' \cap T' = \varnothing$, $f^{-1}(S') \equiv_E f^{-1}(T')$ entails $S' \equiv_{E'} T'$.

**Remark 10.** The Anonymity-Consistency and Minimal Discriminatory Power requirements are mild and uncontroversial. Thus the crucial requirement for Taylor-Zwicker-normality is Merging Invariance which amounts to 'closure under Rudin-Keisler pullbacks' as defined in [TayZwi99]. The intuition underlying Merging Invariance (which also motivates our choice of terminology) is that if two equally desirable coalitions are affected by internal merging processes, then the resulting post-merger coalitions should also be equally desirable.

It turns out that the symmetric components of strong coalitional $\boldsymbol{L}$-*desirability relations* for effectivity functions as defined above are indeed a Taylor-Zwicker-normal class.

**Proposition 11.** Let $X$ be a non-empty set, and $N$ a non-empty finite set. Then the class of $\boldsymbol{L}$-desirability indifference relations

$$\{\sim^*_{\boldsymbol{L}(E)} \ : \ \text{there exist } N' \subseteq N, X' \subseteq X \text{ such that } (N', X', E) \text{ is a monotonic effectivity function } \}$$

is Taylor-Zwicker-normal.

and $T \not\succcurlyeq S$.

*Proof.* (*Anonymity-Consistency*): Let $(N', X', E)$ be an anonymous effectivity function for some $N' \subseteq N, X' \subseteq X$, and $S, T \subseteq N'$ with $\#S = \#T$. Then for any $V \subseteq N'$ such that $V \cap (S \cup T) = \varnothing$, $E_S(V) = E_T(V)$, while for any $U \subseteq N'$ such that $E_S(U) \subset E_S(V)$ it must be —by monotonicity of $E$ and definition of $E_S$— that $\#(S \cup U) < \#(S \cup V)$ hence there exists $U' \subseteq N'$ such that $\#(T \cup U') = \#(S \cup U)$ and consequently —by anonymity of $E$ and definition of $E_T$— $E_T(U') = E_S(U) \subset E_S(V) = E_T(V)$. It follows that —by definition of $\mathrm{r}_{E_S}$ and $\mathrm{r}_{E_T}$— $\mathrm{r}_{E_S}(V) = \mathrm{r}_{E_T}(V)$.

(*Minimal Discriminatory Power*): Just consider $\{4\}$. Clearly, $E_{\{2,3\}}(\{4\}) = \{A \subseteq X : A \neq \varnothing\} \supset \{X\} = E_{\{2\}}(\{4\}) = E_{\{2,3\}}(\{2\}) = E_{\{2,3\}}(\{3\})$ (recall that $\#X \geq 2$ by hypothesis). Hence $\mathrm{r}_{E_{\{2,3\}}}(\{4\}) > 0 = \mathrm{r}_{E_{\{2\}}}(\{4\})$, and *not*$(\{2,3\} \sim^*_{\mathbf{L}(E)} \{2\})$.

(*Merging Invariance*): Let $\mathbf{G} = (N, X, E)$ be a coalitional game form, $f : N \to N'$ a surjective function and $\mathbf{G}' = (N', X, E')$ the coalitional game form defined as follows:

For any $S' \subseteq N'$ ,$A \subseteq X$, $A \in E'(S')$ if and only if $A \in E(f^{-1}(S'))$. Hence, by definition, $E' = E \circ f^{-1}$.

Now, take $S', T' \subseteq N'$ such that $f^{-1}(S') \equiv_{\mathrm{r}_E} f^{-1}(T')$. Thus, by definition, for any $V \subseteq N$ with $V \cap (f^{-1}(S') \cup f^{-1}(T')) = \varnothing$ there exists a positive integer $k$ such that $k$ is the common height of $V$ in $\mathbf{L}((N, X, E_{f^{-1}(S')}))$ and in $\mathbf{L}((N, X, E_{f^{-1}(T')}))$, *i.e.*, $k = \mathrm{r}_{E_{f^{-1}(S')}}(V) = \mathrm{r}_{E_{f^{-1}(T')}}$. It follows that, by definition of such concept lattices, for any $k' \leqslant k$ there exist $\{S_1, .., S_{k'-2}\} \subseteq \mathcal{P}(N)$, $\{T_1, .., T_{k'-2}\} \subseteq \mathcal{P}(N)$ such that

$$E(f^{-1}(S') \cup V) = E_{f^{-1}(S')}(V) \supset E_{f^{-1}(S')}(S_1) \supset \dots$$
$$\dots \supset E_{f^{-1}(S'')}(S_{k'-2}) \supset E_{f^{-1}(S'')}(\varnothing)$$

and

$$E(f^{-1}(T') \cup V) = E_{f^{-1}(T')}(V) \supset E_{f^{-1}(S')}(T_1) \supset \dots$$
$$\dots \supset E_{f^{-1}(T'')}(T_{k'-2}) \supset E_{f^{-1}(T'')}(\varnothing).$$

Next, let us assume that *not* $S' \equiv_{\mathrm{r}_{E'}} T'$. Then one may also assume without loss of generality that there exist $U' \subseteq N'$, a positive integer $k^*$, and $\{S'_1, .., S'_{k^*-2}\} \subseteq \mathcal{P}(N')$ such that

$$U' \cap (S' \cup T') = \varnothing,$$

$$E(f^{-1}(S' \cup U')) = E'_{S'}(U') \supset E'_{S'}(S'_1) \supset .. \supset E'_{S'}(S'_{k^*-2}) \supset E'_{S'}(\varnothing),$$

while $m < k^*$ for any positive integer $m$ such that there exist $\{T'_1, .., T'_{m-2}\} \subseteq \mathcal{P}(N')$ with

$$E(f^{-1}(T' \cup U')) = E'_{T'}(U') \supset E'_{T'}(T'_1) \supset .. \supset E'_{T'}(T'_{m-2}) \supset E'_{T'}(\varnothing).$$

But notice that $f^{-1}(T' \cup U') = f^{-1}(T') \cup f^{-1}(U')$, whence the foregoing statements on nested sequences may be rewritten as follows:

$$E(f^{-1}(S') \cup f^{-1}(U')) = E'_{S'}(U') \supset E'_{S'}(S'_1) \supset \ldots \supset E'_{S'}(S'_{k^*-2}) \supset E'_{S'}(\varnothing)$$

and

$$E(f^{-1}(T') \cup f^{-1}(U')) = E'_{T'}(U') \supset E'_{T'}(T'_1) \supset \ldots \supset E'_{T'}(T'_{m-2}) \supset E'_{T'}(\varnothing)$$

entails $m < k^*$. Moreover, $T' \cap U' = \varnothing$ entails $f^{-1}(T') \cap f^{-1}(U') = \varnothing$ and similarly $S' \cap U' = \varnothing$ entails $f^{-1}(S') \cap f^{-1}(U') = \varnothing$. Therefore $f^{-1}(U') \subseteq N$ is such that

$$f^{-1}(U') \cap (f^{-1}(S') \cup f^{-1}(T')) = \varnothing$$

while

$$\mathrm{r}_{E_{f^{-1}(S')}}(f^{-1}(U')) \geqslant k^* > \mathrm{r}_{E_{f^{-1}(T')}}(f^{-1}(U')),$$

hence *not* $f^{-1}(S') \equiv_{\mathrm{r}_E} f^{-1}(T')$, a contradiction. q.e.d.

Thus, strong coalitional $\boldsymbol{L}$-desirability relations —while distinct from the most straightforward extensions to effectivity functions of the strong coalition desirability relations for simple games— are indeed regular enough to be members of the Taylor-Zwicker-normal class. In particular, it follows that strong coalitional $\boldsymbol{L}$-desirability relations may well be intransitive as stated by the following

**Corollary 12.** Let $X, N$ be non-empty finite sets. Then there exists a monotonic effectivity function $(N, X, E)$ such that $\sim^*_{\boldsymbol{L}(E)}$ is not transitive.

*Proof.* Straightforward from Proposition 11 above, Theorem 4.2.2 in Taylor and Zwicker's [TayZwi99], as combined with the fact that a simple game $(N, W)$ can always be re-interpreted as an effectivity function on $(N, X)$ with $\#X = 2$. q.e.d.

Corollary 12 suggests the need for a detailed analysis of conditions on an effectivity function that imply transitivity or acyclicity (as well as totality) of its strong coalitional $\boldsymbol{L}$-desirability relation. This is however not the place for a full-fledged analysis of $\boldsymbol{L}$-*desirability relations* which is indeed best left as a topic for further research.

# 3 Computing Concept Lattices of Coalitional Game Forms: Some Examples

Computing the concept lattice of a coalitional game form may well involve a heavy computational burden (but *cf.* [GanWil99, Chapter 2] for a short discussion of some feasible algorithms). The present section is devoted to the computation of the (bulks of) concept lattices of some prominent and well-known game forms that are simple enough to allow for "manual" calculations. Comparing such concept lattices, and their parameters, will enable us to classify the underlying game forms according to the properties introduced in the previous section.

### 3.1 Modular-Arithmetical Game Forms for Deterministic Social Lotteries

A strategic game form for a deterministic social lottery on $(N, X)$ is a tuple $G^{\mathrm{L}} = (N, X, (S_i)_{i \in N}, h)$ with $S_i = (X \times \mathbb{Z}_+)$ for any $i \in N$, and $h((x_i, z_i)_{i \in N}) = x_{i^*}$ where $i^* = \sum_{i \in N} x_i (\mathrm{mod}\, n)$ and $n = \#N$ (*cf.*, *e.g.*, [DanSot02] for an extensive discussion of game forms of this sort). Clearly, $E_\alpha(G^{\mathrm{L}})$ is a consensual effectivity function, while $E_\beta(G^{\mathrm{L}})(S) = \mathcal{P}(X) \setminus \{\varnothing\}$ for any $S \subseteq N$.

**Claim 13.** Let $G^{\mathrm{L}}$ be a lottery game form as defined above. Then $\mathbf{L}(G^{\mathrm{L}}_\alpha) = \mathbf{4}$, while $\mathbf{L}(G^{\mathrm{L}}_\beta) = \mathbf{3}$ (hence in particular $B(\mathbf{L}(G^{\mathrm{L}}_\beta)) = \mathbf{1}$).

*Proof.* Straightforward: just observe that the concepts comprising the bulk of $\mathbf{L}(\mathbf{G}^{\mathrm{L}}_\alpha)$ are $(\{N\}, \mathcal{P}(X) \setminus \{\varnothing\})$ and $(\mathcal{P}(N) \setminus \{\varnothing\}, \{X\})$, while the only concept in the bulk of $\mathbf{L}(\mathbf{G}^{\mathrm{L}}_\beta)$ is $(\mathcal{P}(N) \setminus \{\varnothing\}, \mathcal{P}(X) \setminus \{\varnothing\})$. q.e.d.

Thus constancy of $E_\beta(G^{\mathrm{L}})$ is reflected by flatness of its concept lattice. Moreover, flatness of the concept lattice also implies that concept-latticial coalitional desirabilities relations as defined above reduce to a single indifference class.

### 3.2 Demand Game Forms

A (strategic) demand game form on $(N, X)$ is a tuple $G^{\mathrm{D}} = (N, (X_i)_{i \in N}, x^*)$ where $X \subseteq \Pi_{i \in N} X_i$, and $x^* \in X$ denotes the conflict outcome. The players in $N$ may agree on any outcome in $X$ by choosing $x = ((x_i)_{i \in N}) \in X$. The conflict outcome $x^*$ obtains if the players fail to agree on any other outcome. Therefore $E_\alpha(G^{\mathrm{D}})$ is given by the following rule: for any $S, B$ ,$\varnothing \neq S \subseteq N$, $\varnothing \neq B \subseteq X$, and $B \in (E_\alpha(G^{\mathrm{D}}))(S)$ if and only if either $S = N$ or $x^* \in B$. The validity of the following claim is easily established:

**Claim 14.** Let $G^{\mathrm{D}}$ be a demand game form as defined above. Then $E_\alpha(G^{\mathrm{D}}) = E_\beta(G^{\mathrm{D}})$ and $\mathbf{L}(G^{\mathrm{D}}) = \mathbf{4}$. Thus $G^{\mathrm{D}}$ is unspecialized and consensual (but not simple, fully distributed or strictly hierarchical).

*Proof.* It is well-known that for any strategic game form $E_\alpha(G) = E_\beta(G)$ if and only if $E_\alpha(G)$ is maximal (*cf.* [Pel84, Lemma 5.1.17]). Now take $\varnothing \neq A \subseteq X, \varnothing \neq S \subseteq N$ such that $A \notin E_\alpha(G^{\mathrm{D}})(S)$. Then —by definition of $G^{\mathrm{D}}$— $S \neq N$ and $x^* \notin A$: hence $N \backslash S \neq \varnothing$ and $x^* \in X \backslash A$, *i.e.*, $X \backslash A \in E_\alpha(G^{\mathrm{D}})(N \backslash S)$. Moreover the definition of $G^{\mathrm{D}}$ clearly entails that $S = N$ is the sole coalition such that $A \in E_\alpha(G^{\mathrm{D}})(S)$ for each $A \in \mathcal{P}(X) \backslash \{\varnothing\}$ (this fact also implies that $G^{\mathrm{D}}$ is consensual). Indeed, $S = N$ is the sole coalition such that $A \in E_\alpha(G^{\mathrm{D}})(S)$ for *some* $A \in \{B \subseteq X : x^* \notin B\}$ while —by definition of $G^{\mathrm{D}}$— $A \in E_\alpha(G^{\mathrm{D}})(S)$ for any coalition $S \neq \varnothing$ and any $A \subseteq X$ with $x^* \in A$ (hence in particular $\mathrm{r}_{E_\alpha(G^{\mathrm{D}})}(\{i\}) = 1$ for any $i \in N$). It follows that $B(\mathbf{L}(G^{\mathrm{D}})) = ((\{N\}, \mathcal{P}(X) \backslash \{\varnothing\}), (\mathcal{P}(N) \backslash \{\varnothing\}, \{B \subseteq X : x^* \in B\})) = \mathbf{2}$ (modulo isomorphisms). q.e.d.

Thus a demand game form provides an interesting elementary example of an unspecialized and not strictly hierarchical game form which is not simple while being endowed with a *simple* Galois-latticial bulk.

### 3.3 Voting-by-Veto Game Forms 1: The Finite Case

Majoritarian voting game forms rely on a sharp distinction between winning, *i.e.*, all powerful coalitions and losing, *i.e.*, powerless coalitions. It is well-known that unfortunately majoritarian-like voting procedures are, generally speaking, *unstable* in that at many preference profiles their *core*[10] is *empty*. By contrast, certain *voting-by-limited-veto* procedures (as briefly introduced in section 2 above and thoroughly analyzed elsewhere, *e.g.*, in [MouPel82, Mou83, Pel84, DanSot93]) enjoy several nice stability properties (including core-stability[11] on standard large preference domains) and rely on a considerably more complex allocation of decision power. This is neatly reflected by the properties of their concept lattices. As a prominent example of a voting-by-limited-veto procedure that shares anonymity and neutrality properties with majoritarian-like schemes we shall focus on a version of the *proportional veto procedure*, first introduced by Moulin (*cf.*, *e.g.*, [Mou83, AbdKei91]). Namely, we consider a *proportional veto procedure with endogenous agenda formation* that can be described as follows. A distinguished outcome $x^*$ —the *"status quo"*— is identified. Then each player *makes k proposals*, is informed on the resulting set of outcomes, and *issues*

[10] An outcome $x \in X$ is dominated within a game in coalitional form $G = (N, X, E, (\succcurlyeq_i)_{i \in N})$, where $(N, X, E)$ is a coalitional game form and $(\succcurlyeq_i)_{i \in N}$ is the profile of total preference preorders on $X$, if there exist $A \subseteq X$ and $S \subseteq N$ such that $A \in E(S)$ and for any $y \in A$ and $i \in S$ both $y \succcurlyeq_i x$ and *not* $x \succcurlyeq_i y$. The *core* of $G$ is the set of outcomes of $X$ which are *not* dominated in $G$.

[11] A coalitional game form $\mathbf{G} = (N, X, E)$ is core-stable on a certain domain $D$ of preference profiles if for any preference profile $\succcurlyeq$ in $D$ the core of the game induced by $\mathbf{G}$ at $\succcurlyeq$ is non-empty. Convexity implies core-stability on most standard preference-domains. Voting-by-veto as defined below in the text is indeed additive hence convex and therefore core-stable.

*k vetos* —according to a *prefixed order*— on non-vetoed alternatives. The unique non-vetoed outcome is selected. The corresponding effectivity function $E^{\text{PV}}$ (that is regular and maximal, hence unambiguously determined) is defined by the following rule:
For any $S \subseteq N$, $A \subseteq X$, $A \in E^{\text{PV}}(S)$ if and only if

$$\lceil (kn+1)\frac{s}{n} \rceil > kn + 1 - a,$$

where $s = \#S$, $n = \#N$, and $a = \#A$.

Since each coalition-size corresponds to a distinctive "degree" of decision power, the concept lattice $\mathbf{L}(\mathbf{G}^{\text{PV}})$ is easily computed. Thus it is straightforward to establish validity of the following

**Claim 15.** Let $\mathbf{G}^{\text{PV}} = (N, X, E^{\text{PV}})$ be the proportional veto effectivity function as defined above. Then

(i) $\mathbf{L}(\mathbf{G}^{\text{PV}}) = \mathbf{1} \oplus \mathbf{n} \oplus \mathbf{1}$ (where $\mathbf{n}$ denotes the chain of size $n$). Hence in particular
(ii) $\mathbf{G}^{PV}$ is consensual and unspecialized (but not simple, fully distributed, or strictly hierarchical).

*Proof.* (i): Notice that for any $S \subseteq N$ such that $\#S = n - h$, $\lceil (kn+1)\frac{s}{n} \rceil > kn+1-a$ if and only if $a > kh$, *i.e.*, $A \in E^{\text{PV}}(S)$ iff $\#A > kh$. It follows that $B(\mathbf{L}(N, X, E^{\text{PV}})) = ((\{S \subseteq N : \#S \geq n-h\}, \{A \subseteq X : \#A > kh\}), h = 0, .., n-1) = \mathbf{n}$ (modulo isomorphisms). In particular, $(N, X, E^{\text{PV}})$ turns out to be consensual ($N$ is the only coalition of maximum height) and unspecialized (since its Galois lattice is a chain). q.e.d.

It turns out that the foregoing construct is amenable to a generalization which provides a nice example of a coalitional game form with an *infinite* concept lattice, as presented below.

### 3.4 Voting-by-Veto Game Forms 2: A Coalitional Game Form with an Infinite Concept Lattice

Let us now consider the following extension of the proportional voting-by-veto *coalitional* game form to an infinite setting. Let $(N, X, E_{\mu,\nu})$ be a coalitional game form, where $N = \mathbb{N}$ is a countably infinite set, $X = [0, 1]$ and $E_{\mu,\nu} : \mathcal{P}(\mathbb{N}) \to \mathcal{P}(\mathbb{X})$ is defined by the following rule: there exist positive and permutation-invariant probability measures $\mu : \mathcal{P}(\mathbb{N}) \to \mathbb{R}_+$, $\nu : \mathcal{P}(\mathbb{X}) \to \mathbb{R}_+$ such that for any $S \subseteq \mathbb{N}, A \subseteq \mathbb{X}$, $A \in E_{\mu,\nu}(S)$ if and only if $\mu(S) + \nu(A) > 1$. It is readily checked that even this infinite version $(N, X, E_{\mu,\nu})$ of the voting-by-veto coalitional game form is indeed an effectivity function. Moreover, $(N, X, E_{\mu,\nu})$ satisfies both superadditivity and monotonicity, therefore it can be regarded as the effectivity function of a (non-deterministic) strategic game form (*cf.*, *e.g.*, [Bor+95] whose proof for finite $N$ and $X$ can be easily adapted to the present infinite setting).

**Claim 16.** Let $\mathbf{G}^{\mu,\nu} = (N, X, E_{\mu,\nu})$ be a coalitional game form as defined above. Then $\mathbf{L}(\mathbf{G}^{\mu,\nu}) = \mathbf{1} \oplus \mathbb{N} \oplus \mathbf{1}$.

*Proof.* From positivity and permutation-invariance of $\mu$ and $\nu$ it follows that for any $S \subseteq N$ and $i \in N \backslash S$, $\mu(S \cup \{i\}) > \mu(S)$, and $(\curlywedge_E \circ \curlyvee_E)(\{S \cup \{i\}\}) \subset (\curlywedge_E \circ \curlyvee_E)(S)$, whence $B(\mathbf{L}(N, X, E_{\mu,\nu})) = \mathbb{N}$. q.e.d.

It should be noticed that concept lattices of both demand and (neutral) voting game forms as considered above are invariably *chains*. The following example shows that there also exist elementary game forms whose concept lattices are *not* chains.

### 3.5 $2 \times 2$ Strategic Game Forms

Let us consider a $2 \times 2$ *strategic game form* $G^2 = (N, X, (S_1, S_2), h)$, where $N = \{1, 2\}$, $X = \{a, b, c, d\}$, $S_1 = \{s_1, t_1\}$, $S_2 = \{s_2, t_2\}$, and $h(s_1, s_2) = a$, $h(s_1, t_2) = b$, $h(t_1, s_2) = c$, $h(t_1, t_2) = d$. Clearly enough $E_\alpha(G^2) \neq E_\beta(G^2)$. The following Claim summarizes the situation:

**Claim 17.** Let $G^2$ be a $2 \times 2$ strategic game form as defined above. Then

(i) $B(\mathbf{L}(N, X, E_\alpha(G^2))) = B(\mathbf{L}(N, X, E_\beta(G^2))) = \mathbf{2}^2$, the 4-valued Boolean lattice if $G^2$ is "generic", *i.e.*, $\#X = 4$. Hence in particular both $E_\alpha(G^2)$ and $E_\beta(G^2)$ are consensual, fully distributed, specialized (and, of course, neither simple nor strictly hierarchical);

(ii) $B(\mathbf{L}(N, X, E_\alpha(G^2))) = B(\mathbf{L}(N, X, E_\beta(G^2))) = \mathbf{3}$ if $\#X = 3$ and the replicated outcome is *not* on a diagonal, whereas $B(\mathbf{L}(E_\alpha(G^2))) = B(\mathbf{L}(E_\beta(G^2))) = \mathbf{2}$ if $\#X = 3$ and the replicated outcome is on a diagonal;

(iii) $B(\mathbf{L}(N, X, E_\alpha(G^2))) = B(\mathbf{L}(N, X, E_\beta(G^2))) = \mathbf{2}$ if $\#X = 2$ and there exists a diagonal which does not include replicated outcomes, whereas $B(\mathbf{L}(E_\alpha(G^2))) = \mathbf{2}$ and $B(\mathbf{L}(E_\beta(G^2))) = \mathbf{1}$ if $\#X = 2$ and each diagonal includes replicated outcomes.

*Proof.* Let $G^2$ be a $2 \times 2$ strategic game form with outcome function $h$ as described in the text. Clearly enough by choosing rows and columns, respectively, 1 can ($\alpha$- and $\beta$-)enforce $\{a, b\}$, $\{c, d\}$ (and supersets), whereas 2 can ($\alpha$- and $\beta$-)enforce $\{a, c\}$, $\{b, d\}$ (and supersets).

(i): It is easily checked by direct inspection that the only difference between $E_\alpha(G^2)$ and $E_\beta(G^2)$ —in the "generic" case, *i.e.*, with $\#X = 4$— reduces to the fact that

$$\{\{a,d\},\{b,c\}\} \in [E_\beta(\{1\}) \cap E_\beta(\{2\})] \backslash [E_\alpha(\{1\}) \cup E_\alpha(\{2\})].$$

It follows that $B(\mathbf{L}(\mathbf{G}^2_\alpha)) =$

$$\begin{aligned}\{\ & (\{N\}, \{A \subseteq X : A \neq \varnothing\}),\\ & (\{N,\{1\}\}, \{A \subseteq X : A = \{a,b\}, A = \{c,d\} \text{ or } \#A \geq 3\}),\\ & (\{N,\{2\}\}, \{A \subseteq X : A = \{a,c\}, A = \{b,d\} \text{ or } \#A \geq 3\}),\\ & (\{N,\{1\},\{2\}\}, \{A \subseteq X : \#A \geq 3\}) \qquad \}\end{aligned}$$

and $B(\mathbf{L}(\mathbf{G}^2_\beta)) =$

$$\begin{aligned}\{\ & (\{N\}, \{A \subseteq X : A \neq \varnothing\}),\\ & (\{N,\{1\}\}, \{A \subseteq X : \#A \geq 2, A \neq \{a,c\}, A \neq \{b,d\}\}),\\ & (\{N,\{2\}\}, \{A \subseteq X : \#A \geq 2, A \neq \{a,b\}, A \neq \{c,d\}\}),\\ & (\{N,\{1\},\{2\}\}, \{A \subseteq X : A = \{a,d\}, A = \{b,c\}, \text{ or } \#A \geq 3\}) \ \},\end{aligned}$$

where $\mathbf{G}^2_\alpha = (N, X, E_\alpha(G^2)$ and $\mathbf{G}^2_\beta = (N, X, E_\beta(G^2))$. But then, modulo (latticial) isomorphisms, $B(\mathbf{L}(\mathbf{G}^2_\alpha)) = B(\mathbf{L}(\mathbf{G}^2_\beta)) = \mathbf{2}^2$.

(ii): When $\#X = 3$, three subcases may be distinguished: a) $a = b$ or $c = d$ (*i.e.*, the identical outcomes are in the same row), b) $a = c$ or $b = d$ (*i.e.*, the identical outcomes are in the same column), c) $a = d$ or $b = c$ (*i.e.*, the identical outcomes are on a diagonal). It should also be noticed that under both a) and b) $E_\alpha(G^2) = E_\beta(G^2)$: this is so because whenever a (2-dimensional) row or column has (two) identical entries —say $x$— the following facts are easily checked by direct inspection of the game(-form) matrix: 1) $x$ is an element of both diagonals, 2) each diagonal replicates a distinct row (column) if the double $x$ is on a column (a row). But then $\{\{a,d\},\{c,d\}\} \subseteq E_\alpha(G^2)(\{1\}) \cap E_\alpha(G^2)(\{2\})$. Thus, $E_\alpha(G^2) = E_\beta(G^2) = E(G^2)$ follows from our previous observation on the relationship between $E_\alpha(G^2)$ and $E_\beta(G^2)$ in the "generic" case: we denote by $\mathbf{L}(N, X, E(G^2))$ their concept lattice, which is determined as follows. Under case a) it is checked by direct inspection that if we denote by $x$ the replicated outcome, and by $y, z$ the remaining outcomes then

$$E(G^2)(\{1\}) = \{A \subseteq X : A \supseteq \{x\}\} \cup \{\{y,z\}\} \text{ and}$$

$$E(G^2)(\{2\}) = \{A \subseteq X : A \supseteq \{x,y\} \text{ or } A \supseteq \{x,z\}\}.$$

Hence, $B(\mathbf{L}(N, X, G^2))) =$

$$\{\ (\{N\}, \{A \subseteq X : A \neq \varnothing\}),$$
$$(\{N, \{1\}\}, \{A \subseteq X : A = \{y, z\}\ \text{, or } A \supseteq \{x\}\}),$$
$$(\{N, \{1\}, \{2\}\}, \{A \subseteq X : A \supseteq \{x, y\} \text{ or } A \supseteq \{x, z\}\})\ \}.$$

Similarly, under case b) $B(\mathbf{L}(\mathbf{N}, \mathbf{X}, G^2))) =$

$$\{\ (\{N\}, \{A \subseteq X : A \neq \varnothing\}),$$
$$(\{N, \{2\}\}, \{A \subseteq X : A = \{y, z\}, \text{ or } A \supseteq \{x\}\}),$$
$$(\{N, \{1\}, \{2\}\}, \{A \subseteq X : A \supseteq \{x, y\} \text{ or } A \supseteq \{x, z\}\})\ \},$$

where again $x$ denotes the replicated outcome. Thus, under cases a) and b) $B(\mathbf{L}(N, X, E(G^2))) = \mathbf{3}$.

Under case c) it is immediately seen that for each row there is (exactly) one column that includes the same elements, and vice versa. Moreover, no row or column amounts to a pair of duplicated outcomes, and none of the diagonals can possibly be replicated by a row or a column ( because that would entail $\#X < 3$). Hence if we denote by $x$ the replicated outcomes and by $y, z$ the other outcomes

$$E_\alpha(G^2)(\{1\}) = E_\alpha(G^2)(\{2\}) = \{A \subseteq X : \#A \geq 2, A \neq \{y, z\}\} \text{ and}$$

$$E_\beta(G^2)(\{1\}) = E_\beta(G^2)(\{2\}) = \{A \subseteq X : A \neq \varnothing, A \neq \{y\}, A \neq \{z\}\}.$$

Thus, $B(\mathbf{L}(N, X, E_\alpha(G^2))) = B(\mathbf{L}(N, X, E_\beta(G^2))) = \mathbf{2}$.

(iii): If $\#X = 2$ four cases may be distinguished: a) an outcome $x \in X$ and a strategy profile $(u, v) \in S_1 \times S_2$ exist such that $x = h(r, w)$ if and only if $(r, w) \neq (u, v)$; b) each row is made up of replicated outcomes; c) each column is made up of replicated outcomes; d) each diagonal is made up of replicated outcomes.

Under case a) it is immediately checked that

$$E_\alpha(G^2)(\{1\}) = E_\alpha(G^2)(\{2\}) = E_\beta(G^2)(\{1\}) = E_\beta(G^2)(\{2\}) =$$
$$= \{A \subseteq X : A \neq \varnothing, A \neq \{x\}\}.$$

It follows that $B(\mathbf{L}(N, X, E_\alpha(G^2))) = B(\mathbf{L}(N, X, E_\beta(G^2))) = \mathbf{2}$.

Under case b),

$$E_\alpha(G^2)(\{1\}) = E_\beta(G^2)(\{1\}) = \{A \subseteq X : A \neq \varnothing\} \text{ and}$$

$$E_\alpha(G^2)(\{2\}) = E_\beta(G^2)(\{2\}) = \{X\}.$$

Hence $B(\mathbf{L}(N, X, E_\alpha(G^2))) = B(\mathbf{L}(N, X, E_\beta(G^2))) = \mathbf{2}$.
Similarly, under case c),

$$E_\alpha(G^2)(\{2\}) = E_\beta(G^2)(\{2\}) = \{A \subseteq X : A \neq \varnothing\} \text{ and}$$

$$E_\alpha(G^2)(\{1\}) = E_\beta(G^2)(\{1\}) = \{X\}.$$

Thus, again, $B(\mathbf{L}(N, X, E_\alpha(G^2))) = B(\mathbf{L}(N, X, E_\beta(G^2))) = \mathbf{2}$.
Under case d),

$$E_\alpha(G^2)(\{1\}) = E_\alpha(G^2)(\{2\}) = \{X\} \text{ and}$$

$$E_\beta(G^2)(\{1\}) = E_\beta(G^2)(\{2\}) = \{A \subseteq X : A \neq \varnothing\}.$$

It follows that $B(\mathbf{L}(N, X, E_\alpha(G^2))) = \mathbf{2}$, while $B(\mathbf{L}(N, X, E_\beta(G^2))) = \mathbf{1}$.

q.e.d.

Therefore a generic $2 \times 2$ strategic game form provides a first elementary example of an effectivity function whose concept lattice has a non-trivial width. Typical two players perfect information strictly competitive extensive game (forms) —either finite or infinite— exhibit the very same conceptual structure.

## 4 Concluding Remarks

The results presented in the previous sections suggest that our understanding of games in terms of coalitional power may be significantly enhanced if lattice- and order-theoretic methods are deployed within game-theoretic models. In particular it has been shown that concept lattices of game forms provide a few natural 'complexity' measures for the latter and a notion of coalitional rank which in turn suggests a further extension of desirability relations from simple games to effectivity functions.

In a sense all that is scarcely surprising: the concept lattice of *any* binary relation may be regarded as one particular way of encoding the information provided by the latter. Indeed, such an encoding of the interaction structure is likely to contribute new ways of organizing the relevant data and help elicit new patterns: as a matter of fact, concept lattices have become in the last two decades the most rapidly growing strand of applied lattice theory, with applications to an impressive array of fields ranging from retrieval systems and analysis of lexical data bases to color perception (*cf.* Ganter and Wille's [GanWil99] and the Appendix by the same authors in [Grä98]). In that connection, finding applications of concept lattices to the analysis of game-theoretic models is indeed something to be expected.

The present paper was focused on *classifying* game forms as coalitional-power-classifying data structures from a purely structural perspective. Classifications of this structural sort may in fact contribute to game analysis in at least two ways. First, concept-latticial classifications can improve our understanding of game-theoretic objects as such. Second, make it possible to identify which structural features of games interfere (or do not interfere) with their behaviour. In that connection, a most interesting issue is whether —and to what extent— various types of requirements on the allocation of coalitional power thus classified interfere with *implementability/stability* properties. In particular, one might ask, for any pair $(N, X)$ of player- and outcome-sets, what are the possible widths and/or lenghts of the concept lattices of coalitional game forms $(N, X, E)$ enjoying some behaviourally prominent property such as, say, being convex or core-stable on a suitably large domain of preference profiles (that issue is addressed in [Van06], where it is shown that —essentially— restricting to *convex* coalitional game forms does not introduce any additional constraint on the set of achievable widths or lenghts of their concept lattices)?

All in all, the results obtained so far are in our view promising enough to suggest that indeed —when it comes to the tasks of classifying and understanding coalitional power— concept-latticial representations of game forms are a quite powerful tool which deserves further careful study.

## References.


[AbdKei91] Joseph **Abdou** and Hans **Keiding**, Effectivity Functions in Social Choice, Kluwer 1991

[BarMon70] Marc **Barbut** and Bernard **Monjardet**, Ordre et Classification, Algèbre et Combinatoire, 2 volumes, Hachette 1970

[Bir67] Garrett **Birkhoff**, Lattice Theory, 3rd edition. American Mathematical Society 1967

[DanSot93] Vladimir I. **Danilov** and Alexander I. **Sotskov**, On Strongly Consistent Social Choice Functions, **Journal of Mathematical Economics** 22 (1993), p. 327–346

[DanSot02] Vladimir I. **Danilov** and Alexander I. **Sotskov**, Social Choice Mechanisms, Springer 2002

[DavPri90] Brian A. **Davey** and Hilary A. **Priestley**, Introduction to Lattices and Order, Cambridge University Press 1990

[dSw99] Harrie **de Swart** (*ed.*), Logic, Game Theory, and Social Choice, Proceedings of the International Conference LGS 99, Tilburg University Press 1999

[Ein85] Ezra **Einy**, The Desirability Relation of Simple Games, **Mathematical Social Sciences** 10 (1985), p. 155–168

[GanWil99] Bernhard **Ganter** and Rudolf **Wille**, Formal Concept Analysis, Springer 1999

[Grä98] George A. **Grätzer**, General Lattice Theory, Birkhäuser 1998

[Mou83] Hervé **Moulin**, The Strategy of Social Choice, North Holland 1983 [Advanced textbooks in economics]

[MouPel82] Hervé **Moulin** and Bezalel **Peleg**, Cores of Effectivity Functions and Implementation Theory, **Journal of Mathematical Economics** 10 (1982), p. 115–145

[Bor+95] Peter **Borm**, Gert-Jan **Otten**, Ton **Storcken**, and Stef **Tijs**, Effectivity Functions and Associated Claim Game Correspondences, **Games and Economic Behavior** 9 (1995), p. 172–190

[Pel84] Bezalel **Peleg**, Game Theoretic Analysis of Voting in Committees, Cambridge University Press 1984 [Econometric Society Monographs]

[TayZwi99] Alan D. **Taylor** and William S. **Zwicker**, Simple Games: Desirability Relations, Trading, Pseudoweightings, Princeton University Press 1999

[Van99] Stefano **Vannucci**, On a Lattice-Theoretic Representation of Coalitional Power in Game Correspondences, *in:* [dSw99, p. 575–588]

[Van06] Stefano **Vannucci**, Concept Lattices and Convexity of Coalitional Game Forms, DEP Siena Working Paper Series 2006



**Received**: March 1st, 2005;
**In revised version**: October 16th, 2005;
**Accepted by the editors**: June 6th, 2007.

Stefan **Bold**, Benedikt **Löwe**,
Thoralf **Räsch**, Johan **van Benthem** (*eds.*)
**Foundations of the Formal Sciences V**
Infinite Games

# Sets with Large Intersections: A Game Theoretic Approach

Enrico Zoli*

Facoltà di Architettura
Università degli Studi di Bologna
via Cavalcavia 55
47023 Cesena (FC), Italy
e.zoli@unibo.it

ABSTRACT. We say that a family $\mathcal{M}$ of subsets of $\mathbb{R}$ has the large intersection property if any intersection of countably many diffeomorphic images of members of $\mathcal{M}$ has Hausdorff dimension one. Here it is shown, via very simple proofs, that the family of winning sets in the sense of Schmidt's $(\alpha, \beta)$-games has the large intersection property. Incidentally, we point out an alternative proof of Jarník's theorem stating that the set of badly approximable numbers has Hausdorff dimension one.

## 1 Introduction

*Notation.* Throughout, $\mathbb{N}$ and $\mathbb{R}$ stand for the set of nonnegative integers (or natural numbers) and the set of reals, respectively. A natural number $n$ is identified with the set of its predecessors: $n = \{0, 1, \ldots, n-1\}$.

The intervals considered in this paper are assumed to be nontrivial (*i.e.*, of positive length) and compact. The length of an interval $I$ is denoted by $\ell(I)$. Given $\delta \in (0,1)$, the symbol $B^\delta(I)$ denotes the set of all subintervals $J$ of $I$ such that $\ell(J) = \delta\ell(I)$. For $x \in \mathbb{R}$ and $r > 0$, by $I[x, r]$ we mean the interval with center $x$ and length $2r$.

*Background: Hausdorff dimensions.* Given a subset $F$ of $\mathbb{R}$, the Hausdorff dimension $\dim F$ of $F$ is defined as the infimum of the positive reals $s$ such that, for every $\varepsilon > 0$, there is a sequence $(I_n)_{n\in\mathbb{N}}$ of intervals covering $F$ and satisfying $\ell(I_n) < \varepsilon$ and $\sum_{n\in\mathbb{N}} \ell^s(I_n) < \infty$.

*The author sincerely thanks the referee for many precious suggestions.

The following are easy consequences of the definition (a fuller account on Hausdorff dimensions may be found in Falconer's book [Fal90]):

*(i)* If $E \subseteq F \subseteq \mathbb{R}$, then $0 = \dim \varnothing \leq \dim E \leq \dim F \leq \dim \mathbb{R} = 1$.

*(ii)* If $(F_n)_{n\in\mathbb{N}}$ is a sequence of sets in $\mathbb{R}$, then $\dim \bigcup_{n\in\mathbb{N}} F_n = \sup_{n\in\mathbb{N}} \dim F_n$.

*(iii)* If $F \subseteq \mathbb{R}$ and $\varphi : \mathbb{R} \to \mathbb{R}$ is a diffeomorphism (namely, a surjection of class $C^1$ with a never vanishing derivative), then $\dim F = \dim \varphi[F]$.

*(iv)* If $F$ is a subset of positive Lebesgue outer measure, then $\dim F = 1$.

*Sets with large intersections.* Let $(F_n)_{n\in\mathbb{N}}$ be a sequence of subsets of $\mathbb{R}$, and $(\varphi_n)_{n\in\mathbb{N}}$ a sequence of diffeomorphisms $\varphi_n : \mathbb{R} \to \mathbb{R}$. It follows from *(ii)* and *(iii)* that the Hausdorff dimension of $F := \bigcup_{n\in\mathbb{N}} \varphi_n[F_n]$ is

$$\dim F = \sup_{n\in\mathbb{N}} \dim \varphi_n[F_n] = \sup_{n\in\mathbb{N}} \dim F_n.$$

In contrast to this, it is generally impossible to determine the dimension of $E := \bigcap_{n\in\mathbb{N}} \varphi_n[F_n]$ in terms of the dimensions of the $F_n$, the only rough (and obvious) estimate being derived from *(i)* and *(iii)*:

$$\dim E \leq \inf_{n\in\mathbb{N}} \dim \varphi_n[F_n] = \inf_{n\in\mathbb{N}} \dim F_n. \tag{1}$$

It turns out that the inequality in (1) can be "dramatically" strict: it is easy to construct two subsets of $\mathbb{R}$ of Hausdorff dimension one, even dense in $\mathbb{R}$, with empty intersection.

It is therefore useful to determine conditions on the sets $F_n$ guaranteeing the *equality* in (1), *i.e.*, the largest (in the sense of Hausdorff) possible intersection. Indeed, sets fulfilling such large intersection properties occur naturally in number theory (*e.g.*, in the theory of diophantine approximations) and in dynamical systems (in problems of small divisors resonances, for example).

Several authors have addressed this problem and introduced systems of sets of Hausdorff dimension $s$ such that countable intersections of the sets also have Hausdorff dimension $s$. We mention the "regular sets" in [BakSch70], the "$\mathcal{M}^s_\infty$-dense" construction in [Fal85], and the constructions using "ubiquitous systems" in [DodRynVic90]. All these approaches, however, have in the author's opinion two disadvantages: they are somewhat technical; emphasis is not given on the sets *per se*, but rather on the recipe for their construction.

A very good model is proposed by Falconer in [Fal94], where he actually unifies the aforementioned constructions by giving a simple and direct definition of sets with large intersection properties. Falconer's approach suggests the following

**Definition 1.** A family $\mathcal{M}$ of subsets of $\mathbb{R}$ has the large intersection property if, for any sequence $(F_n)_{n\in\mathbb{N}}$ of members of $\mathcal{F}$ and any sequence $(\varphi_n)_{n\in\mathbb{N}}$ of diffeomorphisms $\varphi_n : \mathbb{R} \to \mathbb{R}$, we have $\dim \bigcap_{n\in\mathbb{N}} \varphi_n[F_n] = 1$.

It is easy to deduce that all members $F$ of $\mathcal{F}$ are everywhere of Hausdorff dimension one, *i.e.*, $\dim F \cap O = 1$ for any nonempty open subset $O$ of $\mathbb{R}$. The family consisting of the full Lebesgue measure subsets of $\mathbb{R}$ obviously has the large intersection property, by *(iv)*.

Aim of this note is to indicate another (less trivial) class of sets with the large intersection property. More precisely:

We will show in Theorem 5 that the winning sets in the sense of Schmidt's $(\alpha, \beta)$-games have Hausdorff dimension one. We shall verify in Theorem 7 that the family $\mathcal{W}$ consisting of all winning sets is closed under taking diffeomorphic images. Combined with Schmidt's result quoted in Theorem 2, *i.e.*, that the intersection of countably many members of $\mathcal{W}$ in its turn belongs to $\mathcal{W}$, Theorems 5 and 7 immediately imply Corollary 8 stating that $\mathcal{W}$ has the large intersection property.

Incidentally, as a direct consequence of Schmidt's analysis, we also point out an alternative proof of the classical theorem of Jarník quoted in Corollary 6: the set of badly approximable numbers has Hausdorff dimension one.

While Theorems 5 and 7 are not new, for they are included in Schmidt's [Sch66, Theorems 6 and 1], their proofs —*ad hoc* for the real line case— are particularly short and simpler than the original ones.

## 2 Schmidt's $(\alpha, \beta)$-games and winning sets

Let $\alpha, \beta \in \left(0, \frac{1}{2}\right)$ and $W \subseteq \mathbb{R}$. The $(\alpha, \beta)$-game relative to $W$ between the two players Adam and Eve is defined as follows. Adam selects an interval $A_0$ in $\mathbb{R}$. Then Eve chooses $E_0 \in B^\alpha(A_0)$. Adam, in his turn, selects $A_1 \in B^\beta(E_0)$. Eve responds with $E_1 \in B^\alpha(A_1)$, and so on. In general, for any $n \in \mathbb{N}$ the rules of the game impose $E_n \in B^\alpha(A_n)$ and $A_{n+1} \in B^\beta(E_n)$.

A winning $(\alpha, \beta)$-strategy (for Eve) relative to $W$ is a map $\sigma_W = \sigma_W(\alpha, \beta)$ such that to every finite sequence of Adam's moves $A_0, A_1, \ldots, A_n$, it associates $E_n := \sigma_W(\alpha, \beta; n; A_0, A_1, \ldots, A_n) \in B^\alpha(A_n)$ in such a way that

$$\bigcap_{n\in\mathbb{N}} E_n = \bigcap_{n\in\mathbb{N}} A_n \subseteq W.$$

$W$ is said to be $(\alpha, \beta)$-winning if there exists a winning $(\alpha, \beta)$-strategy relative to $W$. To put it another way: if Eve can always "hit" an element of $W$, independently of Adam's sequence $(A_n)_{n\in\mathbb{N}}$ of legal moves. $W$ is $\alpha$-winning if it is $(\alpha, \beta)$-winning for every $\beta \in \left(0, \frac{1}{2}\right)$. Moreover, $W$ is said to be winning if it is $\alpha$-winning for every $\alpha \in \left(0, \frac{1}{2}\right)$.

**Theorem 2 (Schmidt, [Sch66, Theorem 2]).** For each $\alpha \in \left(0, \frac{1}{2}\right)$, the intersection of countably many $\alpha$-winning sets is $\alpha$-winning. Hence, the intersection of countably many winning sets is winning.

It is worth remarking that the system of winning sets neither contains, nor is contained in, the family consisting of the sets of full Lebesgue measure. The set of badly approximable numbers, *i.e.*, of those reals $x$ fulfilling

$$\left|x - \frac{p}{q}\right| > \frac{c}{q^2}$$

for some $c > 0$ and all rationals $\frac{p}{q}$, serves as a remarkable example:

**Theorem 3 (Schmidt, [Sch66, Theorem 3]; Khintchine, [Khi63, Theorems 23, 29]).** The set $B$ of badly approximable numbers is winning and of Lebesgue measure zero. Hence, $\mathbb{R} \setminus B$ is of full Lebesgue measure but it is not winning.

*Proof.* Schmidt proves that $B$ is winning. On the other hand, $B$ has Lebesgue measure zero, according to Khintchine's celebrated theorem. The first assertion is thus verified.

To prove the second, observe that, for $\alpha, \beta \in \left(0, \frac{1}{2}\right)$, Eve knows a winning $(\beta, \alpha)$-strategy $\sigma_B(\beta, \alpha)$ relative to $B$. Certainly Eve cannot win the $(\alpha, \beta)$-game relative to $\mathbb{R}\setminus B$, if Adam (judiciously) chooses his $(n+2)$th move $A_{n+1}$ according to Eve's winning $(\beta, \alpha)$-strategy $\sigma_B(\beta, \alpha)$ (Adam's first choice $A_0$ being immaterial).

As an alternative proof, simply apply Theorem 2: the intersection of two winning sets is winning, but $\varnothing = B \cap (\mathbb{R} \setminus B)$ is not. q.e.d.

## 3 Results

It follows from [Sch66, Corollary 2] that any winning subset of $\mathbb{R}$ has Hausdorff dimension one. Below we indicate a short and simple proof of Schmidt's result valid for the real line case. Differently from Schmidt's, our proof is based on the following direct consequence of the so-called "mass distribution principle" [Fal90, Section 4.1]:

**Lemma 4 (Falconer, [Fal90, Example 4.6]).** Let $(\varepsilon_n)_{n\in\mathbb{N}}$ be a strictly decreasing sequence of positive reals, and $(C_n)_{n\in\mathbb{N}}$ a sequence of subsets of $\mathbb{R}$

with $C_0 = [0,1]$ and each $C_{n+1}$ a union of intervals of equal length separated by gaps of length at least $\varepsilon_n$. Suppose that for every $n \in \mathbb{N}$ each interval of $C_n$ contains $m_n \geq 2$ intervals of $C_{n+1}$. Then, defined $F := \bigcap_{n\in\mathbb{N}} C_n$, we have

$$\dim F \geq \liminf_{n\to\infty} \frac{\log(m_0 \cdots m_n)}{-\log(m_{n+1}\varepsilon_{n+1})}.$$

**Theorem 5 (Schmidt, [Sch66, Corollary 2]).** If $W$ is a winning subset of $\mathbb{R}$, then $\dim W = 1$.

*Proof.* We shall verify that, for every $\alpha \in \left(0, \frac{1}{2}\right)$ and every $\alpha$-winning subset $W$ of $\mathbb{R}$, we have $\dim W = 1$ (this obviously implies the statement above). For this purpose, it suffices to show that $W$ contains a family of Cantor-like sets of Hausdorff dimensions arbitrarily close to 1.

Fix arbitrarily a natural $m \geq 2$, and let $\beta_m := \frac{1}{2m+1}$. Put $C_0 := [0,1]$ and $E_0 := \sigma_W(\alpha, \beta_m; 0; C_0)$, being $\sigma_W(\alpha, \beta_m)$ a winning $(\alpha, \beta_m)$-strategy relative to $W$. Divide $E_0$ into $2m+1$ subintervals of equal length, and denote with $I_i$, the index $i$ varying in $m$, the second, fourth, ..., $2m$th of them. Then, let $C_1 := \bigcup_{i\in m} I_i$ and, for all $i \in m$, let $E_{0i} := \sigma_W(\alpha, \beta_m; 1; C_0, I_i)$.

This done, divide each $E_{0i}$ into $2m+1$ subintervals of equal length and call $I_{ij}$, with $j \in m$, the one occupying the $2(j+1)$th place. Again, put $C_2 := \bigcup_{i\in m} \bigcup_{j\in m} I_{ij}$ and, for any $i, j \in m$, let $E_{0ij} := \sigma_W(\alpha, \beta_m; 2; C_0, I_i, I_{ij})$.

Proceeding inductively, for any $n \geq 1$ and any $i_1, i_2, \ldots, i_n \in m$, divide each $E_{0i_1\cdots i_n}$ into $2m+1$ subintervals of equal length; label with $I_{i_1\cdots i_n j}$, the index $j$ ranging over $m$, the one in the $2(j+1)$th place; let

$$C_{n+1} := \bigcup_{i_1,\ldots,i_n\in m} \bigcup_{j\in m} I_{i_1\cdots i_n j}$$

and, for all $j \in m$,

$$E_{0i_1\cdots i_n j} := \sigma_W(\alpha, \beta_m; n+1; C_0, I_{i_1}, \ldots, I_{i_1\cdots i_n j}).$$

Finally, define $F := \bigcap_{n\in\mathbb{N}} C_n$. By construction, $F$ is contained in $W$. Moreover, for every $n \in \mathbb{N}$, the set $C_{n+1}$ is a union of finitely many intervals separated each other by gaps of length at least $\varepsilon_n := (\alpha\beta_m)^{n+1}$, and each interval of $C_n$ contains precisely $m_n := m$ intervals of $C_{n+1}$. Therefore, in view of the previous lemma we obtain

$$\dim W \geq \dim F \geq \liminf_{n\to\infty} \frac{\log m^n}{-\log m\left(\frac{\alpha}{2m+1}\right)^{n+1}} = \frac{\log m}{\log(2m+1) - \log\alpha}.$$

The conclusion follows by letting $m \to \infty$. q.e.d.

As a first corollary of Theorems 3 and 5, we immediately obtain the following noteworthy theorem of Jarník:

**Corollary 6 (Pollington, Velani, [PolVel00, Section 2.2]).** The set of badly approximable numbers has Hausdorff dimension one.

Jarník's proof makes essential use of the classical result that a number is badly approximable if and only if the partial quotients of its continued fraction expansion are uniformly bounded from above [Khi63, Theorem 23]. The game theoretic approach just indicated avoids the theory of continued fractions, depending only on geometric arguments.

That the system of winning sets is invariant under taking diffeomorphic images follows as a particular case from [Sch66, Theorem 1]. Here we give a very simple proof of Schmidt's theorem, *ad hoc* for the real line case — we add, however, that the key idea of playing a suitable "virtual" game, in combination with the "real" one, is to ascribe to him.

**Theorem 7 (Schmidt, [Sch66, Theorem 1]).** Let $W$ be a winning subset of $\mathbb{R}$, and $\varphi : \mathbb{R} \to \mathbb{R}$ a diffeomorphism. Then $\varphi[W]$ is winning.

*Proof.* Fix a winning set $W$, two constants $\alpha, \beta \in \left(0, \frac{1}{2}\right)$, and a diffeomorphism $\varphi : \mathbb{R} \to \mathbb{R}$. We have to check that $\varphi[W]$ is $(\alpha, \beta)$-winning.

In view of the continuity of the derivative $\varphi'$ of $\varphi$, and by neglecting finitely many moves if necessary, without any loss of generality we may assume Adam's first choice $A_0$ so small that there exist two constants $k, K > 0$ fulfilling $\frac{K}{k}\alpha < \frac{1}{2}$ and both the following equivalent conditions:

$$k \leq |\varphi'(x)| \leq K \text{ for every } x \in \varphi^{-1}(A_0); \tag{2}$$

$$\frac{1}{K} \leq |(\varphi^{-1})'(y)| \leq \frac{1}{k} \text{ for every } y \in A_0. \tag{3}$$

With the $(n+1)$th move $A_n := I[a_n, \varrho_n]$ done by the first player Adam, let us associate $A'_n := I\left[\varphi^{-1}(a_n), \frac{\varrho_n}{K}\right]$. Then, let us determine Eve's choice $E'_n$ accordingly to a winning $\left(\frac{K}{k}\alpha, \frac{k}{K}\beta\right)$-strategy relative to $W$ (it does exist, by assumption). Thus

$$E'_n := I\left[e'_n, \frac{\varrho_n}{k}\alpha\right] = \sigma_W\left(\frac{K}{k}\alpha, \frac{k}{K}\beta; n; A'_0, A'_1, \ldots, A'_n\right) \in B^{\frac{K}{k}\alpha}(A'_n)$$

for a certain $e'_n \in \mathbb{R}$. Finally, put $E_n := I[\varphi(e'_n), \alpha\varrho_n]$.

By (2), (3), the theorem of Lagrange from elementary analysis, and the self-evident observation that, for any $x, y \in \mathbb{R}$, $r > 0$, and $\delta \in (0,1)$, we have $I[y, \delta r] \in B^{\delta}(I[x,r])$ if and only if $|x - y| \leq r(1 - \delta)$, for all $n \in \mathbb{N}$ we have:

$$E'_n \in B^{\frac{K}{k}\alpha}(A'_n) \text{ implies } E_n \in B^{\alpha}(A_n);$$
$$A_{n+1} \in B^{\beta}(E_n) \text{ implies } A'_{n+1} \in B^{\frac{k}{K}\beta}(E'_n).$$

It then follows: $\bigcap_{n\in\mathbb{N}} E'_n = \{\lim_{n\to\infty} e'_n\} \subseteq W$, which clearly gives, being $\varphi$ continuous, $\bigcap_{n\in\mathbb{N}} E_n = \{\lim_{n\to\infty} \varphi(e'_n)\} \subseteq \varphi[W]$. q.e.d.

The main (and last) result of this note is now deduced by simply combining Theorems 2, 5 and 7:

**Corollary 8.** The family of winning sets has the large intersection property.

# References.

[BakSch70] Alan **Baker** and Wolfgang M. **Schmidt**, Diophantine Approximation and Hausdorff Dimension, **Proceedings of the London Mathematical Society** 21 (1970), p. 1–11

[DodRynVic90] Maurice M. **Dodson**, Bryan P. **Rynne**, and James A. **Vickers**, Diophantine Approximation and a Lower Bound for Hausdorff Dimension, **Mathematika** 37 (1990), p. 59–73

[Fal85] Kenneth J. **Falconer**, Classes of Sets with Large Intersection, **Mathematika** 32 (1985), p. 191–205

[Fal90] Kenneth J. **Falconer**, Fractal Geometry: Mathematical Foundations and Applications, John Wiley & Sons 1990

[Fal94] Kenneth J. **Falconer**, Sets with Large Intersection Properties, **Journal of the London Mathematical Society** 49 (1994), p. 267–280

[Khi63] Alexander Y. **Khintchine**, Continued Fractions, Noordhoff 1963

[PolVel00] Andrew D. **Pollington** and Sanju L. **Velani**, On a Problem in Simultaneous Diophantine Approximation: Littlewood's Conjecture, **Acta Mathematica** 185 (2000), p. 287–306

[Sch66] Wolfgang M. **Schmidt**, On Badly Approximable Numbers and Certain Games, **Transactions of the American Mathematical Society** 123 (1966), p. 178–199

**Received**: January 20th, 2005;
**In revised version**: May 23rd, 2005;
**Accepted by the editors**: July 29, 2005.